Tenth Edition

Basic Statistics

Tales of Distributions

Tenth Edition

Basic Statistics

Tales of Distributions

Chris Spatz
Hendrix College

WADSWORTH
CENGAGE Learning™

Australia • Brazil • Canada • Mexico • Singapore • Spain
United Kingdom • United States

Basic Statistics: Tales of Distributions, Tenth Edition
Chris Spatz

Acquisitions Editor: Jane Potter
Assistant Editor: Paige Leeds
Editorial Assistant: Alicia Mclaughlin
Technology Project Manager: Andy Yap
Marketing Manager: Tierra Morgan
Marketing Assistant: Anna Andersen
Marketing Communications Manager: Talia Wise
Content Project Management: Pre-Press PMG
Art Director: Vernon Boes
Manufacturing Buyer: Judy Inouye
Permissions Editor Text: Bob Kauser
Production Service: Pre-Press PMG
Cover Designer: Bill Stanton
Compositor: Pre-Press PMG

Library of Congress Control Number: 2009939372

Student Edition:
ISBN-13: 978-0-495-80891-6
ISBN-10: 0-495-80891-1

Paper Edition:
ISBN-13: 978-0-495-90968-2
ISBN-10: 0-495-90968-8

Wadsworth
20 Davis Drive
Belmont CA 94002
USA

Printed in the United States of America
3 4 5 6 7 13 12

About the Author

Chris Spatz is at Hendrix College where he twice served as chair of the Psychology Department. Spatz, whose Ph.D. degree is in experimental psychology from Tulane University, was awarded postdoctoral fellowships in animal behavior at the University of California, Berkeley, and the University of Michigan in Ann Arbor. He held faculty positions at The University of the South and the University of Arkansas, Monticello.

Spatz is a co-author of *Research Methods in Psychology: Ideas, Techniques, and Reports* (with Edward P. Kardas). He wrote the chapter on statistical techniques and analysis for *21st Century Psychology: A Reference Handbook* and was a section editior for the four-volume work *Encyclopedia of Statistics in Behavioral Science.* He has contributed chapters and entries to other references works and serves as a manuscript reviewer for journals that focus on teaching.

Spatz is married to Thea Siria Spatz; they have three children and seven grandchildren. Besides writing, Spatz enjoys outdoor activities such as canoeing, camping, and gardening, especially with family members. He swims several times a week and is active in the United Methodist Church.

With love and affection,

this textbook is dedicated to

Thea Siria Spatz, Ed.D., CHES.

Brief Contents

Contents

Preface

Even if our statistical appetite is far from keen, we all of us should like to know enough to understand, or to withstand, the statistics that are constantly being thrown at us in print or conversation—much of it pretty bad statistics. The only cure for bad statistics is apparently more and better statistics. All in all, it certainly appears that the rudiments of sound statistical sense are coming to be an essential of a liberal education.

—Robert Sessions Woodworth

Basic Statistics: Tales of Distributions, Tenth Edition, is a textbook for a one-term statistics course in the social or behavioral sciences, education, or the allied health/nursing field. Its focus is conceptualization, understanding, and interpretation, rather than computation. Although designed to be comprehensible and complete for students who take only one statistics course, it includes many elements that prepare students for additional statistics courses. Basic experimental design terms such as *independent* and *dependent variables* are explained and used in examples so that students can be expected to write fairly complete interpretations of their analyses. In many places, the student is invited to stop and think or stop and do an exercise. Some problems simply ask the student to decide which statistical technique is appropriate. In sum, this book's approach reinforces instructors who emphasize critical thinking in their course.

This textbook has been remarkably successful for 35 years, at times being a Wadsworth "best-seller" among statistics texts. Reviewers have praised the book as have students and professors. A common refrain is that the book has a distinctive, androgynous voice and a conversational style that is engaging, encouraging, and even endearing. Other features that distinguish this textbook from others include:

- Problems are interspersed throughout the chapter rather than grouped at the end
- Answers to problems are extensive; there are more than 50 pages of detailed answers
- Examples and problems come from a variety of disciplines and everyday life
- Most problems are based on actual studies rather than fabricated scenarios

- Interpretation is emphasized; interpretation headings in the answers are highlighted
- Empirical explanations are provided for theoretical concepts
- The effect size index is treated as a descriptive statistic and not just an add-on to hypothesis-testing problems
- Computer software analyses are illustrated with SPSS printouts
- Important words and phrases are defined in the margin when they first occur
- *Objectives* at the beginning of each chapter serve first as an orientation list and later as a review list
- *Clues to the Future* alert students to concepts that will be repeated
- *Error Detection* boxes tell ways to detect or prevent mistakes
- *Transition Passages* alert students to changes in focus that are part of the chapters that follow
- *Comprehensive Problems* encompass all (or most) of the techniques in a chapter
- *What Would You Recommend?* problems require choices from among the techniques in several chapters
- A final chapter, *Choosing Tests and Writing Interpretations* provides consolidation activities
- The Book Companion website has a variety of student aids for each chapter

For this tenth edition, I amplified my earlier emphasis that data sets should be approached with an attitude of exploration. To reflect this increased emphasis, the titles of three of the descriptive statistics chapters include the phrase "Exploring Data." As for other changes, the chapter that covered central tendency and variability is now two separate chapters, which reflects how many instructors assign that material. All of the problems and examples based on contemporary data (height, family income, tennis rankings, and such) are updated. Several scenarios and their hypothetical data have been replaced with data based on actual studies. Words, sentences, paragraphs, and examples were revised to improve clarity or reflect current practice. The 360 problems and their sometimes extensive answers were given special attention to be sure they matched the text. Digital object identifiers (doi numbers), when available, were added to references to make their electronic retrieval easier.

For students, the book companion website (www.cengage.com/psychology/ spatz) has multiple-choice questions and definition flashcards for each chapter. For professors, the *Instructor's Manual with Test Bank*, which is available in print and electronically, has teaching suggestions and almost 2000 test items, most of which have been classroom tested.

Students who engage themselves in this book and in their course can expect to

- Understand and explain statistical reasoning
- Choose correct statistical techniques for the data from simple experiments
- Solve statistical problems
- Write explanations that are congruent with statistical analyses

A gentle revolution is going on in the practice of statistics. In the past, statistics moved toward more sophisticated ways to test a null hypothesis. Recently, the direction shifted to an emphasis on descriptive statistics and graphs, simpler analyses, and less reliance on null hypothesis statistical testing (NHST). (See the initial report and the follow-up report of the APA Task Force on Statistical Inference at *www.apa.org/science/bsaweb-tfsi.html.*) This book reflects many of those changes.

Acknowledgments

I am grateful for all the help I received over three-plus decades from students, faculty, and staff at Hendrix College. I especially acknowledge my helpful colleagues at Hendrix's Bailey Library. For this edition, Peggy Morrison was the go-to person. Roger E. Kirk, my consulting editor for the first six editions, deserves special thanks for identifying errors and teaching me some statistics. Rob Nichols wrote sampling programs, and Bob Eslinger produced accurate graphs of the F, t, χ^2, and normal distributions. Paige Leeds and Beth Kluckhohn shepherded the manuscript through production. I especially want to acknowledge James O. Johnston, my friend and former co-author, who suggested that we do a statistics book, dreamed up the subtitle, and worked with me on the first three editions.

I also want to acknowledge the help of reviewers for this edition:

Daniel J. Calcagnetti, Fairleigh Dickinson University
Thomas J. Capo, University of Maryland
Robert G. Jones, Missouri State University
Marcel Satsky Kerr, Texas Wesleyan University
Elizabeth Mason, California University of Pennsylvania
Ed McAllister, The Sage Colleges
Jennifer Peszka, Hendrix College
Joseph H. Porter, Virginia Commonwealth University
Robert D. Westbrook, Ohio University

In addition to the reviewers for this edition, I have benefited from the criticism of more than 75 other professors who formally reviewed previous editions of this book, including Adansi Amankwaa, Albany State College; Michael Baird, San Jacinto College; Evelyn Blanch-Payne, Albany State University; Chris Bloom, University of Southern Indiana; Curtis Brant, Baldwin-Wallace College; Daniel Calcagnetti, Fairleigh Dickinson University; Thomas Capo, University of Maryland; Sky Chafin, Palomar College; Yong Dai, Louisiana State University; Kathleen Dillon, Western New England College; Beverly Dretzke, University of Wisconsin–Eau Claire; Alexis Grosofsky, Beloit College; Laura Heinze, University of Kansas; Marcel Satsky Kerr, Tarleton State University–Central Texas; Gerald Lucker, University of Texas, El Paso; Sandra McIntire, Rollins College; Craig Nagoshi, Arizona State University; Jennifer Peszka, Hendrix College; Marilyn Pugh, Texas Weslyan University; David Schwebel, University of Alabama at Birmingham; Christy Scott, Pepperdine University; Francisco Silva, University of Redlands; Elizabeth Ann Spatz, Hendrix College; Boyd Spencer, Eastern Illinois University; Greg Streib, Georgia State University; Philip Tolin, Central Washington University; and Anthony Walsh, Salve Regina University.

My most important acknowledgment goes to my wife and family, who helped and supported me in many ways over the life of this project. Words are not adequate here.

I've always had a touch of the teacher in me—first as an older sibling, then as a parent and professor, and now as a grandfather. Education is a first-class task, in my opinion. I hope this book conveys my enthusiasm for teaching and also my philosophy of teaching. (By the way, if you are a student who is so thorough as to read the whole preface, you should know that I included phrases and examples in a number of places that reward your kind of diligence.)

If you find an error in this book, please report it to me at spatz@hendrix.edu. I will post corrections at *www.hendrix.edu/statistics10thED.*

Tenth Edition

Basic Statistics

Tales of Distributions

Introduction

CHAPTER

1

OBJECTIVES FOR CHAPTER 1

After studying the text and working the problems in this chapter, you should be able to:

1. Distinguish between descriptive and inferential statistics
2. Define the words *population, sample, parameter, statistic,* and *variable* as these terms are used in statistics
3. Distinguish between quantitative and qualitative variables
4. Identify the lower and upper limits of a quantitative measurement
5. Identify four scales of measurement and distinguish among them
6. Distinguish between statistics and experimental design
7. Define some experimental-design terms—*independent variable, dependent variable,* and *extraneous variable*—and identify these variables in the description of an experiment
8. Describe the relationship of statistics to epistemology
9. Describe actions to take to analyze a data set
10. Identify a few events in the history of statistics

THE REPORTER PACED back and forth outside the yellow tape trying to catch the eye of any police officer who might talk about what happened inside the large house beyond the tape barrier. As a dark, unmarked sedan pulled up, the reporter watched and then smiled as a woman carrying a bag emerged from the vehicle.

"Hello, Detective Drew, I'm glad you are here. Can you tell me what's going on?"

"Well, George, I don't have anything for you officially except that I'm here to gather evidence and tell the story of what happened in there. I'm keen to explore the situation and see what I can determine."

Lifting the yellow tape, the detective walked quickly toward the mansion.

I think Detective Drew would probably be a good statistician. Her approach strategy, exploration, is the best attitude to have when you inspect data. Her goal, a story supported by evidence, is the statistician's goal as well. The only other thing she needs is a mental collection of statistical tools (which she could get by studying this textbook and taking a college course in statistics). With an attitude of exploration and a story supported by a statistical analysis, Ms. Drew could contribute to any project that depends on quantitative data.

What projects and disciplines depend on quantitative data? The list is long and variable; the disciplines include psychology, biology, sociology, education, medicine, politics, business, economics, forestry, and others. Examples and problems in this textbook come from all these disciplines. Statistics is a powerful method for getting answers from data, and that makes it popular with investigators in a wide variety of endeavors.

Statistics is used in areas that might surprise you. As examples, statistics has been used to determine the effect of cigarette taxes on smoking among teenagers, the safety of a new surgical anesthetic, and the memory of young school-age children for pictures (which is as good as that of college students). Statistics show what diseases have an inheritance factor, how to improve short-term weather forecasts, and why giving intentional walks in baseball is a poor strategy. All these examples come from *Statistics: A Guide to the Unknown,* a book edited by Judith M. Tanur and others (1989). Written for those "without special knowledge of statistics," this book has 29 essays on topics as varied as those above.

The importance of statistical literacy is becoming more and more apparent. As one bit of evidence, Gigerenzer, et al. (2007), in their public interest article on health statistics, point out that statistical illiteracy among both patients and physicians undermines the information exchange necessary for informed consent and shared decision making. The result is anxiety, confusion, and undue enthusiasm for testing and treatment.

Whatever your current interests or thoughts about your future as a statistician, I believe you will benefit from this course. When a statistics course is successful, students learn to identify the questions that a set of data can answer, determine the statistical procedures that will provide the answers, carry out the procedures, and then, using plain English and graphs, tell the story the data reveal. Also, they find statistical thinking helpful in other arenas of their lives.

The best way for you to acquire all these skills (especially the part about telling the story) is to *engage* statistics. Engaged students are easily recognized; they are prepared for exams, not easily distracted while studying, and generally finish assignments on time. *Becoming* an engaged student may not be so easy, but many have achieved it. Here are my recommendations. Read with the goal of understanding. Attend class. Do all the assignments (on time). Write down questions. Ask for explanations. Expect to understand. (Disclaimer: I'm not suggesting that you marry statistics, but just engage for this one course.)

Are you uncertain about your arithmetic and algebra skills? Appendix A in the back of this book may help. It consists of a pretest (to see if you need to refresh your memory) and a review (to provide that refresher).

What Do You Mean, "Statistics"?

The *Oxford English Dictionary* says that the word *statistics* came into use more than 200 years ago. At that time, *statistics* referred to a country's quantifiable political characteristics—characteristics such as population, taxes, and area. Statistics meant "state numbers." Tables and charts of those numbers turned out to be a very satisfactory way to compare different countries and to make projections about the future. Later, tables and charts proved useful to people studying trade (economics) and natural phenomena (science). Statistical thinking spread because it helped.

descriptive statistic
A number that conveys a particular characteristic of a set of data.

mean
Arithmetic average; sum of scores divided by number of scores.

inferential statistics
Method that uses sample evidence and probability to reach conclusions about unmeasurable populations.

Today two different techniques are called *statistics.* One technique, **descriptive statistics,**[1] produces a number or a figure that summarizes or describes a set of data. You are already familiar with some descriptive statistics. For example, you know about the arithmetic average, called the **mean.** You have probably known how to compute a mean since elementary school—just add up the numbers and divide the total by the number of entries. As you already know, the mean describes the central tendency of a set of numbers. The basic idea is simple: A descriptive statistic summarizes a set of data with one number or graph. This book will cover about a dozen descriptive statistics.

The other statistical technique is **inferential statistics.** With inferential statistics, you use measurements from a sample to reach conclusions about a larger, *unmeasured* population. There is, of course, a problem with samples. Samples always depend *partly* on the luck of the draw; chance helps determine the particular measurements you get. If you have the measurements for the entire population, chance doesn't play a part—all the variation in the numbers is "true" variation. But with samples, some of the variation is the true variation in the population and some is just the chance ups and downs that go with a sample. Inferential statistics was developed as a way to account for the effects of chance that come with sampling. This book will cover about a dozen and a half inferential statistics.

Here is a textbook definition: Inferential statistics is a method that takes chance factors into account when samples are used to reach conclusions about populations. Like most textbook definitions, this one condenses many elements into a short sentence. Because the idea of using samples to understand populations is perhaps the *most important concept* in this course, please pay careful attention when elements of inferential statistics are explained.

Inferential statistics has proved to be a very useful method in scientific disciplines. Many other fields use inferential statistics, too, so I selected examples and problems from a variety of disciplines for this text and its auxiliary materials such as the book companion website, *www.cengage.com/psychology/spatz.*

Here is an example of inferential statistics from psychology. Today there is a lot of evidence that people remember the tasks they fail to complete better than the tasks they complete. This is known as the *Zeigarnik effect.* Bluma Zeigarnik asked the participants in her experiment to do about 20 tasks, such as work a puzzle, make a clay figure, and construct a box from cardboard.[2] For each participant, half the tasks were interrupted before completion. Later, when the participants were asked to recall the tasks they worked on, they listed more of the interrupted tasks (about 7) than the completed tasks (about 4).

So, should you conclude that interruption improves memory? Not yet. It might be that interruption actually has no effect but that several *chance factors* happened to favor the interrupted tasks in Zeigarnik's particular experiment. One way to meet this objection is to conduct the experiment again. Similar results would lend support to the

[1] Boldface words and phrases are defined in the margin and also in Appendix D, Glossary of Words.

[2] A summary of this study can be found in Ellis (1938). The complete reference and all others in the text are listed in the References section at the back of the book.

conclusion that interruption improves memory. A less expensive way to meet the objection is to use an inferential statistics test.

An inferential statistics test begins with the actual data from the experiment. It ends with a probability—the probability of obtaining data like those actually obtained if it is true that interruption *does not* affect memory. If the probability is very small, then you can conclude that interruption *does* affect memory.[3]

For the Zeigarnik memory experiment, the conclusion might be written as: "The greater recall of the interrupted tasks, compared to those that were completed, is most likely due to interruption because chance by itself would rarely produce this large a difference between two samples." The words *chance* and *rarely* tell you that probability is an important element of inferential statistics.

My more complete answer to what I mean by "statistics" is Chapter 6 in *21st Century Psychology: A Reference Handbook* (Spatz, 2008). This 8-page chapter summarizes in words (no formulas) the statistical concepts usually covered in statistics courses. The chapter can orient you as you begin your study of statistics and later provide a review after you finish your course.

Statistics: A Dynamic Discipline

Many people continue to think of statistics as a collection of techniques that were developed long ago, that have not changed, and that will be the same in the future. That view is mistaken. Statistics is a dynamic discipline characterized by more than a little controversy. New techniques in both descriptive and inferential statistics have been developed in recent years. As for controversy, a number of statisticians recently made a strong case for *banning* the use of a very popular inferential statistics technique (null hypothesis significance tests). Other statisticians disagreed, although they acknowledged that the tests are sometimes misused and that other techniques yield more information. For fairly nontechnical summaries of this issue see Dillon (1999) or Spatz (2000). For a more technical summary, see Nickerson (2000) or go to *www.apa.org/science/bsaweb-tfsi.html* for a report from the American Psychological Association. For alternatives; see Erceg-Hurn and Mirosevich, 2008.

In addition to out-and-out controversy over techniques, attitudes toward data analysis have shifted in recent years. The shift has been toward the idea of exploring data to see what it reveals and away from using statistical analyses to nail down a conclusion. This shift owes much of its impetus to John Tukey (1915–2000), who promoted Exploratory Data Analysis (Lovie, 2005). Tukey invented techniques such as the boxplot (Chapter 5) that reveal several characteristics of a data set simultaneously.

Today, statistics is used in a wide variety of fields. Researchers start with a phenomenon, event, or process that they want to understand better. They make measurements that produce numbers. The numbers are manipulated according to the rules and conventions of statistics. Based on the outcome of the statistical analysis, researchers draw conclusions and then write the story of their new understanding of the phenomenon, event, or process. Statistics is just one tool that researchers use, but it is often an essential tool.

---

[3] If it really is the case that interruption doesn't affect memory, then differences between the samples are the result of chance.

clue to the future

The first part of this book is devoted to descriptive statistics (Chapters 2–6) and the second part to inferential statistics (Chapters 7–15). Inferential statistics is the more comprehensive of the two because it combines descriptive statistics, probability, and logic.

What's in It for Me?

What's in it for me? is a reasonable question. Decide which of the following answers apply to you and then pencil in others that are true for you.

One thing that is in it for you is that, when you have successfully completed this course, you will understand how statistical analyses are used to *make decisions.* You will understand both the elegant beauty and the ugly warts of the process. H. G. Wells, a novelist who wrote about the future, said, "Statistical thinking will one day be as necessary for efficient citizenship as the ability to read and write." So chalk this course up under the heading "general education about decision making, to be used throughout life."

Another result of successfully completing this course is that you will be able to use a tool employed in many disciplines. Although the people involved in these disciplines are interested in different phenomena, they all use statistics. Lawyers, for example, sometimes use statistics to provide evidence for their clients. I have assisted lawyers at times, occasionally testifying in court about the proper interpretation of data. (You will analyze and interpret some of these data in later chapters.)

In American history the authorship of 12 of *The Federalist* papers was disputed for a number of years. (*The Federalist* papers were 85 short essays written under the pseudonym "Publius" and published in New York City newspapers in 1787 and 1788. Written by James Madison, Alexander Hamilton, and John Jay, the essays were designed to persuade the people of the state of New York to ratify the Constitution of the United States.) To determine authorship of the 12 disputed papers, each was graded with a quantitative *value analysis* in which the importance of such values as national security, a comfortable life, justice, and equality was assessed. The value analysis scores were compared with value analyses of papers known to have been written by Madison and Hamilton (Rokeach, Homant, and Penner, 1970). Another study, by Mosteller and Wallace, analyzed *The Federalist* papers using the frequency of words such as *by* and *to* (reported in Tanur et al., 1989). Both studies concluded that Madison wrote all 12 essays.

Here is another example from law. Rodrigo Partida was convicted of burglary in Hidalgo County, a border county in southern Texas. A grand jury rejected his motion for a new trial. Partida's attorney filed suit, claiming that the grand jury selection process discriminated against Mexican-Americans. In the end (*Castaneda* v. *Partida,* 430 U.S. 482 [1976]), Justice Harry Blackmun of the U.S. Supreme Court wrote, regarding the number of Mexican-Americans on grand juries, "If the difference between the expected and the observed number is greater than two or three standard deviations, then the hypothesis that the jury drawing was random (is) suspect." In Partida's case the difference was approximately 12 standard deviations, and the Supreme Court ruled that Partida's attorney had presented *prima facie* evidence. (*Prima facie* evidence is so good that one side wins the case unless the other side rebuts the evidence, which in this case did not happen.) *Statistics: A Guide to the Unknown* includes two essays on the use of statistics by lawyers.

I hope that this course will encourage you or even teach you to ask questions about the statistics you hear and read. Consider the questions, one good and one better, asked of a restaurateur in Paris who served delicious rabbit pie. He served rabbit pie even when other restaurateurs could not obtain rabbits. A suspicious customer asked this restaurateur if he was stretching the rabbit supply by adding horse meat.

"Yes, a little," he replied.

"How much horse meat?"

"Oh, it's about 50–50," the restaurateur replied.

The suspicious customer, satisfied, began eating his pie. Another customer, who had learned to be more thorough in asking questions about statistics, said, "50–50? What does *that* mean?"

"One rabbit and one horse" was the reply.

This section began with a question from you: What's in it for me? My answer is that when you learn the basics of statistics, you will understand data better. You will be able to communicate with others who use statistics, and you'll be better able to persuade others. But what about your answer? What do you expect of yourself and of this course? (Writing your answers in the margin is an example of engagement.)

Some Terminology

Like most courses, statistics introduces you to many new words. In statistics, most of the terms are used over and over again. Your best move, when introduced to a new term, is to *stop, read* the definition carefully, and *memorize* it. As the term continues to be used, you will become more and more comfortable with it. Making notes is helpful.

Populations and Samples

population
All measurements of a specified group.

sample
Measurements of a subset of a population.

A **population** consists of all the scores of some specified group. A **sample** is a subset of a population. The population is the thing of interest, is defined by the investigator, and includes all cases. The following are some populations:

Family incomes of college students in the fall of 2008

Weights of crackers eaten by obese male students

Depression scores of Alaskans

Errors in maze running by rats with brain damage

Gestation times for human beings

Memory scores of human beings[4]

---

[4] I didn't pull these populations out of thin air; they are all populations that researchers have gathered data on. Studies of these populations will be described in this book.

Investigators are always interested in populations. However, as you can determine from these examples, populations can be so large that not all the members can be studied. The investigator must often resort to measuring a sample that is small enough to be manageable. A sample taken from the population of incomes of families of college students might include only 40 students. From the last population on the list, Zeigarnik used a sample of 164.

Most authors of research articles carefully explain the characteristics of the *samples* they use. Often, however, they do not identify the *population,* leaving that task to the reader. For example, consider a study of a drug therapy program based on a sample of 14 men who were acute schizophrenics at Hospital A. Suppose the report of the study doesn't identify the population. What is the population? Among the possibilities are: all male acute schizophrenics at Hospital A, all male acute schizophrenics, all acute schizophrenics, and all schizophrenics. The answer to the question, What is the population? depends on the specifics of a research area, but many researchers generalize generously. For example, for some topics it is reasonable to generalize from the results of a study on rats to "all mammals." In all cases, the reason for gathering data from a sample is to generalize the results to a larger population even though sampling introduces some uncertainty into the conclusions.

Parameters and Statistics

A **parameter** is some numerical (number) or nominal (name) characteristic of a population. An example is the mean reading readiness score of all first-grade pupils in the United States. A **statistic** is some numerical or nominal characteristic of a sample. The mean reading readiness score of 50 first-graders is a statistic and so is the observation that 45 percent are girls.

parameter
Numerical or nominal characteristic of a population.

statistic
A numerical or nominal characteristic of a sample.

A parameter is constant; it does not change unless the population itself changes. The mean of a population is exactly one number. Unfortunately, the parameter often cannot be computed because the population is unmeasurable. So, a statistic is used as an estimate of the parameter, although, as suggested before, statistics tend to differ from one sample to another. If you have five samples from the same population, you will probably have five different sample means. In sum, parameters are constant; statistics are variable.

Variables

A **variable** is something that exists in more than one amount or in more than one form. Height and gender are both variables. The variable height is measured using a standard scale of feet and inches. For example, the notation 5′7″ is a numerical way to identify a group of persons who are similar in height. Of course, there are many other groups, each with an identifying number. Gender is also a variable. With gender, there are (usually) only two groups of people. Again, we may identify each group by assigning a number to it. All participants represented by 0 have the same gender. I will often refer to numbers like 5′7″ and 0 as *scores* or *test scores.* A score is simply the result of a measurement.

variable
Something that exists in more than one amount or in more than one form.

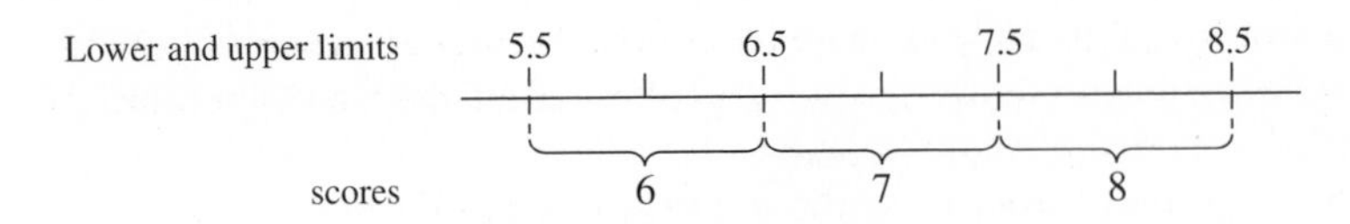

FIGURE 1.1 The lower and upper limits of recall scores of 6, 7, and 8

Quantitative Variables

quantitative variable
Variable whose levels indicate different amounts.

Many of the variables you will work with in this text are **quantitative variables.** When a quantitative variable is measured, the scores tell you something about the amount or degree of the variable. At the very least, a larger score indicates more of the variable than a smaller score does.

Take a closer look at the quantitative memory scores in Zeigarnik's experiment. *Number of tasks recalled* comes in whole numbers such as 4 or 7. However, it is reasonable to think that of two persons who scored 7, one just barely got 7 and another almost scored 8. Picture the quantitative variable, recall, as **Figure 1.1**.

lower limit
Bottom of the range of possible values that a measurement on a quantitative variable can have.

upper limit
The top of the range of possible values that a measurement on a quantitative variable can have.

Figure 1.1 shows that a score of 7 is used for a range of possible recall values—the range from 6.5 to 7.5. The number 6.5 is the **lower limit** and 7.5 is the **upper limit** of the score of 7. The idea is that recall can be any value between 6.5 and 7.5 but that all the recall values in this range are expressed as 7. In a similar way, a score of 42 seconds stands for all the values between 41.5 and 42.5; 41.5 is the lower limit and 42.5 is the upper limit of 42 seconds.

Sometimes scores are expressed in tenths, hundredths, or thousandths. Like integers, these scores have lower and upper limits that extend halfway to the next value on the quantitative scale.

Qualitative Variables

qualitative variable
Variable whose levels are different kinds, not different amounts.

Qualitative variables do not have the continuous nature that quantitative variables have. For example, gender is a qualitative variable. The two measurements on this variable, female and male, are different in a qualitative way. Using a 0 and a 1 instead of names doesn't change this fact. Another example of a qualitative variable is political affiliation, which has measurements such as Democrat, Republican, Independent, and Other.

Some qualitative variables have the characteristic of *order.* College year classification is a qualitative variable with ordered measurements of senior, junior, sophomore, and freshman. Another qualitative variable is military rank, which has measurements such as sergeant, corporal, and private.

Problems and Answers

At the beginning of this chapter, I urged you to engage statistics. Have you? For example, did you read the footnotes? Have you looked up any words you weren't sure of? (How near are you to dictionary definitions when you study?) Have you read a paragraph a second time, wrinkled your brow in concentration, made notes in the book margin, or promised yourself to ask your instructor or another student about something you aren't sure of? *Engagement* shows up as activity. Best of all, the activity at times is a nod to yourself and a satisfied, "Now, I understand."

From time to time, I will use my best engagement tactic: I'll give you a set of problems so that you can practice what you have just been reading about. Working these problems correctly is additional evidence that you have been engaged. You will find the answers at the end of the book in Appendix G. Here are some suggestions for *efficient* learning.

1. Buy yourself a notebook for statistics. Work all the problems for this course in it because I sometimes refer back to a problem you worked previously. When you make an error, don't erase it—put an X through it and work the problem correctly below. Seeing your error later serves as a reminder of what not to do on a test. If you find that *I* have made an error, write to me with a reminder of what not to do in the next edition.
2. Never, *never* look at an answer before you have worked the problem (or at least tried twice to work the problem).
3. For each set of problems, work the first one and then immediately check your answer against the answer in the book. If you make an error, find out why you made it—faulty understanding, arithmetic error, or whatever.
4. Don't be satisfied with just doing the math. If a problem asks for an interpretation, write out your interpretation.
5. When you finish a chapter, go back over the problems immediately, reminding yourself of the various techniques you have learned.
6. Use any blank spaces near the end of the book for your special notes and insights.

Now, here is an opportunity to see how actively you have been reading.

PROBLEMS

1.1. For the quantitative variables that follow, give the lower and upper limits. For the qualitative variables, write "qualitative."
  a. Seconds required to work puzzle—65
  b. Identification number for mild mental retardation in the American Psychiatric Association manual—317
  c. Category of daffodils in a flower show—5
  d. High School Advanced Placement Examination score—4
  e. Milligrams of aspirin—81

1.2. Write a paragraph that gives the definitions of *population, sample, parameter,* and *statistic* and the relationships among them.

1.3. Two different techniques are called *statistics* today. Fill in the blank with one of them.
  a. To reach a conclusion about an unmeasured population, use __________ statistics.
  b. __________ statistics take chance into account to reach a conclusion.
  c. __________ statistics are numbers or graphs that summarize a set of data.

Scales of Measurement

Numbers mean different things in different situations. Consider three answers that appear to be identical but are not:

What number were you wearing in the race?	"5"
What place did you finish in?	"5"
How many minutes did it take you to finish?	"5"

The three 5s all look the same. However, the three variables (identification number, finish place, and time) are quite different. Because of the differences, each 5 has a different interpretation.

To illustrate this difference, consider another person whose answers to the same three questions were 10, 10, and 10. If you take the first question by itself and know that the two people had scores of 5 and 10, what can you say? You can say that the first runner was different from the second, but *that is all.* (Think about this until you agree.) On the second question, with scores of 5 and 10, what can you say? You can say that the first runner was faster than the second and, of course, that they are different. Comparing the 5 and 10 on the third question, you can say that the first runner was twice as fast as the second runner (and, of course, was faster and different).

The point of this discussion is to draw the distinction between the *thing* you are interested in and the *number* that stands for the thing. Much of your experience with numbers has been with pure numbers or with measurements such as time, length, and amount. "Four is twice as much as two" is true for the pure numbers themselves and for time, length, and amount, but it is not true for finish places in a race. Fourth place is not twice anything in relation to second place—not twice as *slow* or twice as *far behind* the second runner.

S. S. Stevens's 1946 article is probably the most widely referenced effort to focus attention on these distinctions. He identified four different *scales of measurement,* each of which carries a different set of information. Each scale uses numbers, but the information that can be inferred from the numbers differs. The four scales are nominal, ordinal, interval, and ratio.

nominal scale
Measurement scale in which numbers serve only as labels and do not indicate any quantitative relationship.

The information in the **nominal scale** is quite simple. In the nominal scale, numbers are used simply as names and have no real quantitative value. Numerals on sports uniforms are an example. Thus, 45 is *different* from 32, but that is all you can say. The person represented by 45 is not "more than" the person represented by 32, and certainly it would be meaningless to calculate a mean from the two scores. Examples of nominal variables include psychological diagnoses, personality types, and political parties. Psychological diagnoses, like other nominal variables, consist of a set of categories. People are assessed and then classified into one of the categories. The categories have both a name (such as posttraumatic stress disorder or autistic disorder) and a number (309.81 and 299.00, respectively). On a nominal scale, the numbers mean only that the categories are different. In fact, for a nominal scale variable, the numbers could be assigned to categories at random. Of course, all things that are alike must have the same number.

ordinal scale
Measurement scale in which numbers are ranks; equal differences between numbers do not represent equal differences between the things measured.

A second kind of scale, the **ordinal scale,** has the characteristic of the nominal scale (different numbers mean different things) plus the characteristic of indicating *greater than* or *less than.* In the ordinal scale, the object with the number 3 has less or more of something than the object with the number 5. Finish places in a race are an example of an ordinal scale. The runners finish in rank order, with 1 assigned to the winner, 2 to the runner-up, and so on. Here, 1 means less time than 2. Judgments about anxiety, quality, and recovery often correspond to an ordinal scale. "Much improved," "improved," "no change," and "worse" are levels of an ordinal recovery variable. Ordinal scales are characterized by *rank order.*

The third kind of scale is the **interval scale,** which has the properties of both the nominal and ordinal scales plus the additional property that *intervals between the numbers are equal.* "Equal interval" means that the distance between the things represented by 2 and 3 is the same as the distance between the things represented by 3 and 4. Temperature is measured on an interval scale. The difference in temperature between 10°C and 20°C is the same as the difference between 40°C and 50°C. The Celsius thermometer, like all interval scales, has an arbitrary zero point. On the Celsius thermometer, this zero point is the freezing point of water at sea level. Zero degrees on this scale does not mean the complete absence of heat; it is simply a convenient starting point. With interval data, there is one restriction: You may not make simple ratio statements. You may not say that 100° is twice as hot as 50° or that a person with an IQ of 60 is half as intelligent as a person with an IQ of 120.[5]

interval scale
Measurement scale in which equal differences between numbers represent equal differences in the thing measured. The zero point is arbitrarily defined.

The fourth kind of scale, the **ratio scale,** has all the characteristics of the nominal, ordinal, and interval scales plus one other: It has a *true zero point,* which indicates a complete absence of the thing measured. On a ratio scale, zero means "none." Height, weight, and time are measured with ratio scales. Zero height, zero weight, and zero time mean that no amount of these variables is present. With a true zero point, you can make ratio statements such as 16 kilograms is four times heavier than 4 kilograms.[6] **Table 1.1** summarizes the major differences among the four scales of measurement.

ratio scale
Measurement scale with characteristics of interval scale; also, zero means that none of the things measured is present.

Knowing the distinctions among the four scales of measurement will help you in two tasks in this course. The kind of *descriptive statistics* you can compute from numbers

TABLE 1.1 Characteristics of the four scales of measurement

	Scale characteristics			
Scale of measurement	Different numbers for different things	Numbers convey greater than and less than	Equal differences mean equal amounts	Zero means none of what was measured was detected
Nominal	Yes	No	No	No
Ordinal	Yes	Yes	No	No
Interval	Yes	Yes	Yes	No
Ratio	Yes	Yes	Yes	Yes

[5] Convert 100°C and 50°C to Fahrenheit ($F = 1.8C + 32$) and suddenly the "twice as much" relationship disappears.

[6] Convert 16 kilograms and 4 kilograms to pounds (1 kg = 2.2 pounds) and the "four times heavier" relationship is maintained.

depends, in part, on the scale of measurement the numbers represent. For example, it is senseless to compute a mean of numbers on a nominal scale. Calculating a mean Social Security number, a mean telephone number, or a mean psychological diagnosis is either a joke or evidence of misunderstanding numbers.

Understanding scales of measurement is sometimes important in choosing the kind of *inferential statistic* that is appropriate for a set of data. If the dependent variable (see next section) is a nominal variable, then a chi square analysis is appropriate (Chapter 14). If the dependent variable is a set of ranks (ordinal data), then a nonparametric statistic is required (Chapter 15). Most of the data analyzed with the techniques described in Chapters 7–13 are interval and ratio scale data.

The topic of scales of measurement is controversial among statisticians. Part of the controversy involves viewpoints about the underlying thing you are interested in and the number that represents the thing (Wuensch, 2005). In addition, it is sometimes difficult to classify some of the variables used in the social and behavioral sciences. Often they appear to fall between the ordinal scale and the interval scale. For example, a score may provide more information than simply rank, but equal intervals cannot be proved. Examples include aptitude and ability tests, personality measures, and intelligence tests. In such cases, researchers generally treat the scores as if they were interval scale data.

Statistics and Experimental Design

Here is a story that will help you distinguish between statistics (applying straight logic) and experimental design (observing what actually happens). This is an excerpt from a delightful book by E. B. White, *The Trumpet of the Swan* (1970, pp. 63–64).

> The fifth-graders were having a lesson in arithmetic, and their teacher, Miss Annie Snug, greeted Sam with a question.
>
> "Sam, if a man can walk three miles in one hour, how many miles can he walk in four hours?"
>
> "It would depend on how tired he got after the first hour," replied Sam. The other pupils roared. Miss Snug rapped for order.
>
> "Sam is quite right," she said. "I never looked at the problem that way before. I always supposed that man could walk twelve miles in four hours, but Sam may be right: that man may not feel so spunky after the first hour. He may drag his feet. He may slow up."
>
> Albert Bigelow raised his hand. "My father knew a man who tried to walk twelve miles, and he died of heart failure," said Albert.
>
> "Goodness!" said the teacher. "I suppose *that* could happen, too."
>
> "Anything can happen in four hours," said Sam. "A man might develop a blister on his heel. Or he might find some berries growing along the road and stop to pick them. That would slow him up even if he wasn't tired or didn't have a blister."
>
> "It would indeed," agreed the teacher. "Well, children, I think we have all learned a great deal about arithmetic this morning, thanks to Sam Beaver."
>
> Everyone had learned how careful you have to be when dealing with figures.

Statistics involves the manipulation of numbers and the conclusions based on those manipulations (Miss Snug). Experimental design (also called research methods) deals with all the things that influence the numbers you get (Sam and Albert). **Figure 1.2** illustrates these two approaches to getting an answer. This text could have been a "pure" statistics book, from which you would learn to analyze numbers without knowing

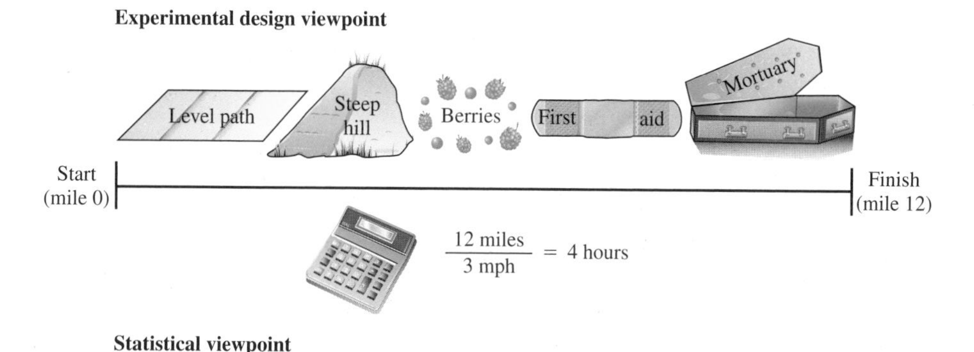

FIGURE 1.2 Travel time from an experimental design viewpoint and a statistical viewpoint

where they came from or what they referred to. You would learn about statistics, but such a book would be dull, dull, dull. On the other hand, to describe procedures for collecting numbers is to teach experimental design—and this book is for a statistics course. My solution to this conflict is generally to side with Miss Snug but to include some aspects of experimental design throughout the book. Knowing experimental design issues is especially important when it comes time to interpret a statistical analysis. Here's a start on experimental design.

Experimental Design Variables

The overall task of an experimenter is to discover relationships among variables. Variables are things that vary, and researchers have studied personality, health, gender, anger, caffeine, memory, beliefs, age, skill.... (I'm sure you get the picture—almost anything can be a variable.)

Independent and Dependent Variables

A simple experiment has two major variables, the **independent variable** and the **dependent variable.** In the simplest experiment, the researcher selects two values of the independent variable for investigation. Values of the independent variable are usually called **levels** and sometimes called **treatments.**

The basic idea is that the researcher finds or creates two groups of participants that are similar except for the independent variable. These individuals are measured on the dependent variable. The question is whether the data will allow the experimenter to claim that the values on the dependent variable *depend* on the level of the independent variable.

The values of the dependent variable are found by *measuring* or *observing* the participants in the investigation. The dependent variable might be scores on a personality test, the number of items remembered, or whether or not a passerby offered assistance. For the independent variable, the two groups might have been selected because they were

independent variable
Variable controlled by the researcher; changes in this variable may produce changes in the dependent variable.

dependent variable
The observed variable that is expected to change as a result of changes in the independent variable in an experiment.

level
One value of the independent variable.

treatment
One value (or level) of the independent variable.

already different—in age, gender, personality, and so forth. Alternatively, the experimenter might have produced the difference in the two groups by an experimental manipulation such as creating different amounts of anxiety or providing different levels of practice.

An example might help. Suppose for a moment that as a budding gourmet cook you want to improve your spaghetti sauce. One of your buddies suggests adding marjoram. To investigate, you serve spaghetti sauce at two different gatherings. For one group of guests the sauce is spiced with marjoram; for the other it is not. At both gatherings you count the number of favorable comments about the spaghetti sauce. *Stop reading; identify the independent and the dependent variables.*[7]

The dependent variable is the number of favorable comments, which is a measure of the taste of the sauce. The independent variable is marjoram, which has two levels: present and absent.

Extraneous Variables

One of the pitfalls of experiments is that every situation has other variables besides the independent variable that might possibly be responsible for the changes in the dependent variable. These other variables are called **extraneous variables.** In the story, Sam and Albert noted several extraneous variables that could influence the time to walk 12 miles.

extraneous variable
Variable other than the independent variable that may affect the dependent variable.

Are there any extraneous variables in the spaghetti sauce example? Oh yes, there are many, and just one is enough to raise a suspicion about a conclusion that relates spaghetti sauce taste to marjoram. Extraneous variables include the amount and quality of the other ingredients in the sauce, the spaghetti itself, the "party moods" of the two groups, and how hungry everybody was. If any of these extraneous variables was actually operating, it weakens the claim that a difference in the comments about the sauce is the result of the presence or absence of marjoram.

The simplest way to remove an extraneous variable is to be sure that all participants are equal on that variable. For example, you can ensure that the sauces are the same except for marjoram by mixing up one pot of ingredients, dividing it into two batches, adding marjoram to one batch but not the other, and then cooking. The "party moods" variable can be controlled (equalized) by conducting the taste test in a laboratory. Controlling extraneous variables is a complex topic that is covered in courses that focus on research methods and experimental design.

In many experiments it is impossible or impractical to control all the extraneous variables. Sometimes researchers think they have controlled them all, only to find that they did not. The effect of an uncontrolled extraneous variable is to prevent a simple cause-and-effect conclusion. Even so, if the dependent variable changes when the independent variable changes, something is going on. In this case, researchers can say that the two variables are *related* but that other variables may play a part, too.

At this point you can test your understanding by engaging yourself with these questions: What were the independent and dependent variables in the Zeigarnik experiment? How many levels of the independent variable were there?

---

[7] Try for answers. Then, if need be, here's a hint: First, identify the dependent variable; for the dependent variable, you don't know values until data are gathered. Next, identify the independent variable; you can tell what the values of the independent variable are just from the description of the design.

How well did Zeigarnik control extraneous variables? For one thing, each participant was tested at both levels of the independent variable. That is, the recall of each participant was measured for interrupted tasks and for uninterrupted tasks. One advantage of this technique is that it naturally controls many extraneous variables. Thus, extraneous variables such as age and motivation were exactly the same for tasks that were interrupted as for tasks that were not because the same people contributed scores to both levels.

At various places in the following chapters, I will explain experiments and the statistical analyses using the terms *independent* and *dependent variables*. These explanations usually assume that all extraneous variables were controlled; that is, you may assume that the experimenter knew how to design the experiment so that changes in the dependent variable could be attributed correctly to changes in the independent variable. However, I present a few investigations (like the spaghetti sauce example) that I hope you recognize as being so poorly designed that conclusions cannot be drawn about the relationship between the independent variable and dependent variable. Be alert.

Here's my summary of the relationship between statistics and experimental design. Researchers suspect that there is a relationship between two variables. They design and conduct an experiment; that is, they *choose* the levels of the independent variable (treatments), *control* the extraneous variables, and then *measure* the participants on the dependent variable. The measurements (data) are analyzed using statistical procedures. Finally, the researcher tells a story that is consistent with the results obtained and the procedures used.

Statistics and Philosophy

The two previous sections directed your attention to the relationship between statistics and experimental design; this section will direct your thoughts to the place of statistics in the grand scheme of things.

Explaining the grand scheme of things is the task of philosophy and, over the years many schemes have been advanced. For a scheme to be considered a grand one, it has to propose answers to questions of **epistemology**—that is, answers to questions about the nature of knowledge.

epistemology
The study or theory of the nature of knowledge.

One of the big questions in epistemology is: How do we acquire knowledge? The answer *reason* and the answer *experience* have gotten a lot of attention.[8] For those who emphasize the importance of reason, mathematics has been a continuing source of inspiration. Classical mathematics starts with a set of axioms that are assumed to be true. Theorems are thought up and are then proved by giving axioms as reasons. Once a theorem is proved, it can be used as a reason in a proof of other theorems.

Statistics has its foundations in mathematics. Statistics, therefore, is based on reason. As you go about the task of memorizing definitions of terms such as $\overline{X}$, σ, and ΣX, calculating their values from data, and telling the story of what they mean, know deep down that you are using a technique in logical reasoning. Using logical reasoning is called rationalism, which is one approach to questions of epistemology. (Experimental design is more complex; it includes experience and observation as well as reasoning.)

During the 20th century, statistical methods of analyzing data revolutionized the philosophy of science. The 19th century was characterized by observation and

[8] In philosophy, those who emphasize reason are rationalists and those who emphasize experience are empiricists.

description. The underlying philosophy was that natural phenomena had an objective reality and the variation that always accompanied a set of observations was the result of imprecise observing or imprecise instruments. In the 20th century, science used a newly developed set of techniques called statistics to analyze the variation that was always present. The result was that particular amounts of variation could be associated with particular causes. Many philosophers of science adopted the belief that the underlying reality was variation rather than some static, Platonic ideal (Salsburg, 2001).

Let's move from these formal descriptions of philosophy to a more informal one. A very common task of most human beings can be described as *trying to understand.* Statistics has helped many in their search for better understanding, and it is such people who have recommended (or demanded) that statistics be taught in school. A reasonable expectation is that you, too, will find statistics useful in your future efforts to understand and persuade.

Speaking of persuasion, you have probably heard it said, "You can prove anything with statistics." The implied message is that a conclusion based on statistics is suspect because statistical methods are unreliable. Well, it just isn't true that statistical methods are unreliable, but it is true that people can misuse statistics (just as any tool can be misused). One of the great advantages of studying statistics is that you get better at recognizing statistics that are used improperly.

Statistics: Then and Now

Statistics began with counting. The beginning of counting, of course, was prehistory. The origin of the mean is almost as obscure. That statistic was in use by the early 1700s, but no one is credited with its discovery. Graphs, however, began when J. H. Lambert, a Swiss-German scientist and mathematician, and William Playfair, an English political economist, invented and improved graphs in the period 1765 to 1800 (Tufte, 2001).

In 1834, a group of Englishmen in London formed the Royal Statistical Society. Just 5 years later, on November 27, 1839, at 15 Cornhill in Boston, a group of Americans founded the American Statistical Society. Less than 3 months later, for a reason that you can probably figure out, the group changed its name to the American Statistical Association, which continues today (*www.amstat.org*).

According to Walker (1929), the first university course in statistics in the United States was probably "Social Science and Statistics," taught at Columbia University in 1880. The professor was a political scientist, and the course was offered in the economics department. In 1887, at the University of Pennsylvania, separate courses in statistics were offered by the departments of psychology and economics. By 1891, Clark University, the University of Michigan, and Yale had been added to the list of schools that taught statistics, and anthropology had been added to the list of departments. Biology was added in 1899 (Harvard) and education in 1900 (Columbia).

You might be interested in when statistics was first taught at your school and in what department. College catalogs are probably your most accessible source of information.

This course provides you with the opportunity to improve your ability to understand and use statistics. Kirk (2008) identifies four levels of statistical sophistication:

Category 1—those who understand statistical presentations

Category 2—those who understand, select, and apply statistical procedures

Category 3—applied statisticians who help others use statistics

Category 4—mathematical statisticians who develop new statistical techniques and discover new characteristics of old techniques

I hope that by the end of your statistics course, you will be well along the path to becoming a category 2 statistician.

How to Analyze a Data Set

The end point of analyzing a data set is a story that explains the relationships among the variables in the data set. I recommend that you analyze a data set in three steps. The first step is exploratory. Read all the information and examine the data. Calculate descriptive statistics and focus on the differences that are revealed. In this textbook, descriptive statistics are emphasized in Chapters 2 through 6 and include graphs, percentages, and means. Calculating descriptive statistics helps you develop preliminary ideas for your story (step 3). The second step is to answer the question, Could the differences observed be due to chance?[9] Often, an inferential statistic such as a null hypothesis statistical test will provide an answer. Inferential statistics are covered in Chapters 7 through 15. The third step is to write the story the data reveal. Incorporate the descriptive and inferential statistics to support the conclusions in the story. Of course, the skills you've learned and taught yourself about composition will be helpful as you compose and write your story. Don't worry about length; most good statistical stories about simple data sets can be told in one paragraph.

Write your story using journal style, which is quite different from textbook style. Textbook style, at least this textbook, is chatty, redundant, and laced with footnotes.[10] Journal style, on the other hand, is terse, formal, and devoid of footnotes. For examples of journal style, see Appendix G, which has answers to the textbook problems. Look at paragraphs that have Interpretation highlighted.

Helpful Features of This Book

At various points in the chapter, I have encouraged your engagement in statistics. Your active participation is *necessary* if you are to learn statistics. For my part, I worked to organize this book and write it in a way that encourages active participation. Here are some of the features you should find helpful.

Objectives

Each chapter begins with a list of skills the chapter is designed to help you acquire. Read this list of objectives first to find out what you are to learn to do. Then thumb through the chapter and read the headings. Next, study the chapter, working all the problems *as you come to them.* Finally, reread the objectives. Can you meet each one? If so, put a check mark beside that objective.

[9] The question is whether the *difference* could be due to chance, not whether the data are due to chance.

[10] You are reading the footnotes, aren't you? Your answer — "Well, yes, it seems I am."

Problems and Answers

The problems in this text are in small groups within the chapter rather than clumped together at the end. This encourages you to read a little and work problems, followed by more reading and problems. Psychologists call this pattern *spaced practice*. Spaced practice patterns lead to better performance than massed practice patterns.

Because working problems and writing answers are *necessary* to learn statistics, I have used interesting problems and have sometimes written interesting answers for the problems; in some cases, I even put new information in the answers. The problems come from a variety of disciplines; the answers are in Appendix G.

Many of the problems are conceptual questions that do not require any arithmetic. Think these through and *write* your answers. Being able to calculate a statistic is almost worthless if you cannot explain in English what it means. Writing reveals how thoroughly you understand. To emphasize the importance of explanation, I've highlighted the Interpretation portion of each answer in Appendix G.

On several occasions, problems or data sets are used more than once, either later in the chapter or even in another chapter. If you do not work the problem when it is first presented, you are likely to be frustrated when it appears again. To alert you, I have put an asterisk (*) beside problems that are used again.

At the end of many chapters, comprehensive problems are marked with a ✵. Working these problems requires knowing most of the material in the chapter. For most students, it is best to work all the problems, but be sure you can work those marked with a ✵.

Sometimes you may find a minor difference between your answer and mine (in the last decimal place, for example). This discrepancy will probably be the result of rounding errors and does not deserve further attention.

clues to the future

Often a concept is presented that will be used again in later chapters. These ideas are separated from the rest of the text in a box labeled "Clue to the Future." You have already seen one of these "Clues" in this chapter. Attention to these concepts will pay dividends later in the course.

error detection

I have boxed in, at various points in the book, ways to detect errors. Some of these "Error Detection" tips will also help you better understand the concept. Because many of these checks can be made early, they can prevent the frustrating experience of getting an impossible answer when the error could have been caught in step 2.

Figure and Table References

Sometimes the words *Figure* and *Table* are in boldface print. This means that you should examine the figure or table at that point. Afterward, it will be easy for you to return to your place in the text—just find the boldface type.

Transition Passages

At six places in this book, there are major differences between the material you just finished and the material in the next section. "Transition Passages," which describe the differences, separate these sections.

Glossaries

This book has three separate glossaries of words, symbols, and formulas.

1. *Words.* The first time an important word is used in the text, it appears in boldface type accompanied by a definition in the margin. In later chapters, the word may be boldfaced again, but margin definitions are not repeated. Appendix D is complete glossary of words (page 405). I suggest you mark this appendix.
2. *Symbols.* Statistical symbols are defined in Appendix E (page 409). Mark it too.
3. *Formulas.* Formulas for all the statistical techniques used in the text are printed in Appendix F (page 411), in alphabetical order according to the name of the technique.

Computers, Calculators, and Pencils

Computer programs, calculators, and pencils with erasers are all tools used at one time or another by statisticians. Any or all of these devices may be part of the course you are taking. Regardless of the calculating aids that you use, however, your task is the same:

- Read a problem.
- Decide what statistical procedure to use.
- Apply that procedure using the tools available to you.
- Write an interpretation of the results.

In addition, you should be able to detect gross errors by comparing your statistical computations to the raw data.

Pencils, calculators, and computers represent, in ascending order, tools that are more and more error-free. People who routinely do statistics use computers to calculate answers. You may or may not use one at this point. Remember, though, whether you are using computer programs or not, your principal task is to understand and describe.

For many of the worked examples in this book, I included the output of a popular statistical software program, SPSS.[11] If your course includes SPSS, these tables should help familiarize you with the program. VassarStats, a user-friendly program maintained by Richard Lowry, is available free on the web at *http://faculty.vassar.edu/lowry/VassarStats.html.*

[11] The original name of the program was Statistical Package for Social Sciences.

Concluding Thoughts for This Introductory Chapter

Most students find that this book works well for them as a *textbook* in their statistics course. Those who keep their book often find it a very useful *reference* book after the course is over. In courses that follow statistics and even after leaving school, they find themselves looking up a definition or reviewing a procedure.[12] I hope that you not only learn from this book but also join those students who keep the book as part of their personal library.

This book is a fairly complete introduction to elementary statistics. There is more to the study of statistics—lots, lots more—but there is a limit to what you can do in one term. Some of you, however, will learn the material in this book and want to know more. If you are such a student, I suggest that you "forage for yourself." Encyclopedias, both general and specialized, are good places to forage. Try the *International Encyclopedia of the Social and Behavioral Sciences* (2001) or the *Encyclopedia of Statistics in Behavioral Science* (2005).

I also recommend that when you finish this course (but before any final examination), you study and work the problems in Chapter 16, the last chapter in the book. It is designed to be an overview/integrative chapter.

For me, studying statistics and using them to understand the world around me has been both helpful and satisfying. I hope you come to a similar conclusion.

PROBLEMS

1.4. Name the four scales of measurement identified by S. S. Stevens.

1.5. Give the properties of each of the scales of measurement.

1.6. Identify the scale of measurement in each of the following cases.

- **a.** Geologists have a "hardness scale" for identifying different rocks, called Mohs' scale. The hardest rock (diamond) has a value of 10 and will scratch all others. The second hardest will scratch all but the diamond, and so on. Talc, with a value of 1, can be scratched by every other rock. (A fingernail, a truly handy field-test instrument, has a value between 2 and 3.)
- **b.** The volumes of three different cubes are 40, 64, and 65 cubic inches.
- **c.** Three different highways are identified by their numbers: 40, 64, and 65.
- **d.** Republicans, Democrats, Independents, and Others are identified on the voters' list with the numbers 1, 2, 3, and 4, respectively.
- **e.** The winner of the Miss America contest was Miss California; the four runners-up were Miss Ohio, Miss Illinois, Miss Pennsylvania, and Miss Michigan.[13]
- **f.** The prices of the three items are $3.00, $10.00, and $12.00.
- **g.** She earned three degrees: B.A., M.S., and Ph.D.

---

[12] This text's index is unusually extensive. If you make margin notes, they will help, too.

[13] Contest winners have come most frequently from these states, which have had six, six, five, five, and five winners, respectively.

1.7. Undergraduate students conducted the three studies that follow. For each study identify the dependent variable, the independent variable, the number of levels of the independent variable, and the names of the levels of the independent variable.

a. Becca had students in a statistics class rate a resume, telling them that the person had applied for a position that included teaching statistics at their college. The students rated the resume on a scale of 1 (not qualified) to 10 (extremely qualified). All the students received identical resumes, except that the candidate's first name was Jane on half the resumes and John on the other half.

b. Michael's participants filled out the Selfism scale, which measures narcissism. (Narcissism is neurotic self-love.) In addition, students were classified as first-born, second-born, and later-born.

c. Johanna had participants read a description of a crime and "Mr. Anderson," the person convicted of the crime. For some participants, Mr. Anderson was described as a janitor. For others, he was described as a vice president of a large corporation. For still others, no occupation was given. After reading the description, participants recommended a jail sentence (in months) for Mr. Anderson.

1.8. Researchers who are now well known conducted the three classic studies that follow. For each study, identify the dependent variable, the independent variable, and the number and names of the levels of the independent variable. Complete items i and ii.

a. Theodore Barber hypnotized 25 different people, giving each a series of suggestions. The suggestions included arm rigidity, hallucinations, color blindness, and enhanced memory. Barber counted the number of suggestions that the hypnotized participants complied with (the mean was 4.8). For another 25 people, he simply asked them to achieve the best score they could (but no hypnosis was used). This second group was given the same suggestions, and the number complied with was counted (the mean was 5.1). (See Barber, 1976.)

i. Identify a nominal variable and a statistic.

ii. In a sentence, describe what Barber's study shows.

b. Elizabeth Loftus had participants view a film clip of a car accident. Afterward, some were asked, How fast was the car going? and others were asked, How fast was the car going when it passed the barn? (There was no barn in the film.) A week later, Loftus asked the participants, Did you see a barn? If the barn had been mentioned earlier, 17 percent said yes; if it had not been mentioned, 3 percent said yes. (See Loftus, 1979.)

i. Identify a population and a parameter.

ii. In a sentence, describe what Loftus's study shows.

c. Stanley Schachter and Larry Gross gathered data from obese male students for about an hour in the afternoon. At the end of this time, a clock on the wall was correct (5:30 p.m.) for 20 participants, slow (5:00 p.m.) for 20 others, and fast (6:00 p.m.) for 20 more. The actual time, 5:30, was the usual dinnertime for these students. While participants filled out a final questionnaire, Wheat

Thins® were freely available. The weight of the crackers each student consumed was measured. The means were: 5:00 group—20 grams; 5:30 group—30 grams; 6:00 group—40 grams. (See Schachter and Gross, 1968.)
i. Identify a ratio scale variable.
ii. In a sentence, describe what this study shows.

1.9. There are uncontrolled extraneous variables in the study described here. Name as many as you can. Begin by identifying the dependent and independent variables. An investigator concluded that statistics Textbook A was better than Textbook B, after comparing two statistics classes. One class, which met MWF at 10:00 A.M., used Textbook A and was taught by Professor X. The other class, which met for 3 hours on Wednesday evening, used Textbook B and was taught by Professor Y. At the end of the term, all students took the same comprehensive test. The mean score for the Textbook A students was higher than the mean score for the Textbook B students.

1.10. In philosophy, the study of the nature of knowledge is called __________.

1.11. a. The two approaches to epistemology identified in the text are __________ and __________.
b. Statistics has its roots in __________.

1.12. Your textbook recommends a three-step approach to analyzing a data set. Summarize the steps.

1.13. Read the nine objectives at the beginning of this chapter. Responding to them will help you consolidate what you have learned.

ADDITIONAL HELP FOR CHAPTER 1

Visit *cengage.com/psychology/spatz*. At the Student Companion Site, you'll find multiple-choice tutorial quizzes flashcards with definitions, and workshops. For this chapter, there is a Statistical Workshop on Scale of Measurement and a Research Methods Workshop on Experimental Methods (dependent and independent variables).

KEY TERMS

Dependent variable (p. 13)
Descriptive statistics (p. 3)
Epistemology (p. 15)
Extraneous variable (p. 14)
Independent variable (p. 13)
Inferential statistics (p. 3)
Interval scale (p. 11)
Level (p. 13)
Lower limit (p. 8)
Mean (p. 3)
Nominal scale (p. 10)
Ordinal scale (p. 10)
Parameter (p. 7)
Population (p. 6)
Qualitative variable (p. 8)
Quantitative variable (p. 8)
Ratio scale (p. 11)
Sample (p. 6)
Statistic (p. 7)
Treatment (p. 13)
Upper limit (p. 8)
Variable (p. 7)

transition passage
to descriptive statistics

THE MOST COMMON way to divide the statistics pie is into descriptive statistics and inferential statistics. The next five chapters are about descriptive statistics. You are already familiar with some of these descriptive statistics, such as the mean, range, and bar graphs. Others may be less familiar—the standard deviation, correlation coefficient, and boxplot. All of these and others that you will study will be helpful in your efforts to understand data.

The phrase: *Exploring Data* appears in three of the chapter titles that follow. The phrase is a reminder to approach a data set with the attitude of an explorer, an attitude of *What can I find here*? Descriptive statistics are especially valuable in the early stages of an analysis as you explore what the data have to say. Later, descriptive statistics are essential when you convey your story of the data to others. In addition, many descriptive statistics have important roles in the inferential statistical techniques that are covered in later chapters. Let's get started.

CHAPTER

2

Exploring Data: Frequency Distributions and Graphs

OBJECTIVES FOR CHAPTER 2

After studying the text and working the problems in this chapter, you should be able to:

1. Arrange a set of scores into simple and grouped frequency distributions
2. Describe the characteristics of frequency polygons, histograms, and bar graphs and explain the information provided in each one
3. Name certain distributions by looking at their shapes
4. Describe the characteristics of a line graph
5. Comment on the recent use of graphics

YOU HAVE NOW invested some time and effort learning the preliminaries of elementary statistics. Concepts have been introduced and you have an overview of the course. The chapters that follow immediately are about descriptive statistics. To get an idea of the specific topics, read this chapter title and the next four chapter titles thoughtfully.

raw score
Score obtained by observation or from an experiment.

You'll begin your study of descriptive statistics with a group of **raw scores.** Raw scores, or raw data, can be obtained in many ways. For example, if you administer a questionnaire to a group of college students, the scores from the questionnaire are raw scores. I will illustrate several of the concepts in this chapter and the next three by using raw scores that are representative of 100 college students.

If you would like to engage fully in this chapter, take 3 minutes to complete the five-item questionnaire in **Figure 2.1,** and calculate your score. Score yourself now, before you read further. (If you read the five chapter titles, I'll bet that you answered the five questions. Engagement always helps.)

The investigation of subjective well-being is an increasingly hot topic in psychology. Subjective well-being is a person's evaluation of his or her life. Recently, Diener and Seligman (2004) marshaled evidence that governments would be better served to focus

Instructions: Five statements with which you may agree or disagree are shown below. Using the scale that follows, indicate your agreement with each item by placing the appropriate number on the line preceding that item. Please be open and honest in your responding.

1 = Strongly disagree	5 = Slightly agree
2 = Disagree	6 = Agree
3 = Slightly disagree	7 = Strongly agree
4 = Neither agree nor disagree	

____ 1. In most ways my life is close to my ideal.

____ 2. The conditions of my life are excellent.

____ 3. I am satisfied with my life.

____ 4. So far I have gotten the important things I want in life.

____ 5. If I could live my life over, I would change almost nothing.

Scoring instructions: Add the numbers in the blanks to determine your score.

FIGURE 2.1 Questionnaire

on policies that increase subjective well-being rather than continuing with their traditional concern, which is policies of economic well-being.

As a concept, subjective well-being consists of emotional components and cognitive components. An important cognitive component is satisfaction with life. To measure this cognitive component, Ed Diener and his colleagues (Diener et al., 1985) developed the Satisfaction With Life Scale (SWLS). The SWLS is the short, five-item questionnaire in Figure 2.1. It is a reliable, valid measure of a person's global satisfaction with life. **Table 2.1** is an unorganized collection of SWLS scores of 100 representative college students.

TABLE 2.1 **Representative scores of 100 college students on the Satisfaction With Life Scale**

15	31	22	26	19	27	33	24	27	25
20	26	29	32	21	13	35	9	25	25
28	23	17	27	30	16	11	29	26	20
23	30	16	24	27	29	26	10	23	34
5	19	28	29	27	30	32	22	17	13
35	28	27	25	26	25	23	21	29	28
20	27	30	22	22	12	32	25	24	23
20	24	26	26	29	33	29	24	20	25
19	25	9	21	32	30	27	24	10	5
22	26	26	28	23	27	25	28	27	31

Simple Frequency Distributions

Table 2.1, which is just a jumble of numbers, is not very interesting or informative. (My guess is that you glanced at it and went quickly on.) A much more informative presentation of the 100 scores is an arrangement called a **simple frequency distribution. Table 2.2** is a simple frequency distribution—an ordered arrangement that shows the frequency of each score. Look at Table 2.2. (I would guess that you spent more time on Table 2.2 than on Table 2.1 and that you got more information from it.)

simple frequency distribution
Scores arranged from highest to lowest, with the frequency shown for each score.

Look again at Table 2.2. The SWLS score column tells the name of the variable that is being measured. The generic name for any variable is *X*, which is the symbol used in formulas. The Frequency (*f*) column shows how frequently a score occurred. The tally marks are used when you construct a rough-draft version and are not usually included in the final form. *N* is the number of scores and is found by summing the numbers in the *f* column. You will have a lot of opportunities to construct simple frequency distributions, so here are the steps. General instructions are given first, followed by their application to the data in Table 2.1 (italics).

1. Find the highest and lowest scores. *Highest score is 35; lowest is 5.*
2. In column form, write in descending order all the numbers. *35 to 5.*
3. At the top of the column, name the variable being measured. *Satisfaction With Life Scale* scores.
4. Start with the number in the upper left-hand corner of the scores, draw a line under it, and place a tally mark beside that number in the column of numbers. *Underline 15 in Table 2.1, and place a tally mark beside 15 in Table 2.2.*

TABLE 2.2 Rough draft of a simple frequency distribution of Satisfaction With Life Scale scores for a representative sample of 100 college students

SWLS score (*X*)	Tally marks	Frequency (*f*)	SWLS score (*X*)	Tally marks	Frequency (*f*)
35	//	2	19	///	3
34	/	1	18		0
33	//	2	17	//	2
32	////	4	16	//	2
31	//	2	15	/	1
30	~~////~~	5	14		0
29	~~////~~ //	7	13	//	2
28	~~////~~ /	6	12	/	1
27	~~////~~ ~~////~~	10	11	/	1
26	~~////~~ ////	9	10	//	2
25	~~////~~ ////	9	9	//	2
24	~~////~~ /	6	8		0
23	~~////~~ /	6	7		0
22	~~////~~	5	6		0
21	///	3	5	//	2
20	~~////~~	5			$N = 100$
	(continued above)				

5. Continue underlining and tallying for all the unorganized scores.
6. Add a column labeled f (frequency).
7. Count the number of tallies by each score and enter the count in the f column. *2, 1, 2, 4, . . . , 0, 2.*
8. Add the numbers in the f column. If the sum is equal to N, you haven't left out any scores. *Sum = 100.*

error detection

Underlining numbers is much better than crossing them out. Easy-to-read numbers are appreciated later when you check your work or do an additional analysis.

A simple frequency distribution is a useful way to explore a set of data because you pick up valuable information with just a glance. For example, the highest and lowest scores are readily apparent in any frequency distribution. In addition, after some practice with frequency distributions, you can ascertain the general shape of the distribution and make an informed guess about measures of central tendency and variability.

Table 2.3 shows a formal presentation of the data in Table 2.1. Formal presentations are used to present data to colleagues, professors, supervisors, editors, and others. Formal presentations usually do not include tally marks and often do not include zero-frequency scores.

TABLE 2.3 Formal simple frequency distribution of Satisfaction With Life Scale scores for a representative sample of 100 college students

SWLS score (X)	Frequency (f)	SWLS score (X)	Frequency (f)
35	2	22	5
34	1	21	3
33	2	20	5
32	4	19	3
31	2	17	2
30	5	16	2
29	7	15	1
28	6	13	2
27	10	12	1
26	9	11	1
25	9	10	2
24	6	9	2
23	6	5	2
(continued above)			$N = 100$

Grouped Frequency Distributions

Many researchers would condense Table 2.3 even more. The result is a **grouped frequency distribution. Table 2.4** is a rough-draft example of such a distribution.[1] Raw data are usually condensed into a grouped frequency distribution when researchers want to present the data as a graph or as a table.

grouped frequency distribution
Scores compiled into equal-sized intervals. Includes the frequency of scores in each interval.

class interval
A range of scores in a grouped frequency distribution.

In a grouped frequency distribution, scores are grouped into equal-sized ranges called **class intervals.** In Table 2.4, the entire range of scores, from 35 to 5, is reduced to 11 class intervals. Each interval covers three scores; the symbol i indicates the size of the interval. In Table 2.4, $i = 3$. The midpoint of each interval represents all the scores in that interval. For example, five students had scores of 15, 16, or 17. The midpoint of the class interval 15–17 is 16.[2] The midpoint, 16, represents all 5 scores. There are no scores in the interval 6–8, but zero-frequency intervals are included in formal grouped frequency distributions if they are within the range of the distribution.

Class intervals have lower and upper limits, much like simple scores obtained by measuring a quantitative variable. A class interval of 15–17 has a lower limit of 14.5 and an upper limit of 17.5.

The only difference between grouped frequency distributions and simple frequency distributions is class intervals. The details of establishing class intervals are described in Appendix B. For problems in this chapter, I will give you the class intervals to use.

TABLE 2.4 Rough draft of a grouped frequency distribution of Satisfaction With Life Scale scores ($i = 3$)

SWLS scores (class interval)	Midpoint (X)	Tally marks	f
33–35	34	~~||||~~	5
30–32	31	~~||||~~ ~~||||~~ |	11
27–29	28	~~||||~~ ~~||||~~ ~~||||~~ ~~||||~~ |||	23
24–26	25	~~||||~~ ~~||||~~ ~~||||~~ ~~||||~~ ||||	24
21–23	22	~~||||~~ ~~||||~~ ||||	14
18–20	19	~~||||~~ |||	8
15–17	16	~~||||~~	5
12–14	13	|||	3
9–11	10	~~||||~~	5
6–8	7		0
3–5	4	||	2
			$N = 100$

[1] A formal grouped frequency distribution, much like a formal simple frequency distribution, does not include tally marks.

[2] When the scores are whole numbers, make i an odd number. With i odd, the midpoint of the class interval is a whole number.

PROBLEMS

***2.1.** (This is the first of many problems with an asterisk. See the footnote.) The following numbers are the heights in inches of two groups of Americans in their 20s. Choose the group that interests you more, and organize the 50 numbers into a simple frequency distribution using the rough-draft form. For the group you choose, your result will be fairly representative of the whole population of 20- to 29-year-olds (*Statistical Abstract of the United States: 2009, 2008*).

Women

64	67	63	65	59
66	66	62	65	65
60	72	64	61	65
69	64	64	66	60
65	67	65	62	68
60	65	63	64	60
66	64	59	63	65
63	63	66	64	65
66	67	62	62	63
64	70	64	63	65

Men

72	65	72	68	70
69	73	71	69	67
77	67	72	73	70
73	64	72	69	69
70	71	71	70	75
72	68	62	68	74
66	70	72	66	75
69	71	68	73	69
71	69	69	65	76
71	72	65	70	70

***2.2.** (The frequency distribution you will construct for this problem is a little different. The "scores" are names and thus nominal data.) A political science student traveled on every street in a voting precinct on election day morning, recording the yard signs (by initial) for the five candidates. The five candidates were Attila (A), Bolivar (B), Gandhi (G), Lenin (L), and Mao (M). Construct an appropriate frequency distribution from her observations. (She hoped to find the relationship between yard signs and actual votes.)

G A M M M G G L A G B A A G G B L M M
A G G B G L M A A L M G G M G L G A A
B L G G A G A M L M G B A G L G M A

***2.3.** You may have heard or read that the normal body temperature (oral) is 98.6°F. The numbers that follow are temperature readings from healthy adults, aged 18–40. Arrange the data into a grouped frequency distribution, using 99.3–99.5 as the highest class interval and 96.3–96.5 as the lowest. (Based on Mackowiak, Wasserman, and Levine, 1992.)

98.1	97.5	97.8	96.4	96.9	98.9	99.5	98.6	98.2	98.3
97.9	98.0	97.2	99.1	98.4	98.5	97.4	98.0	97.9	98.3
98.8	99.5	98.7	97.9	97.7	99.2	98.0	98.2	98.3	97.0
99.4	98.9	97.9	97.4	97.8	98.6	98.7	97.9	98.4	98.8

***2.4.** An experimenter read 60 related statements to a class. He then asked the students to indicate which of the next 20 statements were among the first 60.

* An asterisk means that the information in the problem will be used in other problems later in the book. If you do all the problems in a notebook, you will have a handy reference when a problem turns up again.

Due to the relationships among the concepts in the sentences, many seemed familiar but, in fact, none of the 20 had been read before. The following scores indicate the number (out of 20) that each student had "heard earlier." (See Bransford and Franks, 1971.) Arrange the scores into a simple frequency distribution. Based on this description and your frequency distribution, write a sentence of interpretation.

14	11	10	8	12	13	11	10	16	11
11	9	9	7	14	12	9	10	11	6
13	8	11	11	9	8	13	16	10	11
9	9	8	12	11	10	9	7	10	

Graphs of Frequency Distributions

You have no doubt heard the saying *A picture is worth a thousand words.* When it comes to numbers, a not-yet-well-known saying is *A graph is better than a thousand numbers.* Actually, as long as I am rewriting sayings, I would also like to say *The more numbers you have, the more valuable graphics are.* Graphics are becoming more and more important in data analysis and persuasion.

abscissa
The horizontal axis of a graph; *X* axis.

ordinate
The vertical axis of a graph; *Y* axis.

Pictures that present statistical data are called graphics. The most common graphic is a graph composed of a horizontal axis (variously called the baseline, *X* axis, or **abscissa**) and a vertical axis (called the *Y* axis, or **ordinate**). To the right and upward are both positive directions; to the left and downward are both negative directions. The axes cross at the *origin*. **Figure 2.2** is a picture of these words.

This section presents three kinds of graphs that are used to present frequency distributions—*frequency polygons, histograms,* and *bar graphs.* Frequency distribution graphs present an entire set of observations from a sample or a population. If the variable being graphed is a *quantitative* variable, use a frequency polygon or a

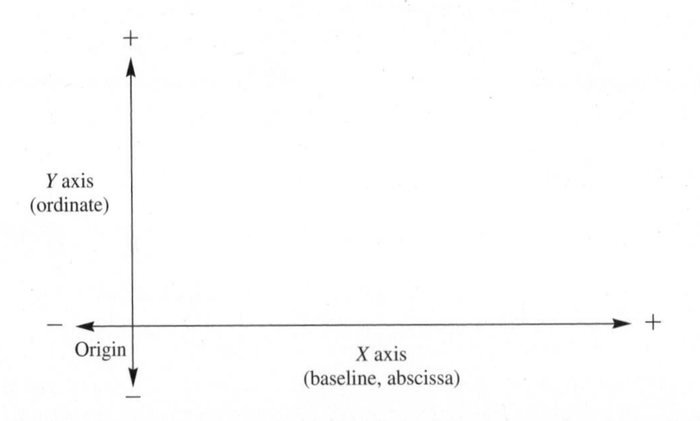

FIGURE 2.2 The elements of a graph

FIGURE 2.3 Frequency polygon of Satisfaction With Life Scale scores of 100 college students

histogram; if the variable is *qualitative,* use a bar graph. (Think about the SWLS scores in Table 2.4 and the yard sign data in problem 2.2. Which variable is quantitative and which is qualitative?)

Frequency Polygon

A **frequency polygon** is used to graph quantitative variables. The SWLS scores in Table 2.4 are quantitative data, so a frequency polygon is appropriate. The result is **Figure 2.3.** An explanation of Figure 2.3 will serve as your introduction to constructing polygons.

frequency polygon
Frequency distribution graph of a quantitative variable; frequency points are connected by lines.

Each point of the frequency polygon represents two numbers: the class midpoint directly below it on the *X* axis and the frequency of that class directly across from it on the *Y* axis. By looking at the data points in Figure 2.3, you can see that five students are represented by the midpoint 34, zero students by 7, and so on.

The name of the variable being graphed goes on the *X* axis (Satisfaction With Life Scale scores). The *X* axis is marked off in equal intervals, with each tick mark indicating the midpoint of a class interval. Low scores are on the left. The lowest class interval midpoint, 1, and the highest, 37, have zero frequencies. Frequency polygons are closed at both ends. In cases where the lowest score in the distribution is well above zero, it is conventional to replace numbers smaller than the lowest score on the *X* axis with a slash mark, which indicates that the scale to the origin is not continuous.

Histogram

A **histogram** is another graphing technique that is appropriate for quantitative variables. **Figure 2.4** shows the SWLS data of Table 2.4 graphed as a histogram. A histogram is constructed by raising bars from the *X* axis to the appropriate frequencies. The lines that separate the bars intersect the *X* axis at the lower and upper limits of the class intervals.

histogram
Frequency distribution graph of a quantitative variable with frequencies indicated by contiguous vertical bars.

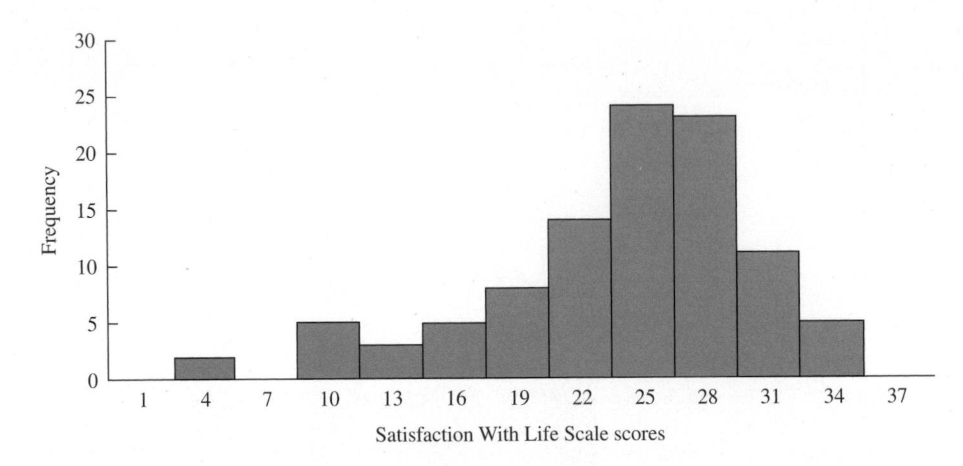

FIGURE 2.4 Histogram of Satisfaction With Life Scale scores of 100 college students

Here are some considerations for deciding whether to use a frequency polygon or a histogram. On the one hand, if you are displaying two overlapping distributions on the same axes, a frequency polygon is less cluttered than a histogram. On the other hand, it is easier to read frequencies from a histogram, and histograms are the better choice when you are presenting discrete data. (*Discrete data* are quantitative data that do not have intermediate values; the number of children in a family is an example.)

Bar Graph

A **bar graph** is used to present the frequencies of the categories of a qualitative variable. A conventional bar graph looks exactly like a histogram except for the wider spaces between the bars. The space is usually a signal that a qualitative variable is being graphed. Conventionally, bar graphs also have the name of the variable being graphed on the *X* axis and frequency on the *Y* axis.

bar graph
Graph of the frequency distribution of nominal or qualitative data.

If an ordinal scale variable is being graphed, the order of the values on the *X* axis follows the order of the variable. If, however, the variable is a nominal scale variable, then *any* order on the *X* axis is permissible. Alphabetizing might be best. Other considerations may lead to some other order.

Figure 2.5 is a bar graph of the six most common liberal arts majors among U.S. college graduates in 2006–2007. The majors are listed on the *X* axis; the number of graduates is on the *Y* axis. I ordered the majors from most to fewest graduates, but other orders would be more appropriate in other circumstances. For example, moving the bar for English majors to the first position would be better for a presentation about English majors.

PROBLEMS

2.5. Answer the following questions for Figure 2.3.
 a. What is the meaning of the number 25 on the *X* axis?
 b. What is the meaning of the number 5 on the *Y* axis?
 c. How many students had scores in the class interval 6–8?

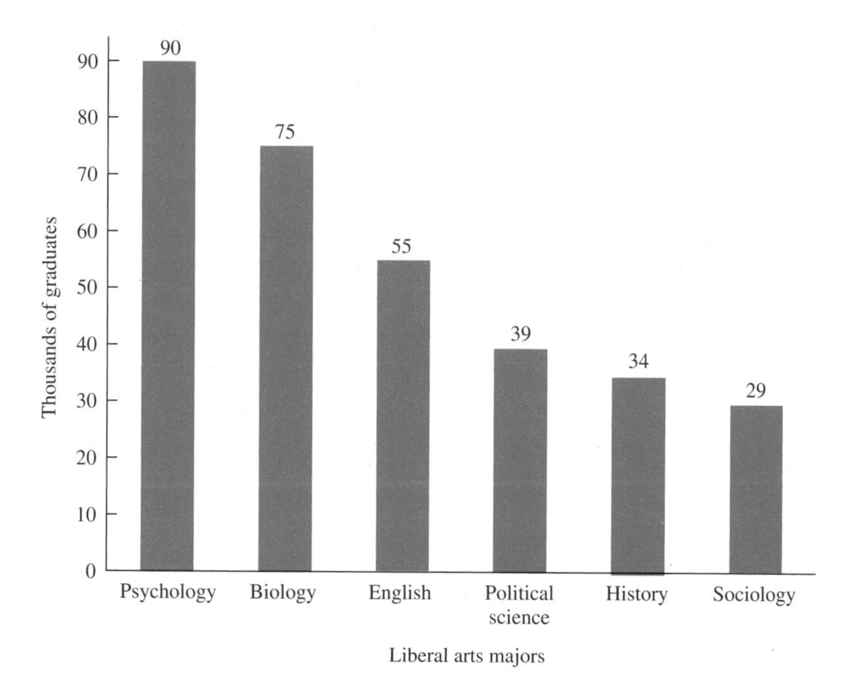

FIGURE 2.5 The six most common liberal arts majors among college graduates for the academic year 2006–2007

2.6. For the height data in problem 2.1 you constructed a frequency distribution. Graph it, being careful to include the zero-frequency scores.

2.7. The average hourly workweek varies from country to country for industrial workers. In 2009 the most recent figures available were: Brazil, 43.9; Canada, 38.5; Finland, 37.7; India, 46.8; Japan, 38.7; Korea, 45.4; Mexico, 45.5; United Kingdom, 40.9; United States, 41.2. What kind of graph is appropriate for these data? Create a graph and execute it.

***2.8.** Decide whether the following distributions should be graphed as frequency polygons or as bar graphs. Graph both distributions.

X.

Class interval	f
48–52	1
43–47	1
38–42	2
33–37	4
28–32	5
23–27	7
18–22	10
13–17	12
8–12	6
3–7	2

Y.

Class interval	f
54–56	3
51–53	7
48–50	15
45–47	14
42–44	11
39–41	8
36–38	7
33–35	4
30–32	5
27–29	2
24–26	0
21–23	0
18–20	1

2.9. Look at the frequency distribution that you constructed for the yard-sign observations in problem 2.2. Which kind of graph should be used to display these data? Compose and graph the distribution.

Describing Distributions

There are three ways to describe the form or shape of a distribution: words, pictures, and mathematics. In this section you will use the first two ways. The only mathematical method I cover is in Chapter 14 on chi square.

Symmetrical Distributions

Symmetrical distributions have two halves that more or less mirror each other; the left half looks pretty much like the right half. In many cases symmetrical distributions are bell-shaped; the highest frequencies are in the middle of the distribution, and scores on either side occur less and less frequently. The distribution of heights in problem 2.6 is an example.

normal distribution (normal curve)
A mathematically defined, theoretical distribution or a graph of scores with a particular bell shape.

There is a special case of a bell-shaped distribution that you will soon come to know very well. It is the **normal distribution** (or **normal curve**). The left panel in **Figure 2.6** is an illustration of a normal distribution.

A rectangular distribution (also called a uniform distribution) is a symmetrical distribution that occurs when the frequency of each value on the *X* axis is the same. The right panel of Figure 2.6 is an example. You will see these distributions again in Chapter 7.

Skewed Distributions

skewed distribution
Asymmetrical distribution; may be positive or negative.

positive skew
Graph with a great preponderance of low scores.

In some distributions, the scores that occur most frequently are near one end of the scale, which leaves few scores at the other end. Such distributions are **skewed.** Skewed distributions, like a skewer, have one end that is thin and narrow. On a graph, if the thin point is to the right—the positive direction—the curve has a **positive skew.** If the thin

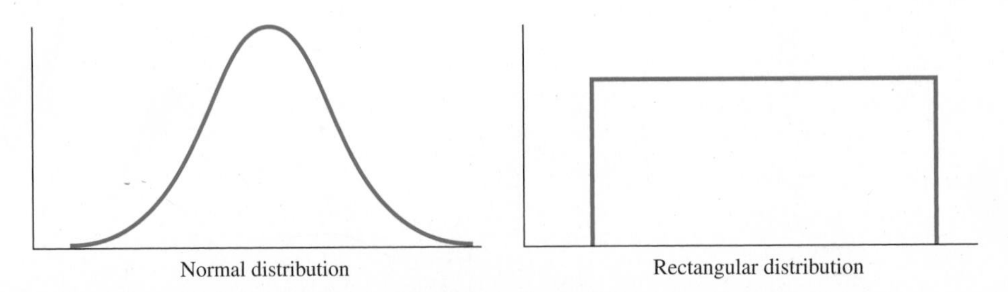

FIGURE 2.6 A normal distribution and a rectangular distribution

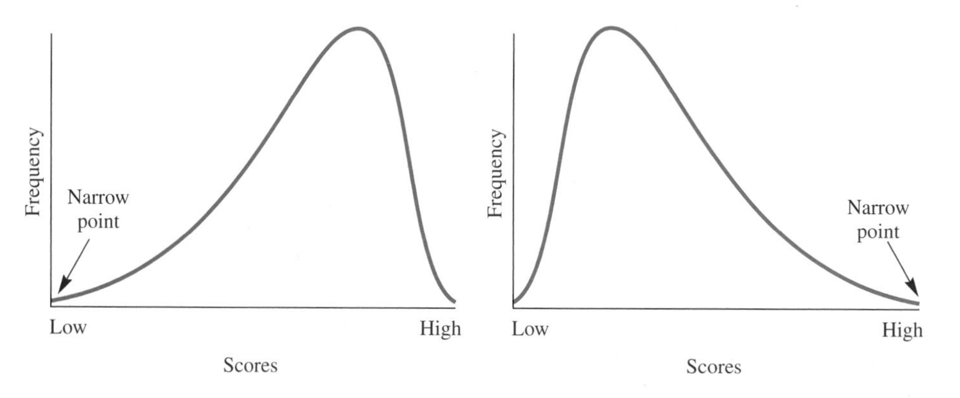

FIGURE 2.7 A negatively skewed curve and a positively skewed curve

point is to the left, the curve is **negatively skewed. Figure 2.7** shows a negatively skewed curve on the left and a positively skewed curve on the right. The data in Table 2.4 are negatively skewed.

negative skew
Graph with a great preponderance of large scores.

Bimodal Distributions

bimodal distribution
Distribution with two modes.

A graph with two distinct humps is called a **bimodal distribution.**[3] Both distributions in **Figure 2.8** would be referred to as bimodal even though the humps aren't the same height in the distribution on the right. For both distributions and graphs, if two high-frequency scores are separated by scores with lower frequencies, the name *bimodal* is appropriate.

Any set of measurements of a phenomenon can be arranged into a distribution; scientists and others find them most helpful. Sometimes measurements are presented in a frequency distribution table and sometimes in a graph (and sometimes both ways).

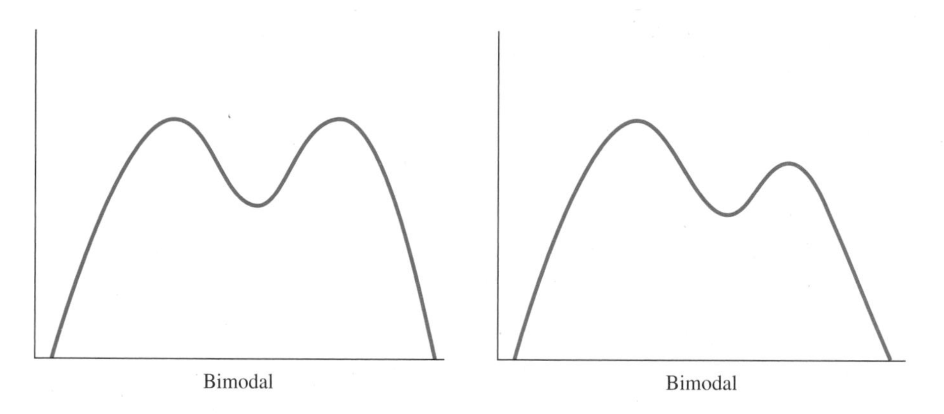

FIGURE 2.8 Two bimodal curves

[3] The mode of a distribution is the score that occurs most frequently.

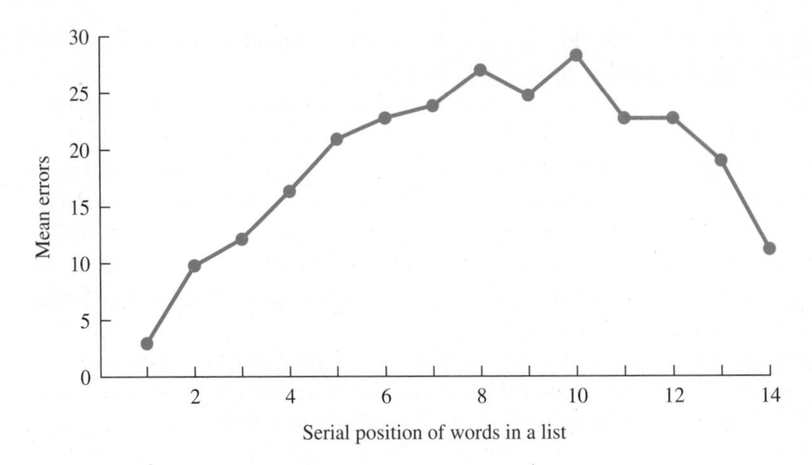

FIGURE 2.9 The serial position effect

In the first chapter I said that by learning statistics you could become more persuasive. Frequency distributions and graphs (but especially graphs) will be valuable to you as you face audiences large and small over the years.

The Line Graph

Perhaps the graph most frequently used by scientists is the **line graph.** A line graph is a picture of the relationship between *two* variables. A point on a line graph represents the value on the *Y* variable that goes with the corresponding value on the *X* variable. The point might represent two scores by one person or the mean score of a group of people.

line graph
Graph that shows the relationship between two variables with lines.

Line graphs are very popular with those who convey information. Peden and Hausmann (2000) found that two-thirds of the graphs in psychology textbooks were line graphs. Likewise, Boehner and Howe (1996) report that line graphs accounted for two-thirds of the data graphs in a random selection of psychology-related journal articles.

Figure 2.9 is an example of a line graph. It shows the *serial position effect.* If you practice an ordered list of words until you can repeat it without error, you'll find that you make the most errors on the words just past the middle. The serial position effect applies to any sequence of material that is learned in order. Knowing about the serial position effect, you should devote extra study time to items in the middle of a list.

More on Graphics

Almost 100 graphs and figures appear in the chapters, problems, and answers that follow. Some graphs are frequency distributions; others are line graphs. In addition, you will find boxplots and scatterplots (with explanations). Group means and the variability of the scores about those means are pictured with bar graphs. Although this chapter is the only one with *graphs* in the title, this textbook and scientific research in general are full of graphs. New designs are being created regularly.

Recognition of the importance of graphs has been emerging in the past two decades or so. L. D. Smith et al. (2002) described psychologists' use of graphs in their work, pointing out that researchers use graphs throughout a project. Even before data are collected, graphs provide a way to convey expected results and compare them to previous results. Wainer and Velleman (2001) show that graphs serve as a guide to the future in addition to their traditional role as a picture of past discoveries. Certainly, a scene of several scientists scribbling a graph on a paper napkin or a blackboard is almost a stereotype of how scientists work.

Graphs are one of our most powerful ways of persuading others. When a person is confused or is of the opposite opinion, a graph can be convincing. One particular champion of graphics is Edward Tufte, an emeritus professor at Yale University. His book, *The Visual Display of Quantitative Information* (2nd edition, 2001), celebrates and demonstrates the power of graphs, and also includes a reprint of "(perhaps) the best statistical graphic ever drawn."[4] Tufte says that regardless of your field, when you construct a quality graphic, it improves your understanding of the phenomenon you are studying. So, if you find yourself somewhere on that path between confusion and understanding, you should try to construct a graph. A graph communicates information with the simultaneous presentation of words, numbers, and pictures. In the heartfelt words of a sophomore engineering student I know, "Graphs sure do help."

Designing an effective and pleasing graph requires planning and rough drafts. Sometimes conventional advice is helpful and sometimes not. For example, "Make the height 60 to 75 percent of the width" works for some data sets and not for others. Graphic designers should heed Tufte's admonition (2001, epilogue), "It is better to violate any principle than to place graceless or inelegant marks on paper."

So, how can you learn to put graceful, elegant marks on paper? Here are some books that can help. *Statistical Graphics for Univariate and Bivariate Data* (Jacoby, 1997) is an 88-page primer with examples from many disciplines. Nicol and Pexman (2003) provide a detailed guide that illustrates a wide variety of graphs and figures; it also covers poster presentations. *The Elements of Graphing Data* (Cleveland, 1994) is a thorough textbook that includes some theory about graphs.

A Moment to Reflect

At this point in your statistics course, I have a question for you: How are you doing? I know that questions like this are better asked by a personable human than by an impersonal textbook. Nevertheless, please give your answer. I hope you can say OK or perhaps something better. If your answer isn't OK or better, now is the time to make changes. Take paper and pen and write down changes that you think (or know!) will help. Develop a plan. Share it with someone who will support you. Implement your plan.

[4] I'll give you just a hint about this graphic. A French engineer drew it well over a century ago to illustrate a disastrous military campaign by Napoleon against Russia in 1812–13. This graphic was nominated by Wainer (1984) as the "World's Champion Graph."

PROBLEMS

***2.10.** Determine the direction of the skew for the two curves in problem 2.8 by examining the curves or the frequency distributions (or both).

2.11. Describe how a line graph is different from a frequency polygon.

2.12. From the data for the five U.S. cities that follow, construct a line graph that shows the relationship between elevation above sea level and mean January temperature. The latitudes of the five cities are almost equal; all are within 1 degree of 35°N latitude. Write a sentence of interpretation.

City	Elevation (feet)	Mean January temperature (°F)
Albuquerque, NM	5000	34
Amarillo, TX	3700	35
Flagstaff, AZ	6900	29
Little Rock, AR	350	39
Oklahoma City, OK	1200	36

2.13. Without looking at Figures 2.6 and 2.8, sketch a normal distribution, a rectangular distribution, and a bimodal distribution.

2.14. Is the narrow point of a positively skewed distribution directed toward the right or the left?

2.15. For each frequency distribution listed, tell whether it is positively skewed, negatively skewed, or approximately symmetrical (bell-shaped or rectangular).

a. Age of all people alive today
b. Age in months of all first-graders
c. Number of children in families
d. Wages in a large manufacturing plant
e. Age at death of everyone who died last year
f. Shoe size

2.16. Write a paragraph on graphs.

***2.17.** **a.** Problem 2.3 presented data on oral body temperature. Use those data to construct a simple frequency distribution.
b. Using the grouped frequency distribution you created for your answer to problem 2.3, construct an appropriate graph, and describe its form.

2.18. Read and respond to the five objectives at the beginning of the chapter. Engaged responding will help you remember what you learned.

ADDITIONAL HELP FOR CHAPTER 2

Visit *cengage.com/psychology/spatz*. At the Student Companion Site, you'll find multiple-choice tutorial quizzes and flashcards with definitions.

KEY TERMS

Abscissa (p. 30)
Bar graph (p. 32)
Bimodal distribution (p. 35)
Class intervals (p. 28)
Frequency (p. 26)
Frequency polygon (p. 31)
Grouped frequency distribution (p. 28)
Histogram (p. 31)
Line graph (p. 36)
Negative skew (p. 35)
Normal distribution (normal curve) (p. 34)
Ordinate (p. 30)
Positive skew (p. 34)
Raw scores (p. 24)
Rectangular distribution (p. 34)
Simple frequency distribution (p. 26)
Skewed distribution (p. 34)
Symmetrical distribution (p. 34)

CHAPTER 3

Exploring Data: Central Tendency

OBJECTIVES FOR CHAPTER 3

After studying the text and working the problems in this chapter, you should be able to:

1. Find the mean, median, and mode of a simple frequency distribution
2. Determine whether a measure of central tendency is a statistic or a parameter
3. Detect bimodal distributions
4. Determine the central tendency measure that is most appropriate for a set of data
5. Estimate the direction of skew of a frequency distribution from the relationship of the mean to the median
6. Calculate a weighted mean

IN THE PREVIOUS chapter, you learned to present a distribution of scores using frequency distributions and graphs. These two descriptive methods show the *form* of a distribution. This chapter covers measures of **central tendency,** which are the most common way to describe a set of data. Measures of central tendency give you one number or descriptor that represents, or is typical of, a distribution.

central tendency
Descriptive statistics that indicate a typical or representative score.

Recall from Chapter 1 that when you have a population of scores, parameters can be calculated. In most research situations, however, only samples are available, so the calculations produce statistics. Naturally, a good statistic is one that mimics its parameter. Fortunately, the common formula for the sample mean produces the best estimate of its corresponding population parameter. For measures of variability, the topic of the next chapter, the situation is not so straightforward.

clue to the future

The distributions you work with in this chapter are *empirical distributions* based on observed scores. This chapter and the next three are about these empirical frequency distributions. Starting with Chapter 7 and throughout the rest of the book, you will also use *theoretical distributions*—distributions based on mathematical formulas and logic rather than on actual observations.

Measures of Central Tendency

There are a number of measures of central tendency; the most popular are the mean, median, and mode.

The Mean

The symbol for the **mean** of a sample is $\overline{X}$ (pronounced "mean" or "X-bar"). The symbol for the mean of a population is μ (a Greek letter, pronounced "mew"). Of course, an $\overline{X}$ is only one of many possible means from a population. Because other samples from that same population produce somewhat different $\overline{X}$'s, a degree of uncertainty goes with $\overline{X}$.

mean
The arithmetic average; the sum of the scores divided by the number of scores.

Of course, if you had an entire population of scores, you could calculate μ and it would carry no uncertainty with it. Most of the time, however, the population is not available and you must make do with a sample. Fortunately, mathematical statisticians have shown that the formula for $\overline{X}$ produces a value that is the best estimator of μ. It also turns out that this formula for $\overline{X}$ is the same as the formula for μ. The difference in $\overline{X}$ and μ, then, is in the interpretation. $\overline{X}$ carries some uncertainty with it; μ does not.

Here's a very simple example of a sample mean. Suppose a college freshman arrives at school in the fall with a promise of a monthly allowance for spending money. Sure enough, on the first of each month, there is money to spend. However, 3 months into the school term, our student discovers a recurring problem: too much month left at the end of the money.

Rather quickly, our student zeros in on money spent at the Student Center. So, for a 2-week period, he records everything bought at the center, a record that includes coffee, both regular and cappuccino Grande, bagels (with cream cheese), chips, soft drinks, ice cream, and the occasional banana. His data are presented in **Table 3.1**. You already know how to compute the mean of these numbers, but before you do, eyeball the data and then write down your *estimate* of the mean in the space provided. The formula for the mean is

$$\overline{X} = \frac{\Sigma X}{N}$$

where $\overline{X}$ = the mean[1]
Σ = an instruction to add (Σ is uppercase Greek sigma)
X = a score or observation; ΣX means to add all the X's
N = number of scores or observations

[1] Many scientific journals use a capital M to represent the mean.

TABLE 3.1 Expenditures at the Student Center during a 2-week period

Day	Money spent
1	\$4.25
2	2.50
3	5.25
4	0.00
5	4.90
6	0.85
7	0.00
8	0.00
9	5.70
10	3.00
11	0.00
12	0.00
13	8.90
14	5.25
	$\Sigma = \$40.60$

Your estimate of the mean ______

For the data in Table 3.1,

$$\overline{X} = \frac{\Sigma X}{N} = \frac{\$40.60}{14} = \$2.90$$

These data are for a 2-week period, but our freshman is interested in his expenditures for at least 1 month and, more likely, for many months. Thus, the result is a sample mean and the $\overline{X}$ symbol is appropriate. The amount, \$2.90, is an estimate of the amount that our friend spends at the Student Center each day. \$2.90 may seem unrealistic to you. If so, go back and explore the data; you can find an explanation.

Now we come to an important part of any statistical analysis, which is to answer the question, *So what?* Calculating numbers or drawing graphs is a part of almost every statistical problem, but unless you can tell the story of what the numbers and pictures mean, you won't find statistics worthwhile.

The first use you can make of Table 3.1 is to estimate the student's monthly Student Center expenses. This is easy to do. Thirty days times \$2.90 is \$87.00. Now, let's suppose our student decides that this \$87.00 is an important part of the "monthly money problem." The student has three apparent options. The first is to get more money. The second is to spend less at the Student Center. The third is to justify leaving things as they are. For this third option, our student might perform an economic analysis to determine what he gets in return for his almost \$90 a month. His list might be pretty impressive: lots of visits with friends, information about classes, courses, and professors, a borrowed book that was just super, thousands of calories, and more.

The point of all this is that part of the attack on the student's money problem involved calculating a mean. However, an answer of \$2.90 doesn't have much meaning by itself. Interpretations and comparisons are called for.

Characteristics of the mean Two characteristics of the mean are important for you to know. Both characteristics will come up again later.

First, if the mean of a distribution is subtracted from each score in that distribution and the differences are added, the sum will be zero; that is, $\Sigma(X - \overline{X}) = 0$. The statistic, $X - \overline{X}$, is called a deviation score. To *demonstrate* to yourself that $\Sigma(X - \overline{X}) = 0$, you might pick a few numbers to play with (numbers 1, 2, 3, 4, and 5 are easy to work with). In addition, if you know the rules that govern algebraic operations of summation (Σ) notation, you can *prove* the relationship $\Sigma(X - \overline{X}) = 0$. (See Kirk, 2008, p. 81.)

Second, the mean is the point about which the sum of the squared deviations is minimized. If we subtract the mean from each score, square each deviation, and add the squared deviations together, the resulting sum will be smaller than if any number other than the mean had been used; that is, $\Sigma(X - \overline{X})^2$ is a minimum. You can demonstrate this relationship for yourself by playing with some numbers.

The Median

median
Point that divides a distribution of scores into equal halves.

The **median** is the *point* that divides a distribution of scores into two parts that are equal in size. To find the median of the Student Center expense data, begin by arranging the daily expenditures from highest to lowest. The result is **Table 3.2**, which is called an *array*. Because there are 14 scores, the halfway point, or median, will have 7 scores above it and 7 scores below it. The seventh score from the bottom is \$2.50. The seventh score from the top is \$3.00. The median, then, is halfway between these two scores, or \$2.75.[2] Remember, the median is a hypothetical *point* in the distribution; it may or may not be an actual score.

TABLE 3.2 **Data of Table 3.1 arrayed in descending order**

X	
\$8.90	
5.70	
5.25	
5.25	7 scores
4.90	
4.25	
3.00	
	Median = \$2.75
2.50	
0.85	
0.00	
0.00	7 scores
0.00	
0.00	
0.00	

[2] The halfway point between two numbers is the mean of the two numbers. Thus, (\$3.00 + \$2.50)/2 = \$2.75.

What is the interpretation of a median of \$2.75? The simplest interpretation is that on half the days our student spends less than \$2.75 in the Student Center and on the other half he spends more.

What if there had been an odd number of days in the sample? Suppose the student chose to sample half a month, or 15 days. Then the median would be the eighth score. The eighth score has seven scores above and seven below. For example, if an additional day was included, during which \$5.00 was spent, the median would be \$3.00. If the additional day's expenditure was zero, the median would be \$2.50.

The reasoning you just went through can be facilitated using a formula that locates the median in a distribution. The formula is:

$$\text{Median location} = \frac{N + 1}{2}$$

The location may be at an actual score (as in the second example) or a point between two scores (the first example).

The Mode

mode
Score that occurs most frequently in a distribution.

The third central tendency statistic is the **mode.** As mentioned earlier, the mode is the most frequently occurring score—the score with the highest frequency. For the Student Center expense data, the mode is \$0.00. Table 3.2 shows the mode most clearly. The zero amount occurred five times, and all other amounts occurred only once.

When a mode is given, it is often helpful to tell the percentage of times it occurred. You will probably agree that "The mode was \$0.00, which occured on 36 percent of the days" is more informative than "The mode was \$0.00."

Finding Central Tendency of Simple Frequency Distributions

Mean

Table 3.3 is an expanded version of Table 2.3, the frequency distribution of Satisfaction With Life Scale (SWLS) scores in Chapter 2. The steps for finding the mean from a simple frequency distribution follow, but first (looking only at the data and not at the summary statistics at the bottom) *estimate* the mean of the SWLS scores in the space at the bottom of Table 3.3.[3]

The first step in calculating the mean from a simple frequency distribution is to multiply each score in the X column by its corresponding f value, so that all the people who make a particular score are included. Next, sum the fX values and divide the total by N. (N is the sum of the f values.) The result is the mean.[4] In terms of a formula,

$$\mu \text{ or } \overline{X} = \frac{\Sigma fX}{N}$$

[3] If the data are arranged in a frequency distribution, you can estimate the mean by selecting the most frequent score (the mode) or by selecting the score in the middle of the list.
[4] When you work with whole numbers, calculating a mean to two decimal places is usually sufficient.

TABLE 3.3 Calculating the mean of the simple frequency distribution of the Satisfaction With Life Scale scores

SWLS score (X)	f	fX	SWLS score (X)	f	fX
35	2	70	22	5	110
34	1	34	21	3	63
33	2	66	20	5	100
32	4	128	19	3	57
31	2	62	17	2	34
30	5	150	16	2	32
29	7	203	15	1	15
28	6	168	13	2	26
27	10	270	12	1	12
26	9	234	11	1	11
25	9	225	10	2	20
24	6	144	9	2	18
23	6	138	5	2	10
(continued above)				$\Sigma = 100$	2400

Your estimate of the mean _______

For the data in Table 3.3,

$$\mu \text{ or } \bar{X} = \frac{\Sigma fX}{N} = \frac{2400}{100} = 24.00$$

How did 24.00 compare to your estimate?

To answer the question of whether 24.00 is $\bar{X}$ or μ, you need more information. Here's the question to ask: Is there any interest in a group larger than these 100? If the answer is *no,* the 100 scores are a population and $24.00 = \mu$. If the answer is *yes,* the 100 scores are a sample and $24.00 = \bar{X}$.

Median

The formula for finding the location of the median that you used earlier works for a simple frequency distribution, too.

$$\text{Median location} = \frac{N + 1}{2}$$

Thus, for the scores in Table 3.3,

$$\text{Median location} = \frac{N + 1}{2} = \frac{100 + 1}{2} = 50.5$$

To find the 50.5th position, begin adding the frequencies in Table 3.3 from the bottom (2 + 2 + 2 + 1 + · · ·). The total is 43 by the time you include the score of 24. Including 25 would make the total 52—more than you need. So the 50.5th score is among those nine scores of 25. The median is 25.

Suppose you start the quest for the median at the top of the distribution rather than at the bottom. Again, the location of the median is at the 50.5th position in the distribution. To get to 50.5, add the frequencies from the top (2 + 1 + 2 + · · ·). The sum of the frequencies of the scores from 35 down to and including 26 is 48. The next score, 25, had a frequency of 9. Thus, the 50.5th position is among the scores of 25. The median is 25.

error detection

Calculating the median by starting from the top of the distribution produces the same result as calculating the median by starting from the bottom.

Mode

It is easy to find the mode from a simple frequency distribution. In Table 3.3, the score with the highest frequency, 10, is the mode. So 27 is the mode.

In Chapter 2 (page 35) you learned to recognize curves with two distinct humps as *bimodal*. Such curves mirror distributions that have two high-frequency scores (modes) separated by one or more low-frequency scores.

PROBLEMS

3.1. Find the median for the following sets of scores.
- **a.** 2, 5, 15, 3, 9
- **b.** 9, 13, 16, 20, 12, 11
- **c.** 8, 11, 11, 8, 11, 8

3.2. Which of the following distributions is bimodal?
- **a.** 10, 12, 9, 11, 14, 9, 16, 9, 13, 20
- **b.** 21, 17, 6, 19, 23, 19, 12, 19, 16, 7
- **c.** 14, 18, 16, 28, 14, 14, 17, 18, 18, 6

***3.3.** Refer to problem 2.2, the political scientist's yard-sign data.
- **a.** Decide which measure of central tendency is appropriate and find it.
- **b.** Is this central tendency measure a statistic or a parameter?
- **c.** Write a sentence of interpretation.

***3.4.** Refer to problem 2.1. Find the mean, median, and mode of the heights of both groups of Americans in their 20s. Work from the Appendix G answers to problem 2.1.

3.5. For distribution **a,** find the median starting from the bottom. For distribution **b,** find the median starting from the top. The median of distribution **c** can probably be solved by inspection.

a.

X	f
15	4
14	3
13	5
12	4
11	2
10	1

b. 4, 0, −1, 2, 1, 0, 3, 1, −2, −2, −1, 2, 1, 3, 0, −2, 1, 2, 0, 1

c. 28, 27, 26, 21, 18, 10

3.6. What two mathematical characteristics of the mean were covered in this section?

* An asterisk indicates that later problems are based on this one. Your answer and your understanding of an asterisked problem will help when the data set is presented again.

error detection

Estimating (also called eyeballing) is a valuable way to avoid big mistakes. Begin work by quickly making an estimate of the answers. If your calculated answers differ from your estimates, wisdom dictates that you reconcile the difference. You have either overlooked something when estimating or made a computation mistake.

When to Use the Mean, Median, and Mode

A common question is: Which measure of central tendency should I use? The general answer is, given a choice, use the mean. Sometimes, however, the data limit your choice. Here are three considerations.

Scale of Measurement

A mean is appropriate for ratio or interval scale data, but not for ordinal or nominal distributions. A median is appropriate for ratio, interval, and ordinal scale data, but not for nominal data. The mode is appropriate for any of the four scales of measurement.

You have already thought through part of this issue in working problem 3.3. In it you found that the yard-sign names (very literally, a nominal variable) could be characterized with a mode, but it would be impossible to try to add up the names and divide by N or to find the median of the names.

For an ordinal scale such as class standing in college, either median or mode makes sense. The median would probably be sophomore, and the mode would be freshman.

Skewed Distributions

Even if you have interval or ratio data, the mean may be a misleading choice if the distribution is severely skewed. Here's a story to illustrate.

The developer of Swampy Acres Retirement Homesites is attempting to sell building lots in a southern "paradise" to out-of-state buyers. The marks express concern about flooding. The developer reassures them: "The average elevation of the lots is 78.5 feet and the water level in this area has never ever exceeded 25 feet." The developer tells the truth, but this average truth is misleading. The actual lay of the land is shown in **Figure 3.1**. Now look at **Table 3.4**, which shows the elevations of the 100 lots arranged in a grouped frequency distribution. (Grouped frequency distributions are explained in Appendix B.)

To calculate the mean of a grouped frequency distribution, multiply the midpoint of each interval by its frequency, add the products, and divide by the total frequency. Thus, in Table 3.4 the mean is 78.5 feet, exactly as the developer said. However, only the 20 lots on the bluff are out of the flood zone; the other 80 lots are, on the average, under water. The mean, in this case, is misleading. What about the median? The median of the distribution in Table 3.4 is 12.5 feet, well under the high-water mark, and a better overall descriptor of Swampy Acres Retirement Homesites.

The distribution in Table 3.4 is severely skewed. For distributions that are moderately or severely skewed, the median is often preferred over the mean because the median is unaffected by extreme scores.

FIGURE 3.1 Elevation of Swampy Acres

Open-Ended Class Intervals

Even if you have interval or ratio data and the distribution is fairly symmetrical, there is a situation for which you *cannot* calculate a mean. If the highest interval or the lowest interval of a grouped frequency distribution is open-ended, it has no midpoint and so you cannot calculate a mean. Age data are sometimes reported with the oldest being "75 and over." *The Statistical Abstract of the United States* reports household income with the largest category as $200,000 and over. Because there is no midpoint to "75 and over" or to "$200,000 or more," you cannot calculate a mean. Medians and modes are appropriate measures of central tendency when one or both of the extreme class intervals are open-ended.

In summary, use the mean if it is appropriate. To follow this advice you must recognize data for which the mean is not appropriate. Perhaps **Table 3.5** will help.

TABLE 3.4 **Frequency distribution of lot elevations at Swampy Acres**

Elevation, in feet	Midpoint (X)	Number of lots (f)	fX
348–352	350	20	7000
13–17	15	30	450
8–12	10	30	300
3–7	5	20	100
		$\Sigma = 100$	7850

$$\mu = \frac{\Sigma fX}{N} = \frac{7850}{100} = 78.5 \text{ feet}$$

TABLE 3.5 Data characteristics and recommended central tendency statistic

	Recommended statistic		
Data characteristic	Mean	Median	Mode
Nominal scale data	No	No	Yes
Ordinal scale data	No	Yes	Yes
Interval scale data	Yes	Yes	Yes
Ratio scale data	Yes	Yes	Yes
Open-ended category(ies)	No	Yes	Yes
Skewed distribution	No	Yes	Yes

Determining Skewness from the Mean and Median

There is a rule of thumb that uses the relationship of the mean to the median to help determine skewness. This rule works most of the time. It usually works for continuous data such as SWLS scores but is less trustworthy for discrete data such as the number of adult residents in U.S. households (von Hippel, 2005).

The rule is that when the mean is larger than the median, you can expect the distribution to be positively skewed. If the mean is smaller than the median, expect negative skew. **Figure 3.2** shows the relationship of the mean to the median for a positively skewed and a negatively skewed distribution of continuous data.

I'll illustrate by changing the slightly skewed data in Table 3.1 into more severely skewed data. The original expenditures in Table 3.1 have a mean of $2.90 and a median of $2.50, a difference of $0.40. If I add an expenditure of $100.00 to the 14 scores, the mean jumps to $9.37; the median moves up to $3.00. The difference now is $6.37. This example follows the general rule that the greater the difference between the mean and median, the greater the skew.[5] Note also that the mean of the new distribution ($9.37)

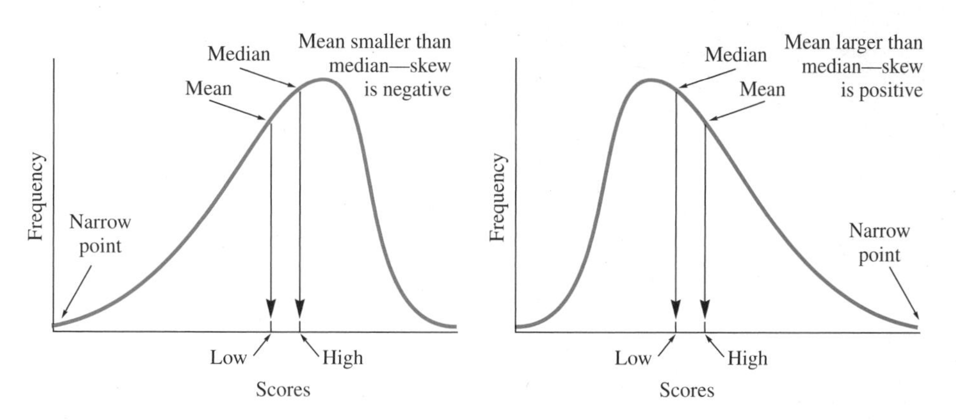

FIGURE 3.2 The effect of skewness on the relative position of the mean and median for continuous data

[5] For a mathematical measure of skewness, see Kirk, 2008, p. 112.

is greater than every score in the distribution except for $100.00. For severely skewed distributions, the mean is not a typical score; the median is usually a better measure of central tendency.

The Weighted Mean

weighted mean
Overall mean calculated from two or more samples with different *N*'s.

Sometimes several sample means are available from the same or similar populations. In such cases, a **weighted mean,** $\overline{X}_W$, is the best estimate of the population parameter, μ. If every sample has the same N, you can compute a weighted mean by adding the means and dividing by the number of means. If the sample means are based on N's of different sizes, however, you cannot use this procedure. Here is a story that illustrates the right way and the wrong way to calculate a weighted mean.

At my undergraduate college, a student with a cumulative grade point average (GPA) of 3.25 in the middle of the junior year was eligible to enter a program to "graduate with honors." [In those days (1960) usually fewer than 10 percent of a class had GPAs greater than 3.25.] Discovering this rule after my sophomore year, I decided to figure out if I had a chance to qualify. Calculating a cumulative GPA seemed easy enough to do: Four semesters had produced GPAs of 3.41, 3.63, 3.37, and 2.16. Given another GPA of 3.80, the sum of the five semesters would be 16.37, and dividing by 5 gave an average of 3.27, well above the required 3.25.

Graduating with honors seemed like a great ending for college, so I embarked on a goal-oriented semester—a GPA of 3.80 (a B in German and an A in everything else). And, at the end of the semester I had accomplished the goal. Unfortunately, "graduating with honors" was not to be.

There was a flaw in my method of calculating my cumulative GPA. My method assumed that all of the semesters were equal in weight, that they had all been based on the same number of credit hours. My calculations based on this assumption are shown on the left side of **Table 3.6**.

Unfortunately, all five semesters were not the same; the semester with the GPA of 2.16 was based on 19 hours, rather than the usual 16 or so.[6] Thus, that semester should have been weighted more heavily than semesters with fewer hours.

The formula for a weighted mean is

$$\overline{X}_W = \frac{N_1\overline{X}_1 + N_2\overline{X}_2 + \cdots + N_K\overline{X}_K}{N_1 + N_2 + \cdots + N_K}$$

where $\overline{X}_W$ = the weighted mean

$\overline{X}_1, \overline{X}_2, \overline{X}_K$ = sample means

N_1, N_2, N_K = sample sizes

K = number of samples

[6] That semester was more educational than a GPA of 2.16 would indicate. I read a great deal of American literature that spring, but unfortunately, I was not registered for any courses in American literature!

TABLE 3.6 Two methods of calculating a mean from five semesters' GPAs; the method on the left is correct only if all semesters have the same number of credit hours

Flawed method	Correct method		
Semester GPA	Semester GPA	Credit hours	GPA × hours
3.41	3.41	17	58
3.63	3.63	16	58
3.37	3.37	19	64
2.16	2.16	19	41
3.80	3.80	16	61
$\Sigma = 16.37$		$\Sigma = 87$	$\Sigma = 282$
$\overline{X} = \frac{16.37}{5} = 3.27$	$\overline{X}_W = \frac{282}{87} = 3.24$		

The right side of Table 3.6 shows the steps for a weighted mean, which is required to correctly calculate a cumulative GPA. Each semester's GPA is multiplied by its number of credit hours. These products are summed and that total is divided by the sum of the hours. As you can see from the numbers on the right, the actual cumulative GPA was 3.24, not high enough to qualify for the honors program.

More generally, to find the mean of a set of means, multiply each separate mean by its N, add these products together, and divide the total by the sum of the N's. As an example, three means of 2.0, 3.0, and 4.0, calculated from the scores of samples with N's of 6, 3, and 2, produce a weighted mean ($\overline{X}_W$) of 2.64. Do you agree?

Estimating Answers

You may have noticed that I have (subtly?) worked in the advice: *As your first step, estimate an answer*. Here's why I think you should begin a problem by estimating the answer: (1) Taking a few seconds to estimate keeps you from plunging into the numbers before you fully understand the problem. (2) An estimate is especially helpful when a calculator or a computer does the bulk of your number crunching. Although these wonderful machines don't make errors, the people who enter the numbers or give instructions do, occasionally. Your initial estimate helps you catch these mistakes. (3) Noticing that an estimate differs from a calculated value gives you a chance to correct an error before anyone else sees it.

It can be exciting to look at an answer, whether it is on paper, a calculator display, or a computer screen, and say, "That *can't* be right!" and then to find out that, sure enough, the displayed answer is wrong. If you develop your ability to estimate answers, I promise that you will sometimes experience this excitement.

PROBLEMS

3.7. For each of the following situations, tell which measure of central tendency is appropriate and why.

a. As part of a study on prestige, an investigator sat on a corner in a high-income residential area and classified passing automobiles according to color: black, gray, white, silver, green, and other.

b. In a study of helping behavior, an investigator pretended to have locked himself out of his car. Passersby who stopped were classified on a scale of 1 to 5 as (1) very helpful, (2) helpful, (3) slightly helpful, (4) neutral, and (5) discourteous.

c. In a study of household income in a city, the following income categories were established: $0–$20,000, $20,001–$40,000, $40,001–$60,000, $60,001–$80,000, $80,001–$100,000, and $100,001 and more.

d. In a study of *per capita* income in a city, the following income categories were established: $0–$20,000, $20,001–$40,000, $40,001–$60,000, $60,001–$80,000, $80,001–$100,000, $100,001–$120,000, $120,001–$140,000, $140,001–$160,000, $160,001–$180,000, and $180,001–$200,000.

e. First admissions to a state mental hospital for 5 years were classified by disorder: schizophrenic, delusional, anxiety, dissociative, and other.

f. A teacher gave her class an arithmetic test; most of the children scored in the range 70–79. A few scores were above this, and a few were below.

3.8. A senior psychology major performed the same experiment on three groups and obtained means of 74, 69, and 75 percent correct. The groups consisted of 12, 31, and 17 participants, respectively. What is the overall mean for all participants?

3.9. **a.** By inspection, determine the direction of skew of the expenditure data in Table 3.2 and the elevation data in Table 3.4. Verify your judgment by comparing the means and medians.

b. For problem 2.10, you determined by inspection the direction of skew of two distributions (**X** and **Y**). To verify your judgment, calculate and compare the mean and median of each distribution. The data are in problem 2.8.

3.10. A 3-year veteran of the local baseball team was calculating his lifetime batting average. The first year he played for half the season and batted .350 (28 hits in 80 at-bats). The second year he had about twice the number of at-bats and his average was .325. The third year, although he played even more regularly, he was in a slump and batted only .275. Adding the three season averages and dividing the total by 3, he found his lifetime batting average to be .317. Is this correct? Explain.

3.11. The data in Table 3.1 (money spent at the Student Center) were created so I could illustrate characteristics of central tendency statistics.

The data that follow are based on a study of actual expenditures for food by freshmen students at Central Michigan University. (See http://www.westga.edu/~bquest/2008/local08.pdf). The scores are for a two week period and are from students who reported positive expenditures. Find the mean, median, and mode and determine the direction of skew from the relationship of the mean to the median.

$20, $30, $9, $22 $18, $54, $24, $2
$81, $24, $20, $33, $13, $28, $20

3.12. This chapter begins with a list of objectives. Read each one and judge yourself.

ADDITIONAL HELP FOR CHAPTER 3

Visit *cengage.com/psychology/spatz*. At the Student Companion Site, you'll find multiple-choice tutorial quizzes and flashcards with definitions.

KEY TERMS

Array (p. 43)
Bimodal (p. 46)
Central tendency (p. 40)
Empirical distribution (p. 41)
Estimating (p. 47, 51)
Mean (p. 41)
Median (p. 43)
Mode (p. 44)
Open-ended class intervals (p. 48)
Scale of measurement (p. 47)
Skewed distributions (pp. 47, 49)
Theoretical distribution (p. 41)
Weighted mean (p. 50)

CHAPTER

4

Exploring Data: Variability

OBJECTIVES FOR CHAPTER 4

After studying the text and working the problems in this chapter, you should be able to:

1. Explain the concept of variability
2. Find and interpret the range of a distribution
3. Find and interpret the interquartile range of a distribution
4. Distinguish among the standard deviation of a population, the standard deviation of a sample used to estimate a population standard deviation, and the standard deviation used to describe a sample
5. Know the meaning of σ, $\hat{s}$, and S
6. For grouped and ungrouped data, calculate a standard deviation and interpret it
7. Calculate the variance of a distribution

variability
Having more than one value.

A DISTRIBUTION OF scores has three features that are independent of each other—*form*, *central tendency*, and *variability*. Knowledge of all three features leads you to a fairly complete understanding of a distribution. You studied form (frequency distributions and graphs) in Chapter 2 and central tendency in Chapter 3. Measures of **variability** (this chapter) tell separate story and deserve more attention than they get.

The value of knowing about variability is illustrated by a story of two brothers who, on a dare, water skied on Christmas Day (the temperature was about 35°F). On the *average* each skier finished his turn 5 feet from the shoreline (where one may step off the ski into only 1 foot of very cold water). This bland central tendency of the two actual stopping places, however, does not convey the excitement of the day.

The first brother, determined to avoid the cold water, held onto the towrope too long. Scraped and bruised, he finally stopped rolling at a spot 35 feet up on the rocky shore. The second brother, determined not to share the same fate, released the towrope too soon. Although he swam the 45 feet to shore very quickly, his lips were very blue. No, to hear that the average stopping place was 5 feet from the shoreline doesn't capture the excitement of the day. To get the full story, you should ask about variability. Here are some other situations in which knowing the variability is important.

1. You are an elementary school teacher preparing for your first year teaching the fourth grade. Looking over the summary material for your students, you are delighted to discover that the class average for reading is the 71st percentile on the Stanford 10 Achievement Test. You begin thinking up some sophisticated reading projects for the year.

Caution: *If the variability around that average of 71st percentile is low, your projects with the class will probably succeed. However, if the variability is great, the projects will be too complicated for some, but for others, even these projects will not be challenging enough.*

2. Suppose that your temperature, taken with a thermometer under your tongue, is 97.5°F. You begin to worry. This is below even the average of 98.2°F that you learned in the previous chapter (based on Mackowiak, Wasserman, and Levine, 1992).

Caution: *There is variability around that mean of 98.2°F. Is 97.5°F below the mean by just a little or by a lot? Measuring variability is necessary if you are to answer this question.*

3. Having graduated from college, you are considering two offers of employment, one in sales and the other in management. The pay is about the same for both. Using the library to check out the statistics for salespeople and managers, you find that those who have been working for 5 years in each type of job also have similar averages. You conclude that the pay for the two occupations is equal.

Caution: *Pay is more variable for those in sales than for those in management. Some in sales make much more than the average and some make much less, whereas the pay of those in management is clustered together. Your reaction to this difference in variability might help you choose.*

The information that measures of variability provide is completely independent of that provided by the mean, median, and mode. **Table 4.1** shows three distributions, each with a mean of 15. As you can see, however, the actual scores are quite different.

TABLE 4.1 **Illustration of three different distributions that have equal means**

	X_1	X_2	X_3
	25	17	90
	20	16	30
	15	15	15
	10	14	0
	5	13	−60
Mean	15	15	15

Fortunately, measures of variability reveal the differences in the three distributions, giving you information that the mean does not.

This chapter is about statistics and parameters that measure the variability of a distribution. The *range* is the first measure I will describe; the second is the *interquartile range.* The third, the *standard deviation,* is the most important. Most of this chapter is about the standard deviation. A fourth way to measure variability is the *variance.*

The Range

range
Highest score minus the lowest score.

The **range** of a quantitative variable is the highest score minus the lowest score.

$$\text{Range} = X_H - X_L$$

where X_H = highest score
X_L = lowest score

The range of the Satisfaction With Life Scale scores you worked with in previous chapters (see Table 2.2) was 30. The highest score was 35; the lowest was 5. Thus, the range is $35 - 5 = 30$. In Chapter 2 you worked with oral body temperatures. (See problem 2.3 and its answer.) The highest temperature was 99.5°F; the lowest was 96.4°F. The range is 3.1°F. Knowing that the range of normal body temperature is more than 3 degrees tends to soften a strict interpretation of a particular temperature. In manufacturing, the range is used in some quality-control procedures. From a small sample of whatever is being manufactured, inspectors calculate a range and compare it to an expected figure. A large range means there is too much variability in the process and adjustments are called for.

The range is a quickly calculated, easy-to-understand statistic that is just fine in some situations. However, you can probably imagine two distributions of scores that have the same mean and the same range but are very different. (Go ahead, imagine them.) If you are able to dream up two such distributions, it follows that some other measures of variability are needed if you are to distinguish among different distributions using just one measure of central tendency and one measure of variability.

Interquartile Range

interquartile range
Range of scores that contain the middle 50 percent of a distribution.

percentile
Point below which a specified percentage of the distribution falls.

The next measure of variability, the **interquartile range,** tells the range of scores that enclose the middle 50 percent of the distribution. It is an important element of *boxplots,* which you will study in Chapter 5. (A boxplot is a graphic that conveys the data of a distribution and some of its statistical characteristics with *one picture.*)

To find the interquartile range, you must have the 25th percentile score and the 75th percentile score. You may already be familiar with the concept of **percentile** scores. The 10th percentile score has 10 percent of the distribution below it; it is near the bottom. The 95th percentile score is near the top; 95 percent of the scores in the distribution are smaller. The 50th percentile divides the distribution into equal halves.

TABLE 4.2 Finding the 25th and 75th percentiles

Score	f	
31	1	} 9 scores
30	3	
29	5	
28	4	
27	5	
26	4	
25	6	
24	7	
23	3	} 5 scores
22	2	
	$N = 40$	

Determining percentiles is something you practiced when you calculated medians. The median is the point that divides a distribution into equal halves. Thus, the median is the 50th percentile. Finding the 25th and 75th percentile scores involves the same kind of reasoning that you used to find the median.

The 25th percentile score is the one that has 25 percent of the scores below it. Look at **Table 4.2**, which shows a frequency distribution of 40 scores. To find the 25th percentile score, multiply 0.25 times N. Thus, $0.25 \times 40 = 10$. When $N = 40$, the 10th score *up* from the *bottom* is the 25th percentile score. You can see in Table 4.2 that there are 5 scores of 23 or lower. The 10th score is among the 7 scores of 24. Thus, the 25th percentile is a score of 24.

The 75th percentile score has 75 percent of the scores below it. The easiest way to find it is to work from the top, using the same multiplication procedure, 0.25 times N ($0.25 \times 40 = 10$). The 75th percentile score is the 10th score *down* from the *top* of the distribution. In Table 4.2 there are 9 scores of 29 or higher. The 10th score is among the 4 scores of 28, so the 75th percentile score is 28.

The interquartile range (IQR) is the 75th percentile minus the 25th percentile:

$$\text{IQR} = \text{75th percentile} - \text{25th percentile}$$

Thus, for the distribution in Table 4.2, $\text{IQR} = 28 - 24 = 4$. The interpretation is that the middle 50 percent of the scores have values from 24 to 28.

PROBLEMS

4.1. Find the range for the two distributions.
  a. 17, 5, 1, 1
  b. 0.45, 0.30, 0.30

***4.2.** Find the interquartile range of the Satisfaction With Life Scale scores in Table 3.3. Write a sentence of interpretation.

***4.3.** Find the interquartile range of the heights of both groups of those 20- to 29-year-old Americans. Use the frequency distributions you constructed for problem 2.1. (These distributions may be in your notebook of answers; they are also in Appendix G.)

The Standard Deviation

standard deviation
Descriptive measure of the dispersion of scores around the mean.

The most widely used measure of variability is the **standard deviation.** It is popular because it is very reliable and it provides information about the proportions within a distribution if you know the distribution's form. Once you understand standard deviations, you can express quantitatively the difference between the two distributions you imagined previously in the section on the range. (You did do the imagining, didn't you?)

The standard deviation (or its close relatives) turn up in every chapter after this one. Of course, in order to understand, you'll have to read all the material, work the problems, and do the interpretations. But what you get for all this work is a lifetime of understanding the most popular yardstick of variability.

Your principal task in this section is to learn the distinctions among three different standard deviations. Which standard deviation you use in a particular situation will be determined by your purpose. **Table 4.3** lists symbols, purposes, and descriptions. It will be worth your time to study Table 4.3 thoroughly now.

Distinguishing among these three standard deviations is sometimes a problem for beginning students. Be alert in situations where a standard deviation is used. With each situation, you will acquire more understanding. I will first discuss the calculation of σ and S and then deal with $\hat{s}$.

The Standard Deviation as a Descriptive Index of Variability

Both σ and S are used to *describe* the variability of a set of data. σ is a parameter of a population; S is a statistic of a sample. Both are calculated with similar formulas. I will show you two ways to arrange the arithmetic for these formulas, the deviation-score formula and the raw-score formula.

TABLE 4.3 **Symbols, purposes, and descriptions of three standard deviations**

Symbol	Purpose	Description
σ	Measure a population's variability	Lowercase Greek *sigma*. A parameter. Describes variability when a population of data is available.
$\hat{s}$	Estimate a population's variability	Lowercase $\hat{s}$ (s-hat). A statistic. An *estimate* of σ (in the same way that $\overline{X}$ is an estimate of μ). The variability statistic you will use most often in this book.
S	Measure a sample's variability	Capital S. A statistic. Describes the variability of a sample when there is no interest in estimating σ.

You can best learn what a standard deviation is actually measuring by working through the steps of the *deviation-score* formula. *The raw-score* formula, however, is quicker (and sometimes more accurate, depending on rounding). Algebraically, the two formulas are identical. My suggestion is that you learn both methods. By the time you get to the end of the chapter, you should have both the understanding that the deviation-score formula produces and the efficiency that the raw-score formula gives. After that, you may want to use a calculator or computer to do all the arithmetic for you.

Deviation Scores

A **deviation score** is a raw score minus the mean of the distribution, whether the distribution is a sample or a population.

deviation score
Raw score minus the mean of its distribution.

$$\text{Deviation score} = X - \overline{X} \qquad \text{or} \qquad X - \mu$$

Raw scores that are greater than the mean have positive deviation scores, raw scores that are less than the mean have negative deviation scores, and raw scores that are equal to the mean have a deviation score of zero.

Table 4.4 provides an illustration of how to compute deviation scores for a small sample of data. In Table 4.4, I first computed the mean, 8, and then subtracted it from each score. The result is deviation scores, which appear in the right-hand column.

A deviation score tells you the number of points that a particular score deviates from, or differs from, the mean. In Table 4.4, the $X - \overline{X}$ value for Alex, 6, tells you that he scored six points above the mean. Luke, 0, scored at the mean, and Stephen, −5, scored five points below the mean.

error detection

Notice that the sum of the deviation scores is always zero. Add the deviation scores; if the sum is not zero, you have made an error. You have studied this concept before. In chapter 3 you learned that $\Sigma(X - \overline{X}) = 0$.

TABLE 4.4 The computation of deviation scores from raw scores

Name	Score	$X - \overline{X}$	Deviation score
Alex	14	14 − 8	6
Ian	10	10 − 8	2
Luke	8	8 − 8	0
Zachary	5	5 − 8	−3
Stephen	3	3 − 8	−5
	$\Sigma X = 40$		$\Sigma(X - \overline{X}) = 0$

$$\overline{X} = \frac{\Sigma X}{N} = \frac{40}{5} = 8$$

TABLE 4.5 Using the deviation-score formula to compute *S* for cookie sales by a sample of six Girl Scouts

Boxes of cookies X	Deviation scores $X - \overline{X}$	$(X - \overline{X})^2$
28	18	324
11	1	1
10	0	0
5	–5	25
4	–6	36
2	–8	64
$\Sigma X = 60$	$\Sigma(X - \overline{X}) = 0$	$\Sigma(X - \overline{X})^2 = 450$

$$\overline{X} = \frac{\Sigma X}{N} = \frac{60}{6} = 10 \text{ boxes}$$

$$S = \sqrt{\frac{\Sigma(X - \overline{X})^2}{N}} = \sqrt{\frac{450}{6}} = \sqrt{75} = 8.66 \text{ boxes}$$

Computing S and σ Using Deviation Scores

The deviation-score formula for computing the standard deviation as a descriptive index is

$$S = \sqrt{\frac{\Sigma(X - \overline{X})^2}{N}} \quad \text{or} \quad \sigma = \sqrt{\frac{\Sigma(X - \mu)^2}{N}}$$

where S = standard deviation of a sample
σ = standard deviation of a population
N = number of scores (same as the number of deviations)

The numerator of standard deviation formulas, $\Sigma(X - \overline{X})^2$, is shorthand notation that tells you to find the deviation score for each raw score, square the deviation score, and then add all the squares together. This sequence illustrates one of the rules for working with summation notation: Perform the operations within the parentheses first.

How can a standard deviation add to your understanding of a phenomenon? Let's take the sales of Girl Scout cookies. Table 4.5 presents some imaginary (but true-to-life) data on boxes of cookies sold by six Girl Scouts. I'll use these data to illustrate the calculation of the standard deviation, S. If these data were a population, the value of σ would be identical.

The numbers of boxes sold are listed in the X column of **Table 4.5**. The rest of the arithmetic needed for calculating S is also given. To compute S by the deviation-score formula, first find the mean.[1] Obtain a deviation score for each raw score by subtracting the mean from the raw score. Square each deviation score and sum the squares to obtain $\Sigma(X - \overline{X})^2$. Divide $\Sigma(X - \overline{X})^2$ by N. Take the square root. The result is $S = 8.66$ boxes.

Now, what does $S = 8.66$ boxes mean? How does it help your understanding? The 8.66 boxes is a measure of the variability in the number of boxes the six Girl Scouts

[1] For convenience I arranged the data so the mean is an integer. If you are confronted with decimals, it usually works to carry three decimals in the problem and then round the final answer to two decimal places.

sold. If S was zero, you would know that each girl sold the same number of boxes. The closer S is to zero, the more confidence you can have in predicting that the number of boxes any girl sold was equal to the mean of the group. Conversely, the further S is from zero, the less confidence you have. With $S = 8.66$ and $\overline{X} = 10$, you know that the girls varied a great deal in cookie sales.

I realize that, so far, my interpretation of the standard deviation has not given you any more information than the range does. (A range of zero means that each girl sold the same number of boxes, and so forth.) The range, however, has no additional information to give you—the standard deviation does, as you will see in Chapter 7.

Now look again at Table 4.5 and the formula for S. Notice what is happening. The mean is subtracted from each score. This difference, whether positive or negative, is squared and these squared differences are added together. This sum is divided by N and the square root is found. Every score in the distribution contributes to the final answer, but they don't all contribute equally.

Notice the contribution made by a score such as 28, which is far from the mean; its contribution to $\Sigma(X - \overline{X})^2$ is large. This makes sense because the standard deviation is a yardstick of variability. Scores that are far from the mean cause the standard deviation to be greater. Take a moment to think through the contribution to the standard deviation made by a score near the mean.[2]

error detection

All standard deviations are positive numbers. If you find yourself trying to take the square root of a negative number, you've made an error.

PROBLEMS

4.4. Give the symbol and purpose of each of the three standard deviations.

4.5. Using the deviation-score method, compute S for the three sets of scores.

- ***a.** 7, 6, 5, 2
- **b.** 14, 11, 10, 8, 8
- ***c.** 107, 106, 105, 102

4.6. Compare the standard deviation of problem 4.5a with that of problem 4.5c. What conclusion can you draw about the effect of the size of the scores on the following?

- **a.** standard deviation
- **b.** mean

***4.7.** The temperatures listed are averages for March, June, September, and December. Calculate the mean and standard deviation for each city. Summarize your results in a sentence.

San Francisco, CA	54°F	59°F	62°F	52°F
Albuquerque, NM	46°F	75°F	70°F	36°F

---

[2] If you play with a formula, you will become more comfortable with it and understand it better. Make up a small set of numbers and calculate a standard deviation. Change one of the numbers, or add a number, or leave out a number. See what happens. Besides teaching yourself about standard deviations, you may learn more efficient ways to use your calculator.

4.8. No computation is needed here; just eyeball the scores in set I and set II and determine which set is more variable or whether the two sets are equally variable.

a. *set I:* 1, 2, 4, 1, 3 — *set II:* 9, 7, 3, 1, 0
b. *set I:* 9, 10, 12, 11 — *set II:* 4, 5, 7, 6
c. *set I:* 1, 3, 9, 6, 7 — *set II:* 14, 15, 14, 13, 14
d. *set I:* 8, 4, 6, 3, 5 — *set II:* 4, 5, 7, 6, 15
e. *set I:* 114, 113, 114, 112, 113 — *set II:* 14, 13, 14, 12, 13

Computing S and σ with the Raw-Score Formula

The deviation-score formula helps you understand what is actually going on when you calculate a standard deviation. It is the formula to use until you do understand. Unfortunately, except in textbook examples, the deviation-score formula almost always has you working with decimal values. The raw-score formula, which is algebraically equivalent, involves far fewer decimals. It also produces answers more quickly, especially if you are working with a calculator.

The raw-score formula is

$$S \text{ or } \sigma = \sqrt{\frac{\Sigma X^2 - \frac{(\Sigma X)^2}{N}}{N}}$$

where ΣX^2 = sum of the squared scores
$(\Sigma X)^2$ = square of the sum of the raw scores
N = number of scores

Although the raw-score formula may appear more forbidding, it is actually easier to use than the deviation-score formula because you don't have to compute deviation scores. The numbers you work with will be larger, but your calculator won't mind.

Table 4.6 shows the steps for calculating S or σ by the raw-score formula. The data are for boxes of cookies sold by the six Girl Scouts. The arithmetic of Table 4.6 can be

TABLE 4.6 Using the raw-score formula to compute *S* for cookie sales by a sample of six Girl Scouts

Boxes of cookies X	X^2
28	784
11	121
10	100
5	25
4	16
2	4
$\Sigma X = 60$	$\Sigma X^2 = 1050$

Note: $(\Sigma X)^2 = (60)^2 = 3600$

$$S = \sqrt{\frac{\Sigma X^2 - \frac{(\Sigma X)^2}{N}}{N}} = \sqrt{\frac{1050 - \frac{(60)^2}{6}}{6}} = \sqrt{\frac{1050 - 600}{6}} = \sqrt{\frac{450}{6}} = \sqrt{75} = 8.66 \text{ boxes}$$

expressed in words: Square the sum of the values in the X column and divide the total by N. Subtract this result from the sum of the values in the X^2 column. Divide this difference by N and find the square root. The result is S or σ. Notice that the value of S in Table 4.6 is the same as the one you calculated in Table 4.5. In this case, the mean is an integer, so the deviation scores introduced no rounding errors.[3]

ΣX^2 and $(\Sigma X)^2$: Did you notice the difference in these two terms when you were working with the data in Table 4.6? If so, congratulations. You cannot calculate a standard deviation correctly unless you understand the difference. Reexamine Table 4.6 if you aren't sure of the difference between ΣX^2 and $(\Sigma X)^2$. Be alert for these two sums in the problems that are coming up.

error detection

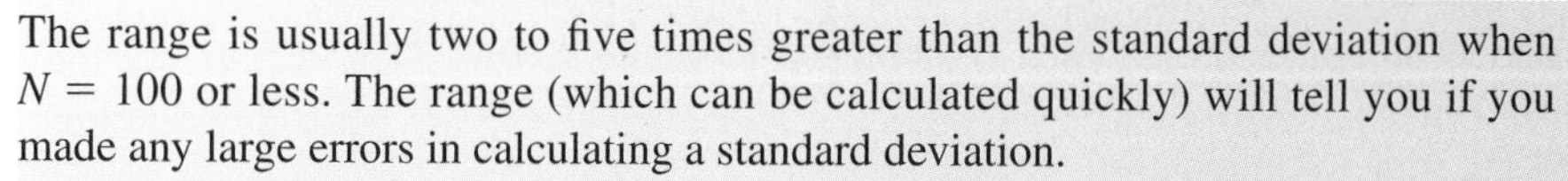

The range is usually two to five times greater than the standard deviation when $N = 100$ or less. The range (which can be calculated quickly) will tell you if you made any large errors in calculating a standard deviation.

PROBLEMS

***4.9.** Look at the following two distributions. Without calculating (just look at the data), decide which one has the larger standard deviation and estimate its size. (You may wish to calculate the range before you estimate.) Finally, make a choice between the deviation-score and the raw-score formulas and compute S for each distribution. Compare your computation with your estimate.

a. 5, 4, 3, 2, 1, 0 **b.** 5, 5, 5, 0, 0, 0

4.10. For each of the distributions in problem 4.9, divide the range by the standard deviation. Is the result between 2 and 5?

4.11. By now you can look at the following two distributions and see that **a** is more variable than **b.** The difference in the two distributions is in the lowest score (2 and 6). Calculate σ for each distribution, using the raw-score formula. Notice the difference in σ that is produced by the change of just one score.

a. 9, 8, 8, 7, 2 **b.** 9, 8, 8, 7, 6.

---

[3] Some textbooks and statisticians prefer the algebraically equivalent formula

$$S \text{ or } \sigma = \sqrt{\frac{N\Sigma X^2 - (\Sigma X)^2}{N^2}}$$

I am using the formula in the text because the same form, or parts of it, will be used in other procedures. Yet another formula is often used in the field of testing. To use this formula, you must first calculate the mean:

$$S = \sqrt{\frac{\Sigma X^2}{N} - \bar{X}^2} \quad \text{and} \quad \sigma = \sqrt{\frac{\Sigma X^2}{N} - \mu^2}$$

All these arrangements of the arithmetic are algebraically equivalent.

$\hat{s}$ as an Estimate of σ

Remember that $\hat{s}$ (ess-hat) is the principal statistic you will learn in this chapter; it will be used again and again throughout the rest of this text. Statisticians often add a "hat" to a symbol to indicate that something is being estimated. If you have sample data and you want to calculate an estimate of σ, use the statistic $\hat{s}$:

$$\hat{s} = \sqrt{\frac{\Sigma(X - \bar{X})^2}{N - 1}}$$

Note that the difference between $\hat{s}$ and σ is that $\hat{s}$ has $N - 1$ in the denominator, whereas σ has N.

This issue of dividing by N or by $N - 1$ sometimes leaves students shrugging their shoulders and muttering, "OK, I'll memorize it and do it however you want." I would like to explain, however, why you use $N - 1$ in the denominator when you have sample data and want to estimate σ.

Because the formula for σ is

$$\sigma = \sqrt{\frac{\Sigma(X - \mu)^2}{N}}$$

it would seem logical just to calculate $\bar{X}$ from the sample data, substitute $\bar{X}$ for μ in the numerator of the formula, and calculate an answer. This solution will, more often than not, produce a numerator that is *too small* [as compared to the value of $\Sigma(X - \mu)^2$].

To explain this surprising state of affairs, remember (page 43) a characteristic of the mean: For any set of scores, the expression $\Sigma(X - \bar{X})^2$ is minimized. That is, for any set of scores, this sum is smaller when $\bar{X}$ is used than it is if some other number (either larger or smaller) is used in place of $\bar{X}$. Thus, for a sample of scores, substituting $\bar{X}$ for μ gives you a numerator that is minimized. However, what you want is a value that is the same as $\Sigma(X - \mu)^2$. Now, if the value of $\bar{X}$ is at all different from μ, then the minimized value you get using $\Sigma(X - \bar{X})^2$ will be too small.

The solution that statisticians have adopted for this underestimation problem is to use $\bar{X}$ and then to divide the too-small numerator by a smaller denominator—namely, $N - 1$. This results in a statistic that is a much better estimator of σ.[4]

---

[4] To illustrate this issue for yourself, use a small population of scores and do some calculations. For a population with scores of 1, 2, and 3, $\sigma = 0.82$. Three different samples with $N = 2$ are possible in the population. For each sample, calculate the standard deviation using $\bar{X}$ and N in the denominator. Find the mean of these three standard deviations. Now, for each of the three samples, calculate the standard deviation using $N - 1$ in the denominator and find the mean of these three. Compare the two means to the σ that you want to estimate.

Mathematical statisticians use the term *unbiased estimator* for statistics whose average value (based on many samples) is exactly equal to the parameter of the population the samples came from. Unfortunately, even with $N - 1$ in the denominator, $\hat{s}$ is not a unbiased estimator of σ, although the bias is not very serious. There is, however, an unbiased measure of variability called the variance. The sample variance, $\hat{s}^2$, is an unbiased estimator of the population variance, σ^2. (See the Variance section that follows.)

You just finished three dense paragraphs and a long footnote—lots of ideas per square inch. You may understand it already, but if you don't, take 10 or 15 minutes to reread, do the exercise in footnote 4, and think.

Note also that as N gets larger, the subtraction of 1 from N has less and less effect on the size of the estimate of variability. This makes sense because the larger the sample size is, the closer $\bar{X}$ will be to μ, on average.

One other task comes with the introduction of $\hat{s}$: the decision whether to calculate σ, S, or $\hat{s}$ for a given set of data. Your choice will be based on your purpose. If your purpose is to estimate the variability of a population using data from a sample, calculate $\hat{s}$. (This purpose is common in inferential statistics.) If your purpose is to describe the variability of a sample or a population, use S or σ, respectively.

Calculating $\hat{s}$

To calculate $\hat{s}$ from raw scores, I recommend this formula:

$$\hat{s} = \sqrt{\frac{\Sigma X^2 - \frac{(\Sigma X)^2}{N}}{N - 1}}$$

This formula is the same as the raw-score formula for σ except for $N - 1$ in the denominator. This raw-score formula is the one you will probably use for your own data.[5]

Sometimes you may need to calculate a standard deviation for data already arranged in a frequency distribution. For a simple frequency distribution or a grouped frequency distribution, the formula is

$$\hat{s} = \sqrt{\frac{\Sigma f X^2 - \frac{(\Sigma f X)^2}{N}}{N - 1}}$$

where f is the frequency of scores in an interval.

Here are some data to illustrate the calculation of $\hat{s}$ both for ungrouped raw scores and for a frequency distribution. Consider puberty. As you know, females reach puberty earlier than males (about 2 years earlier on the average). Is there any difference between the genders in the *variability* of reaching this developmental milestone? Comparing standard deviations will give you an answer.

If you have only a sample of ages for each sex and your interest is in all females and all males, $\hat{s}$ is the appropriate standard deviation. **Table 4.7** shows the calculation of $\hat{s}$ for the females. Work through the calculations. The standard deviation is 2.19 years.

Data for the ages at which males reach puberty are presented in a simple frequency distribution in **Table 4.8**. Work through these calculations, noting that grouping causes two additional columns of calculations. For males, $\hat{s}$ is 1.44 years.

[5] Calculators with standard deviation functions differ. Some use N in the denominator, some use $N - 1$, and some have both standard deviations. You will have to check yours to see how it is programmed.

TABLE 4.7 Calculation of $\hat{s}$ for age at which females reach puberty ungrouped raw scores

Age (X)	X^2
17	289
15	225
13	169
12	144
12	144
11	121
11	121
11	121

$\Sigma X = 102 \qquad \Sigma X^2 = 1334 \qquad (\Sigma X)^2 = 10{,}404$

$\overline{X} = 12.75$ years

$$\hat{s} = \sqrt{\frac{\Sigma X^2 - \frac{(\Sigma X)^2}{N}}{N-1}} = \sqrt{\frac{1334 - \frac{(102)^2}{8}}{7}} = \sqrt{\frac{1334 - 1300.50}{7}} = 2.19 \text{ years}$$

Thus, based on sample data, you can conclude that there is more variability among females in the age of reaching puberty than there is among males. (Incidentally, you would be correct in your conclusion—I chose the numbers so they would produce results that are similar to population figures.)

Here are three final points about simple and grouped frequency distributions.

- ΣfX^2 is found by squaring X, multiplying by f, and then summing.
- $(\Sigma fX)^2$ is found by multiplying f by X, summing, and then squaring.
- For grouped frequency distributions, X is the midpoint of a class interval.

TABLE 4.8 Calculation of $\hat{s}$ for age at which males reach puberty (simple frequency distribution)

Age (X)	f	fX	fX^2
18	1	18	324
17	1	17	289
16	2	32	512
15	4	60	900
14	5	70	980
13	3	39	507
	$N = 16$	$\Sigma fX = 236$	$\Sigma fX^2 = 3512$

$\overline{X} = 14.75$ years

$$\hat{s} = \sqrt{\frac{\Sigma fX^2 - \frac{(\Sigma fX)^2}{N}}{N-1}} = \sqrt{\frac{3512 - \frac{(236)^2}{16}}{15}} = \sqrt{\frac{3512 - 3481}{15}} = 1.44 \text{ years}$$

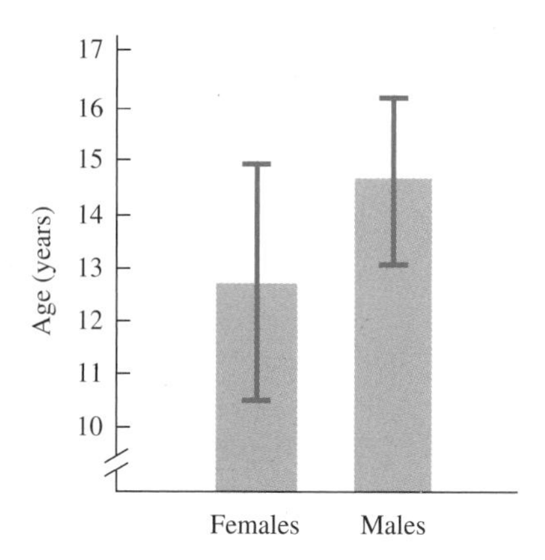

FIGURE 4.1 Bar graph of mean age of onset of puberty for females and males (error bars = I standard deviation)

error detection

ΣfX^2 and $(\Sigma fX)^2$ are similar in appearance, but they tell you to do different operations. The different operations produce different results.

For both grouped and simple frequency distributions, most calculators give you ΣfX and ΣfX^2 if you properly key in X and f for each line of the distribution. Procedures differ depending on the brand. The time you invest in learning will be repaid several times over in future chapters (not to mention the satisfying feeling you will get).

clue to the future

You will be glad to know that in your efforts to calculate a standard deviation, you have produced two other useful statistics along the way. Each has a name, and they turn up again in future chapters. The number that you took the square root of to get the standard deviation is called the variance (also called a *mean square*). The expression in the numerator of the standard deviation, $\Sigma(X - \overline{X})^2$, is called the *sum of squares*, which is an important concept in Chapters 11–13.

Graphing Standard Deviations

The information on variability that a standard deviation conveys can often be added to a graph. **Figure 4.1** is a double-duty bar graph of the puberty data that shows standard deviations as well as means. The lines extend 1 standard deviation above and below the mean.[6] From this graph you can see at a glance that the mean age of puberty for females is younger than for males and that they are more variable in reaching puberty.

[6] Other measures of variability besides standard deviations are also presented as an extended line. The caption or the legend of the graph tells you the measure.

TABLE 4.9 SPSS output of descriptive statistics for the SWLS scores

Statistics		
SWLSscores		
N	Valid	100
	Missing	0
Mean		24.0000
Median		25.0000
Mode		27.00
Std. Deviation		6.41809
Variance		41.192
Range		30.00
Percentiles	25	21.0000
	50	25.0000
	75	28.0000

The Variance

The last step in calculating a standard deviation is to find a square root. The number you take the square root of is the **variance**. The symbols for the variance are σ^2 (population variance) and $\hat{s}^2$ (sample variance used to estimate the population variance). By formula,

variance
Square of the standard deviation.

$$\sigma^2 = \frac{\Sigma(X - \mu)^2}{N} \quad \text{and} \quad \hat{s}^2 = \frac{\Sigma(X - \overline{X})^2}{N - 1} \quad \text{or} \quad \hat{s}^2 = \frac{\Sigma X^2 - \frac{(\Sigma X)^2}{N}}{N - 1}$$

The difference between σ^2 and $\hat{s}^2$ is the term in the denominator. The population variance uses N and the sample variance uses $N - 1$. The variance is not very useful as a *descriptive* statistic. It is, however, of enormous importance in inferential statistics, especially in a technique called the analysis of variance (Chapters 11, 12, and 13).

Statistical Software Programs

There are many computer software programs that calculate statistics. One of the most widely used is called SPSS. At several places in this book, I have included look-alike tables from SPSS. **Table 4.9** has SPSS output of measures of central tendency and variability of the SWLS scores you have been working with since Chapter 2.

PROBLEMS

4.12. Make a rough-sketch bar graph that shows the means and standard deviations of the problem 4.7 temperature data for San Francisco and Albuquerque. Include a caption.

4.13. Here is an interpretation problem. Remember the student who recorded money spent at the Student Center every day for 14 days? The mean was \$2.90 per day. Suppose the student wanted to reduce his Student Center spending. Write a sentence of advice if $\hat{s} = \$0.02$ and a second sentence if $\hat{s} = \$2.50$.

4.14. Describe in words the relationship between the variance and the standard deviation.

4.15. A researcher had a sample of scores from the freshman class on a test that measured attitudes toward authority. She wished to estimate the standard deviation of the scores of the entire freshman class. (She had data from 20 years ago and she believed that current students were more homogeneous than students in the past.) ΣX and ΣX^2 have already been determined. Calculate the proper standard deviation and variance.

$N = 21 \qquad \Sigma X = 304 \qquad \Sigma X^2 = 5064$

4.16. Here are those data on the heights of the Americans in their 20s that you have been working with. For each group, calculate $\hat{s}$. Write a sentence of interpretation.

Women		Men	
Height (in.)	*f*	Height (in.)	*f*
72	1	77	1
70	1	76	1
69	1	75	2
68	1	74	1
67	3	73	4
66	6	72	7
65	10	71	6
64	9	70	7
63	7	69	8
62	4	68	4
61	1	67	2
60	4	66	2
59	2	65	3
		64	1
		62	1

4.17. A high school English teacher measured the attitudes of 11th-grade students toward poetry. After a 9-week unit on poetry, she measured the students' attitudes again. She was disappointed to find that the mean change was zero.

Following are some representative scores. (High scores represent more favorable attitudes.) Calculate $\hat{s}$ for both before and after scores, and write a conclusion based on the standard deviations.

Before	7	5	3	5	5	4	5	6
After	9	8	2	1	8	9	1	2

4.18. In manufacturing, engineers strive for consistency. The following data are the errors, in millimeters, in giant gizmos manufactured by two different processes. Choose S or $\hat{s}$ and determine which process produces the more consistent gizmos.

Process A	0	1	−2	0	−2	3
Process B	1	−2	−1	1	−1	2

***4.19.** Find the interquartile range for normal oral body temperature and write a sentence of interpretation. For data, use your answer to problem 2.17a.

4.20. Estimate the population standard deviation for oral body temperature using data you worked with in problem 2.17a. You may find ΣfX and ΣfX^2 by working from the answer to problem 2.17a, or, if you understand what you need to do to find these values from that table, you can use $\Sigma fX = 3928.0$ and $\Sigma fX^2 = 385{,}749.24$.

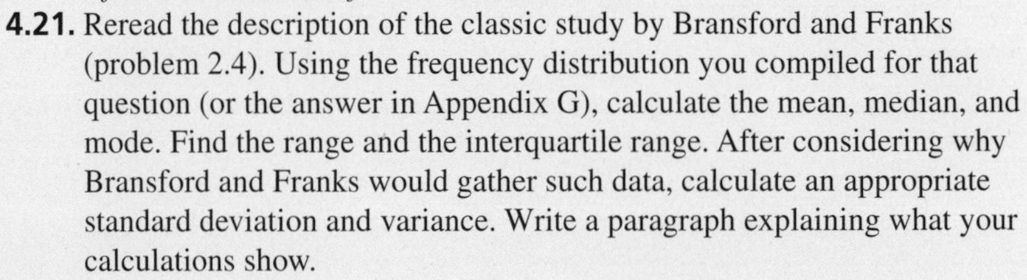

4.21. Reread the description of the classic study by Bransford and Franks (problem 2.4). Using the frequency distribution you compiled for that question (or the answer in Appendix G), calculate the mean, median, and mode. Find the range and the interquartile range. After considering why Bransford and Franks would gather such data, calculate an appropriate standard deviation and variance. Write a paragraph explaining what your calculations show.

4.22. Return to the objectives at the beginning of the chapter. Can you do each one?

ADDITIONAL HELP FOR CHAPTER 4

Visit *cengage.com/psychology/spatz*. At the Student Companion Site, you'll find multiple-choice tutorial quizzes, flashcards with definitions, and workshops. For this chapter, there is a Statistical Workshop on Central Tendency and Variability.

KEY TERMS

Deviation score (p. 59)
Estimate (p. 58)
Interquartile range (p. 56)
Percentile (p. 56)
Range (p. 56)
Standard deviation (p. 58)
Variability (p. 54)
Variance (p. 68)

CHAPTER 5

Other Descriptive Statistics

OBJECTIVES FOR CHAPTER 5

After studying the text and working the problems in this chapter, you should be able to:

1. Use *z* scores to compare two scores in one distribution
2. Use *z* scores to compare scores in one distribution with scores in a second distribution
3. Construct and interpret boxplots
4. Identify outliers in a distribution
5. Calculate an effect size index and interpret it
6. Compile descriptive statistics and an explanation into a Descriptive Statistics Report

THIS CHAPTER INTRODUCES statistical techniques that have two things in common. First, all are descriptive statistics. Second, each one combines two or more statistical components. Fortunately, the components are ones you studied in the preceding two chapters. These statistical techniques should prove quite helpful as you improve your own ability to explore and understand data.

The first statistic, the *z score,* tells you the relative standing of a raw score in its distribution. The formula for a *z* score combines a raw score with its distribution's mean and standard deviation. A *z*-score description of a raw score works regardless of the kind of raw scores or the shape of the distribution. This section also defines *outliers*, which are extreme scores in a distribution, and offers suggestions of what to do about them.

The second section of this chapter covers *boxplots.* A boxplot is a graphic with information on one variable, much like a frequency polygon, but it conveys lots more about a distribution. With just one picture, a boxplot gives you the median, range, interquartile range, and skew of a distribution. Often, it provides additional information as well.

This chapter introduces *d*, the *effect size index*. An effect size index allows you to describe the size of the difference between two distributions as small, medium, or large. If you know that an independent variable makes a difference in the scores on the dependent variable, then *d* indicates how much of a difference the independent variable makes.

The final section of this chapter has no new statistics. It shows you how to put together the ones you have learned into a Descriptive Statistics Report, which is an organized collection of descriptive statistics that helps a reader understand the data that were gathered.

Probably all college graduates are familiar with measures of central tendency; most are familiar with measures of variability. However, only those with a good education in quantitative thinking are familiar with all the techniques presented in this chapter. So, learn this material. You will then understand more than most and you will be better equipped to help others.

Describing Individual Scores

Suppose one of your friends says he got a 95 on a math exam. What does that tell you about his mathematical ability? From your previous experience with tests, 95 may seem like a pretty good score. This conclusion, however, depends on a couple of assumptions, and unless those assumptions are correct, a score of 95 is *meaningless*. Let's return to the conversation with your friend.

After you say, "95! Congratulations," suppose he tells you that 200 points were possible. Now a score of 95 seems like something to hide. "My condolences," you say. But then he tells you that the highest score on that difficult exam was 105. Now 95 has regained respectability and you chortle, "Well, all right!" In response, he shakes his head and tells you that the mean score was 100. The 95 takes a nose dive. As a final blow, you find out that 95 was the lowest score, that nobody scored worse than your friend. With your hand on his shoulder, you cut off further discussion of the test with, "Come on, I'll buy you an ice cream cone."

This example illustrates that the meaning of a score of 95 depends on the rest of the test scores. Fortunately, there are several ways to convert a raw score into a measure that signals its relationship to other scores. Percentiles are one well-known example. I'll explain two others that are important for statistical analyses, *z* scores and outliers.

The *z* score

***z* score**
Score expressed in standard deviation units.

The *z* **score** is a widely used technique. It modifies an individual score so that it conveys the score's relationship to both the mean and the standard deviation of its fellow scores. The formula is

$$z = \frac{X - \bar{X}}{S}$$

Remember that the numerator, $X - \bar{X}$, is an acquaintance of yours, the deviation score. A *z* score describes the relation of X to $\bar{X}$ with respect to the variability of the

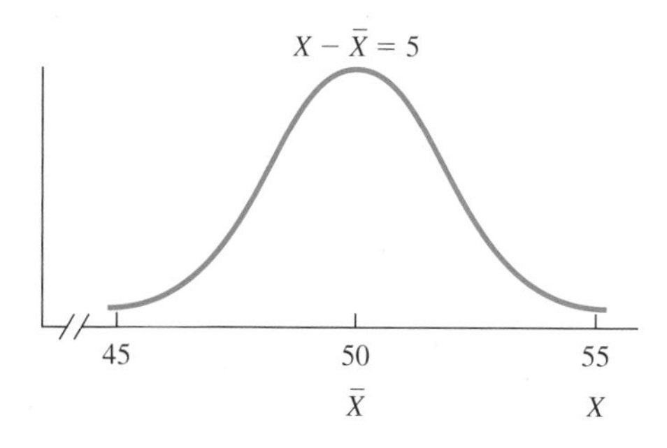

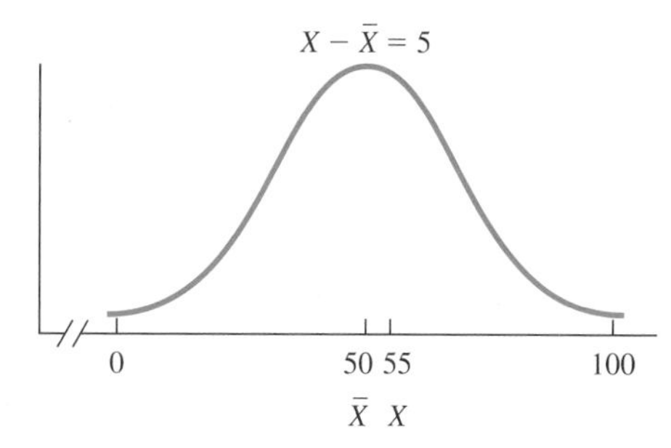

FIGURE 5.1 A comparison of $X - \bar{X} = 5$ for a distribution with a small standard deviation (left) and a large standard deviation (right)

distribution. For example, if you know that a score (X) is 5 units from the mean—that is, $X - \bar{X} = 5$—you know only that the score is better than average, but you have no idea how far above average it is. If the distribution has a range of 10 units and $\bar{X} = 50$, then an X of 55 is a very high score. On the other hand, if the distribution has a range of 100 units, an X of 55 is barely above average. Look at **Figure 5.1**, which is a picture of the ideas in this paragraph.

standard score
Score expressed in standard deviation units; z is one example.

Thus, to know a score's position in a distribution, the variability of the distribution must be taken into account. The way to do this is to divide $X - \bar{X}$ by a unit that measures variability, the standard deviation. The result is a deviation score per unit of standard deviation.[1] A z score is often referred to as a **standard score** because it is a deviation score expressed in standard deviation units.

Any distribution of raw scores can be converted into a distribution of z scores; for each raw score, there is one z score. Positive z scores represent raw scores that are greater than the mean; negative z scores go with raw scores that are less than the mean. In both cases, the absolute value of the z score tells the number of standard deviations the score is from the mean. The mean of the z scores is 0; its standard deviation is 1.

When one raw score is converted into a z score, the z score gives the raw score's position in the distribution. If *two* raw scores are converted, the two z scores tell you their positions relative to *each other* as well as to the distribution. Finally, z scores are also used to compare two scores from different distributions, even when the scores are measuring different things. (If this seems like trying to compare apples and oranges, see problem 5.5.)

In the general psychology course I took as a freshman, the professor returned tests with a z score rather than a percentage score. This z score was the key to figuring out your grade. A z score of $+1.50$ or higher was an A, and -1.50 or lower was an F. (z scores between $+.50$ and $+1.50$ received B's. If you assume the professor practiced symmetry, you can figure out the rest of the grading scale.)

---

[1] This technique is based on the same idea that percents are based on. For example, 24 events at one time and 24 events a second time appear to be the same. But don't stop with appearances. Ask for the additional information that you need, which is, "24 events out of how many chances?" An answer of "50 chances the first time and 200 chances the second" allows you to convert both 24s to "per centum" (per one hundred). Clearly, 48 percent and 12 percent are different, even though both are based on 24 events. Other examples of this technique of dividing to make raw scores comparable include miles per gallon, per capita income, bushels per acre, and points per game.

TABLE 5.1 The *z* scores of selected students on two 100-point tests in general psychology

	Test 1		Test 2	
Student	Raw score	z score	Raw score	z score
Kris	76	+2.20	76	−1.67
Robin	54	.00	86	.00
Marty	58	+.40	82	−.67
Terry	58	+.40	90	+.67
	Test 1: $\bar{X} = 54$		Test 2: $\bar{X} = 86$	
	$S = 10$		$S = 6$	

Table 5.1 lists the raw scores (percentage correct) and z scores for four of the many students who took two of the tests in that class. Begin by noting that for test 1, $\bar{X} = 54$ and $S = 10$. For test 2, $\bar{X} = 86$, $S = 6$. Well, so what? To answer this question, examine the scores of the four individuals. Consider the first student, Kris, who scored 76 on both tests. The two 76s appear to be the same, but the z scores show that they are not. The first 76 was a high A and the second 76 was an F. The second student, Robin, appears to have improved on the second test, going from 54 to 86, but in fact, the scores were the test averages both times (a grade of C). Robin was the same on both tests. Marty also appears to have improved if you examine only the raw scores. However, the z scores reveal that Marty did *worse* on the second test. Finally, comparing Terry's and Robin's raw scores, you can see that although Terry scored 4 points higher than Robin on each test, the z scores show that Terry's improvement was greater than Robin's on the second test.

The reason for these surprising comparisons is that the means and standard deviations are so different for the two tests. Perhaps the second test was easier, or the material was more motivating to students, or the students studied more. Maybe the teacher prepared better. Perhaps all of these reasons were true.

To summarize, z scores give you a way to compare raw scores. The basis of the comparison is the distribution itself rather than some external standard (such as a grading scale of 90–80–70–60 percent for As, Bs, and so on).

A word of caution: z is used as both a descriptive statistic and an inferential statistic. As a descriptive statistic, its range is limited. For a distribution of 100 or so scores, the z scores might range from approximately −3 to +3. For many distributions, especially when N is small, the range is less.

As an inferential statistic, however, z values are not limited to ±3. To illustrate, z score texts are used in Chapter 15 to help decide whether two populations are different. The value of z depends heavily on how different the two populations actually are. When z is used as an inferential statistic, values much greater than 3 can occur.

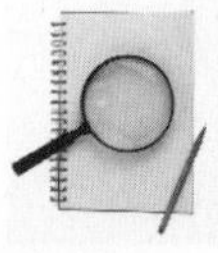

clue to the future

z scores will turn up often as you study statistics. They are prominent in this book in Chapters 5, 7, 8, and 15.

Outliers

Outlier
An extreme score separated from the others and 1.5 × IQR or more beyond the 25th or 75th percentile.

Outliers are scores in a distribution that are unusually small or unusually large. An outlier has a disproportionate influence, compared to any of the others scores, on the mean, standard deviation, and other statistical measures. They can certainly affect the outcome of a statistical analysis and they appear to be common (Wilcox, 2005a).

Although there is no general agreement on how to identify outliers, Hogan and Evalenko (2006) found that the most common definition in statistics textbooks is

Lower outlier = 25th percentile − (1.5 × IQR)

Upper outlier = 75th percentile + (1.5 × IQR)

Using these definitions and the heights of 20–29-year-old women and men, we can determine heights that qualify as outliers. For women,

Lower outlier = 25th percentile − (1.5 × IQR) = 63 − 1.5(3) = 58.5 inches or shorter

Upper outlier = 75th percentile + (1.5 × IQR) = 66 + 1.5(3) = 70.5 inches or taller

For men,

Lower outlier = 25th percentile − (1.5 × IQR) = 68 − 1.5(4) = 62 inches or shorter

Upper outlier = 75th percentile + (1.5 × IQR) = 72 + 1.5(4) = 78 inches or taller

What should you do if you detect an outlier in your data? The answer is to think. Could the outlier score be a recording error? Is there a way to check? The outlier score may not be an error, of course. Each of us probably knows someone who is taller or shorter than the outlier heights identified above. Still outliers distort means, standard deviations, and other statistics. Fortunately, mathematical statisticians have developed statistical techniques for data with outliers (Wilcox, 2005b), but these are typically covered only in advanced courses.

PROBLEMS

5.1. The mean of any distribution has a z score equal to what value?

5.2. What conclusion can you reach about Σz?

5.3. Under what conditions would you prefer that your personal z score be negative rather than positive?

5.4. Harriett and Heslope, twin sisters, were intense competitors, but they never competed against each other. Harriett specialized in long-distance running and Heslope was an excellent sprint swimmer. As you can see from the distributions in the accompanying table, each was the best in her event. Take the analysis one step further and use z scores to determine who is the more outstanding twin. You might start by looking at the data and making an estimate.

10k runners	Time (min)	50m swimmers	Time (sec)
Harriett	37	Heslope	24
Dott	39	Ta-Li	26
Liz	40	Deb	27
Marette	42	Betty	28

5.5. Tobe grows apples and Zeke grows oranges. In the local orchards the mean weight of apples is 5 ounces, with $S = 1.0$ ounce. For oranges the mean weight is 6 ounces, with $S = 1.2$ ounces. At harvest time, each entered his largest specimen in the Warwick County Fair. Tobe's apple weighed 9 ounces and Zeke's orange weighed 10 ounces. This particular year Tobe was ill on the day of judgment, so he sent his friend Hamlet to inquire who had won. Adopt the role of judge and use z scores to determine the winner. Hamlet's query to you is: "Tobe, or not Tobe; that is the question."

5.6. Ableson's anthropology professor announced that the poorest exam grade for each student would be dropped. Ableson scored 79 on the first anthropology exam. The mean was 67 and the standard deviation 4. On the second exam, he made 125. The class mean was 105 and the standard deviation 15. On the third exam, the mean was 45 and the standard deviation 3. Ableson got 51. Which test should be dropped?

5.7. Using your answer to problem 4.19, determine which of the following temperatures qualifies as an outlier.

a. 98.6°F
b. 99.9°F
c. 96.0°F
d. 96.6°F
e. 100.5°F

Boxplots

boxplot
Graph that shows a distribution's range, interquartile range, skew, median, and sometimes other statistics.

So far you have studied three different characteristics of distributions: central tendency, variability, and form. A **boxplot** is a way to present all three characteristics with one graphic.[2] What information does a boxplot provide? You get two measures of central tendency, two measures of variability, and a way to estimate skewness. Skewness, of course, is a description of the form of the distribution.

Figure 5.2 is a boxplot of the Satisfaction With Life Scale scores you first encountered in Chapter 2. The older name for this graphic is a "box-and-whisker plot,"

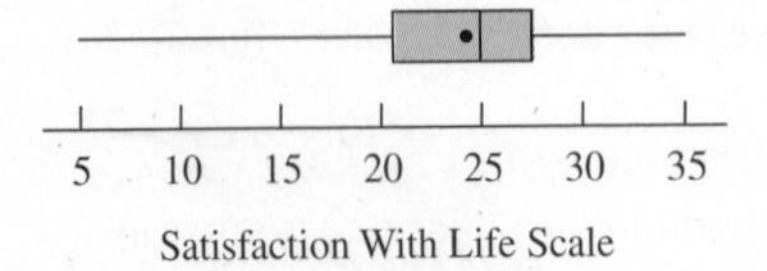

FIGURE 5.2 Boxplot of Satisfaction With Life Scale scores

[2] Boxplots were dreamed up by John Tukey (1915–2000), who invented several statistical techniques that facilitate the exploration of data. See Lovie's (2005) entry on Exploratory Data Analysis.

and as you can see, it consists of a *box* and *whiskers* (and a *line* and a *dot* within the box). The horizontal axis shows the variable that was measured and its values.

Interpreting Boxplots

Central tendency A boxplot gives two measures of central tendency. The vertical line inside the box is at the median. The dot is the mean. In Figure 5.2 you can estimate the median as 25 and the mean as slightly less (it is actually 24).

Variability Both the box and the whiskers in Figure 5.2 tell you about variability. The *box* covers the interquartile range, which you studied in Chapter 4. The left end of the box is the 25th percentile score, and the right end is the 75th percentile score. You can estimate these scores from the horizontal axis. The horizontal width of the box is the **interquartile range.**

The *whiskers* extend to the extreme scores in the distribution. Thus, the whiskers give you a picture of the range. Again, reading from the scale in Figure 5.2, you can see that the highest score is 35 and the lowest is 5.

Skew Finally, both the relationship of the mean to the median and any difference in the lengths of the whiskers help you determine the **skew** of the distribution. You already know that, in general, when the mean is less than the median, the skew is negative; when the mean is greater than the median, the skew is positive. The relationship of the mean to the median is readily apparent in a boxplot.

Skew is also usually indicated by whiskers of different length. The longer whisker tells you the direction of the skew. Thus, if the longer whisker is over the lower scores, the skew is negative; if the longer whisker is over the higher scores, the skew is positive. Given these two rules of thumb to determine skew, you can conclude that the distribution of SWLS scores in Figure 5.2 is negatively skewed.

Variations in boxplots Boxplots are useful during the initial exploratory stages of data analysis and also later for presentation of data to others. Fortunately, the boxplot format can be varied to suit the needs of the researcher. For example, at times the mean is not included in a boxplot. Outliers can be indicated on a boxplot with dots or asterisks beyond the end of a whisker. Also, boxplots are sometimes oriented vertically instead of horizontally.

Boxplot questions Look at **Figure 5.3**, which shows the boxplots of four distributions. Seven questions follow. See how many you get right. If you get all of them right, consider skipping the explanations that follow the answers.

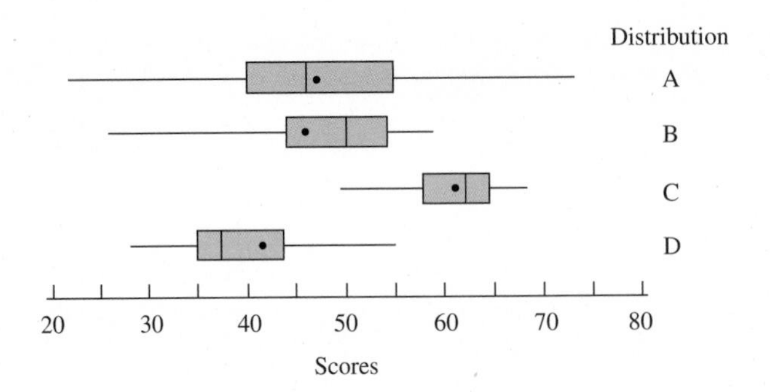

FIGURE 5.3 Boxplots of four distributions

Questions

1. Which distribution has the greatest positive skew?
2. Which distribution is the most compact?
3. Which distribution has a mean closest to 40?
4. Which distribution is most symmetrical?
5. Which distribution has a median closest to 50?
6. Which distribution is most negatively skewed?
7. Which distribution has the largest range?

Answers

1. *Positive skew:* Distribution D. The mean is greater than the median, and the high-score whisker is longer than the low-score whisker.
2. *Most compact:* Distribution C. The range is smaller than other distributions.
3. *Mean closest to 40:* Distribution D.
4. *Most symmetrical:* Distribution A. The mean and median are about the same, and the whiskers are about the same length.
5. *Median closest to 50:* Distribution B.
6. *Most negative skew:* Distribution B. The difference between the mean and median is greater in B than in C, and the difference in whisker length is greater in B than in C.
7. *Largest range:* Distribution A.

error detection

Gross errors from misrecording data are often apparent from a boxplot. Detecting errors is easiest when you use a computer program to generate the boxplot and you are familiar with boxplot interpretation.

PROBLEMS

5.8. Tell the story (mean, median, range, interquartile range, and form) of each of the three boxplots in the figure.

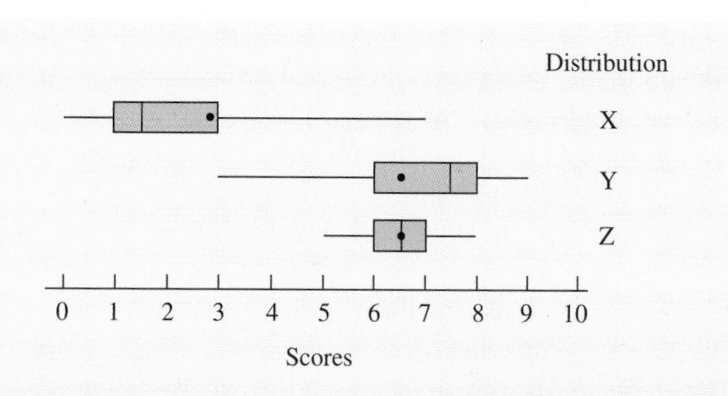

5.9. You have already found the elements needed for boxplots of the heights of 20- to 29-year-old women and men. (See problems 2.1, 3.4, and 4.3.) Draw boxplots of the two groups using one horizontal axis.

***5.10.** Create a boxplot (but without the mean) of the oral body temperature data based on Mackowiak, Wasserman, and Levine (1992). Find the statistics you need from your answer to problems 2.17a and 4.19 (Appendix G).

Effect Size Index

- Men are taller than women.
- First-born children score higher than second-born children on tests of cognitive ability.
- Interrupted tasks are remembered better than uninterrupted tasks.

These three statements follow a pattern that is common when quantitative variables are compared. The pattern is, "The average *X* is greater than the average *Y*." It happens that the three statements are true on average, but they leave an important question unanswered. How *much* taller, higher, or better is the first group than the second? For the heights of men and women, a satisfactory answer is that men, on the average, are almost 6 inches taller than women.

But how satisfactory is it to know that the mean difference in cognitive ability scores of first-born and second-born children is 13 points or that the difference in recall of interrupted and uninterrupted tasks is 3 tasks? Something else is needed. An **effect size index** is the statistician's way to answer the question, How much difference is there?

effect size index
Amount or degree of separation between two distributions.

An *effect size index* gives you a mathematical way to answer the question, How much taller, higher, or better? It works whether you are already familiar with the scale of measurement (such as inches) or you have never heard of the scale before (cognitive ability and recall scores).

The Effect Size Index, *d*

There are several ways to measure effect size (Kirk, 2005). Probably the most common measure is symbolized by *d*, where

$$d = \frac{\mu_1 - \mu_2}{\sigma}$$

Thus, *d* is the difference between means per standard deviation unit.

To calculate d, you must estimate the parameters with statistics. Samples from the μ_1 population and from the μ_2 population produce $\bar{X}_1$ and $\bar{X}_2$. Calculating an estimate of σ requires knowing about degrees of freedom, a concept that is better introduced in connection with hypothesis testing (Chapter 9 and chapters that follow). So, at this point you will have to be content to have σ given to you.

You can easily see that if the difference between means is zero, then $d = 0$. Also, depending on σ, a given difference between two means might produce a small d or a large d. Whether d is positive or negative depends on which group is assigned 1 and which is given 2. Often this decision is arbitrary, which makes the sign of d unimportant. However, in an experiment with an experimental group and a control group, it is conventional to designate the experimental group as group 1.

The Interpretation of *d*

A widely accepted convention of what constitutes small, medium, and large effect sizes was proposed by Jacob Cohen (1969).[3]

Small effect	$d = 0.20$
Medium effect	$d = 0.50$
Large effect	$d = 0.80$

Researchers often use standard deviation language to describe d values. For example, "The independent variable had a large effect, increasing scores on the dependent variable by eight tenths of a standard deviation."

To get a visual idea of these d values, look at **Figure 5.4**. The three components illustrate small, medium, and large values of d for both frequency polygons and boxplots. In the top panel the mean of distribution B is two-tenths of a standard deviation unit greater than the mean of distribution A ($d = 0.20$). You can see that there is a great deal of overlap between the two distributions. Study the other two panels of Figure 5.4, examining the amount of overlap for $d = 0.50$, and $d = 0.80$.

To illustrate further, let's take the heights of women and men. Just intuitively, what adjective would you use to describe the difference in the heights of women and men? A small difference? A large difference?

Well, it certainly is an obvious difference, one that everyone sees. Let's find the effect size index for the difference in heights of women and men. You already have estimates of μ_{women} and μ_{men} from your work on problem 3.4: $\bar{X}_{\text{women}} = 64.2$ inches and $\bar{X}_{\text{men}} = 70.0$ inches. For this problem, $\sigma = 2.8$ inches. Thus,

$$d = \frac{\mu_1 - \mu_2}{\sigma} = \frac{\bar{X}_1 - \bar{X}_2}{\sigma} = \frac{64.2 - 70.0}{2.8} = -2.07$$

The interpretation of $d = 2.07$ is that the difference in heights of women and men is just huge, more than twice the size of a difference that would be judged "large." So, here is a reference point for you for effect size indexes. If a difference is so great that everyone is aware of it, then the effect size index, d, will be greater than large.

Let's take another example. Women, on average, have higher verbal scores than men do. What is the effect size index for this difference? To answer this question,

[3] For an easily accessible source of Cohen's reasoning in proposing these conventions, see Cohen (1992, p. 99).

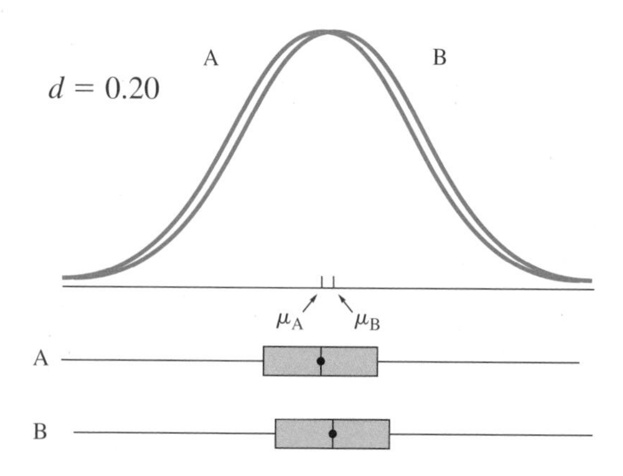

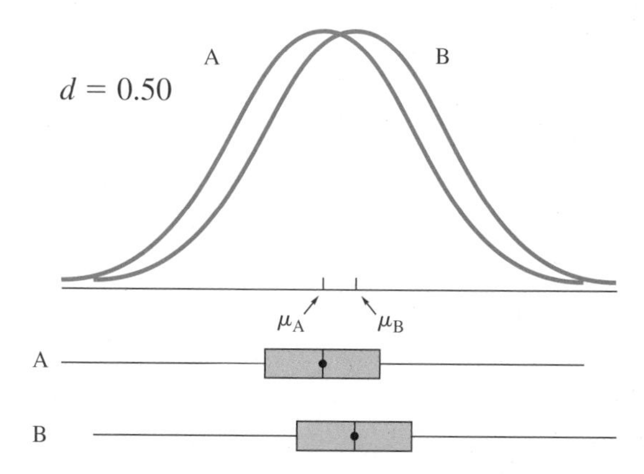

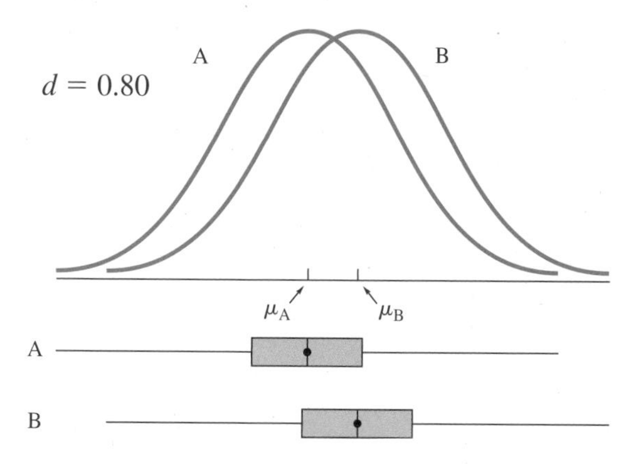

FIGURE 5.4 Frequency polygons and boxplots of two populations that differ by small ($d = 0.20$), medium ($d = 0.50$), and large ($d = 0.80$) amounts

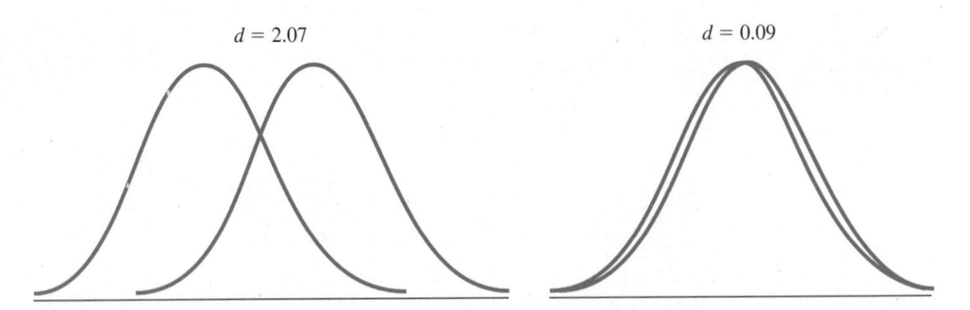

FIGURE 5.5 Frequency polygons that show effect size indexes of 2.07 (heights of men and women) and 0.09 (verbal scores of men and women)

I consulted Hedges and Nowell (1995). Their analysis of six studies with very large representative samples and a total $N > 150{,}000$ revealed that $\bar{X}_{\text{women}} = 513$, $\bar{X}_{\text{men}} =$ 503, and $\sigma = 110$.[4] Thus,

$$d = \frac{\mu_1 - \mu_2}{\sigma} = \frac{\bar{X}_1 - \bar{X}_2}{\sigma} = \frac{513 - 503}{110} = 0.09$$

A *d* value of 0.09 is less than half the size of a value considered "small." Thus, you can say that although the average verbal ability of women is better than that of men, the difference is very small.

Figure 5.5 shows overlapping frequency polygons with *d* values of 2.07 and 0.09, the *d* values found in the two examples in this section.

A common question is: How do you interpret *d* values that are intermediate between 0.20, 0.50, and 0.80? The answer is to use modifiers. So far, I've used the terms "huge" and "half the size of small." Phrases such as "somewhat larger than" and "intermediate between" are often useful.

The Descriptive Statistics Report

Techniques from this chapter and the previous two can be used to compile a Descriptive Statistics Report, which gives you a fairly complete story for a set of data.[5] The most interesting Descriptive Statistics Reports are those that compare two or more distributions of scores. To compile a Descriptive Statistics Report for two groups, (1) construct boxplots, (2) find the effect size index, and (3) tell the story that the data reveal. As for telling the story, cover the following points, arranging them so that your story is told well.

- Form of the distributions
- Central tendency
- Overlap of the two distributions
- Interpretation of the effect size index

[4] I generated these means and standard deviation so the numbers would mimic SAT Critical Reading scores and the outcome would mirror the conclusions of Hedges and Nowell (1995).

[5] A more complete report contains inferential statistics and their interpretation.

TABLE 5.2 Descriptive statistics for a Descriptive Statistics Report of the heights of women and men

	Heights of 20- to 29-year-old Americans	
	Women (in.)	Men (in.)
Mean	64.2	70.0
Median	64	70
Minimum	59	62
Maximum	72	77
25th percentile score	63	68
75th percentile score	66	72
Effect size index	2.07	

To illustrate a Descriptive Statistics Report, let's return to the heights of the men and women that you began working with in Chapter 2. The first steps are to assemble the statistics needed for boxplots and to calculate an effect size index. Look at **Table 5.2**, which shows these statistics. The next step is to construct boxplots (your answer to problem 5.9). The final step is to write a paragraph of interpretation. To write a paragraph, I recommend that you make notes and then organize your points, selecting the most important one to lead with. Write a rough draft. Revise the draft until you are satisfied with it.[6] My version is **Table 5.3**, a Descriptive Statistics Report of the heights of women and men.

I will stop with just one example of a Descriptive Statistics Report. The best way to learn and understand is to create reports of your own. Thus, problems follow shortly.

For the first (but not the last) time, I want to call your attention to the subtitle of this book: *Tales of Distributions.* A Descriptive Statistics Report is a tale of distributions of scores. What on earth would be the purpose of such stories?

TABLE 5.3 A Descriptive Statistics Report on the heights of women and men

The graph shows boxplots of heights of American women and men, aged 20–29. The difference in means produces an effect size index of 2.07.

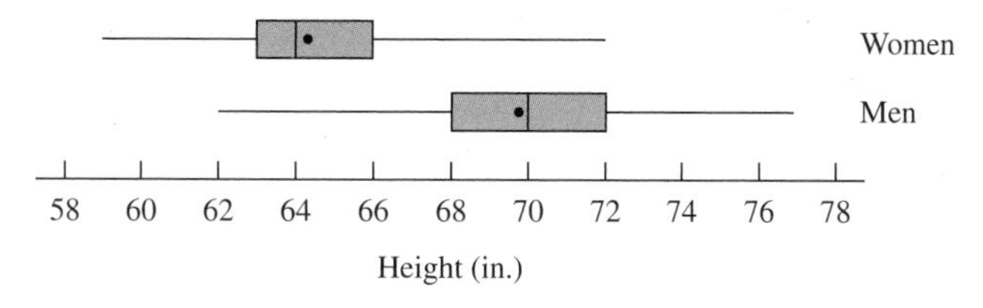

The mean height of women is 64.2 inches; the median is 64 inches. The mean height of men is 70.0 inches; the median is also 70 inches. Men are almost 6 inches taller than women, on average. Although the two distributions overlap, more than 75 percent of the men are taller than 66 inches, a height exceeded by less than 25 percent of the women. This difference in the two distributions is reflected by an effect size index of 2.07, a very large value. (A value of 0.80 is traditionally considered large.) The heights for both women and men are distributed fairly symmetrically.

---

[6] For me, several revisions are needed. This paragraph got nine.

The purpose might be to better understand a scientific phenomenon that you are intensely interested in. The purpose might be to explain to your boss the changes that are taking place in your industry; perhaps it is to convince a quantitative-minded customer to place a big order with you. Your purpose might be to better understand a set of reports on a troubled child (perhaps your own). At this point in your efforts to educate yourself in statistics, you have a good start toward being able to tell the tale of a distribution of data. Congratulations! And, oh yes, here are some more problems so you can get better.

PROBLEMS

5.11. Find the effect size indexes for the three sets of data in the table. Write an interpretation for each d value.

	Group 1 mean	Group 2 mean	Standard deviation
a.	14	12	4
b.	10	19	10
c.	10	19	30

5.12. In hundreds of studies with many thousands of participants, first-born children scored better than second-born children on tests of cognitive ability. Data based on Zajonc (2001) and Zajonc and Bargh (1980) provide mean scores of 469 for first-born and 456 for second-born. For this measure of cognitive ability, $\sigma = 110$. Find d and write a sentence of interpretation.

5.13. Having *the biggest* is a game played far and wide for fun, fortune, and fame (such as being listed in *Guinness World Records*). For example, the biggest cabbage was grown in 1989 by Bernard Lavery of Rhonnda, Wales. It weighed 124 pounds. The biggest pumpkin (1689 pounds) was grown in 2007 by Joseph Jutras of Rhode Island, USA (*Guinness World Records, 2009*, 2008). Calculate $\overline{X}$ and S from the scores below, which are representative of contest cabbages and pumpkins. Use z scores to determine the BIG winner between Lavery and Jutras.

Cabbages	Pumpkins
110	1400
100	1200
90	1000

5.14. For the Satisfaction With Life Score data, determine the highest score and lowest score that qualify as outliers. What scores (if any) in Table 2.3 are outliers? See problem 4.2 for percentiles.

5.15. Is psychotherapy effective? This first-class question has been investigated by many researchers. The data that follow were constructed to mirror the classic findings of Smith and Glass (1977), who analyzed 375 studies.

The *psychotherapy* group received treatment during the study; the *control* group did not. A participant's psychological health score at the beginning of the study was subtracted from the same participant's score at the end of the study, giving the score listed in the table. Thus, each score provides a measure of the *change* in psychological health for an individual. (A negative score means that the person was worse at the end of the study.) For these change scores, $\sigma = 10$. Create a Descriptive Statistics Report.

Psychotherapy	Control	Psychotherapy	Control
7	9	13	−3
11	3	−15	22
0	13	10	−7
13	1	5	10
−5	4	9	−12
25	3	15	21
−10	18	10	−2
34	−22	28	5
7	0	−2	2
18	−9	23	4

5.16. It's time to check yourself. Turn back to the beginning of this chapter and read the objectives. Can you meet each one?

ADDITIONAL HELP FOR CHAPTER 5

Visit *cengage.com/psychology/spatz*. At the Student Companion Site, you'll find multiple-choice tutorial quizzes and flashcards with definitions.

KEY TERMS

Boxplot (p. 76)
Descriptive Statistics Report (p. 82)
Effect size index (p. 79)
Interquartile range (p. 77)
Outliers (p. 75)
Skew (p. 77)
Standard score (p. 73)
z score (p. 71)

transition passage
to bivariate statistics

SO FAR IN this exposition of the wonders of statistics, the examples and problems have been about a single distribution of scores; that is, about one variable. Heights, dollars, satisfaction with life scores, and number of boxes of Girls Scout cookies were all analyzed, but in every case the statistics calculated were on just the one variable. Such one-variable data are called *univariate* distributions.

Another kind of statistical analysis is about two variables (*bivariate* distributions). The analysis of bivariate distributions reveals answers to questions about the relationship between the two variables. For example, the questions might be

What is the relationship between a person's verbal ability and mathematical ability?

or

Knowing a person's verbal aptitude score, what should we predict as his or her freshman grade point average?

Other pairs of variables that might be related include:

Height of daughters and height of their fathers
Wealth and violent crime
Size of groups taking college entrance examinations and scores obtained
Stress and infectious diseases

By the time you finish Chapter6, you will know whether or not these pairs of variables are related and to *what degree*. You will also know how to predict a score on one variable given a score on the other variable.

Chapter 6, "Correlation and Regression," explains two statistical methods. *Correlation* is a method that is used to determine the *degree* of relationship between two variables when you have bivariate data. *Regression* is a method that is used to *predict* scores for one variable when you have measurements on a second variable.

CHAPTER

6

Correlation and Regression

OBJECTIVES FOR CHAPTER 6

After studying the text and working the problems in this chapter, you should be able to:

1. Explain the difference between univariate and bivariate distributions
2. Explain the concept of correlation and the difference between positive and negative correlation
3. Draw scatterplots
4. Compute a Pearson product-moment correlation coefficient (r)
5. Use correlation coefficients to assess *reliability, common variance,* and *effect size*
6. Identify situations in which the Pearson r does not accurately reflect the degree of relationship
7. Name and explain the elements of the regression equation
8. Compute regression coefficients and fit a regression line to a set of data
9. Interpret the appearance of a regression line
10. Predict scores on one variable based on scores from another variable

CORRELATION AND REGRESSION: My guess is that you have some understanding of the concept of correlation and that you are not as comfortable with the word *regression.* Speculation aside, correlation is simpler. Correlation is a statistical technique that describes the degree of relationship between two variables.

Regression is more complex. In this chapter you will use the regression technique to accomplish two tasks, *drawing* the line that best fits the data and *predicting* a person's score on one variable when you know that person's score on a second, correlated variable. Regression has other, more sophisticated uses, but you will have to put those off until you study more advanced statistics.

The ideas identified by the terms *correlation* and *regression* were developed by Sir Francis Galton in England more than 100 years ago. Galton was a genius (he could read at age 3) who had an amazing variety of interests, many of which he actively pursued during his 89 years. He once listed his occupation as "private gentleman," which meant that he had inherited money and did not have to work at a job. Lazy, however, he was not. Galton traveled widely and wrote prodigiously (17 books and more than 200 articles).

From an early age, Galton was enchanted with counting and quantification. Among the many things he tried to quantify were weather, individuals, beauty, characteristics of criminals, boringness of lectures, and effectiveness of prayers. Often he was successful. For example, it was Galton who discovered that atmospheric pressure highs produce clockwise winds around a calm center, and his efforts at quantifying individuals resulted in ways to classify fingerprints that are in use today. Because it worked so well for him, Galton actively promoted the philosophy of **quantification,** the idea that you can understand a phenomenon much better if you translate its essential parts into numbers.

quantification
Concept that translating a phenomenon into numbers promotes a better understanding of the phenomenon.

Many of the variables that interested Galton were in the field of heredity. Although it was common in the 19th century to comment on physical similarities within a family (height and facial characteristics, for example), Galton thought that psychological characteristics, too, tended to run in families. Specifically, he thought that characteristics such as genius, musical talent, sensory acuity, and quickness had a hereditary basis. Galton's 1869 book, *Hereditary Genius,* listed many families and their famous members, including Charles Darwin, his cousin.[1]

Galton wasn't satisfied with the list in that early book; he wanted to express the relationships in quantitative terms. To get quantitative data, he established an anthropometric (people-measuring) laboratory at a health exposition (a fair) and later at a museum in London. Approximately 17,000 people who stopped at a booth paid 3 pence to be measured. They left with self-knowledge; Galton left with quantitative data and a pocketful of coins. For one summary of Galton's results, see Johnson et al., 1985.

Galton's most important legacy is probably his invention of the concepts of correlation and regression. Correlation permits you to express the degree of relationship between any two paired variables. (The relationship between the height of fathers and the height of their adult sons was Galton's classic example.)

Galton was not enough of a mathematician to work out the theory and formulas for his concepts. This task fell to Galton's friend and protégé, Karl Pearson, Professor of Applied Mathematics and Mechanics at University College in London.[2] Pearson's 1896 *product-moment correlation coefficient* and other correlation coefficients that he and his students developed were quickly adopted by researchers in many fields and are widely used today in psychology, sociology, education, political science, the biological sciences, and other areas.

Finally, although Galton and Pearson's fame is for their statistical concepts, their principal goal was to develop recommendations that would improve the human condition. Making recommendations required a better understanding of heredity and evolution, and they saw statistics as the best way to arrive at this better understanding.

---

[1] Galton and Darwin had the same famous grandfather, Erasmus Darwin, but not the same grandmother. For both the personal and intellectual relationships between the famous cousins, see Fancher, 2009.
[2] I have some biographical information on Pearson in Chapter 14, the chapter on chi square. Chi square is another statistical invention of Karl Pearson.

In 1889, Galton described how valuable statistics are (and also let us in on his emotional feelings about statistics):

> Some people hate the very name of statistics, but I find them full of beauty and interest. . . . Their power of dealing with complicated phenomena is extraordinary. They are the only tools by which an opening can be cut through the formidable thicket of difficulties that bars the path of those who pursue the Science of [Humankind].[3]

My plan in this chapter is for you to read about bivariate distributions (necessary for both correlation and regression), to learn to compute and interpret Pearson product-moment correlation coefficients, and to use the regression technique to draw a best-fitting straight line and predict outcomes.

Bivariate Distributions

In the chapters on central tendency and variability, you worked with one variable at a time (**univariate distributions**). Height, time, test scores, and errors all received your attention. If you look back at those problems, you'll find a string of numbers under one heading (see, for example, Table 3.2). Compare those distributions with the one in **Table 6.1.** In Table 6.1, there are scores under the variable *Humor test* and other scores under a second variable, *Intelligence test.* You could find the mean and standard deviation of either of these variables. The characteristic of the data in this table that makes it a **bivariate distribution** is that the scores on the two variables are *paired.* The 50 and the 8 go together; the 20 and the 4 go together. They are paired, of course, because the same person made the two scores. As you will see, there are also other reasons for pairing scores. All in all, bivariate distributions are fairly common.

univariate distribution
Frequency distribution of one variable.

bivariate distribution
Joint distribution of two variables; scores are paired.

The essential idea of a bivariate distribution (which is required for correlation and regression techniques) is that two variables have values that are paired for some logical reason. A bivariate distribution may show positive correlation, negative correlation, or zero correlation.

TABLE 6.1 A bivariate distribution of scores on two tests taken by the same individuals

	Humor test *X* variable	Intelligence test *Y* variable
Larry	50	8
Shep	40	9
Curly	30	5
Moe	20	4

[3] For a short biography of Galton, I recommend Thomas (2005) or Waller (2001).

Positive Correlation

In the case of a *positive correlation* between two variables, high measurements on one variable tend to be associated with high measurements on the other variable, and low measurements on one variable with low measurements on the other. For example, tall fathers tend to have sons who grow up to be tall men. Short fathers tend to have sons who grow up to be short men.

In the case of the manufactured data in **Table 6.2,** fathers have sons who grow up to be *exactly* their height. The data in Table 6.2 represent an extreme case in which the correlation coefficient is 1.00, which is referred to as *perfect correlation.* (Table 6.2 is ridiculous of course; mothers and environments have their say, too.)

Figure 6.1 is a graph of the bivariate data in Table 6.2. One variable (height of father) is plotted on the X axis; the other variable (height of son) is on the Y axis. Each data point in the graph represents a pair of scores, the height of a father and the height of his son. The points in the graph constitute a **scatterplot.** Incidentally, it was when Galton cast his data as a scatterplot graph that the idea of a co-relationship began to become clear to him.

scatterplot
Graph of the scores of a bivariate frequency distribution.

regression line
A line of best fit for a scatterplot.

The line that runs through the points in Figure 6.1 (and in Figures 6.2, 6.3, and 6.5) is called a **regression line.** It is a "line of best fit." When there is perfect correlation ($r = 1.00$), all points fall exactly on the regression line. It is Galton's use of the term *regression* that gives the symbol r for correlation.[4]

TABLE 6.2 **Manufactured data on two variables: heights of fathers and their sons ***

Father	Height (in.) X	Son	Height (in.) Y
Michael Smith	74	Mike, Jr.	74
Christopher Johnson	72	Chris, Jr.	72
Matthew Williams	70	Matt, Jr.	70
Joshua Jones	68	Josh, Jr.	68
Daniel Brown	66	Dan, Jr.	66
David Davis	64	Dave, Jr.	64

* The first names are, in order, the six most common for baby boys born in 1990 in the United States. Rounding out the top ten are Andrew, James, Justin, and Joseph (*www.ssa.gov/cgi-bin/popularnames.cgi*). The surnames are also the six most common in the 1990 U.S. census. Completing the top ten are Miller, Wilson, Moore, and Taylor (*www.census.gov/genealogy/names/dist.all.last*).

[4] The term *regression* can be confusing because it has two separate meanings. As you already know, regression is a *statistical method* that allows you to draw the line of best fit and to make predictions with bivariate data. Regression also refers to a *phenomenon* that occurs when a select group is tested a second time. The phenomenon occurs when those with extreme scores in a distribution (those who did very well or very poorly) are tested a second time. Those who scored high will, on average, score lower on the retest. The mean of those who scored low the first time will increase on the second test.

Galton found that the mean height of sons of extremely tall men was less than the mean height of their fathers and also that the mean height of sons of extremely short men was greater than the mean height of their fathers. In both cases, the sons' mean is closer to the population mean. Because the mean of an extreme group, when measured a second time, tended to regress toward the population mean, Galton named this phenomenon *regression.* The statistical technique he developed to assess regression toward the mean was called regression as well.

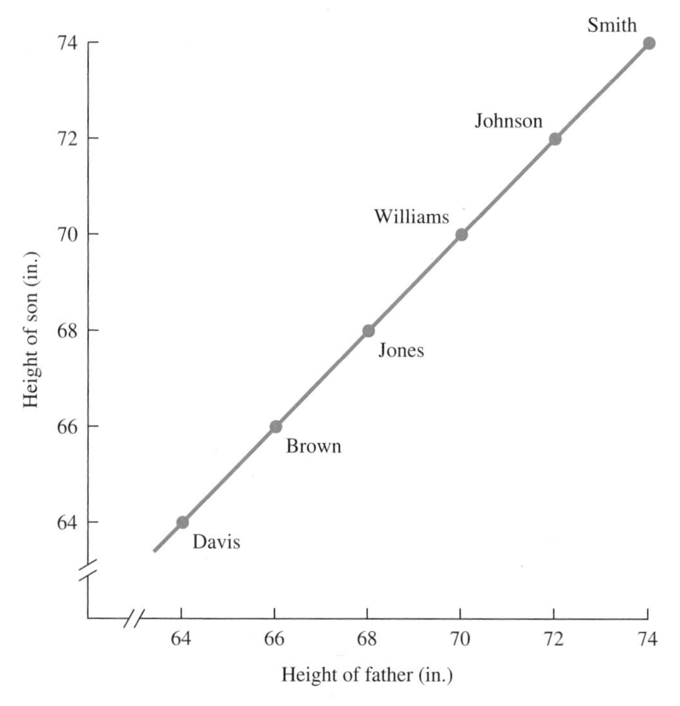

FIGURE 6.1 A scatterplot and regression line for a perfect positive correlation (r = 1.00)

In recent generations, changes in nongenetic factors such as nutrition resulted in sons who tend to be somewhat taller than their fathers, except for extremely tall fathers. If every son grew to be exactly 2 inches taller than his father (or 1 inch or 6 inches, or even 5 inches shorter), the correlation would still be perfect, and the coefficient would still be 1.00. **Figure 6.2** demonstrates this point: You can have a perfect correlation even if the paired numbers aren't the same. The only requirement for perfect correlation is that the *differences* between pairs of scores all be the same. If they are the same, then all the points of a scatterplot lie on the regression line, correlation is perfect, and an exact prediction can be made.

Of course, people cannot predict their sons' heights precisely. The correlation is not perfect and the points do not all fall on the regression line. As Galton found, however, there is a positive relationship; the correlation coefficient is about .50. The points do tend to cluster around the regression line.

In your academic career you have taken an untold number of aptitude and achievement tests. For several of these tests, separate scores were computed for verbal aptitude and mathematics aptitude. Here is a question for you. In the general case, what is the relationship between verbal aptitude and math aptitude? That is, are people who are good in one also good in the other, or are they poor in the other, or is there no relationship? Stop for a moment and compose an answer.

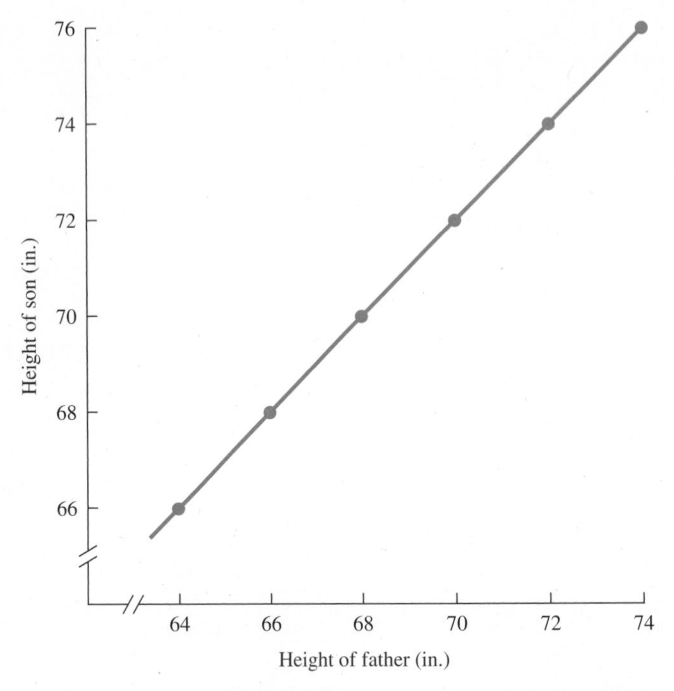

FIGURE 6.2 A scatterplot and regression line with every son 2 inches taller than his father ($r = 1.00$)

As you may have suspected, the next graph shows data that begin to answer the question. **Figure 6.3** shows the scores of eight high school seniors who took the SAT college admissions test. The scores are for the *critical reading* portion of the SAT (CR SAT) and the *mathematics* portion of the test (Math SAT). As you can see in Figure 6.3, there is a positive relationship, though not a perfect one. As the critical reading scores vary upward, mathematics scores *tend* to vary upward. If the score on one is high, the other score tends to be high, and if one is low, the other tends to be low. Later in this chapter you'll learn to calculate the precise *degree* of relationship, and because there is a relationship, you can use a regression equation to predict students' math scores if you know their verbal score.

(Examining Figure 6.3, you might complain that the graph is out of balance; all the data points are stuck up in one corner. It looks ungainly, but I was in a dilemma, which I'll explain later in the chapter.)

Negative Correlation

Here is a scenario that leads to another bivariate distribution. Recall a time when you sat for a college entrance examination (SAT and ACT are the two most common ones.) How many others took the exam at the same testing center that day? Next, imagine your motivation that day. Did you feel motivated to be in the top 10 percent,

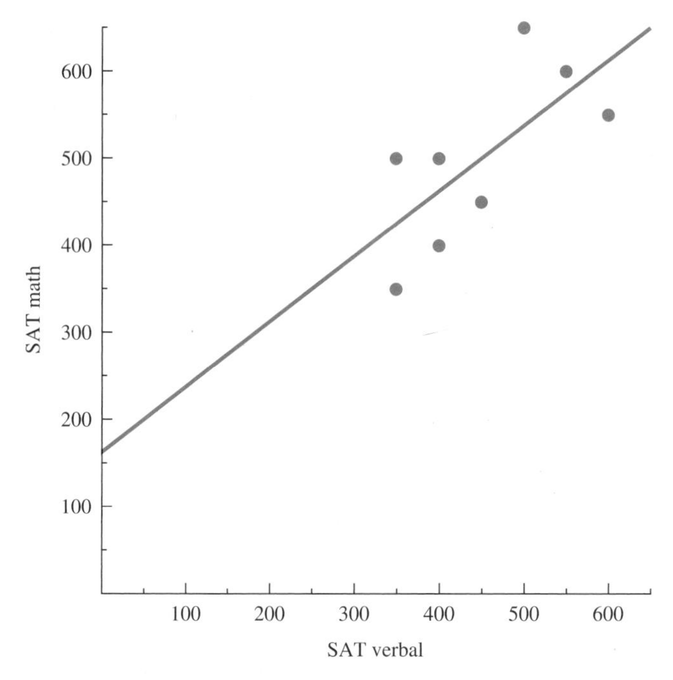

FIGURE 6.3 Scatterplot and regression line for CR SAT and Math SAT scores for eight high school seniors (r = .72). Data derived from *2008 College-Bound Seniors*. Copyright © 2008, the College Board. www.collegeboard.com. Reproduced with permission.

top 50 percent, top 90 percent? With these two numbers in mind (number of fellow test takers and your motivation), you have the ingredients of a bivariate distribution.

Do you think that the relationship between the two variables is positive, like that of critical reading scores and math scores, or that there is no relationship, or that the relationship is negative?

As you probably figured out from this section heading, the answer to the preceding question is—*negative*. **Figure 6.4** is a scatterplot of SAT scores and density of test-takers (state averages for 50 U.S. states). High SAT scores are associated with low densities of test-takers and low SAT scores are associated with high densities of test-takers.[5] This phenomenon is an illustration of the *N-Effect*, the finding that increase in number of competitors goes with a decrease in competitive motivation and thus, test scores (Garcia and Tor, 2009). The cartoon illustrates the *N*-Effect.

When a correlation is negative, increases in one variable are accompanied by decreases in the other variable (an inverse relationship). With negative correlation, the regression line goes from the upper left corner of the graph to the lower right corner. As you may recall from algebra, such lines have a negative slope. Some other

[5] The correlation coefficient is −.68. When Garcia and Tor (2009) statistically removed the effects of confounding variables such as state percentage of high school students who took the SAT, state population density, and other variables, the correlation coefficient was −.35.

examples of variables with negative correlation are highway driving speed and gas mileage, daily rain and daily sunshine, and grouchiness and friendships. As was the case with perfect positive correlation, there is such a thing as perfect negative correlation ($r = -1.00$). In cases of perfect negative correlation also, all the data points of the scatterplot fall on the regression line.

Although some correlation coefficients are positive and some are negative, one is *not more valuable than the other.* The algebraic sign simply tells you the *direction* of the relationship (which is important when you are describing how the variables are related). The absolute size of *r*, however, tells you the *degree* of the relationship. A strong relationship (either positive or negative) is usually more informative than a weaker one.

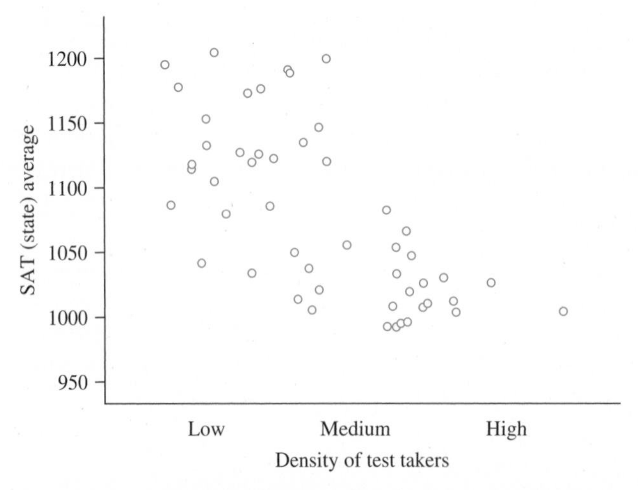

FIGURE 6.4 Scatterplot of state SAT averages and density of test-takers. Courtesy of Stephen Garcia

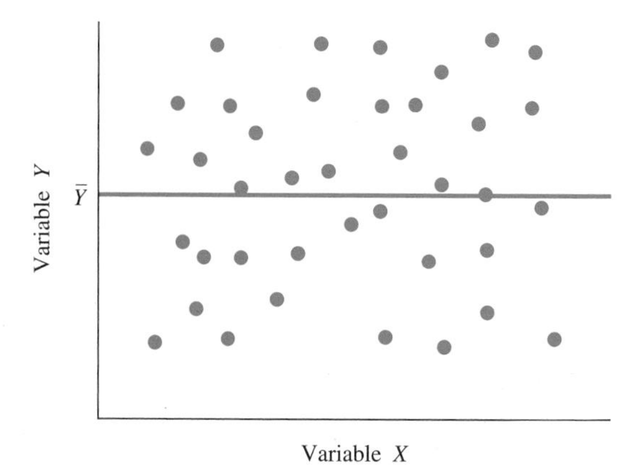

FIGURE 6.5 Scatterplot and regression line for a zero correlation

Zero Correlation

A *zero correlation* means there is no linear relationship between two variables. High and low scores on the two variables are not associated in any predictable manner.

The 50 American states differ in personal wealth; these differences are expressed as per capita income, which ranged from $28,845 (Mississippi) to $54,117 (Connecticut) in 2007 (*Statistical Abstract of the United States: 2009,* 2008). The states also differ in violent crime per capita. Is there a relationship between wealth and violent crime? The correlation coefficient between per capita income and violent crime rate is .03. There is no relationship between the two variables.

Figure 6.5 shows a scatterplot that produces a zero correlation coefficient. When $r = 0$, the regression line is a horizontal line at a height of $\overline{Y}$. This makes sense; if $r = 0$, then your best estimate of Y for any value of X is $\overline{Y}$.

clue to the future

Correlation comes up again in future chapters. The correlation coefficient between two variables whose scores are ranks is explained in Chapter 15. In part of Chapter 10 and in all of Chapter 12, correlation ideas are involved.

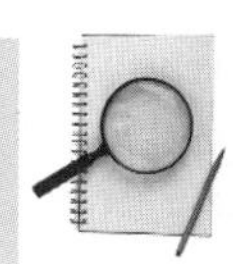

PROBLEMS

6.1. What are the characteristics of a bivariate distribution?

6.2. What is meant by the statement "Variable X and variable Y are correlated"?

6.3. Tell how X and Y vary in a positive correlation. Tell how they vary in a negative correlation.

6.4. Can the following variables be correlated and, if so, would you expect the correlation to be positive or negative?

- **a.** Height and weight of adults
- **b.** Weight of first graders and weight of fifth graders
- **c.** Average daily temperature and cost of heating a home

d. IQ and reading comprehension
e. The first and second quiz scores of students in two sections of General Biology
f. The section 1 scores and the section 2 scores of students in General Biology on the first quiz

The Correlation Coefficient

correlation coefficient
Descriptive statistic that expresses the degree of relationship between two variables.

A **correlation coefficient** provides a quantitative way to express the *degree* of relationship that exists between two variables. The definition formula is

$$r = \frac{\Sigma (z_X z_Y)}{N}$$

where r = Pearson product-moment correlation coefficient
z_X = a z score for variable X
z_Y = the corresponding z score for variable Y
N = number of pairs of scores

Think through the z-score formula to discover what happens when high scores on one variable are paired with high scores on the other variable (positive correlation). The large positive z scores are paired and the large negative z scores are paired. In each case, the multiplication produces large *positive* products, which, when added together, make a large positive numerator. The result is a large positive value of r. Think through for yourself what happens in the formula when there is a negative correlation and when there is a zero correlation.

Though valuable for understanding r, the z-score formula is difficult if you are *calculating* a value for r by hand. Fortunately, however, there are many equivalent formulas for r. I'll describe two and give you some guidance on when to use each.

Computational Formulas for *r*

One formula for computing r is referred to as the *blanched* formula; a second one is called the *raw-score* formula. The blanched formula requires that you first partially "cook" the data to yield means and standard deviations that give you a better feel for the data. For the raw-score formula, you enter the scores directly into the formula. The raw-score formula doesn't have intermediate steps. If all you want is r, then the raw-score formula is the quicker method.

Blanched formula Because researchers often use means and standard deviations when telling the story of the data, this formula is used by many:

$$r = \frac{\frac{\Sigma XY}{N} - (\bar{X})(\bar{Y})}{(S_X)(S_Y)}$$

where X and Y are paired observations

XY = product of each X value multiplied by its paired Y value

$\overline{X}$ = mean of variable X

$\overline{Y}$ = mean of variable Y

S_X = standard deviation of variable X (use N in the denominator)

S_Y = standard deviation of variable Y (use N in the denominator)

N = number of pairs of observations

The expression ΣXY is called the "sum of the cross-products." All formulas for Pearson r include ΣXY. To find ΣXY, multiply each X value by its paired Y value, and then sum those products. Note that one of the terms in the formula has a new meaning. In correlation problems, N is the number of *pairs* of scores.

error detection

ΣXY is not $(\Sigma X)(\Sigma Y)$. To find ΣXY, you do as many multiplications as you have pairs. Afterward, sum the products that you calculated.

As for which variable to call X and which to call Y, it doesn't make any difference for a correlation coefficient. It may, however, make a *big* difference in the regression problems you will work later in this chapter.

Table 6.3 illustrates the steps you use to compute r by the blanched procedure. The data are those that were used to draw Figure 6.3, the scatterplot of the CR SAT and Math SAT scores for the eight students. I made up the numbers so that they would produce the same correlation coefficient as reported by the College Board Seniors Report. Work through the numbers in Table 6.3, paying careful attention to the calculation of ΣXY.

Raw-score formula With the raw-score formula, you calculate r from the raw scores without computing means and standard deviations. The formula is

$$r = \frac{N\Sigma XY - (\Sigma X)(\Sigma Y)}{\sqrt{[N\Sigma X^2 - (\Sigma X)^2][N\Sigma Y^2 - (\Sigma Y)^2]}}$$

You have already learned what all the terms of this formula mean. A reminder: N is the number of *pairs* of values.

Calculators with two or more memory storage registers—and some with one—allow you to accumulate simultaneously the values for ΣX and ΣX^2. After doing so, compute values for ΣY and ΣY^2 simultaneously. This leaves only ΣXY to compute.

Many calculators have a built-in function for r. When you enter X and Y values and press the r key, the coefficient is displayed. If you have such a calculator, I recommend that you use this labor-saving device *after* you have used the computation formulas a number of times. Working directly with terms like ΣXY leads to an understanding of what goes into r.

If you calculate sums that reach above the millions, your calculator may switch into scientific notation. A display such as 3.234234 08 might appear. To convert this number back to familiar notation, just move the decimal point to the right the number of places

TABLE 6.3 Calculation of *r* between critical reading SAT and mathematics SAT aptitude scores by the blanched formula

Student	CR SAT X	Math SAT Y	X^2	Y^2	XY
1	350	400	122,500	160,000	140,000
2	500	420	250,000	176,400	210,000
3	400	470	160,000	220,900	188,000
4	500	450	250,000	202,500	225,000
5	450	520	202,500	270,400	234,000
6	650	590	422,500	348,100	383,500
7	600	600	360,000	360,000	360,000
8	550	670	302,500	448,900	368,500
Σ	4000	4120	2,070,000	2,187,200	2,109,000

$$\overline{X} = \frac{\Sigma X}{N} = \frac{4000}{8} = 500 \qquad \overline{Y} = \frac{\Sigma Y}{N} = \frac{4120}{8} = 515$$

$$S_X = \sqrt{\frac{\Sigma X^2 - \frac{(\Sigma X)^2}{N}}{N}} = \sqrt{\frac{2{,}070{,}000 - \frac{(4000)^2}{8}}{8}} = 93.54$$

$$S_Y = \sqrt{\frac{\Sigma Y^2 - \frac{(\Sigma Y)^2}{N}}{N}} = \sqrt{\frac{2{,}187{,}200 - \frac{(4120)^2}{8}}{8}} = 90.42$$

$$r = \frac{\frac{\Sigma XY}{N} - (\overline{X})(\overline{Y})}{S_X S_Y} = \frac{\frac{2{,}109{,}000}{8} - (500)(515)}{(93.54)(90.42)} = .724 = .72$$

Data derived from *2008 College-Bound Seniors*.

indicated by the number on the right. Thus, 3.234234 08 becomes 323,423,400. The display 1.23456789 12 becomes 1,234,567,890,000.

Table 6.4 illustrates the raw-score procedure for computing r. The data are the same as those in Table 6.3. Note that the value of r is the same with both methods.

TABLE 6.4 Calculation of *r* for critical reading SAT and math SAT aptitude scores by the raw-score formula

$\Sigma X = 4000$ $\quad \Sigma Y = 4120$ $\quad \Sigma X^2 = 2{,}070{,}000$ $\quad \Sigma Y^2 = 2{,}187{,}200$ $\quad \Sigma XY = 2{,}109{,}000$

$$r = \frac{N\Sigma XY - (\Sigma X)(\Sigma Y)}{\sqrt{[N\Sigma X^2 - (\Sigma X)^2][N\Sigma Y^2 - (\Sigma Y)^2]}}$$

$$= \frac{(8)(2{,}109{,}000) - (4000)(4120)}{\sqrt{[(8)(2{,}070{,}000) - (4000)^2][(8)(2{,}187{,}200) - (4120)^2]}}$$

$$= \frac{392{,}000}{541{,}287} = .724 = .72$$

Data derived from *2008 College-Bound Seniors*.

TABLE 6.5 SPSS output of Pearson *r* for CR SAT verbal and math SAT scores

Correlations

		CR.SAT	Math.SAT
CR.SAT	Pearson Correlation	1	.724*
	Sig. (2-tailed)		.042
	N	8	8
Math.SAT	Pearson Correlation	.724*	1
	Sig. (2-tailed)	.042	
	N	8	8

* Correlation is significant at the 0.05 level (2-tailed)..

Table 6.5 shows the SPSS output for a Pearson correlation of the two variables. Again, $r = .72$. The designation *Sig. (2-tailed)* means "the significance level for a two-tailed test," a concept explained in Chapter 9.

The final step in any statistics problem is interpretation. What story goes with a correlation coefficient of .72 between CR SAT verbal scores and Math SAT scores? An r of .72 is a fairly substantial correlation coefficient. Students who have high SAT critical reading scores certainly *tend* to have high SAT math scores. Note, however, that if the correlation had been near zero, you could say that the two abilities are unrelated. If the coefficient had been strong and negative, you could say, "Good in one, poor in the other."

Correlation coefficients should be based on an "adequate" number of pairs of observations. As a general rule of thumb, "adequate" means 30 or more. My SAT example, however, had an N of 8, and most of the problems in the text have fewer than 30 pairs. Small-N problems allow you to spend your time on interpretation and understanding rather than on "number crunching." In Chapter 9, you will learn the reasoning behind my admonition that N be adequate.

The correlation coefficient, r, is a sample statistic. The corresponding population parameter is symbolized by ρ (the Greek letter rho). The formula for ρ is the same as the formula for r (except that parameters such as σ and μ are substituted for the statistics, S and $\bar{X}$). One of the "rules" for statistical names is that parameters are symbolized with Greek letters (μ, σ, ρ) and statistics are symbolized with Latin letters ($\bar{X}$, $\hat{s}$, r). Like many rules, there are exceptions to this one.

clue to the future

The population correlation coefficient, ρ, is important in Chapter 9, where the issue of the reliability of r is addressed. Also, besides the Pearson product-moment correlation coefficient, r, there are other correlation coefficients. I have verbal descriptions of several of these later in this chapter. Chapter 15 explains the Spearman coefficient r_s, which is appropriate for scores that are ranks.

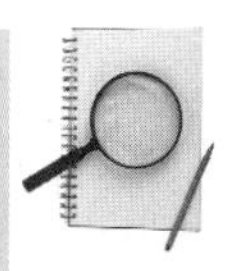

error detection

The Pearson correlation coefficient ranges between −1.00 and +1.00. Values less than −1.00 or greater than +1.00 indicate that you have made an error.

PROBLEMS

***6.5.** This problem is based on data published in 1903 by Karl Pearson and Alice Lee. In the original article, 1376 pairs of father–daughter heights were analyzed. The scores here produce the same means and the same correlation coefficient that Pearson and Lee obtained. For these data, draw a scatterplot and calculate r by both the raw-score formula and the blanched formula.

Father's height, X (in.)	69	68	67	65	63	73
Daughter's height, Y (in.)	62	65	64	63	58	63

***6.6.** The Wechsler Adult Intelligence Scale (WAIS) is an individually administered test that takes over an hour. The Wonderlic Personnel Test can be administered to groups of any size in 15 minutes. X and Y represent scores on the two tests. Summary statistics from a representative sample of 21 adults were $\Sigma X = 2205$, $\Sigma Y = 2163$, $\Sigma X^2 = 235{,}800$, $\Sigma Y^2 = 227{,}200$, $\Sigma XY = 231{,}100$. Compute r and write an interpretation about using the Wonderlic rather than the WAIS.

***6.7.** Is the relationship between stress and infectious disease a strong one or a weak one? Summary values that will produce a correlation coefficient similar to that found by Cohen and Williamson (1991) are: $\Sigma X = 190$, $\Sigma Y = 444$, $\Sigma X^2 = 3940$, $\Sigma Y^2 = 20{,}096$, $\Sigma XY = 8524$, $N = 10$. Calculate r using either method (blanched or raw score).

Scatterplots

You already know something about scatterplots—what their elements are and what they look like when $r = 1.00$, $r = .00$, and $r = -1.00$. In this section I illustrate some intermediate cases and reiterate my philosophy about the value of pictures.

Figure 6.6 shows scatterplots of data with positive correlation coefficients (.20, .40, .80, .90) and negative correlation coefficients (−.60, −.95). If you draw an envelope around the points in a scatterplot, the picture becomes clearer. The thinner the envelope, the larger the correlation. To say this in more mathematical language, the closer the points are to the regression line, the greater the correlation.

Pictures help you understand, and scatterplots are easy to construct. Although the plots require some time, the benefits are worth it. Peden (2001) constructed four data sets that all produce a correlation coefficient of .82. Scatterplots of the data,

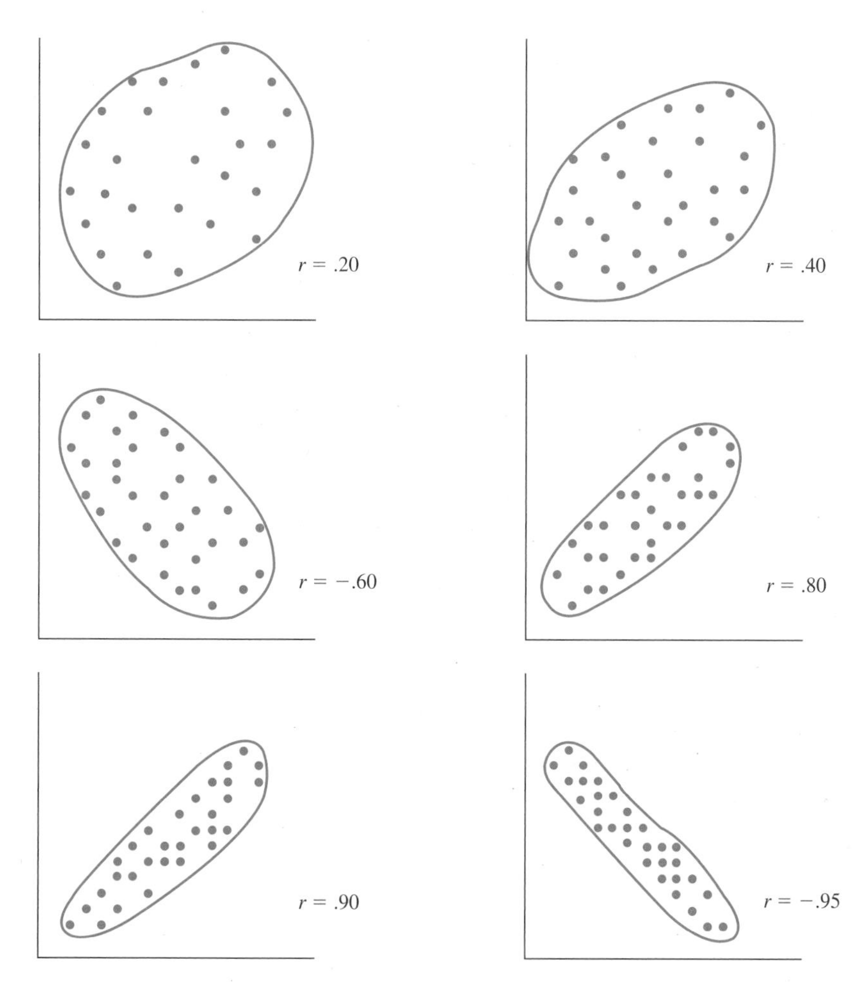

FIGURE 6.6 Scatterplots of data in which $r = .20, .40, -.60, .80, .90$, and $-.95$

however, show four different patterns that lead to four quite different interpretations. So this is a paragraph that encourages you to construct scatterplots—pictures help you understand data.

Interpretations of *r*

The basic interpretation of r is probably familiar to you at this point. A correlation coefficient measures the degree of linear relationship between two variables of a bivariate distribution. Fortunately, additional information about the relationship can be obtained from a correlation coefficient. Over the next few pages, I'll cover some of this additional information. However, a warning is in order—the *interpretation* of r can be a tricky business. I'll alert you to some of the common errors.

Effect Size Index for *r*

What qualifies as a large correlation coefficient? What is small? You will remember that you dealt with similar questions in Chapter 5 when you studied the *effect size index.* In that situation, you had two sample means that were different. The question was: Is the difference big? Jacob Cohen (1969) proposed that the question be answered by calculating an effect size index (d) and that d values of .20, .50, and .80 be considered small, medium, and large effect sizes, respectively. In a similar way, Cohen addressed the question of calculating an effect size index for the correlation coefficient.

You will be delighted with Cohen's calculation of an effect size index for r; the value is r itself. The remaining question is: What is small, medium, and large? Now the story becomes more complicated because correlation coefficients are used to measure many different kinds of relationships. However, Cohen proposed the following:

Small	$r = .10$
Medium	$r = .30$
Large	$r = .50$

These guidelines have not been as acceptable as those that Cohen proposed for d. One of the complaints is that for some problems the guidelines are clearly not appropriate. (See the later section on using r to measure reliability.) Another complaint is that the guidelines do not correspond to the actual distribution of empirical results. For example, Hemphill (2003) examined thousands of correlation coefficients from hundreds of studies and then divided them into thirds. The results were:

Lower third	$< .20$
Middle third	.20 to .30
Upper third	$> .30$

I'll have to leave the issue of "adjectives for correlation coefficients" without being able to give you a simple rule of thumb, but the fact is that the proper adjective to use for a particular r depends on the kind of research being conducted.

Coefficient of Determination

coefficient of determination
Squared correlation coefficient, an estimate of common variance.

The correlation coefficient is the basis of the **coefficient of determination,** which tells the proportion of variance that two variables in a bivariate distribution have in common. The coefficient of determination is calculated by squaring r; it is always a positive value between 0 and 1:

$$\text{Coefficient of determination} = r^2$$

Look back at **Table 6.2,** the heights that produced $r = 1.00$. There is variation among the fathers' heights as well as among the sons' heights. How much of the variation among the sons' heights is associated with the variation in the fathers' heights? All of it! That is, the variation among the sons' heights (74, 72, 70, and so on) is exactly the same variation that is seen among their fathers' heights (74, 72, 70, and so on). In the same way, the variation among the sons' heights in **Figure 6.2** (76, 74, 72, and so on) is the same *variation* that is seen among their fathers' heights (74, 72, 70, and so on). For Table 6.2 and Figure 6.2, $r = 1.00$ and $r^2 = 1.00$.

Now look at **Table 6.3,** the CR SAT and Math SAT scores. There is variation among the CR SAT scores as well as among the Math SAT scores. How much of the variation among the Math SAT scores is associated with the variation among the CR SAT scores? Some of it. That is, the variation among the CR SAT scores (350, 500, 420, 400, and so on) is only partly reflected in the variation among the Math SAT scores (470 and so on). The *proportion* of variance in the CR SAT scores that is associated with the variance in the Math SAT is r^2. In this case, $(.72)^2 = .52$.

What a coefficient of determination of .52 tells you is that 52 percent of the variance in the two sets of scores is common variance. However, 48 percent of the variance is independent variance—that is, variance in one test that is not associated with variance in the other test.

Think for a moment about the many factors that influence CR SAT scores and Math SAT scores. Some factors influence both scores—factors such as motivation, mental sharpness on test day, and, of course, the big one: general intellectual ability. Other factors influence one test but not the other—factors such as anxiety about math tests, chance successes and chance errors, and, of course, the big ones: specific verbal knowledge and specific math knowledge.

Here is another example. The correlation of academic aptitude test scores with first-term college grade point averages (GPAs) is about .50. The coefficient of determination is .25. This means that of all that variation in GPAs (from flunking out to straight A's), 25 percent is associated with aptitude scores. The rest of the variance (75 percent) is related to other factors. Examples of other factors that influence GPA, for good or for ill, include health, roommates, new relationships, and financial situation. Academic aptitude tests cannot predict the variation that these factors produce.

Common variance is often illustrated with two overlapping circles, each of which represents the total variance of one variable. The overlapping portion is the amount of common variance. The left half of **Figure 6.7** shows overlapping circles for the GPA–college aptitude test scores and the right half shows the CR SAT Math SAT data.

Note what a big difference there is between a correlation of .72 and one of .50 when they are interpreted using the common variance terminology. Although .72 and .50 seem fairly close, an r of .72 predicts more than twice the amount of variance that an r of .50 predicts: 52 percent to 25 percent. By the way, common variance is the way professional statisticians interpret correlation coefficients.[6]

---

[6] If you are moderately skilled in algebra and would like to understand more about common variance, I recommend that you finish the material in this chapter on regression and then work through the presentation of Minium and King (2002).

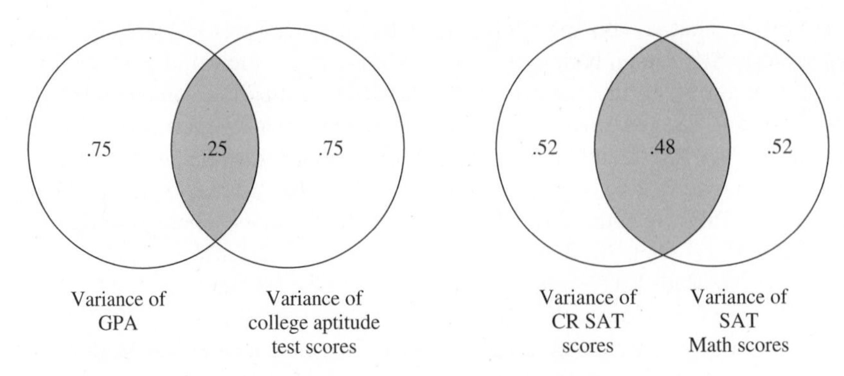

FIGURE 6.7 Two separate illustrations of common variance

Uses of *r*

Reliability of Tests

reliability
Dependability or consistency of a measure.

Correlation coefficients are used to assess the **reliability** of measuring devices such as tests, questionnaires, and instruments. Reliability refers to consistency. Devices that are reliable produce consistent scores that are not subject to chance fluctuations.

Think about measuring a number of individuals and then measuring them a second time. If the measuring device is not influenced by chance, then you get the same measurements both times. If the second measurement is *exactly* the same as the first for every individual, it is easy to conclude that the measuring device is perfectly reliable—that chance has nothing to do with the score you get. However, if the measurements are not *exactly* the same, the disagreement leads to uncertainty. Fortunately, a correlation coefficient tells you the *degree* of agreement between the test and the retest scores. High correlation coefficients mean lots of agreement and therefore high reliability; low coefficients mean lots of disagreement and therefore low reliability. But what size r indicates reliability? The rule of thumb is that an r of .80 or greater indicates reliability for social science measurements.

Here is an example from Galton's data. The heights of 435 adults were measured twice. The correlation was .98. It is not surprising that Galton's method of measuring height was very reliable. The correlation, however, for "highest audible tone," a measure of pitch perception, was only .28 for 349 subjects whom Galton tested a second time within a year (Johnson et al., 1985). One of two interpretations is possible. Either people's ability to hear high sounds changes up and down during a year, or the test was not reliable. In Galton's case, the test was not reliable. Two possible explanations for this lack of reliability are (1) the test environment was not as quiet from one time to the next and (2) the instruments the researchers used were not calibrated exactly the same on both tests.

This section is concerned with the reliability of a *measuring instrument,* which involves measuring one variable twice. The reliability of a *relationship* between two different variables is a different question and will be covered in Chapter 9. When the question is whether a *relationship* is reliable, the .80 rule of thumb does not apply.

To Establish Causation—NOT

A high correlation coefficient does *not* give you the kind of evidence that allows you to make cause-and-effect statements. Therefore, don't do it. Ever.

Jumping to a cause-and-effect conclusion is a cognitively easy leap for humans. For example, Shedler and Block (1990) found that among a sample of 18-year-olds whose marijuana use ranged from abstinence to once a month, there was a positive correlation between use and psychological health. Is this evidence that occasional drug use promotes psychological health?

Because Shedler and Block had followed their participants from age 3 on, they knew about a third variable, the quality of the *parenting* that the 18-year-olds had received. Not surprisingly, parents who were responsive, accepting, and patient and who valued originality had children who were psychologically healthy. In addition, these same children as 18-year-olds had used marijuana on occasion. Thus, two variables—drug use and parenting style—were each correlated with psychological health. Shedler and Block concluded that psychological health and adolescent drug use were both traceable to quality of parenting. (This research also included a sample of frequent users who were not psychologically healthy and who had been raised with a parenting style not characterized by the adjectives above.)

Of course, if you have a sizable correlation coefficient, it *may* be the result of a cause-and-effect relationship between the two variables. For example, early statements about cigarette smoking causing lung cancer were based on simple correlational data. Persons with lung cancer were often heavy smokers. Also, comparisons between countries indicated a relationship (see problem 6.13). However, as careful thinkers—and the cigarette companies—pointed out, both cancer and smoking might have been caused by a third variable; stress was often suggested. That is, stress caused cancer and stress also caused people to smoke. Thus, cancer rates and smoking rates were related (a high correlation), but one did not cause the other; both were caused by a third variable. What was required to establish a cause-and-effect relationship was data from controlled experiments, not correlational data. Experimental data, complete with control groups, finally established the cause-and-effect relationship between cigarette smoking and lung cancer. (Controlled experiments are discussed in Chapter 10, "Hypothesis Testing, Effect Size, and Confidence Intervals: Two-Sample Designs.")

To summarize this section using the language of logic: A sizable correlation is a necessary but not a sufficient condition for establishing causality.

PROBLEMS

6.8. Estimate the correlation coefficients for these scatterplots.

a.
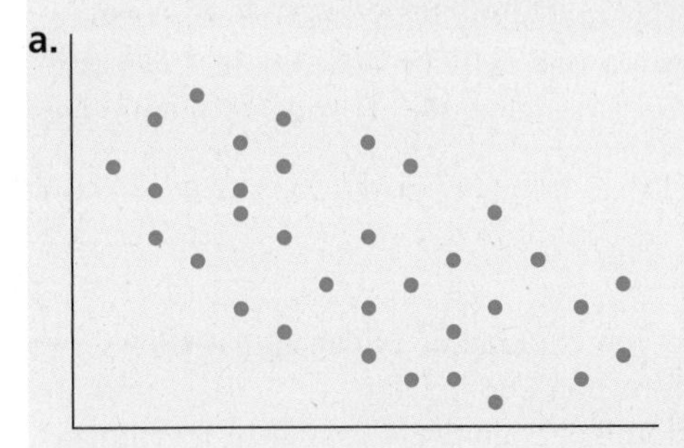

b.
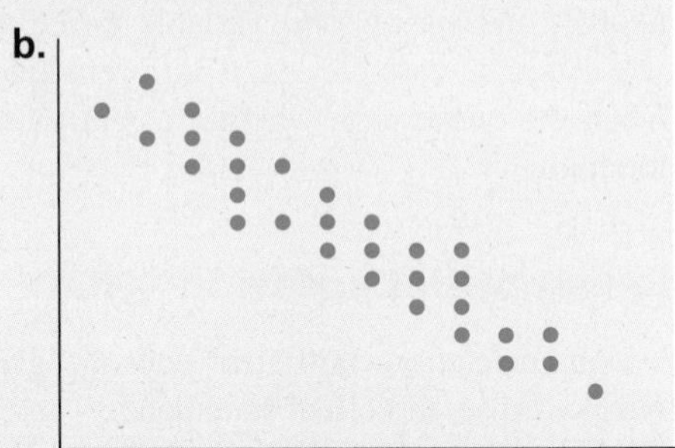

c.
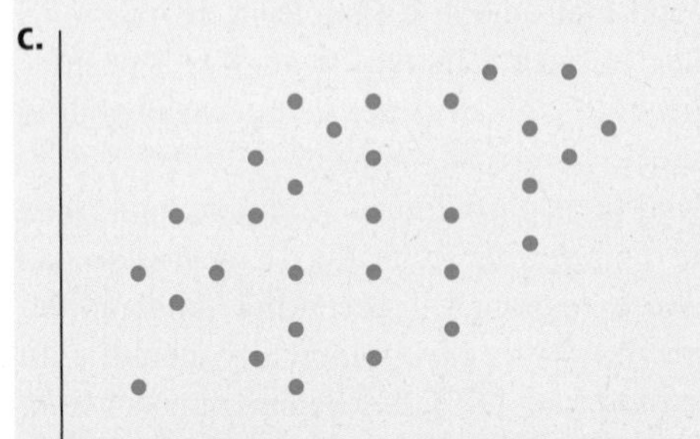

d.
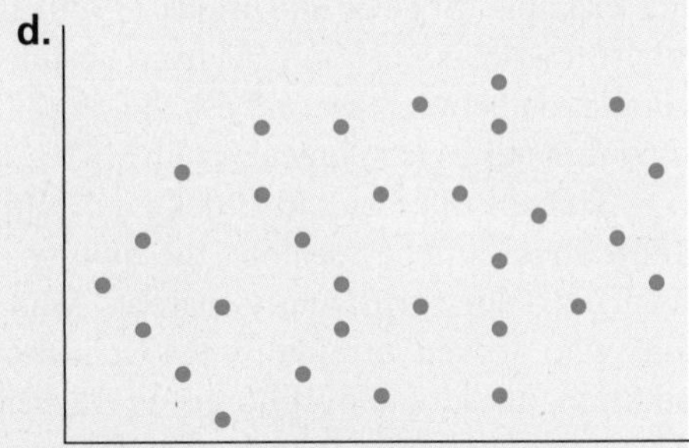

6.9. For the two measures of intelligence in problem 6.6, you found a correlation of .92. What is the coefficient of determination, and what does it mean?

6.10. In problem 6.7, you found that the correlation coefficient between stress and infectious disease was .25. Calculate the coefficient of determination and write an interpretation.

6.11. Examine the following summary statistics (which you have seen before). Can you determine a correlation coefficient? Explain your reasoning.

	Height of women (in.)	Height of men (in.)
ΣX	3210	3500
ΣX^2	206,428	245,470
$N = 50$ pairs		

6.12. What percent of variance in common do two variables have if their correlation is .10? What if the correlation is quadrupled to .40?

6.13. For each of 11 countries, the accompanying table gives the cigarette consumption per capita in 1930 and the male death rate from lung cancer

20 years later in 1950 (Doll, 1955; reprinted in Tufte, 2001). Calculate a Pearson r and write a statement telling what the data show.

Country	Per capita cigarette consumption	Male death rate (per million)
Iceland	217	59
Norway	250	91
Sweden	308	113
Denmark	370	167
Australia	455	172
Holland	458	243
Canada	505	150
Switzerland	542	250
Finland	1112	352
Great Britain	1147	467
United States	1283	191

6.14. Interpret each of these statements.

a. The correlation between vocational-interest scores at age 20 and at age 40 for the 150 participants was .70

b. The correlation between intelligence test scores of identical twins raised together is .86

c. A correlation of $-.30$ between IQ and family size

d. $r = .22$ between height and IQ for 20-year-old men

e. $r = -.83$ between income level and probability of diagnosis of schizophrenia

Strong Relationships but Low Correlation Coefficients

One good thing about understanding something is that you come to know what's going on beneath the surface. Knowing the inner workings, you can judge whether the surface appearance is to be trusted or not. You are about to learn two of the "inner workings" of correlation. These will help you evaluate the meaning of low correlation coefficients. Low correlations do not always mean there is no relationship between two variables.[7]

Nonlinearity

For r to be a meaningful statistic, the best-fitting line through the scatterplot of points must be a *straight line.* If a curved regression line fits the data better than a straight line, r will be low, not reflecting the true relationship between the two variables.

[7] Correlations that do not reflect the true degree of the relationship are said to be *spuriously low* or *spuriously high.*

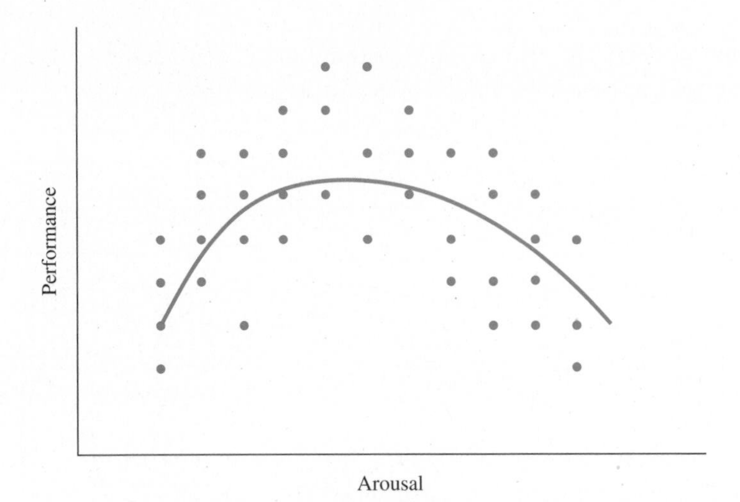

FIGURE 6.8 Generalized relationship between arousal and efficiency of performance

Figure 6.8 is an example of a situation in which r is inappropriate because the best-fitting line is curved. The X variable is arousal and the Y variable is efficiency of performance. At low levels of arousal (sleepy, for example), performance is not very good. Likewise, at very high levels of arousal (agitation, for example), people don't perform well. In the middle range, however, there is a degree of arousal that is optimum; performance is best at moderate levels of arousal.

In Figure 6.8 there is obviously a strong relationship between arousal and performance, but r for the distribution is $-.10$, a value that indicates a very weak relationship. The product-moment correlation coefficient is just not useful for measuring the strength of curved relationships. For curved relationships, researchers often measure the strength of association with the statistic eta (η) or by calculating the formula for a curve that fits the data.

error detection

When a data set produces a low correlation coefficient, a scatterplot is especially recommended. A scatterplot might reveal that a Pearson correlation coefficient is not appropriate for the data set.

Truncated Range

truncated range
Range of the sample is smaller than the range of its population.

Besides nonlinearity, a second situation gives low Pearson correlation coefficients even though there is a strong relationship between the two variables. Spuriously low r values can occur when the range of scores in the sample is much smaller than the range of scores in the population (a **truncated range**).

I'll illustrate with the relationship between GRE scores and grades in graduate school.[8] The relationship graphed in **Figure 6.9** is based on a study by Sternberg and

[8] The Graduate Record Examination (GRE) is used by many graduate schools to help select students for admission.

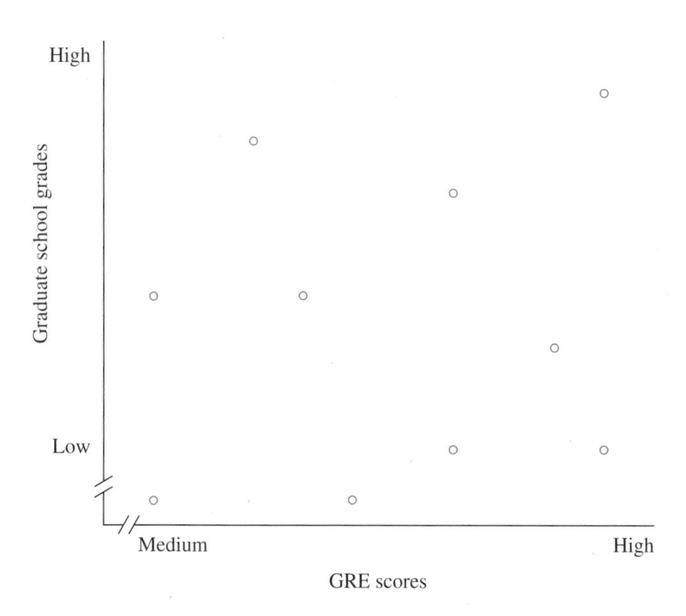

FIGURE 6.9 Scatterplot of GRE scores and graduate school grades in one school

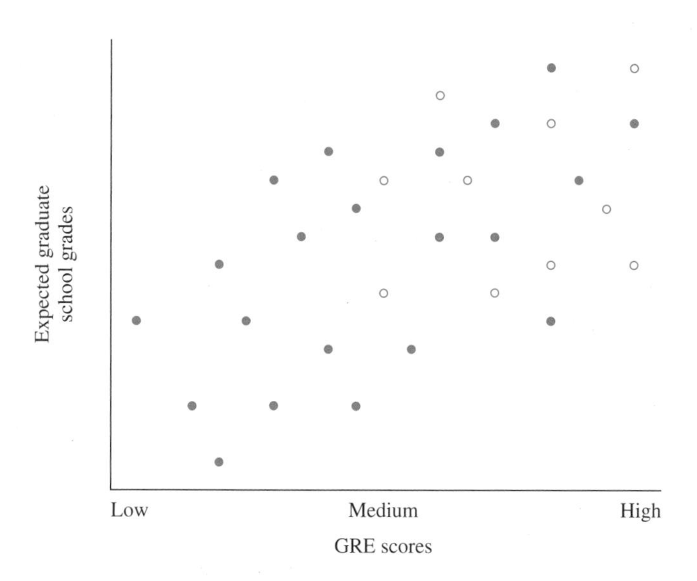

FIGURE 6.10 Hypothetical scatterplot of GRE scores and expected graduate school grades for the population

Williams (1997). These data look like a snowstorm; they lead to the conclusion that there is little relationship between GRE scores and graduate school grades.

However, students in graduate school do not represent the full range of GRE scores; those with low GRE scores are not included. What effect does this restriction of the range have? You can get an answer to this question by looking at **Figure 6.10,** which shows a hypothetical scatterplot of data for the full range of GRE scores. This scatterplot shows a moderate relationship. So, unless you recognized that your sample of graduate students truncated the range of GRE scores, you might be tempted to dismiss GRE scores as "worthless." (A clue that Figure 6.9 illustrates a truncated range is on the horizontal axis. It starts at medium and goes up.)

Other Kinds of Correlation Coefficients

The kind of correlation coefficient you have been learning about—the Pearson product-moment correlation coefficient—is appropriate for measuring the degree of the relationship between two linearly related, continuous variables. Sometimes, however, the data do not consist of two linearly related, continuous variables. What follows is a description of five other situations. In each case, you can express the degree of relationship in the data with a correlation coefficient—but not a Pearson product-moment correlation coefficient. Fortunately, these correlation coefficients can be interpreted much like Pearson product-moment coefficients.

dichotomous variable
A variable that has only two values.

multiple correlation
Correlation coefficient that expresses the degree of relationship between one variable and two or more other variables.

partial correlation
Technique that allows the separation of the effect of one variable from the correlation of two other variables.

1. If one of the variables is **dichotomous** (has only two values), then a biserial correlation (r_b) or a point-biserial correlation (r_{pb}) is appropriate. Variables such as height (recorded as simply tall or short) and gender (male or female) are examples of dichotomous variables.
2. Several variables can be combined, and the resulting combination can be correlated with one variable. With this technique, called **multiple correlation,** a more precise prediction can be made. Performance in school can be predicted better by using several measures of a person rather than one.
3. A technique called **partial correlation** allows you to separate or partial out the effects of one variable from the correlation of two variables. For example, if you want to know the true correlation between achievement test scores in two school subjects, it is probably necessary to partial out the effects of intelligence because cognitive ability and achievement are correlated.
4. When the data are ranks rather than scores from a continuous variable, researchers calculate Spearman r_s, which is covered in Chapter 15.
5. If the relationship between two variables is curved rather than linear, then the correlation ratio eta (η) gives the degree of association (Field, 2005a).

These and other correlational techniques are discussed in intermediate-level textbooks such as Howell (2010).

PROBLEMS

6.15. The correlation between scores on a humor test and a test of insight is .83. Explain what this means. Continue your explanation by interpreting the effect size index and the coefficient of determination. End your explanation with caveats[9] appropriate for r.

6.16. The correlation between number of older siblings and degree of acceptance of personal responsibility for one's own successes and failures is $-.37$. Interpret this correlation. Find the coefficient of determination and explain what it means. What can you say about the cause of the correlation?

6.17. Examine the following data, make a scatterplot, and compute r if appropriate.

Serial position	1	2	3	4	5	6	7	8
Errors	2	5	6	9	13	10	6	4

Linear Regression

linear regression Method that produces a straight line that best fits a bivariate distribution.

What you have learned about correlation will be most helpful as you learn to make quantitative predictions. This technique is called **linear regression.**

A few sections ago I said that the correlation between college entrance examination scores and first-semester grade point averages is about .50. Knowing this correlation, you can predict that those who score high on the entrance examination are more likely to succeed as freshmen than those who score low. This statement is correct, but it is pretty general. Usually you want to predict a *specific* grade point for a *specific* applicant. For example, if you were in charge of admissions at Collegiate U., you want to know the entrance examination score that predicts a GPA of 2.00, the minimum required for graduation. To make specific predictions, you must calculate a regression equation.

Linear regression is a technique that uses the data to produce an equation for a straight line. This equation is then used to make predictions. I'll begin with some background on making predictions from equations.

Making Predictions from a Linear Equation

You are used to making predictions; some you make with a great deal of confidence. "If I get my average up to 80 percent, I'll get a B in this course." "If I spend $15 plus tax on this DVD, I won't have enough left from my $20 to go to a movie."

Often, predictions are based on an assumption that the relationship between two variables is linear, that a straight line will tell the story exactly. Frequently, this assumption is quite justified. For the preceding short-term economics problem, imagine a set of axes with "Amount spent" on the X axis ($0 to $20) and "Amount left" on the Y axis ($0 to $20). A straight line that connects the two $20 marks on the axes tells the whole story. (This is a line with a negative slope that is inclined 45 degrees from horizontal.) Draw this picture. Can you also write the equation for this graph?

[9] Warnings.

Part of your education in algebra was about straight lines. You may recall that the slope-intercept formula for a straight line is

$$Y = mX + b$$

where Y and X are variables representing scores on the Y and X axes
m = slope of the line (a constant)
b = intercept (intersection) of the line with the Y axis (a constant)

Here are some reminders about the *slopes* of lines. The slope of a line is positive if the highest point on the line is to the right of the lowest point. If the highest point is to the left of the lowest point, the slope is negative. Horizontal lines have a slope equal to zero. Lines that are almost vertical have slopes that are either large positive numbers or large negative numbers.

Now, back to the slope-intercept formula. If you are given the two constants $m = 3$ and $b = 6$, the formula becomes $Y = 3X + 6$. Now, if you start with a value for X, you can easily find the value of Y. If $X = 4$, then $Y =$ _____?

The unsolved problem (so far) is how to go from the general formula $Y = mX + b$ to a specific formula like $Y = 3X + 6$—that is, how to find the values for m and b.

A common solution involves a rule and a little algebra. The rule is that if a point lies on the line, then the coordinates of the point satisfy the equation of the line. To use this rule to find an equation, you must know the coordinates for two points. Suppose you are told that one point on the line is where $X = 5$ and $Y = 8$. This point, represented as (5, 8), produces $8 = 5m + b$ when the coordinates are substituted into the general equation. If you are given a second point that lies on the line, you will get another equation with m and b as unknowns. Now you have two equations with two unknowns and you can solve for each unknown in turn, giving you values for m and b. If you would like to check your understanding of this on a simple problem, figure out the formula for the line that tells the story of the $20 problem. For simplicity, use the two points where the line crosses the axes, (0, 20) and (20, 0).

Once you have assumed or have satisfied yourself that a relationship is linear, *any* two points will allow you to draw the line that tells the story of the relationship.

Least Squares—A Line of Best Fit

With this background in place, you are in a position to appreciate the problem Karl Pearson faced at the end of the 19th century when he looked at father–daughter height data (problem 6.5). Look at your scatterplot of those data. It is reasonable to assume that the relationship is linear, but which two points should be used to write the equation for the line? The equation will depend on which two particular points you choose to plug into the formula. The two-point approach produces dozens of different lines. Which one is the best line? What solutions come to your mind?

One solution is to find the mean Y value for each of the X values on the abscissa. Connect those means with straight lines, and then choose two points that make the regression line fall within this narrower range of the means. Pearson's solution, which statistics embraced, is to use a more mathematically sophisticated version of this idea, a method called **least squares.** Look at **Figure 6.11,** which is based on Pearson and Lee's

least squares
Fitting a regression line such that the sum of the squared deviations from the straight regression line is a minimum.

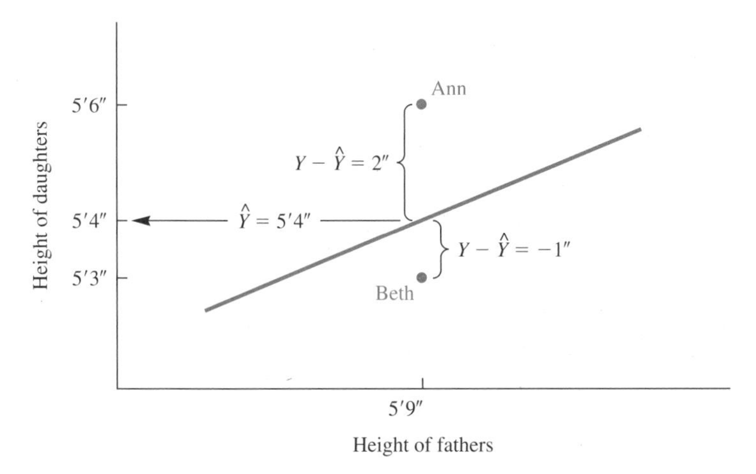

FIGURE 6.11 Hypothetical case of two daughters with a 5′9″ father

1903 data on the heights of fathers and their adult daughters. There is a least squares regression line, which was calculated from all the data points, and two of the data points. The two data points are for 5′6″ Ann and her 5′9″ father and 5′3″ Beth and her father, who is also 5′9″. As you can see from the regression line, 5′4″ is the predicted height for the daughter of a 5′9″ father. (Predicted values are symbolized by $\hat{Y}$, pronounced "Y-hat" or "Y-predicted.") The data show, however, that Ann is 5′6″ and Beth is 5′3″. You can imagine that if the other daughters' heights were plotted on the graph, most of them would not fall on the regression line either. With all this error, why use the least squares method for drawing the line?

The answer is that a least squares regression line *minimizes* error. Calculating error is easy. For each prediction,

$$\text{Error} = Y - \hat{Y}$$

Look again at **Figure 6.11.** Using the two daughters in our example,

Ann: $\text{error} = Y - \hat{Y} = 66 - 64 = 2$

Beth: $\text{error} = Y - \hat{Y} = 63 - 64 = -1$

In a similar way, there is an error for each point in the scatterplot (though for some, the error is zero). The least squares method creates a straight line such that the *sum of the squares* of the errors is a minimum. In symbol form, $\Sigma(Y - \hat{Y})^2$ is a minimum for a straight line calculated by the least squares method.

Not only does the least squares method minimize error, it also produces numerical values for the *slope* and the *intercept.* With a slope and an intercept, you can write the equation for a straight line; this regression line is the one that *best* fits the data.

One more transitional point is necessary. In the language of algebra, the idea of a straight line is expressed as $Y = mX + b$. In the language of statistics, exactly the same idea is expressed as $Y = a + bX$. Y and X are the same in the two formulas, but different letters are used for the slope and the intercept and the order of the terms is different. Unfortunately, the terminology is well established in both fields. However,

regression equation
Equation that predicts values of Y for specific values of X.

the mental translation required of students doesn't cause many problems. Thus, in statistics, b is the slope of the line and a is the intercept of the line with the Y axis.

The Regression Equation

In statistics, the **regression equation** is

$$\hat{Y} = a + bX$$

where $\hat{Y}$ = Y value predicted for a particular X value
a = point at which the regression line intersects the Y axis
b = slope of the regression line
X = value for which you wish to predict a Y value

For correlation problems, the symbol Y can be assigned to either variable, but in regression equations, *Y is assigned to the variable you wish to predict.*

The Regression Coefficients

To use the equation $\hat{Y} = a + bX$ you must have values for a and b, which are called **regression coefficients.** Values for a and b can be calculated from any bivariate set of data. The arithmetic for these calculations comes from the least squares method of line fitting.

regression coefficients
The constants a and b in a regression equation.

If you have already computed r and the standard deviations for both X and Y, then the slope of the regression line, b, is

$$b = r\frac{S_Y}{S_X}$$

where r = correlation coefficient for X and Y
S_Y = standard deviation of the Y variable
S_X = standard deviation of the X variable

Is it clear to you that for positive correlations, b will be a positive number? For negative correlations, b will be negative. The value of b can also be obtained with the formula

$$b = \frac{N\Sigma XY - (\Sigma X)(\Sigma Y)}{N\Sigma X^2 - (\Sigma X)^2}$$

To compute a, the regression line's intercept with the Y axis, use the formula

$$a = \overline{Y} - b\overline{X}$$

where $\overline{Y}$ = mean of the Y scores
b = regression coefficient computed previously
$\overline{X}$ = mean of the X scores

Figure 6.12 is designed to generically illustrate the regression coefficients a and b. In **Figure 6.12,** the regression line crosses the Y axis exactly at 4, so $a = 4.00$. The coefficient b is the slope of the line. To independently determine the slope of a line from a graph,

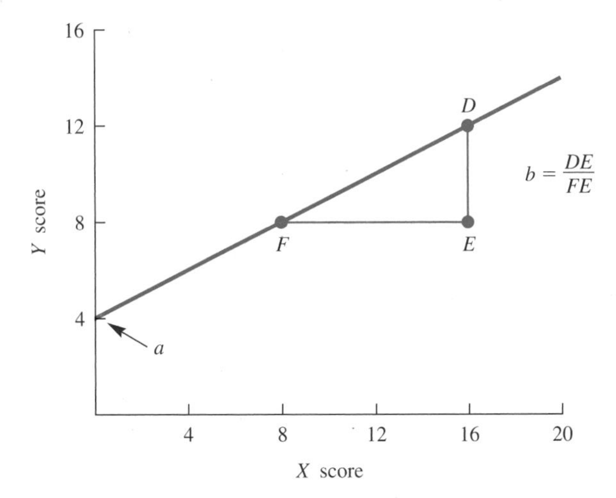

FIGURE 6.12 The regression coefficients *a* and *b*

divide the vertical distance the line rises by the horizontal distance the line covers. In Figure 6.12 the line *DE* (vertical rise of the line *FD*) is half the length of *FE* (the horizontal distance of *FD*). Thus, the slope of the regression line is 0.50 ($DE/FE = b = 0.50$). Put another way, the value of *Y* increases 1/2 point for every one-point increase in *X*.

Calculating a Regression Equation

I'll illustrate calculating a regression equation with the critical reading SAT (CR SAT) scores and first-year college grade point averages (FYGPA). The CR SAT scores come from Table 6.3; the first-year college grade point statistics and the correlation coefficient are from Korbrin, et at. (2008). The task is to predict FYGPA from SAT scores, so I'll designate FYGPA as the *Y* variable.

error detection

Step one in writing a regression equation is to identify the variable whose scores you want to predict. Make that variable *Y*.

To calculate the regression coefficients for a regression equation, you need the correlation coefficient and the mean and standard deviations of both variables.

Correlation coefficient $r = .48$

X	Y
CR SAT	FYGPA
$\overline{X} = 500$	$\overline{Y} = 2.97$
$S_X = 93.54$	$S_Y = 0.71$

The formula for *b* gives

$$b = r\frac{S_Y}{S_X} = (.48)\frac{0.71}{93.54} = (.48)(0.00759) = 0.00364$$

The formula for a gives

$$a = \overline{Y} - b\overline{X} = 2.97 - (0.00364)(500) = 2.97 - 1.822 = 1.148$$

The b coefficient (0.0036) tells you that the slope of the regression line is almost flat. The a coefficient tells you that the regression line intersects the Y axis at 1.148. Entering these regression coefficients values into the regression equation produces a formula that predicts first-year GPA from CR SAT scores:

$$\hat{Y} = a + bX = 1.148 + 0.00364X$$

Predicting a *Y* score–Finding $\hat{Y}$

With this regression equation in hand, you can predict a first-year GPA for any CR SAT score.[10] What GPA does the formula predict for a CR SAT score of 500, which was the mean score for test takers in 2008? In some ways, this is a trick question (but of the good variety). Think through the problem before you look at the arithmetic that follows.

$$\hat{Y} = 1.148 + 0.00364X$$
$$\hat{Y} = 1.148 + (0.00364)(500)$$
$$\hat{Y} = 2.97$$

The "trick" to this problem is that 500 is the mean of the CR SAT scores. The corresponding point for the FYGPA data is its mean, 2.97.

Several sections back I said that if you were in charge of admissions at Collegiate U., you'd want to know the entrance exam score that predicts a graduation GPA of 2.00. With the regression equation above, you can approximate that knowledge by finding the exam score that predicts a first-year GPA of 2.00.

$$\hat{Y} = 1.148 + 0.00364X$$
$$2.00 = 1.148 + 0.00364X$$
$$X = 234$$

SAT scores come in multiples of 10, I'll have to chose between 230 and 240. Because I want a score that predicts applicants who will achieve a FYGPA of *at least* 2.00, I would recommend 240.

To find $\hat{Y}$ values from summary data without calculating a and b, use this formula:

$$\hat{Y} = r\frac{S_y}{S_X}(X - \overline{X}) + \overline{Y}$$

For a CR SAT score of 240, the first year GPA is

$$\hat{Y} = (.48)(0.00759)(-260) + 2.97 = -0.947 = 2.02$$

[10] You may know you own CR SAT score. Using this equation, you can predict your own first-year college grade point average. In addition, you probably already have a first-year college grade point average. How do the two compare?

PROBLEMS

***6.18.** Using the statistics in Table 6.3, write the equation that predicts Math SAT scores from CR SAT scores.

Now you know how to make predictions. Predictions, however, are cheap; anyone can make them. Respect accrues only when your predictions come true. So far, I have dealt with accuracy by simply pointing out that when r is high, accuracy is high, and when r is low, you cannot put much faith in your predicted values of $\hat{Y}$.

standard error of estimate Standard deviation of the differences between predicted outcomes and actual outcomes.

To actually *measure* the accuracy of predictions made from a regression analysis, you need the **standard error of estimate.** This statistic is discussed in most intermediate-level textbooks and in textbooks on testing. (See, for example, Howell, 2008, pp. 221–223.) The materials in Chapters 7 and 8 of this book provide the background needed to understand the standard error of estimate.

Drawing a Regression Line on a Scatterplot

To illustrate drawing a regression line, I'll return to the data in Table 6.3, the two subtests of the SAT. To draw the line, you need a straightedge and two points on the line. Any two points will do. One point that is always on the regression line is $(\bar{X}, \bar{Y})$. Thus, for the SAT data, the two means (500, 515) identify a point. This point is marked on **Figure 6.13** with an open circle.

The second point may take a little more work. For it you need the regression equation. Fortunately, you have that equation from your work on problem 6.18.

$$\hat{Y} = 165 + 0.700X$$

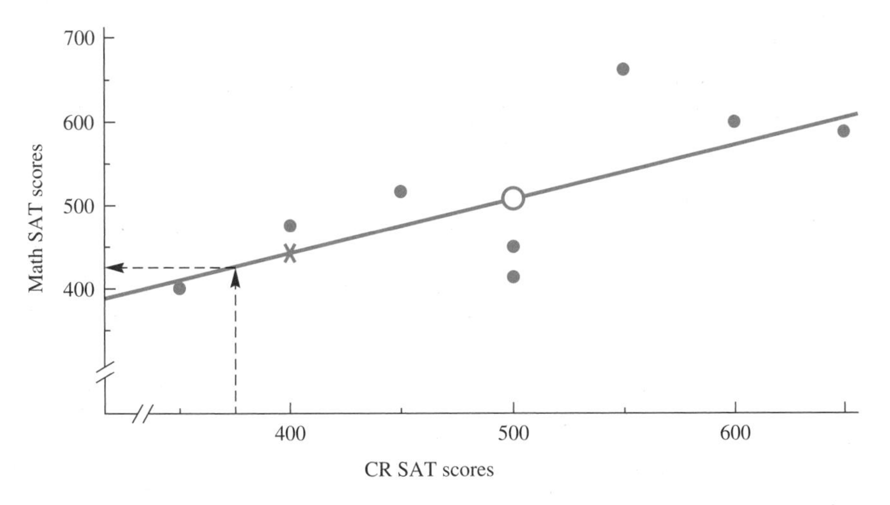

FIGURE 6.13 Scatterplot and regression line for the CR SAT and math SAT scores in Table 6.3

TABLE 6.6 SPSS output of regression coefficients and *r* for the CR SAT and math SAT scores in Table 6.3

		Coefficients[a]				
		Unstandardized Coefficients		Standardized Coefficients		
Model		B	Std. Error	Beta	t	Sig.
1	(Constant)	165.000	138.419		1.192	.278
	CR.SAT	.700	.272	.724	2.572	.042

a. Dependent Variable: M.SAT

To find the second point, choose a value for *X* and solve for *Y*. Any value for *X* within the range of your graph will do; I chose 400 because it made the product of 0.700*X* easy to calculate. (0.700)(400) = 280. Thus, the second point is (400, 445), which is marked on **Figure 6.13** with an X. Finally, line up the straightedge on the two points and extend the line in both directions. Notice that the line crosses the *Y* axis just under 400, which may surprise you because $a = 165$. (I'll come back to this in the next section.)

With a graph such as Figure 6.13, you can make predictions about Math SAT scores from CR SAT scores. From a score on the *X* axis, draw a vertical line up to the regression line. Then draw a horizontal line over to the vertical axis. That *Y* score is $\hat{Y}$, the predicted Math SAT score. On **Figure 6.13,** the Math SAT score predicted for a CR SAT score of 375 is between 400 and 450.

In SPSS, the linear regression program calculates several statistics and displays them in different tables. For the SAT data, the SPSS table *coefficients* is reproduced as **Table 6.6.** The regression coefficients are in the B column under *Unstandardized Coefficients*. The intercept coefficient, the a (165.00), is labeled (*Constant*) and the slope coefficient, *b*(0.700), is labeled *CR.SAT*. The Pearson correlation coefficient (.724) is in the *Beta* column.

The Appearance of Regression Lines

Now, I'll return to the surprise I mentioned in the previous section: The regression line in Figure 6.13 crosses the *Y* axis just below 400 but $a = 165$. The appearance of regression lines depends not only on the calculated values of *a* and *b* also on the units chosen for the *X* and *Y* axes and whether there are breaks in the axes. Look at **Figure 6.14.** Although the two lines appear different, $b = 1.00$ for both. They appear different because the space allotted to each *Y* unit in the right graph is half that allotted to each *Y* unit in the left graph.

I can now explain the dilemma I faced when I composed the "out-of-balance" **Figure 6.3**. The graph is ungainly because it is square (100 *X* units is the same length as 100 *Y* units) and because both axes start at zero, which forces the data points up into a corner. I composed it the way I did because I wanted the regression line to cross the *Y* axis at *a* (note that it does) and because I wanted its slope to *appear* equal to *b* (note that it does).

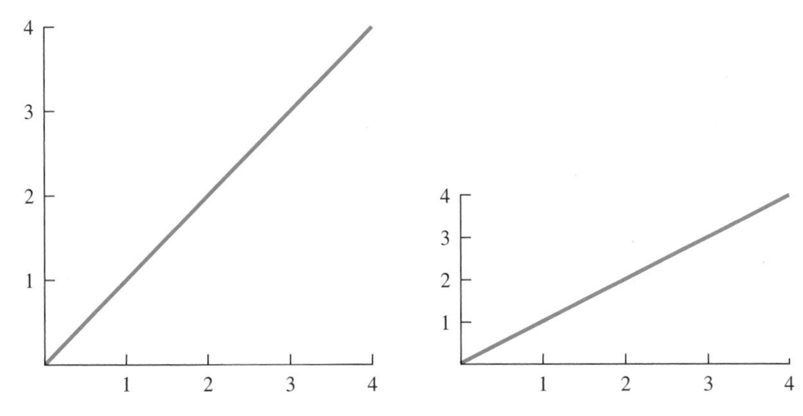

FIGURE 6.14 Two regression lines with the same slope (b = 1.00) but with different appearances. The difference is caused by different-sized units on the Y axis

The more attractive **Figure 6.13** is a scatterplot of the same data as those in Figure 6.3. The difference is that Figure 6.13 has breaks in the axes and 100 Y units is about one half the length of 100 X units. I'm sure by this point you have the message—you cannot necessarily determine a and b by looking at a graph.

Finally, a note of caution: Every scatterplot has two regression lines. One is called the regression of Y onto X, which is what you did in this chapter. The other is the regression of X onto Y. The difference between these two depends on which variable is designated Y. So, in your calculations be sure you assign Y to the variable you want to predict.

PROBLEMS

6.19. In problem 6.5, the father–daughter height data, you found r = .513.
- **a.** Compute the regression coefficients a and b.
- **b.** Use your scatterplot from problem 6.5 and draw the regression line.

6.20. In problem 6.6, the two different intelligence tests with the WAIS test as X and the Wonderlic as Y, you computed r.
- **a.** Compute a and b.
- **b.** What Wonderlic score would you predict for a person who scored 130 on the WAIS?

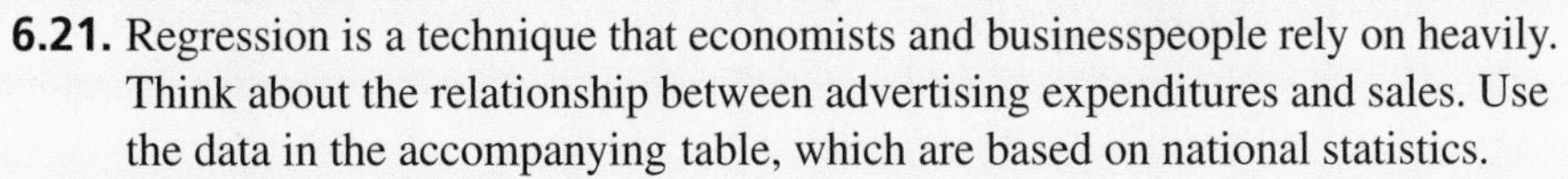

6.21. Regression is a technique that economists and businesspeople rely on heavily. Think about the relationship between advertising expenditures and sales. Use the data in the accompanying table, which are based on national statistics.
- **a.** Find r.
- **b.** Write the regression equation.
- **c.** Plot the regression line on the scatterplot.
- **d.** Predict sales for an advertising expenditure of $10,000.
- **e.** Explain whether any confidence at all can be put in the prediction you made.

Advertising, X ($ thousands)	Sales, Y ($ thousands)
3	70
4	120
3	110
5	100
6	140
5	120
4	100

6.22. The correlation between Stanford–Binet IQ scores and Wechsler Adult Intelligence Scale (WAIS) IQs is about .80. Both tests have a mean of 100. The standard deviation on older versions of the Stanford–Binet was 16. For the WAIS, $S = 15$. What WAIS IQ do you predict for a person who scores 65 on the Stanford–Binet? Write a sentence summarizing these results. (An IQ score of 70 has been used by some schools as a cutoff point between regular classes and special education classes.)

6.23. Many predictions about the future come from regression equations. Use the following data from the *Statistical Abstract of the United States* to predict the number of college graduates with bachelor's degrees in the year 2011. Use the time period numbers rather than years in your calculations and carry four decimal places. Carefully choose which variable to call X and which to call Y.

Time period	1	2	3	4	5
Year	2002	2003	2004	2005	2006
Graduates (millions)	1.29	1.35	1.40	1.44	1.49

6.24. Once again, look over the objectives at the beginning of the chapter. Can you do them?

6.25. Now it is time for *integrative* work on the descriptive statistics you studied in Chapters 2–6. Choose one of the two options that follow.

a. Write an essay on descriptive statistics. Start by jotting down from memory things you could include. Review Chapters 2–6, adding to your list additional facts or other considerations. Draft the essay. Rest. Revise it.

b. Construct a table that summarizes the descriptive statistics in Chapters 2–6. List the techniques in the first column. Across the top of the table, list topics that distinguish among the techniques—topics such as purpose, formula, and so forth. Fill in the table.

Whether you choose option **a** or **b**, save your answer for that time in the future when you are reviewing what you are learning in this course (final exam time?).

ADDITIONAL HELP FOR CHAPTER 6

Visit *cengage.com/psychology/spatz*. At the Student Companion Site, you'll find multiple-choice tutorial quizzes, flashcards with definitions and workshops. For this chapter there are Statistical Workshops on Bivariate Scatter Plots and Correlation.

KEY TERMS

Bivariate distribution (p. 89)
Causation and correlation (p. 105)
Coefficient of determination (p. 102)
Common variance (p. 103–104)
Correlation coefficient (p. 96)
Dichotomous variable (p. 110)
Effect size index for *r* (p. 102)
Intercept (p. 114)
Least squares method (p. 112)
Linear regression (p. 111)
Multiple correlation (p. 110)
Negative correlation (p. 92)
Partial correlation (p. 110)
Positive correlation (p. 90)
Quantification (p. 88)
Regression coefficients (p. 114)
Regression equation (p. 114)
Regression line (p. 117)
Reliability (p. 104)
Scatterplot (p. 100)
Slope (p. 114)
Standard error of estimate (p. 117)
Truncated range (p. 108)
Univariate distribution (p. 89)
Zero correlation (p. 95)

What Would You Recommend? Chapters 1–6

At this point in the text (and at two later points) there is a set of problems titled *What would you recommend?* These problems help you review and integrate your knowledge. For each problem that follows, recommend an appropriate statistic from among those you learned in the first six chapters. Note why you recommend that statistic.

a. Registration figures for the American Kennel Club show which dog breeds are common and which are uncommon. For a frequency distribution for all breeds, what central tendency statistic is appropriate?
b. Among a group of friends, one person is the best golfer. Another person in the group is the best at bowling. What statistical technique allows you to determine that one of the two is better than the other?
c. Tuition at Almamater U. has gone up each of the past 5 years. How can I predict what it will be in 25 years when my child enrolls?
d. Each of the American states has a certain number of miles of ocean coastline (ranging from 0 to 6640 miles). Consider a frequency distribution of these 50 scores. What central tendency statistic is appropriate for this distribution? Explain your choice. What measure of variability do you recommend?

e. Jobs such as "appraiser" require judgments about the value of a unique item. Later, a sale price establishes an actual value. Suppose two applicants for an appraiser's job made judgments about 30 items. After the items sold, an analysis revealed that when each applicant's errors were listed, the average was zero. What other analysis of the data might provide an objective way to decide that one of the two applicants was better?

f. Suppose you study some new, relatively meaningless material until you know it all. If you are tested 40 minutes later, you recall 85 percent; 4 hours later, 70 percent; 4 days later, 55 percent; and 4 weeks later, 40 percent. How can you express the relationship between time and memory?

g. A table shows the ages and the number of voters in the year 2010. The age categories start with "18–20" and end with "65 and over." What statistic can be calculated to describe the age of the typical voter?

h. For a class of 40 students, the study time for the first test ranged from 30 minutes to 6 hours. The grades ranged from a low of 48 to a high of 98. What statistic describes how the variable *Study time* is related to the variable *Grade?*

transition passage
to inferential statistics

YOU ARE NOW through with the part of the book that is clearly about descriptive statistics. You should be able to describe a set of data with a graph, a few choice words, and numbers such as a mean, a standard deviation, and (if appropriate) a correlation coefficient.

The next chapter serves as a bridge between descriptive and inferential statistics. All the problems you will work give you answers that describe something about a person, score, or group of people or scores. However, the ideas about probability and theoretical distributions that you use to work these problems are essential elements of inferential statistics.

So, the transition this time is to concepts that prepare you to plunge into material on inferential statistics. As you will see rather quickly, many of the descriptive statistics that you have been studying are elements of inferential statistics.

CHAPTER 7

Theoretical Distributions Including the Normal Distribution

OBJECTIVES FOR CHAPTER 7

After studying the text and working the problems in this chapter, you should be able to:

1. Distinguish between theoretical and empirical distributions
2. Distinguish between theoretical and empirical probability
3. Predict the probability of certain events from knowledge of the theoretical distribution of those events
4. List the characteristics of the normal distribution
5. Find the proportion of a normal distribution that lies between two scores
6. Find the scores between which a certain proportion of a normal distribution falls
7. Find the number of scores associated with a particular proportion of a normal distribution

THIS CHAPTER HAS more figures than any other chapter, almost one per page. The reason for all these figures is that they are the best way I know to convey to you ideas about theoretical distributions and probability. So, please examine these figures carefully, making sure you understand what each part means. When you are working problems, drawing your own pictures is a big help.

I'll begin by distinguishing between empirical distributions and theoretical distributions. In Chapter 2 you learned to arrange scores in frequency distributions. The scores you worked with were selected because they were representative of scores from actual research. Distributions of such observed scores are **empirical distributions.**

empirical distribution Scores that come from observations.

This chapter has a heavy emphasis on theoretical distributions. Like the empirical distributions in Chapter 2, a theoretical distribution is a presentation of all the scores, usually presented as a graph. **Theoretical distributions,** however, are based on mathematical formulas and logic rather than on empirical observations.

theoretical distribution Hypothesized scores based on mathematical formulas and logic.

Theoretical distributions are used in statistics to determine probabilities. When there is a correspondence between an empirical distribution and a theoretical distribution, you can use the theoretical distribution to arrive at *probabilities* about future empirical events. Probabilities, as you know, are quite helpful in reaching decisions.

This chapter covers three theoretical distributions: rectangular, binomial, and normal. Rectangular and binomial distributions are used to illustrate probability more fully and to establish some points that are true for all theoretical distributions. The third distribution, the normal distribution, will occupy the bulk of your time and attention in this chapter.

Probability

You are already somewhat familiar with the concept of probability. You know, for example, that probability values range from .00 (there is no possibility that an event will occur) to 1.00 (the event is certain to happen).

In statistics, events are sometimes referred to as "successes" or "failures." To calculate the probability of a success using the *theoretical* approach, you first enumerate all the ways a *success* can occur. Then you enumerate all the *events* that can occur (whether successes or failures). Finally, you form a ratio with successes on top (the numerator) and total events on the bottom (the denominator). This fraction, changed to a decimal, is the theoretical probability of a success.

For example, with coin flipping, the theoretical probability of "head" is .50. A head is a success and it can occur in only one way. The total number of possible outcomes is two (head and tail), and the ratio $\frac{1}{2}$ is .50. In a similar way, the probability of rolling a six on a die is $\frac{1}{6} = .167$. For playing cards, the probability of drawing a jack is $\frac{4}{52} = .077$.

The *empirical* approach to finding probability involves observing actual events, some of which are successes and some of which are failures. The ratio of successes to total events produces a probability, a decimal number between .00 and 1.00. To find an empirical probability, you use observations rather than logic to get the numbers.[1]

What is the probability of particular college majors? Remember the data in Figure 2.5, which showed college majors? The probability question can be answered by processing numbers from that figure. Here's how. Choose the major that you are interested in and label the frequency of that major as "number of successes." Divide that number by 1,524,000, which is the total number of baccalaureate degrees granted in 2006–2007. The figure you get answers the probability question. If the major in question is sociology, then 29,000/1,524,000 = .02 is the answer. For English, 55,000/1,524,000 = .04 is the answer.[2] Now, here's a question for you to answer for yourself. Were these probabilities determined theoretically or empirically?

[1] The empirical probability approach is sometimes called the relative frequency approach.

[2] If I missed doing the arithmetic for the major *you* are interested in, I hope you'll do it for yourself.

The rest of this chapter will emphasize theoretical distributions and theoretical probability. You will work with coins and cards next, but before you are finished, I promise you a much wider variety of applications.

A Rectangular Distribution

To show you the relationship between theoretical distributions and theoretical probabilities, I'll use an example of a theoretical distribution that you may be familiar with. **Figure 7.1** is a histogram that shows the distribution of types of cards in an ordinary deck of playing cards. There are 13 kinds of cards, and the frequency of each card is 4. This theoretical curve is a **rectangular distribution.** (The line that encloses a histogram or frequency polygon is called a *curve,* even if it is straight.) The number in the area above each card is the probability of obtaining that card in a chance draw from the deck. That theoretical probability (.077) was obtained by dividing the number of cards that represent the event (4) by the total number of cards (52).

rectangular distribution
Distribution in which all scores have the same frequency.

Probabilities are often stated as "chances in a hundred." The expression $p = .077$ means that there are 7.7 chances in 100 of the event occurring. Thus, from Figure 7.1 you can tell at a glance that there are 7.7 chances in 100 of drawing an ace from a deck of cards. This knowledge might be helpful in some card games.

With this theoretical distribution, you can determine other probabilities. Suppose you want to know your chances of drawing a face card or a 10. These are the shaded events in Figure 7.1. Simply add the probabilities associated with a 10, jack, queen, and king. Thus, $.077 + .077 + .077 + .077 = .308$. This knowledge might be helpful in a game of blackjack, in which a face card or a 10 is an important event (and may even signal "success").

In Figure 7.1 there are 13 kinds of events, each with a probability of .077. It is not surprising that when you add up all the events [(13)(.077)], the result is 1.00.

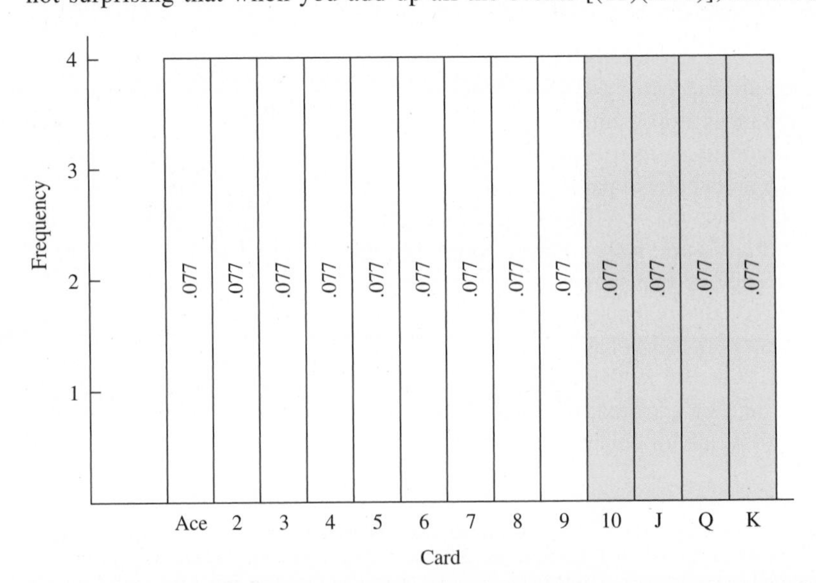

FIGURE 7.1 Theoretical distribution of 52 draws from a deck of playing cards

In addition to the probabilities adding up to 1.00, *the areas add up to 1.00.* That is, by conventional agreement, the *area* under the curve is taken to be 1.00. With this arrangement, any statement about area is also a statement about probability. (If you like to verify things for yourself, you'll find that each slender rectangle has an *area* that is .077 of the area under the curve.) Of the total area under the curve, the proportion that signifies ace is .077, and that is also the probability of drawing an ace from the deck.[3]

clue to the future

The probability of an event or a group of events corresponds to the *area* of the theoretical distribution associated with the event or group of events. This idea will be used throughout this book.

PROBLEMS

7.1. What is the probability of drawing a card that falls between 3 and jack, excluding both?

7.2. If you drew a card at random, recorded the result, and replaced the card in the deck, how many 7s would you expect in 52 draws?

7.3. What is the probability of drawing a card that is higher than a jack *or* lower than a 3?

7.4. If you made 78 draws from a deck, replacing each card, how many 5s and 6s would you expect?

A Binomial Distribution

binomial distribution
Distribution of the frequency of events that can have only two possible outcomes.

The **binomial distribution** is another example of a theoretical distribution. Suppose you take three new quarters and toss them into the air. What is the probability that all three will come up heads? As you may know, the answer is found by multiplying together the probabilities of each of the independent events. For each coin, the probability of a head is $\frac{1}{2}$, so the probability that all three will be heads is $\left(\frac{1}{2}\right)\left(\frac{1}{2}\right)\left(\frac{1}{2}\right) = \frac{1}{8} = .1250$.

Here are two other questions about tossing those three coins. What is the probability of two heads? What is the probability of one head or zero heads? You could answer these questions easily if you had a theoretical distribution of the probabilities. Here's how to construct one. Start by listing, as in **Table 7.1,** the eight possible outcomes of tossing the three quarters into the air. Each of these eight

[3] In gambling situations, uncertainty is commonly expressed in odds. The expression "odds of 5:1" means that there are five ways to fail and one way to succeed; 3:2 means three ways to fail and two ways to succeed. The odds of drawing an ace are 12:1. To convert odds to a probability of success, divide the second number by the sum of the two numbers.

TABLE 7.1 All possible outcomes when three coins are tossed

Outcomes	Number of heads	Probability of outcome
Heads, heads, heads	3	.1250
Heads, heads, tails	2	.1250
Heads, tails, heads	2	.1250
Tails, heads, heads	2	.1250
Heads, tails, tails	1	.1250
Tails, heads, tails	1	.1250
Tails, tails, heads	1	.1250
Tails, tails, tails	0	.1250

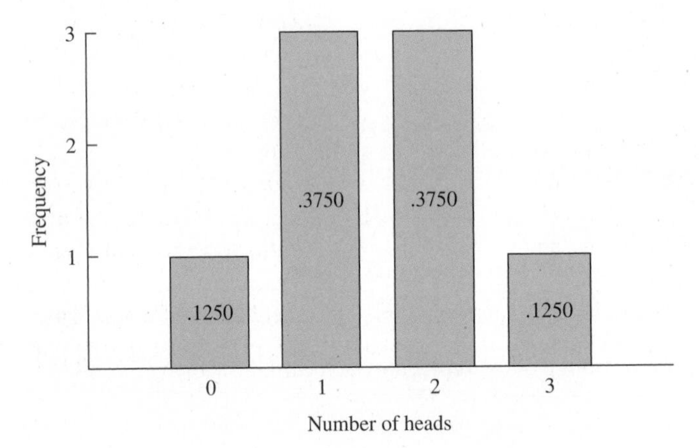

FIGURE 7.2 A theoretical binomial distribution showing the number of heads when three coins are tossed

outcomes is equally likely, so the probability for any one of them is $\frac{1}{8} = .1250$. There are three outcomes in which two heads appear, so the probability of two heads is $.1250 + .1250 + .1250 = .3750$. The probability .3750 is the answer to the first question. Based on Table 7.1, I constructed **Figure 7.2,** which is the theoretical distribution of probabilities you need. You can use it to answer problems 7.5 and 7.6, which follow.[4]

PROBLEMS

7.5. If you toss three coins into the air, what is the probability of a success if success is (a) either one head or two heads? (b) all heads or all tails?

7.6. If you throw the three coins into the air 16 times, how many times would you expect to find zero heads?

[4] The binomial distribution is discussed more fully by Howell (2010) and Pagano (2010, Chap. 9).

Comparison of Theoretical and Empirical Distributions

I have carefully called Figures 7.1 and 7.2 *theoretical* distributions. A theoretical distribution may not reflect *exactly* what would happen if you drew cards from an actual deck of playing cards or tossed quarters into the air. Actual results could be influenced by lost or sticky cards, sleight of hand, uneven surfaces, or chance deviations. Now let's turn to the empirical question of what a frequency distribution of actual draws from a deck of playing cards looks like. **Figure 7.3** is a histogram based on 52 draws from a used deck shuffled once before each draw.

As you can see, Figure 7.3 is not exactly like Figure 7.1. In this case, the differences between the two distributions are due to chance or worn cards and not to lost cards or sleight of hand (at least not conscious sleight of hand). Of course, if I made 52 more draws from the deck and constructed a new histogram, the picture would probably be different from both Figures 7.3 and 7.1. However, if I continued, drawing 520 or 5200 or 52,000 times,[5] and only chance was at work, the curve would be practically flat on the top; that is, the empirical curve would look like the theoretical curve.

The major point here is that a theoretical curve represents the "best estimate" of how the events would actually occur. As with all estimates, a theoretical curve may produce predictions that vary from actual observations, but in the world of real events, it is better than any other estimate.

In summary, then, a theoretical distribution is one based on logic and mathematics rather than on observations. It shows you the probability of each event that is part

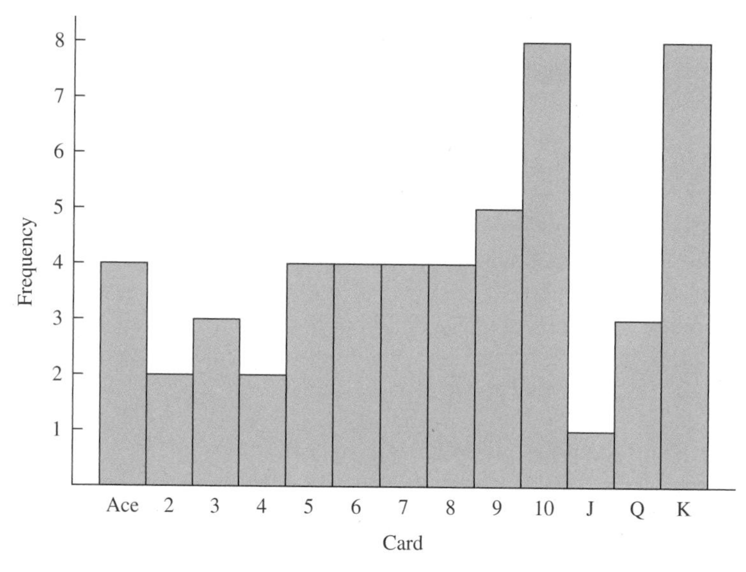

FIGURE 7.3 Empirical frequency distribution of 52 draws from a deck of playing cards

[5] Statisticians use the phrase *the long run* to describe extensive sampling.

of the distribution. When it is similar to an empirical distribution, the probability figures obtained from the theoretical distribution are accurate predictors of actual events.

The Normal Distribution

normal distribution
A bell-shaped, theoretical distribution that predicts the frequency of occurrence of chance events.

One theoretical distribution that has proved to be extremely valuable is the **normal distribution.** With contributions from Abraham de Moivre (1667–1754) and Pierre-Simon Laplace (1749–1827), Carl Friedrich Gauss (1777–1855) worked out the mathematics of the curve and used it to assign precise probabilities to errors in astronomy observations (Wight and Gable, 2005). Because the Gaussian curve was such an accurate picture of the effects of random variation, early writers referred to the curve as the law of error.[6] At the end of the 19th century, Francis Galton called the curve the normal distribution (David, 1995). Perhaps Galton chose the word *normal* based on the Latin adjective *normalis,* which means built with a carpenter's square (and therefore exactly right). Certainly there were statisticians during the 19th century who mistakenly believed that if data were collected without any mistakes, the form of the distribution would be what is today called the normal distribution.

One of the early promoters of the normal curve was Adolphe Quetelet (KA-tle) (1796–1874), a Belgian who showed that many social and biological measurements are distributed normally. Quetelet, who knew about the "law of error" from his work as an astronomer, presented tables showing the correspondence between measurements such as height and chest size and the normal curve. His measure of starvation and obesity was weight divided by height. This index was a precursor of today's BMI (body mass index). During the 19th century Quetelet was widely influential (Porter, 1986). Florence Nightingale, his friend and a pioneer in using statistical analyses to improve health care, said that Quetelet was "the founder of the most important science in the whole world." (See Maindonald and Richardson, 2004, who also include an interview with Nightingale reconstructed from her writings.) Quetelet's work also gave Francis Galton the idea that characteristics we label "genius" could be measured, an idea that led to the concept of correlation.[7]

Although many measurements are distributed approximately normally, it is not the case that data "should" be distributed normally. This unwarranted conclusion has been reached by some scientists in the past.

Finally, the theoretical normal curve has an important place in statistical theory. This importance is quite separate from the fact that empirical frequency distributions often correspond closely to the normal curve.

Description of the Normal Distribution

Figure 7.4 is a normal distribution. It is a bell-shaped, symmetrical, theoretical distribution based on a mathematical formula rather than on empirical observations. (Even so, if you peek ahead to Figures 7.7, 7.8, and 7.9, you will see that empirical

[6] In statistics, *error* means random variation.

[7] Quetelet qualifies as a famous person: A statue was erected in his honor in Brussels, he was the first foreign member of the American Statistical Association, and the Belgian government commemorated the centennial of his death with a postage stamp (1974). For a short intellectual biography of Quetelet, see Faber (2005).

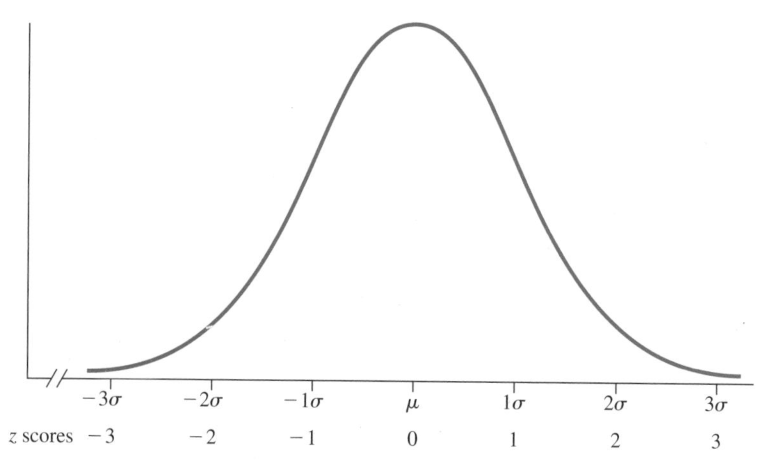

FIGURE 7.4 The normal distribution

curves often look similar to this theoretical distribution.) When the theoretical curve is drawn, the *Y* axis is sometimes omitted. On the *X* axis, ***z* scores** are used as the unit of measurement for the standardized normal curve, where

***z* score**
Score expressed in standard deviation units.

$$z = \frac{X - \mu}{\sigma}$$

where X = a raw score
μ = the mean of the distribution
σ = the standard deviation of the distribution

There are several other things to note about the normal distribution. The mean, the median, and the mode are the same score—the score on the *X* axis where the curve peaks. If a line is drawn from the peak to the mean score on the *X* axis, the area under the curve to the left of the line is half the total area—50 percent—leaving half the area to the right of the line. The tails of the curve are **asymptotic** to the *X* axis; that is, they never actually cross the axis but continue in both directions indefinitely, with the distance between the curve and the *X* axis becoming less and less. Although, in theory, the curve never ends, it is convenient to think of (and to draw) the curve as extending from -3σ to $+3\sigma$. (The *table* for the normal curve in Appendix C, however, covers the area from -4σ to $+4\sigma$.)

asymptotic
Line that continually approaches but never reaches a specified limit.

Another point about the normal distribution is that the two inflection points in the curve are at exactly -1σ and $+1\sigma$. The **inflection points** are where the curve is the steepest—that is, where the curve changes from bending upward to bending over. (See the points above -1σ and $+1\sigma$ in Figure 7.4 and think of walking up, over, and down a bell-shaped hill.)

inflection point
Point on a curve that separates a concave upward arc from a concave downward arc, or vice versa.

To end this introductory section, here's a caution about the word *normal.* The antonym for *normal* is *abnormal. Curves that are not normal distributions, however, are definitely not abnormal.* There is nothing uniquely desirable about the normal

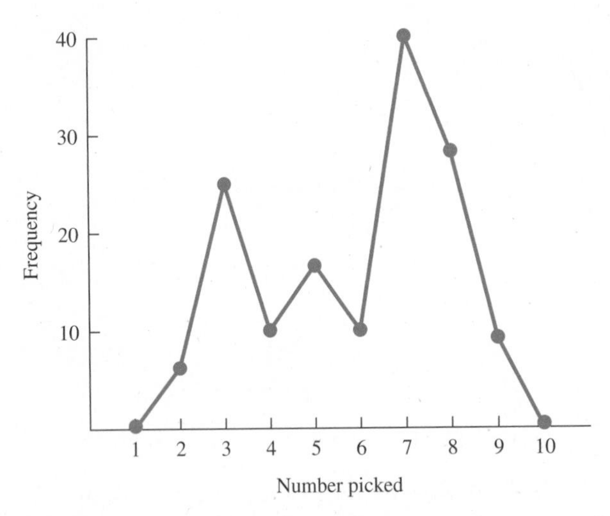

FIGURE 7.5 Frequency distribution of choices of numbers between 1 and 10

distribution. Many nonnormal distributions are also useful to statisticians. Figure 7.1 is an example. It isn't a normal distribution, but it can be very useful. **Figure 7.5** shows what numbers were picked when an instructor asked introductory psychology students to pick a number between 1 and 10. Figure 7.5 is a bimodal distribution with modes at 3 and 7. It isn't a normal distribution, but it will prove useful later in this book.

The Normal Distribution Table

The theoretical normal distribution is used to determine the probability of an event, just as Figure 7.1 was. **Figure 7.6** is a picture of the normal curve, showing the probabilities associated with certain areas. The figure shows that the probability of an event with a z score between 0 and 1.00 is .3413. For events with z scores of 1.00 or larger, the probability is .1587. These probability figures were obtained from Table C in Appendix C. Turn to **Table C** now and insert a bookmark there. Table C is arranged so that you can begin with a z score (column A) and find the following:

1. The area between the mean and the z score (column B)
2. The area from the z score to infinity (∞) (column C)

In column A find the z score of 1.00. The proportion of the curve between the mean and a z score of 1.00 is .3413. The proportion beyond the z score of 1.00 is .1587. Because the normal curve is symmetrical and because the area under the entire curve is 1.00, the sum of .3413 and .1587 will make sense to you. Also, because the curve is symmetrical, these same proportions hold for $z = -1.00$. Thus, all the proportions in Figure 7.6 were derived by finding the proportions associated with a z value of 1.00 in Table C. Don't just read this paragraph; do it. Understanding the normal curve *now* will pay you dividends throughout the book.

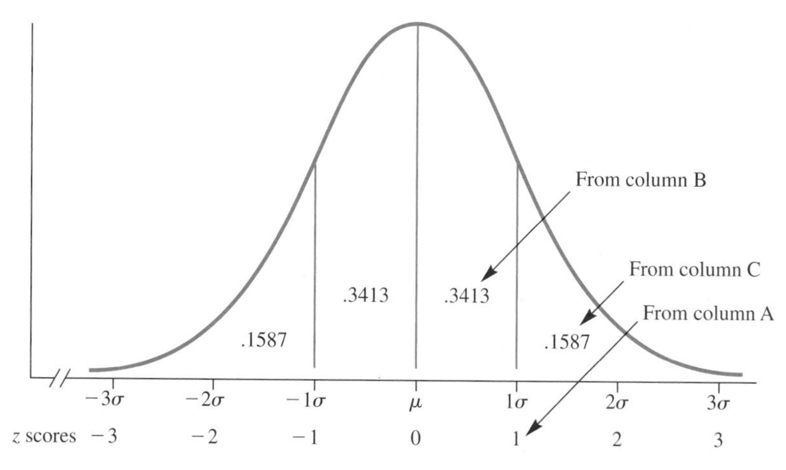

FIGURE 7.6 The normal distribution showing the probabilities of certain *z* scores

Notice that the proportions in Table C are carried to four decimal places and that I used all of them. This is customary practice in dealing with the normal curve because you often want two decimal places when a proportion is converted to a percentage.

PROBLEMS

7.7. In Chapter 5 you read of a professor who gave A's to students with z scores of $+1.50$ or higher.

a. What proportion of a class would be expected to make A's?

b. What assumption must you make to find the proportion in part **a**?

7.8. What proportion of the normal distribution is found in the following areas?

a. Between the mean and $z = .21$

b. Beyond $z = .55$

c. Between the mean and $z = -2.01$

7.9. Is the distribution in Figure 7.5 theoretical or empirical?

As I've already mentioned, many empirical distributions are approximately normally distributed. **Figure 7.7** shows a set of 261 IQ scores, **Figure 7.8** shows the diameter of 199 ponderosa pine trees, and **Figure 7.9** shows the hourly wage rates of 185,822 union truck drivers in the middle of the last century (1944). As you can see, these distributions from diverse fields are similar to Figure 7.4, the theoretical normal distribution. Please note that all of these empirical distributions are based on a "large" number of observations. More than 100 observations are usually required for the curve to fill out nicely.

In this section I made two statistical points: first that Table C can be used to determine areas (proportions) of a normal distribution, and second that many empirical distributions are approximately normally distributed.

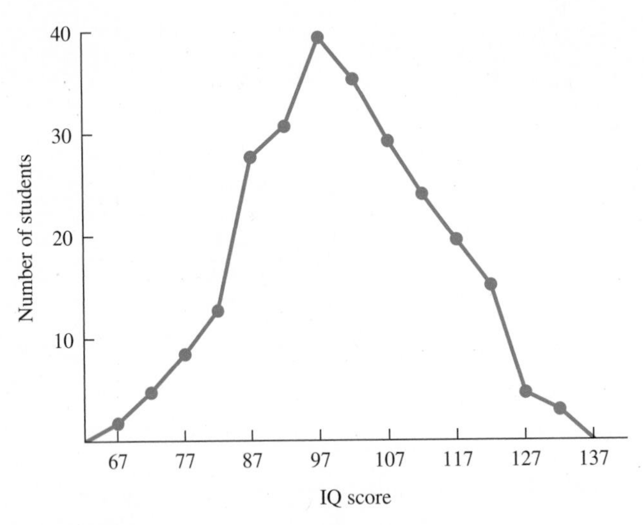

FIGURE 7.7 Frequency distribution of IQ scores of 261 fifth-grade students (unpublished data from J. O. Johnston)

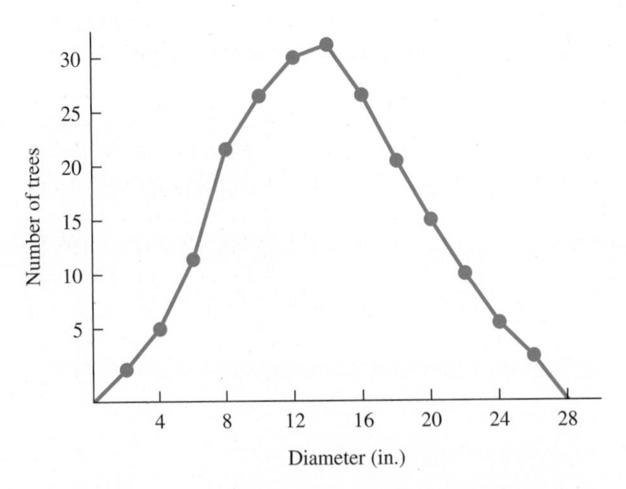

FIGURE 7.8 Frequency distribution of diameters of 100-year-old ponderosa pine trees on 1 acre, $N = 199$ (Forbs and Meyer, 1955)

Converting Empirical Distributions to the Standard Normal Distribution

The point of this section is that any normally distributed empirical distribution can be made to correspond to the theoretical distribution in Table C by using z scores. If the raw scores of an empirical distribution are converted to z scores, the mean of the z scores will be 0 and the standard deviation will be 1. Thus, the parameters of the theoretical normal distribution (which is also called the standardized normal distribution) are: mean = 0, standard deviation = 1.

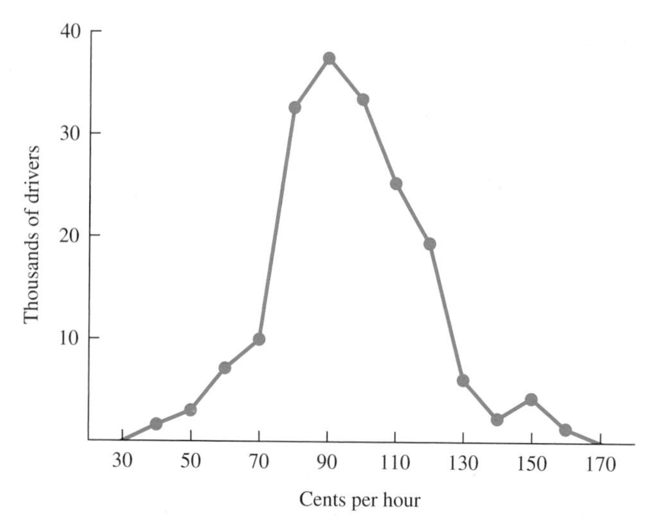

FIGURE 7.9 Frequency distribution of hourly wage rates of union truck drivers on July 1, 1944, N = 185,822 (U.S. Bureau of Labor Statistics, December 1944)

Using z scores calculated from the raw scores of an empirical distribution, you can determine the probabilities of empirical events such as IQ scores, tree diameters, and hourly wages. In fact, with z scores, you can find the probabilities of *any* empirical events that are distributed normally.

Human beings vary from one another in many ways, one of which is cognitive ability. Careful crafted tests such as Wechsler intelligence scales, the Stanford-Binet, and the Wonderlic Personnel Test produce scores (commonly called IQ scores) that are reliable measures of general cognitive ability. These tests have a mean of 100 and a standard deviation of 15.[8] The scores on IQ tests are normally distributed (Micceri, 1989).

Ryan (2008) provides some history and a summary of theories of intelligence, pointing out that ancient Greeks and Chinese used measures of cognitive ability for important personnel decisions. As you have already experienced, college admissions and other academic decisions today are based on tests that measure cognitive ability.

PROBLEM

7.10. Calculate the z scores for IQ scores of

a. 55 **b.** 110 **c.** 103 **d.** 100

Proportion of a Population with Scores of a Particular Size or Greater

Suppose you are faced with finding out what proportion of the population has an IQ of 120 or higher. Begin by sketching a normal curve (either in the margin or on a separate paper). Note on the baseline the positions of IQs of 100 and 120. What is your eyeball estimate of the proportion with IQs of 120 or higher?

---

[8] Older versions of Stanford-Binet tests had a standard deviation of 16. Also, as first noted by Flynn (1987), the actual population mean IQ in many countries is well above 100.

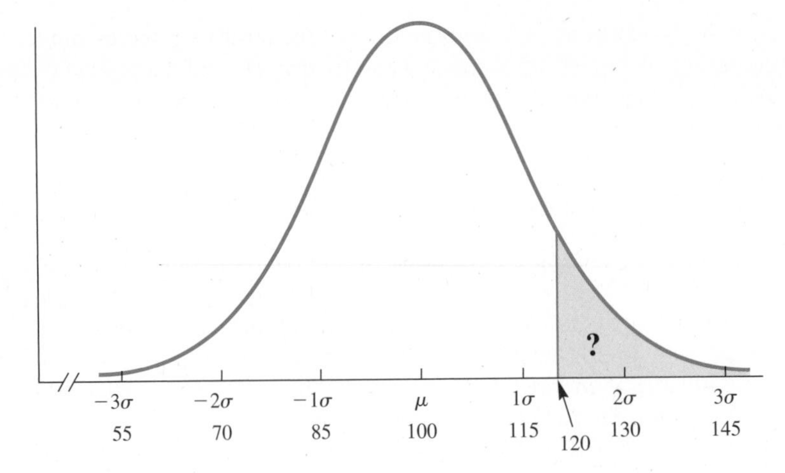

FIGURE 7.10 Theoretical distribution of IQ scores

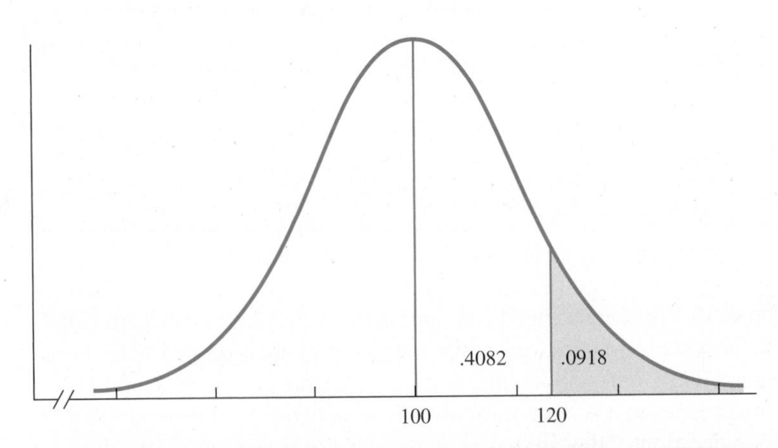

FIGURE 7.11 Proportion of the population with an IQ of 120 or higher

Look at **Figure 7.10.** It is a more formal version of your sketch, giving additional IQ scores on the *X* axis. The proportion of the population with IQs of 120 or higher is shaded. The z score that corresponds with an IQ of 120 is

$$z = \frac{120 - 100}{15} = \frac{20}{15} = 1.33$$

Table C shows that the proportion beyond $z = 1.33$ is .0918. Thus, you expect a proportion of .0918, or 9.18 percent, of the population to have an IQ of 120 or higher. *Because the size of an area under the curve is also a probability statement about the events in that area, there are 9.18 chances in 100 that any randomly selected person will have an IQ of 120 or above.* **Figure 7.11** shows the proportions just determined.

Table C gives the proportions of the normal curve for positive z scores only. However, because the distribution is symmetrical, knowing that .0918 of the population has an IQ of 120 or higher tells you that .0918 has an IQ of 80 or lower. An IQ of 80 has a z score of -1.33.

Questions of "How Many?"

You can answer questions of "how many" as well as questions of proportions using the normal distribution. Suppose 500 first-graders are entering school. How many would be expected to have IQs of 120 or higher? You just found that 9.18 percent of the population would have IQs of 120 or higher. If the population is 500, then calculating 9.18 percent of 500 gives you the number of children. Thus, $(.0918)(500) = 45.9$. So 46 of the 500 first-graders would be expected to have an IQ of 120 or higher.

There are 19 more normal curve problems for you to do in the rest of this chapter. Do you want to maximize your chances of working every one of them correctly the first time? Here's how. For each problem, start by sketching a normal curve. Read the problem and write the givens and the unknowns on your curve. Estimate the answer. Apply the z-score formula. Compare your answer with your estimate; if they don't agree, decide which is in error and make any changes that are appropriate. Confirm your answer by checking the answer in the back of the book. (I hope you decide to go for 19 out of 19!)

error detection

Sketching a normal curve is the best way to understand a problem and avoid errors. Draw vertical lines above the scores you are interested in. Write in proportions.

PROBLEMS

7.11. For many school systems, an IQ of 70 indicates that the child may be eligible for special education. What proportion of the general population has an IQ of 70 or less?

7.12. In a school district of 4000 students, how many would be expected to have IQs of 70 or less?

7.13. What proportion of the population would be expected to have IQs of 110 or higher?

7.14. Answer the following questions for 250 first-grade students.

a. How many would you expect to have IQs of 110 or higher?

b. How many would you expect to have IQs lower than 110?

c. How many would you expect to have IQs lower than 100?

Separating a Population into Two Proportions

Instead of starting with an IQ score and calculating proportions, you can work backward and answer questions about scores if you are given proportions. For example, what IQ score is required to be in the top 10 percent of the population?

My picture of this problem is shown as **Figure 7.12.** I began by sketching a more or less bell-shaped curve and writing in the mean (100). Next, I separated the "top

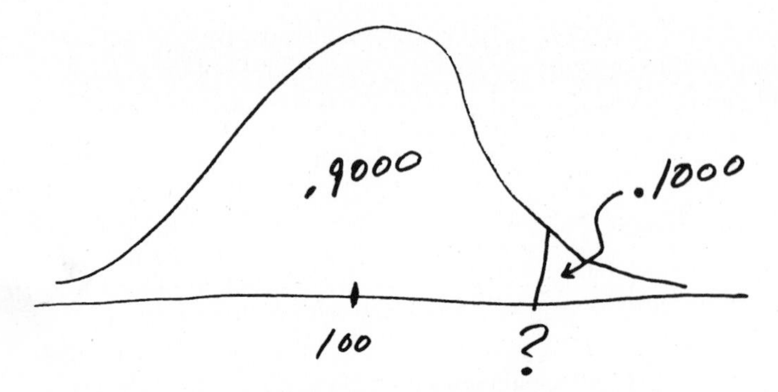

FIGURE 7.12 Sketch of a theoretical distribution of IQ scores divided into an upper 10 percent and a lower 90 percent

10 percent" portion with a vertical line. Because I need to find a score, I put a question mark on the score axis.

With a picture in place, you can finish the problem. The next step is to look in Table C under the column "area beyond z" for .1000. It is not there. You have a choice between .0985 and .1003. Because .1003 is closer to the desired .1000, use it.[9] The z score that corresponds to a proportion of .1003 is 1.28. Now you have all the information you need to solve for X.

To begin, solve the basic z-score formula for X:

$$z = \frac{X - \mu}{\sigma}$$

Multiplying both sides by σ produces

$$(z)(\sigma) = X - \mu$$

Adding μ to both sides isolates X. Thus, when you need to find a score (X) associated with a particular proportion of the normal curve, the formula is

$$X = \mu + (z)(\sigma)$$

Returning to the 10 percent problem and substituting numbers for the mean, the z score, and the standard deviation, you get

$$\begin{aligned} X &= 100 + (1.28)(15) \\ &= 100 + 19.20 \\ &= 119.2 \\ &= 119 \quad \text{(IQs are usually expressed as whole numbers.)} \end{aligned}$$

Therefore, the minimum IQ score required to be in the top 10 percent of the population is 119.

---

[9] You might use interpolation (a method to determine an intermediate score) to find a more accurate z score for a proportion of .1000. This extra precision (and labor) is unnecessary because the final result is rounded to the nearest whole number. For IQ scores, the extra precision does not make any difference in the final answer.

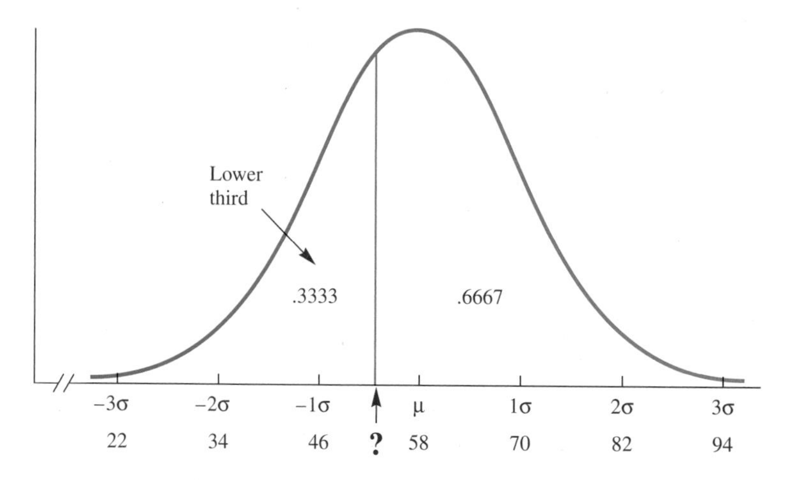

FIGURE 7.13 Distribution of scores on a math achievement exam

Here is a similar problem. Suppose a mathematics department wants to restrict the remedial math course to those who really need it. The department has the scores on the math achievement exam taken by entering freshmen for the past 10 years. The scores on this exam are distributed in an approximately normal fashion, with $\mu = 58$ and $\sigma = 12$. The department wants to make the remedial course available to those students whose mathematical achievement places them in the bottom third of the freshman class. The question is: What score will divide the lower third from the upper two-thirds? Sketch your picture of the problem and check it against **Figure 7.13.**

With a picture in place, the next step is to look in column C of Table C to find .3333. Again, such a proportion is not listed. The nearest proportion is .3336, which has a z value of $-.43$. (This time you are dealing with a z score below the mean, where all z scores are negative.) Applying $z = -.43$, you get

$$
\begin{aligned}
X &= \mu + (z)(\sigma) \\
&= 58 + (-.43)(12) \\
&= 58 - 5.16 \\
&= 52.84 = 53 \text{ points}
\end{aligned}
$$

Using the theoretical normal curve to establish a cutoff score is efficient. All you need are the mean, the standard deviation, and confidence in your assumption that the scores are distributed normally. The empirical alternative for the mathematics department is to sort physically through all scores for the past 10 years, arrange them in a frequency distribution, and calculate the score that separates the bottom one-third.

PROBLEMS

7.15. Mensa is an organization of people who have high IQs. To be eligible for membership, a person must have an IQ "higher than 98 percent of the population." What IQ is required to qualify?

7.16. The mean height of American women aged 20–29 is 64.2 inches, with a standard deviation of 2.5 inches (*Statistical Abstract of the United States: 2009,* 2008).

a. What height divides the tallest 5 percent of the population from the rest?

b. The minimum height required for women to join the U.S. Army is 58 inches. What proportion of the population is excluded?

***7.17.** The mean height of American men aged 20–29 is 70.0 inches, with a standard deviation of 3.0 inches (*Statistical Abstract of the United States: 2009,* 2008).

a. The minimum height required for men to join the U.S. Army is 60 inches. What proportion of the population is excluded?

b. What proportion of the population is taller than Napoleon Bonaparte, who was 5′2″?

7.18. The weight of many manufactured items is approximately normally distributed. For new U.S. pennies, the mean is 3.11 grams and the standard deviation is 0.05 gram (Youden, 1962).

a. What proportion of all new pennies would you expect to weigh more than 3.20 grams?

b. What weights separate the middle 80 percent of the pennies from the lightest 10 percent and the heaviest 10 percent?

Proportion of a Population between Two Scores

Table C in Appendix C can also be used to determine the proportion of the population between two scores. For example, IQ scores that fall in the range from 90 to 110 are often labeled "average." What proportion of the population falls in this range? **Figure 7.14** is a picture of the problem.

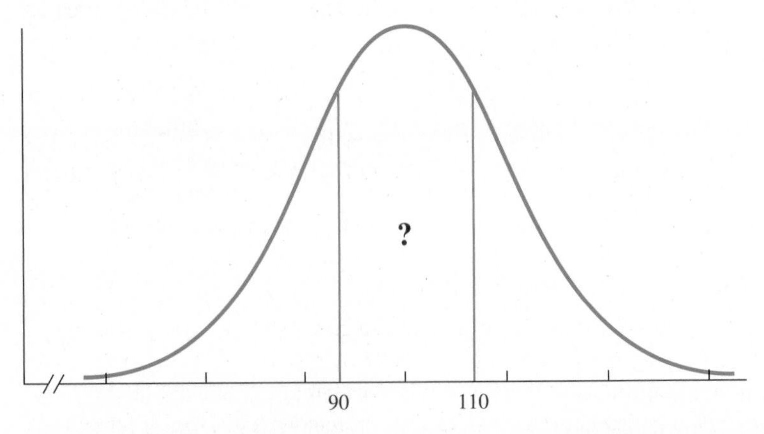

FIGURE 7.14 The normal distribution showing the IQ scores that define the "average" range

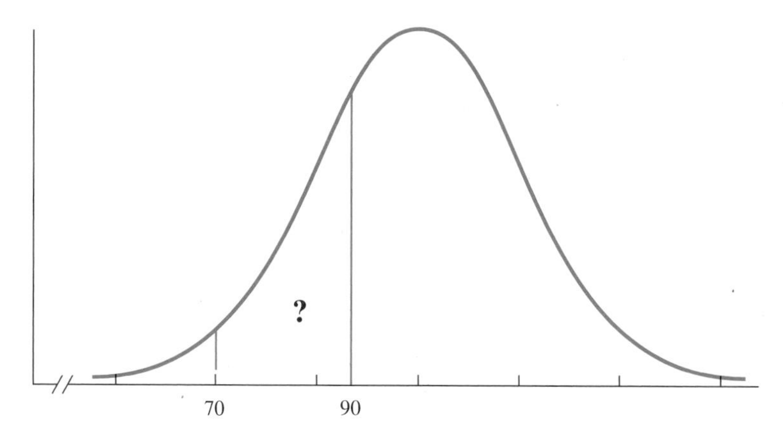

FIGURE 7.15 The normal distribution illustrating the area bounded by IQ scores of 90 and 70

In this problem you must add an area on the left of the mean to an area on the right of the mean. First you need z scores that correspond to the IQ scores of 90 and 110:

$$z = \frac{90 - 100}{15} = \frac{-10}{15} = -.67$$

$$z = \frac{110 - 100}{15} = \frac{10}{15} = .67$$

The proportion of the distribution between the mean and $z = .67$ is .2486, and, of course, the same proportion is between the mean and $z = -.67$. Therefore, (2)(.2486) = .4972, or 49.72 percent. So approximately 50 percent of the population is classified as "average," using the "IQ = 90 to 110" definition.

What proportion of the population would be expected to have IQs between 70 and 90? **Figure 7.15** illustrates this question. There are two approaches to this problem. One is to find the area from 100 to 70 and then subtract the area from 90 to 100. The other way is to find the area beyond 90 and subtract from it the area beyond 70. I'll illustrate with the second approach. The corresponding z scores are

$$z = \frac{90 - 100}{15} = -.67 \quad \text{and} \quad z = \frac{70 - 100}{15} = -2.00$$

The area beyond $z = -.67$ is .2514, and the area beyond $z = -2.00$ is .0228. Subtracting the second proportion from the first, you find that .2286 of the population has an IQ in the range of 70 to 90.

PROBLEMS

***7.19.** The distribution of 800 test scores in an introduction to psychology course was approximately normal, with $\mu = 35$ and $\sigma = 6$.

a. What proportion of the students had scores between 30 and 40?

b. What is the probability that a randomly selected student would score between 30 and 40?

7.20. Now that you know the proportion of students with scores between 30 and 40, would you expect to find the same proportion between scores of 20 and 30? If so, why? If not, why not?

7.21. Calculate the proportion of scores between 20 and 30. Be careful with this one; drawing a picture is especially advised.

7.22. How many of the 800 students would be expected to have scores between 20 and 30?

Extreme Scores in a Population

The extreme scores in a distribution are important in many statistical applications. Most often extreme scores in *either* direction are of interest. For example, many applications focus on the extreme 5 percent of the distribution. Thus, the upper $2\frac{1}{2}$ percent and the lower $2\frac{1}{2}$ percent receive attention. Turn to Table C in Appendix C and find the z score that separates the extreme $2\frac{1}{2}$ percent of the curve from the rest. (Of course, the z score associated with the lowest $2\frac{1}{2}$ percent of the curve will have a negative value.) Please memorize the z score you just looked up. This number will turn up many times in future chapters.

Here is an illustration. What two heart rates (beats per minute) separate the middle 95 percent of the population from the extreme 5 percent? **Figure 7.16** is my sketch of the problem. According to studies summarized by Milnor (1990), the mean heart rate for humans is 71 beats per minute (bpm) and the standard deviation is 9 bpm. To find the two scores, use the formula $X = \mu + (z)(\sigma)$. Using the values given, plus the z score you memorized, you get the following:

Upper score	*Lower score*
$X = \mu + (z)(\sigma)$	$X = \mu - (z)(\sigma)$
$= 71 + (1.96)(9)$	$= 71 - (1.96)(9)$
$= 71 + 17.6$	$= 71 - 17.6$
$= 88.6$, or 89 bpm	$= 53.4$, or 53 bpm

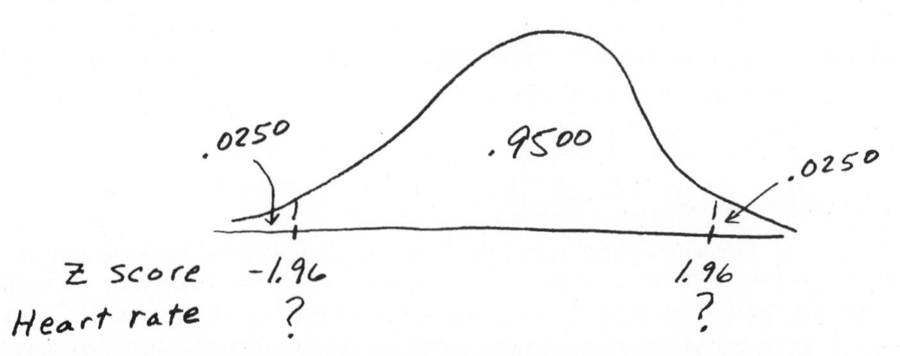

FIGURE 7.16 Sketch showing the separation of the extreme 5 percent of the population from the rest

Thus, 95 percent of the population is expected to have pulse rates between 53 and 89, which leaves 5 percent of the population above 89 or below 53.

clue to the future

The idea of finding scores and proportions that are extreme in either direction will come up again after Chapter 8. In particular, the extreme 5 percent and the extreme 1 percent are important.

PROBLEMS

7.23. What two IQ scores separate the extreme 1 percent of the population from the middle 99 percent? Set this problem up using the "extreme 5 percent" example as a model.

7.24. What is the probability that a randomly selected person has an IQ higher than 139 *or* lower than 61?

7.25. Look at Figure 7.9 and suppose that the union leadership decided to ask for \$0.85 per hour as a minimum wage. For those 185,822 workers, the mean was \$0.99 with a standard deviation of \$0.17. If \$0.85 per hour was established as a minimum, how many workers would get raises?

7.26. Look at Figure 7.8 and suppose that a timber company decided to harvest all trees 8 inches DBH (diameter breast height) or larger from a 100-acre tract. On a 1-acre tract there were 199 trees with $\mu = 13.68$ and $\sigma = 4.83$. How many trees would be expected to be harvested from 100 acres?

Comparison of Theoretical and Empirical Answers

You have been using the theoretical normal distribution to find probabilities and to calculate scores and proportions of IQs, wages, heart rates, and other measures. Earlier in this chapter, I claimed that *if* the empirical observations are distributed like a normal curve, accurate predictions can be made. A reasonable question is: How accurate are all these predictions I've just made? A reasonable answer can be fashioned from a comparison of the predicted proportions (from the theoretical curve) and the actual proportions (computed from empirical data). Figure 7.7 is based on 261 IQ scores of fifth-grade public school students. You worked through examples that produced proportions of people with IQs higher than 120, lower than 90, and between 90 and 110. These actual proportions can be compared with those predicted from the normal distribution. **Table 7.2** shows these comparisons.

As you can see by examining the Difference column of Table 7.2, the accuracy of the predictions ranges from excellent to not so good. Some of this variation can be explained by the fact that the mean IQ of the fifth-grade students was 101 and the standard deviation 13.4. Both the higher mean (101, compared with 100 for the normal curve) and the lower standard deviation (13.4, compared with 15) are due to the systematic exclusion of children with very low IQ scores from regular public schools. Thus, the actual proportion

TABLE 7.2 Comparison of predicted and actual proportions

IQs	Predicted from normal curve	Calculated from actual data	Difference
Higher than 120	.0918	.0920	.0002
Lower than 90	.2514	.2069	.0445
Between 90 and 110	.4972	.5249	.0277

of students with IQs lower than 90 is less than predicted, which is because our school sample is not representative of all 10- to 11-year-old children.

Although IQ scores *are* distributed approximately normally, many other scores are not. Karl Pearson recognized this, as have others. Theodore Micceri (1989) made this point again in an article titled "The Unicorn, the Normal Curve, and Other Improbable Creatures." Caution is always in order when you are using theoretical distributions to make predictions about empirical events. However, don't let undue caution prevent you from getting the additional understanding that statistics offers.

Other Theoretical Distributions

In this chapter you learned a little about rectangular distributions and binomial distributions and quite a bit about normal distributions. Later in this book you will encounter other distributions, such as the *t* distribution, the *F* distribution, and the chi square distribution. (After all, this is a book about tales of distributions.) In addition to the distributions in this book, mathematical statisticians have identified others, all of which are useful in particular circumstances. Some have interesting names such as the Poisson distribution; others have complicated names such as the hypergeometric distribution. In every case, however, a distribution is used because it provides reasonably accurate probabilities about particular events.

PROBLEMS

7.27. For human infants born weighing 5.5 pounds or more, the mean gestation period is 268 days, which is just less than 9 months. The standard deviation is 14 days (McKeown and Gibson, 1951). What proportion of the gestation periods are expected to last 10 months or longer (300 days)?

7.28. The height of residential door openings in the United States is 6′8″. Use the information in problem 7.17 to determine the number of men among 10,000 who have to duck to enter a room.

7.29. An imaginative anthropologist measured the stature of 100 hobbits (using the proper English measure of inches) and found these values:

$\Sigma X = 3600 \qquad \Sigma X^2 = 130{,}000$

Assume that the heights of hobbits are normally distributed. Find μ and σ and answer the following questions.

a. The Bilbo Baggins Award for Adventure is 32 inches tall. What proportion of the hobbit population is taller than the award?
b. Three hundred hobbits entered a cave that had an entrance 39 inches high. The orcs chased them out. How many hobbits could exit without ducking?
c. Gandalf is 46 inches tall. What is the probability that he is a hobbit?

7.30. Please review the objectives at the beginning of the chapter. Can you do what is asked?

ADDITIONAL HELP FOR CHAPTER 7

Visit *cengage.com/psychology/spatz*. At the Student Companion Site, you'll find multiple-choice tutorial quizzes, flashcards with definitions and workshops. For this chapter there is a Statistical Workshop on z Scores.

KEY TERMS

Asymptotic (p. 131)
Binomial distribution (p. 127)
Empirical distribution (p. 124)
Extreme scores (p. 142)
Inflection point (p. 131)
Normal distribution (p. 130)
Probability (p. 125)
Rectangular distribution (p. 126)
Theoretical distribution (p. 125)
z score (p. 131)

CHAPTER 8

Samples, Sampling Distributions, and Confidence Intervals

OBJECTIVES FOR CHAPTER 8

After studying the text and working the problems, you should be able to:

1. Define *random sample* and obtain one if you are given a population of data
2. Define and identify biased sampling methods
3. Distinguish between random samples and the more common research samples
4. Define *sampling distribution* and *sampling distribution of the mean*
5. Discuss the Central Limit Theorem
6. Calculate a standard error of the mean either from σ and N or from sample data
7. Describe the effect of N on the standard error of the mean
8. Use the z-score formula to find the probability that a sample mean or a sample mean more extreme was drawn from a population with a specified mean
9. Describe the t distribution
10. Explain when to use the t distribution rather than the normal distribution
11. Calculate, interpret, and graphically display confidence intervals about sample means

PLEASE BE ESPECIALLY attentive to SAMPLING DISTRIBUTIONS, *a concept that is at the heart of inferential statistics. The subtitle of this book, "Tales of Distributions," is a reference to sampling distributions; every chapter after this one is about some kind of sampling distribution. Thus, you might consider this paragraph to be a* SUPERCLUE *to the future.*

Here is a progression of ideas that lead to what a sampling distribution is and how it is used. In Chapter 2 you studied frequency distributions of scores. If those scores are distributed normally, you can use z scores to determine the probability of the occurrence of any particular score (Chapter 7). Now imagine a frequency distribution, not of scores but of statistics, each calculated from separate samples. This distribution has a form and if it is normal, you could use z scores to determine the probability of the occurrence of any particular value of the statistic. A distribution of sample statistics is called a sampling distribution.

As I've mentioned several times, statistical techniques can be categorized as descriptive and inferential. This is the first chapter that is entirely about inferential statistics, which, as you probably recall, are methods that take *chance factors* into account when *samples* are used to reach conclusions about populations. To take chance factors into account, you must understand sampling distributions and what they tell you. As for samples, I'll describe methods of obtaining them and some of the pitfalls. Thus, this chapter covers topics that are central to inferential statistics.

Here is a story with an inferential statistics problem embedded in it. The problem can be solved using techniques presented in this chapter.

> Late one afternoon, two students were discussing the average family income of students on their campus.
>
> "Well, the average family income for college students is $77,000, nationwide," said the junior, looking up from a book (Pryor et al., 2008). "I'm sure the mean for this campus is at least that much."
>
> "I don't think so," the sophomore replied. "I know lots of students who have only their own resources or come from pretty poor families. I'll bet you a dollar the mean for students here at State U. is below the national average."
>
> "You're on," grinned the junior.
>
> Together the two went out with a pencil and pad and asked ten students how much their family income was. The mean of these ten answers was $69,000.
>
> Now the sophomore grinned. "I told you so; the mean here is $8000 less than the national average."
>
> Disappointed, the junior immediately began to review their procedures. Rather quickly, a light went on. "Actually, this mean of $69,000 is meaningless—here's why. Those ten students aren't representative of the whole student body. They are late-afternoon students, and several of them support themselves with temporary jobs while they go to school. Most students are supported from home by parents who have permanent and better-paying jobs. Our sample was no good. We need results from the whole campus or at least from a representative sample."
>
> To get the results from the whole student body, the two went the next day to the director of the financial aid office, who told them that family incomes for the student body are not public information. Sampling, therefore, was necessary.
>
> After discussing their problem, the two students sought the advice of a friendly statistics professor. The professor explained how to obtain a random sample of students and suggested that 40 replies would be a practical sample size. Forty students, selected randomly, were identified. After three days of phone calls, visits, and callbacks, the 40 responses produced a mean of $73,900, or $3100 less than the national average.
>
> "Pay," demanded the sophomore.
>
> "OK, OK, . . . here!"
>
> Later, the junior began to think. What about another random sample and its mean? It would be different and it *could* be higher. Maybe the mean of $73,900 was just bad luck, just a chance event.
>
> Shortly afterward, the junior confronted the sophomore with thoughts about repeated sampling. "How do we know that the random sample we got told us the truth about the

population? Like, maybe the mean for the entire student body *is* \$77,000. Maybe a sample with a mean of \$73,900 would occur just by chance fairly often. I wonder what the chances are of getting a sample mean of \$73,900 from a population with a mean of \$77,000?"

"Well, that statistics professor told us that if we had any more questions to come back and get an explanation of sampling distributions. In the meantime, why don't I just apply my winnings toward a couple of ice cream cones."

The two statistical points of the story are that random samples are good (the junior paid off only after the results were based on a random sample) and that uncertainty about random samples can be reduced if you know about sampling distributions.

This chapter is about getting a sample, drawing a conclusion about the population the sample came from, and knowing how much faith to put in your conclusion. Of course, there is some peril in this. Even the best methods of sampling produce variable results. How can you be sure that the sample you use will lead to a correct decision about its population? Unfortunately, you *cannot* be absolutely sure.

To use a sample is to agree to accept some uncertainty about the results.

Fortunately, a sampling distribution allows you to measure the uncertainty. On the one hand, if a great deal of uncertainty exists, the sensible thing to do is to say you are uncertain. On the other hand, if there is very little uncertainty, the sensible thing to do is to reach a conclusion about the population, even though there *is* a small risk of being wrong. *Reread this paragraph; it is important.*

In this chapter, the first three sections are about samples. Following that, sampling distributions are explained and one method of drawing a conclusion about a population mean is presented. In the final sections, I will explain the *t* distribution, which is a sampling distribution. The *t* distribution is used to calculate a confidence interval, a statistic that provides information about a population mean.

Random Samples

random sample
Subset of a population chosen so that all samples of the specified size have an equal probability of being selected.

When it comes to finding out about a population, the best sample is a **random sample.** In statistics, *random* refers to the method used to obtain the sample. Any method that allows every possible sample of size N an equal chance to be selected produces a random sample. Random does not mean haphazard or unplanned. To obtain a random sample, you must do the following:

1. Define the population. That is, explain what numbers (scores) are in the population.
2. Identify every member of the population.[1]
3. Select numbers (scores) in such a way that every sample has an equal probability of being selected.

To illustrate, I will use the population of scores in **Table 8.1,** for which $\mu = 9$ and $\sigma = 2$. Using these 20 scores allows us to satisfy requirements 1 and 2. As for requirement 3, I'll describe two methods of selecting numbers so that all the possible samples have an equal probability of being selected. For this illustration, $N = 8$.

[1] Technically, the population consists of the measurements of the members and not the members themselves.

TABLE 8.1 20 scores used as a population; $\mu = 9, \sigma = 2$

9	7	11	13	10	8	10	6	8	8
10	12	9	10	5	8	9	6	10	11

One method of getting a random sample is to write each number in the population on a slip of paper, put the 20 slips in a box, jumble them around, and draw out 8 slips. The numbers on the slips are a random sample. This method works fine if the slips are all the same size, they are jumbled thoroughly, and the population has only a few members. If the population is large, this method is tedious.

A second (usually easier) method of selecting a random sample is to use a table of random numbers, such as Table B in Appendix C. To use the table, you must first assign an identifying number to each of the scores in the population. My version is **Table 8.2.** The population scores do not have to be arranged in any order. Each score in the population is identified by a two-digit number from 01 to 20.

Now turn to **Appendix C, Table B.** Pick an intersection of a row and a column. Any haphazard method will work; close your eyes and put your finger on a spot. Suppose you found yourself at row 80, columns 15–19 (page 388). Find that place. Reading horizontally, the digits are 82279. You need only two digits to identify any member of your population, so you might as well use the first two (columns 15 and 16), which give you 8 and 2 (82). Unfortunately, 82 is larger than any of the identifying numbers, so it doesn't match a score in the population, but at least you are started. From this point you can read two-digit numbers in any direction—up, down, or sideways—but the decision should be made before you look at the numbers. If you decide to read down, you find 04. The identifying number 04 corresponds to a score of 13 in Table 8.2, so 13 becomes the first number in the sample. The next identifying number is 34, which again does not correspond to a population score. Indeed, the next ten numbers are too large. The next usable ID number is 16, which places an 8 in the sample. Continuing the search, you reach the bottom of the table. At this point you can go in any direction; I moved to the right and started back up the two outside columns (18 and 19). The first number, 83, was too large, but the next identifying number, 06, corresponded to an 8, which went into the sample.

TABLE 8.2 Assignment of identifying numbers to a population of scores

ID number	Score	ID number	Score
01	9	11	10
02	7	12	12
03	11	13	9
04	13	14	10
05	10	15	5
06	8	16	8
07	10	17	9
08	6	18	6
09	8	19	10
10	8	20	11

Next, a 15 and a 20 identified population scores of 5 and 11 for the sample. The next number that is between 01 and 20 is 15, but it has already been used, so it should be ignored. The identifying numbers 18, 02, and 14 match scores in the population of 6, 7, and 10. Thus, the random sample of eight consists of these scores: 13, 8, 8, 5, 11, 6, 7, and 10.

PROBLEM

***8.1.** A random sample is supposed to yield a statistic similar to the population parameter. Find the mean of the random sample of eight numbers selected in the text.

What is this table of random numbers? In Table B (and in any set of random numbers), the probability of occurrence of any digit from 0 to 9 at any place in the table is the same: 0.10. Thus, you are just as likely to find 000 as 123 or 397. Incidentally, you cannot generate random numbers out of your head. Certain sequences begin to recur, and (unless warned) you will not include enough repetitions like 000 and 555. If warned, you produce too many.

Here are some suggestions for using a table of random numbers efficiently.

1. In the list of population scores and their ID numbers, check off the ID number when it is chosen for the sample. This helps to prevent duplications.
2. If the population is large (more than 50), it is more efficient to get all the identifying numbers from the table first. As you select them, put them in some rough order to help prevent duplications. After you have all the identifying numbers, go to the population to select the sample.
3. If the population has exactly 100 members, let 00 be the identifying number for 100. By doing this, you can use two-digit identifying numbers, each one of which matches a population score. This same technique works for populations of 10 or 1000 members.

Random sampling has two uses. (1) It is the best method of sampling if you want to generalize to a population. The two sections that follow, biased samples and research samples, both address the issue of generalizing from the sample to the population. (2) As an entirely separate use, random sampling is the mathematical basis for creating sampling distributions, which are central to inferential statistics.

PROBLEMS

8.2. Draw a random sample of 10 from the population in Table 8.1. Calculate $\overline{X}$.

8.3. Draw a random sample with $N = 12$ from the following scores:

76	47	81	70	67	80	64	57	76	81
68	76	79	50	89	42	67	77	80	71
91	72	64	59	76	83	72	63	69	
78	90	46	61	74	74	74	69	83	

Biased Samples

A **biased sample** is one obtained by a method that systematically underselects or overselects from certain groups in the population. Thus, with a biased sampling technique, every sample of a given size does *not* have an equal opportunity of being selected. With biased sampling techniques, you are much more likely to get an unrepresentative sample than you are with random sampling.

biased sample
Sample selected in such a way that not all samples from the population have an equal chance of being chosen.

At times, conclusions based on mailed questionnaires are suspect because the sampling methods were biased. Usually an investigator defines the population, identifies each member, and mails the questionnaire to a randomly selected sample. Suppose that 60 percent of the recipients respond. Can valid results for the population be based on the questionnaires returned? Probably not. There is often good reason to suspect that the 60 percent who responded are different from the 40 percent who did not. Thus, although the population is made up of both kinds of people, the sample reflects only one kind. Therefore, the sample is biased. The probability of bias is particularly high if the questionnaire elicits feelings of pride or despair or disgust or apathy in *some* of the recipients.

A famous case of a biased sample occurred in a poll that was to predict the results of the 1936 election for president of the United States. The *Literary Digest* (a popular magazine) mailed 10 million "ballots" to those on its master mailing list, a list of more than 10 million people compiled from "all telephone books in the U.S., rosters of clubs, lists of registered voters," and other sources. More than 2 million "ballots" were returned and the prediction was clear: Alf Landon by a landslide over Franklin Roosevelt. As you may have learned, the actual results were just the opposite; Roosevelt got 61 percent of the vote.

From the 10 million who had a chance to express a preference, 2 million very interested persons had selected themselves. This 2 million had more than its proportional share of those who were disgruntled with Roosevelt's depression-era programs. The 2 million ballots were a biased sample; the results were not representative of the population. In fairness, it should be noted that the *Literary Digest* used a similar master list in 1932 and predicted the popular vote within 1 percentage point.[2]

Obviously, researchers want to avoid biased samples; random sampling seems like the appropriate solution. The problem, however, is step 2 on page 148. For almost all research problems, it is impossible to identify every member of the population.

What do practical researchers do when they cannot obtain a random sample?

Research Samples

Fortunately, the problem I have posed is not a serious one. For researchers whose immediate goal is to generalize their results directly to a population, techniques besides random sampling produce results that mirror the population in question. Such nonrandom (though carefully controlled) samples are used by a wide variety of people and organizations. The public opinion polls reported by Gallup, Lou Harris, and the

[2] See the *Literary Digest*, August 22, 1936, and November 14, 1936.

Roper organization are based on carefully selected nonrandom samples. ASCAP, the musicians' union, samples the broadcasts of the 12,000 U.S. radio stations and then distributes royalties to members based on sample results. Inventory procedures in large retail organizations rely on sampling. What you know about your blood is based on an analysis of a small sample. (Thank goodness!) And what you think about your friends and family is based on just a sample of their behavior.

Also, the immediate goal of most researchers is not to generalize to a larger population but to determine whether two (or more) treatments produce different outcomes. Their experiments seldom, if ever, involve random samples. Instead, they use convenient, practical samples. If the researchers find a difference, they conduct the experiment again, perhaps varying the independent or dependent variable. If several similar experiments (often called replications) produce a coherent pattern of differences, the researchers and their colleagues conclude that the differences are general, even though no random samples were used at any time. Most of the time, they are correct. For these researchers, the representativeness of their samples is not of immediate concern.

PROBLEMS

8.4. Suppose a large number of questionnaires about educational accomplishments are mailed out. Do you think that some recipients will be more likely to return the questionnaire than others? Which ones? If the sample is biased, will it overestimate or underestimate the educational accomplishments of the population?

8.5. Sometimes newspapers sample opinions of the local population by printing a "ballot" and asking readers to mark it and mail it in. Evaluate this sampling technique.

8.6. Consider as a population the students at State U. whose names are listed alphabetically in the student directory. Are the following samples biased or random?

- **a.** Every fifth name on the list
- **b.** Every member of the junior class
- **c.** 150 names drawn from a box that contains all the names in the directory
- **d.** One name chosen at random from the directory
- **e.** A random sample from those taking the required English course

8.7. Explain why researchers do not use random samples. How are they able to generalize their results?

Sampling Distributions

sampling distribution
Theoretical distribution of a statistic based on all possible random samples drawn from the same population.

A **sampling distribution** is always the sampling distribution of a particular statistic. Thus, there is a sampling distribution of the mean, a sampling distribution of the variance, a sampling distribution of the range, and so forth. Here is a description of an *empirical* sampling distribution of a statistic.

Think about many random samples (each with the same N) all drawn from the same population. The same statistic (for example, the mean) is calculated for each sample. All of these statistics are arranged into a frequency distribution and graphed as a frequency polygon. The mean and the standard deviation of the frequency distribution are calculated. Now, imagine that the frequency polygon is a normal curve. If the distribution is normal and you have its mean and standard deviation, you can calculate *z* scores and find probabilities associated with particular values of the statistic.

The sampling distributions that statisticians and textbooks use are *theoretical* distributions derived from formulas rather than empirical ones. These theoretical distributions are similar, though, because they give you the probability associated with particular values of a statistic.

Sampling distributions are so important that statisticians have special names for their mean and standard deviation. The mean of a sampling distribution is called the **expected value** and the standard deviation is called the **standard error.**[3] I will not have much more to say about expected value, but if you continue your study of statistics, you will encounter it. The standard error, however, will be used many times in this text.

expected value
The mean value of a random variable over an infinite number of samplings.

standard error
Standard deviation of a sampling distribution.

To conclude, a sampling distribution is the theoretical distribution of a statistic based on random samples of size *N*. Sampling distributions are used by all who use inferential statistics, regardless of the nature of their research samples. If you want to see what different sampling distributions look like, peek ahead to Figure 8.3, Figure 8.5, Figure 11.4, and Figure 14.1.

clue to the future

In inferential statistics, decisions are made after comparing actual research outcomes to outcomes predicted by a sampling distribution.

The Sampling Distribution of the Mean

Let's turn to a particular sampling distribution, the sampling distribution of the mean. Remember the population of 20 scores that you sampled from earlier in this chapter? For problem 8.2, you drew a sample of 10 and calculated the mean. For almost every sample in your class, the mean was different. Each, however, was an *estimate* of the population mean.

My example problem in the section on random sampling used $N = 8$; my one sample produced a mean of 8.50. Now, think about drawing 200 samples with $N = 8$ from the population of Table 8.1 scores, calculating the 200 means, and constructing a frequency polygon. You may already be thinking, "That's a job for a computer." Right. The result is **Figure 8.1**.

Look at Figure 8.1. Notice how nicely centered the sampling distribution is about the population mean of 9.00. Most of the sample means ($\bar{X}$'s) are fairly close to the population parameter, μ.

[3] In this and in other statistical contexts, the term *error* means deviations or random variation. The word *error* is left over from the 19th century, when random variation was referred to as the "normal law of error." Of course, *error* sometimes means mistake, so you will have to be alert to the context when this word appears (or you may make an error).

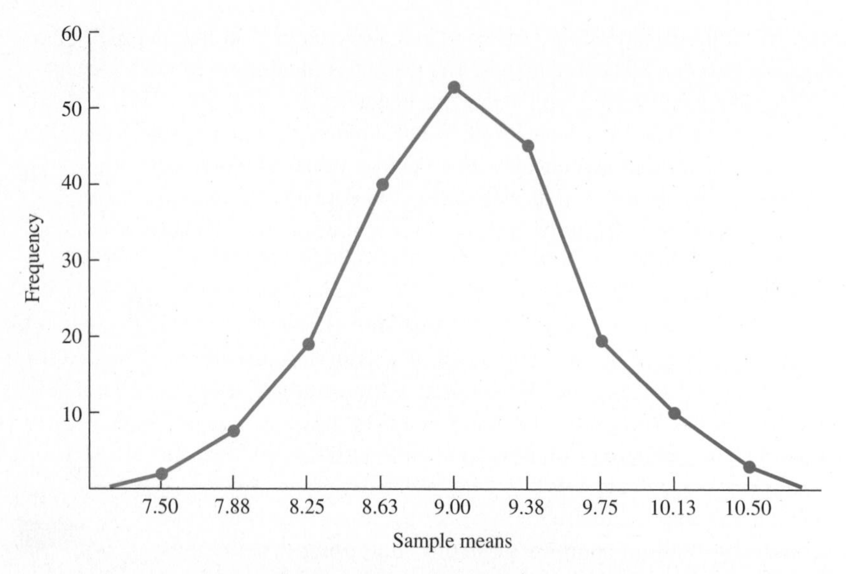

FIGURE 8.1 Empirical sampling distribution of 200 means from the population in Table 8.1. For each sample mean, $N = 8$.

The characteristics of a sampling distribution of the mean are:

1. Every sample is drawn randomly from a specified population.
2. The sample size (N) is the same for all samples.
3. The number of samples is very large.
4. The mean $\overline{X}$ is calculated for each sample.[4]
5. The sample means are arranged into a frequency distribution.

I hope that when you looked at Figure 8.1, you were at least suspicious that it might be the ubiquitous normal curve. It is. Now you are in a position that educated people often find themselves: What you learned in the past, which was how to use the normal curve for scores (X), can be used for a different problem—describing the relationship between $\overline{X}$ and μ.

Of course, the normal curve is a theoretical curve, and I presented you with an empirical curve that only appears normal. I would like to let you prove for yourself that the form of a sampling distribution of the mean is a normal curve, but, unfortunately, that requires mathematical sophistication beyond that assumed for this course. So I will resort to a time-honored teaching technique—an appeal to authority.

Central Limit Theorem

Central Limit Theorem
The sampling distribution of the mean approaches a normal curve as N gets larger.

The authority I appeal to is mathematical statistics, which proved a theorem called the **Central Limit Theorem:**

For any population of scores, regardless of form, the sampling distribution of the mean approaches a normal distribution as N (sample size) gets larger. Furthermore, the sampling distribution of the mean has a mean (the expected value) equal to μ and a standard deviation (the standard error) equal to $\sigma/\sqrt{N}$.

[4] To create a sampling distribution of a statistic other than the mean, substitute that statistic at this step.

This appeal to authority resulted in a lot of information about the sampling distribution of the mean. To put this information into list form:

1. The sampling distribution of the mean approaches a normal curve as N increases.
2. For a population with a mean μ and a standard deviation σ,
   a. The mean of the sampling distribution (expected value) $= \mu$.
   b. The standard deviation of the sampling distribution (standard error) $= \frac{\sigma}{\sqrt{N}}$.

Here are two final points of terminology:

1. The expected value of the mean is symbolized $E(\overline{X})$.
2. The standard error of the mean is symbolized $\sigma_{\overline{X}}$.

The most remarkable thing about the Central Limit Theorem is that it works *regardless* of the form of the original distribution. **Figure 8.2** shows two populations and three sampling distributions from each population. On the left is a rectangular distribution of playing cards (from Figure 7.1) and on the right is the bimodal distribution of number choices (from Figure 7.5). Three sampling distributions of the mean (for $N = 2$, 8, and 30) are below their populations. The take-home message of the Central Limit Theorem is that, regardless of the form of a distribution, the form of the sampling distribution of the mean approaches normal if N is large enough.

How large must N be for the sampling distribution of the mean to be normal? The traditional answer is 30 or more. However, if the population itself is symmetrical, then sampling distributions of the mean will be normal with N's much smaller than 30. In contrast, if the population is severely skewed, N's of more than 30 will be required.

Finally, the Central Limit Theorem does not apply to all sample statistics. Sampling distributions of the median, standard deviation, variance, and correlation coefficient are not normal distributions. The Central Limit Theorem does apply to the mean, which is a most important and popular statistic. (In a frequency count of all statistics, the mean is the mode.)

PROBLEMS

8.8. The standard deviation of a sampling distribution is called the _________ and the mean is called the __________.

8.9. a. Write the steps needed to construct an empirical sampling distribution of the range.

b. What is the name of the standard deviation of this sampling distribution?

c. Think about the expected value of the range compared to the population range. Write a statement about the relationship.

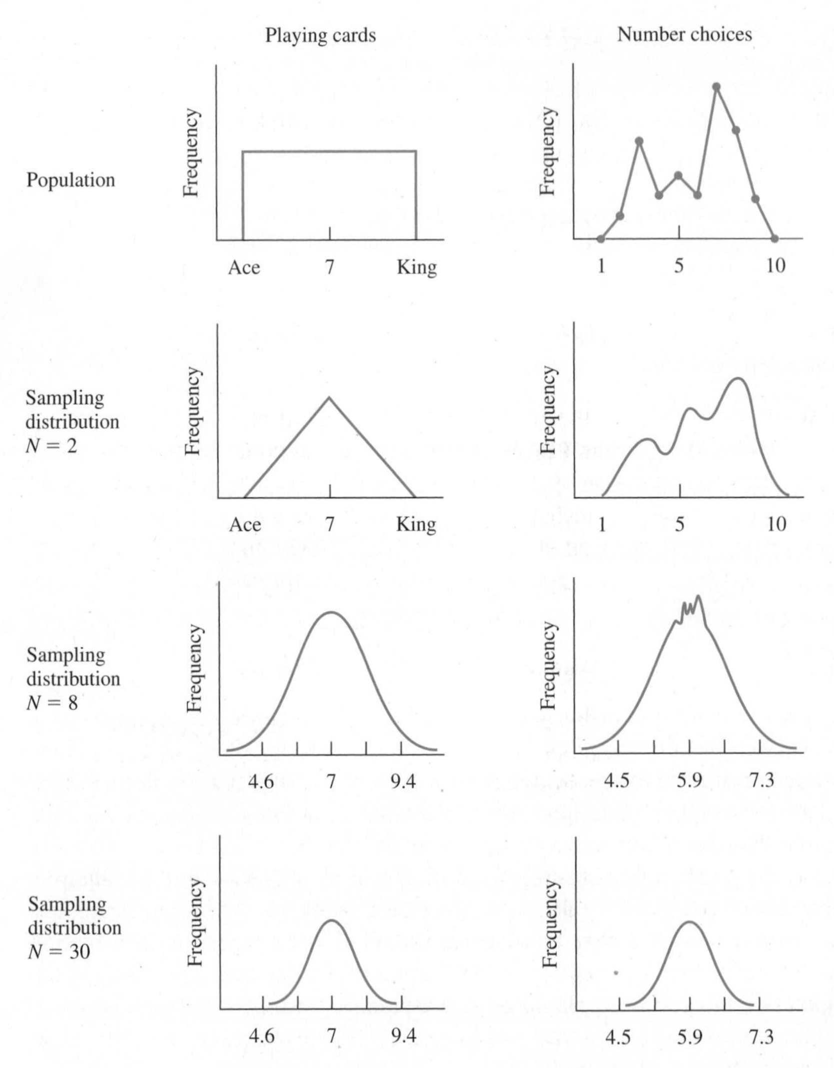

FIGURE 8.2 Populations of playing cards and number choices. Sampling distributions from each population with sample sizes of $N = 2$, $N = 8$, and $N = 30$.

8.10. Describe the Central Limit Theorem in your own words.

8.11. In Chapter 6 you learned how to use a regression equation to predict a score ($\hat{Y}$) if you are given an X score. $\hat{Y}$ is a statistic, so naturally it has a sampling distribution with its own standard error. What can you conclude if the standard error is very small? Very large?

Calculating the Standard Error of the Mean

Calculating the standard error of the mean is fairly simple. I will illustrate with an example that you will use again. For the population of scores in Table 8.1, σ is 2. For a sample size of eight, the standard error of the mean is

$$\sigma_{\overline{X}} = \frac{\sigma}{\sqrt{N}} = \frac{2}{\sqrt{8}} = \frac{2}{2.828} = 0.707$$

The Effect of Sample Size on the Standard Error of the Mean

As you can see by looking at the formula for the standard error of the mean, $\sigma_{\overline{X}}$ becomes smaller as N gets larger. **Figure 8.3** shows four sampling distributions of the mean, all based on the population of numbers in Table 8.1. The sample sizes are 2, 4, 8, and 16. A sample mean of 10 is included in all four figures as a reference point. Notice that as N increases, a sample mean of 10 becomes less and less likely. The importance of sample size will become more apparent as your study progresses.

Determining Probabilities About Sample Means

To summarize where we are at this point: Mathematical statisticians have produced a mathematical invention, the sampling distribution. One particular sampling distribution, the sampling distribution of the mean, is a normal curve, they tell us. Fortunately, having worked problems in Chapter 7 about normally distributed *scores,* we are in a position to check this claim about normally distributed *means.*

One check is fairly straightforward, given that I already have the 200 sample means from the population in Table 8.1. To make this check, I will determine the proportion of sample means that are above a specified point, *using the theoretical normal curve* (Table C in Appendix C). I can then compare this theoretical proportion to the proportion of the 200 sample means that are *actually* above the specified point. If the two numbers are similar, I have evidence that the normal curve can be used to answer questions about sample means.

The z score for a sample mean drawn from a sampling distribution with mean μ and standard error $\sigma/\sqrt{N}$ is

$$z = \frac{\overline{X} - \mu}{\sigma_{\overline{X}}}$$

Any sample mean will do for this comparison; I will use 10.0. The mean of the population is 9.0 and the standard error (for $N = 8$) is 0.707. Thus,

$$z = \frac{\overline{X} - \mu}{\sigma_{\overline{X}}} = \frac{10.0 - 9.0}{0.707} = 1.41$$

By consulting Table C, I see that the proportion of the curve above a z value of 1.41 is .0793. **Figure 8.4** shows the sampling distribution of the mean for $N = 8$. Sample means are on the horizontal axis; the .0793 area above $\overline{X} = 10$ is shaded.

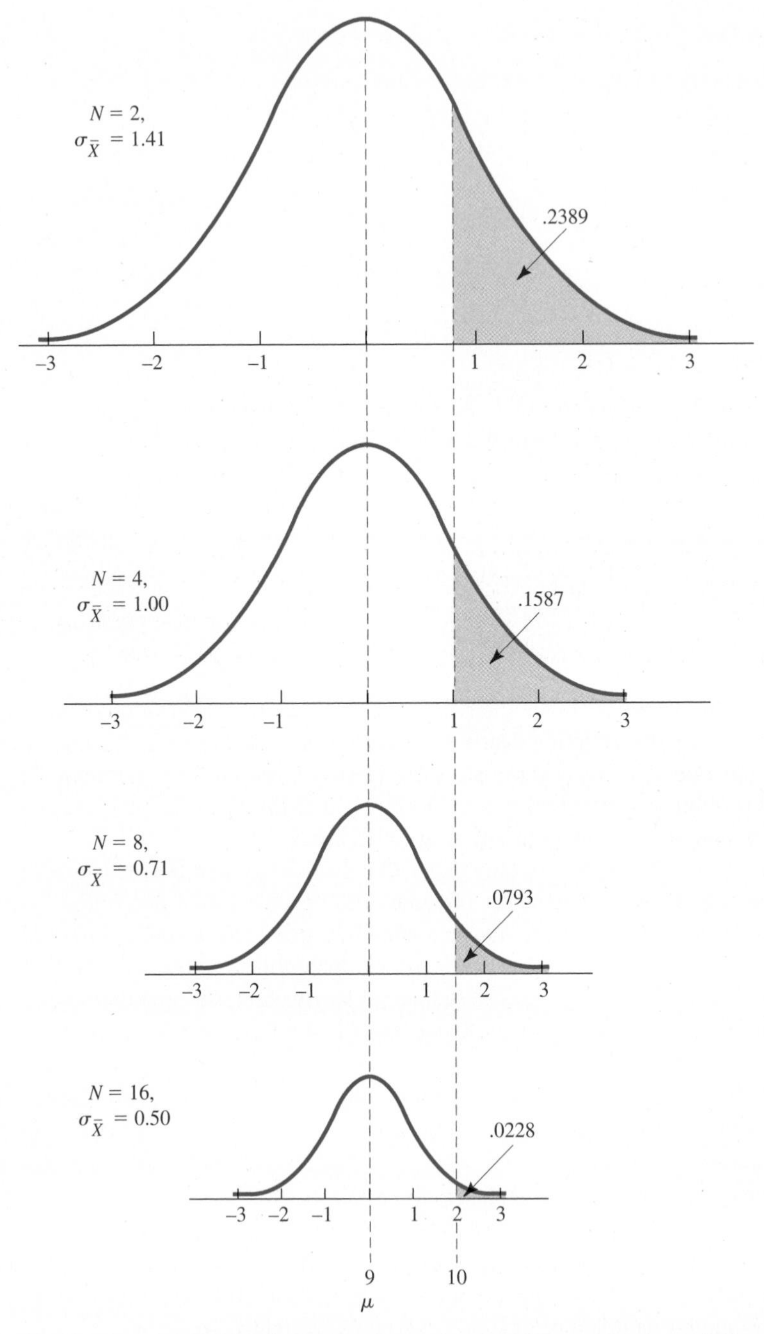

FIGURE 8.3 Sampling distributions of the mean for four different sample sizes. All samples are drawn from the population in Table 8.1. Note how a sample mean of 10 becomes rarer and rarer as $\sigma_{\bar{X}}$ becomes smaller.

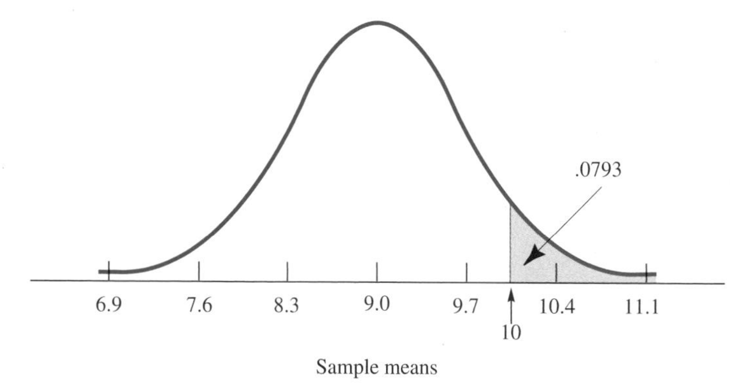

FIGURE 8.4 Theoretical sampling distribution of the mean from the population in Table 8.1. For each sample, $N = 8$.

How accurate is this theoretical prediction of .0793? When I looked at the distribution of sample means that I used to construct Figure 8.1, I found that 13 of the 200 had means of 10 or more, a proportion of .0650. Thus, the theoretical prediction is off by less than $1\frac{1}{2}$ percent. That's not bad; the normal curve model passes the test.[5]

PROBLEMS

***8.12.** When the population parameters are known, the standard error of the mean is $\sigma_{\bar{X}} = \sigma/\sqrt{N}$. The following table gives four σ values and four N values. For each combination, calculate $\sigma_{\bar{X}}$ and enter it in the table.

	σ			
N	1	2	4	8
1				
4				
16				
64				

8.13. On the basis of the table you constructed in problem 8.12, write a precise verbal statement about the relationship between $\sigma_{\bar{X}}$ and N.

8.14. To reduce $\sigma_{\bar{X}}$ to one-fourth its size, you must increase N by how much?

8.15. For the population in Table 8.1, and for samples with $N = 8$, what proportion of the sample means will be 8.5 or less?

8.16. For the population in Table 8.1, and for samples with $N = 16$, what proportion of the sample means will be 8.5 or less? 10 or greater?

[5] As you might suspect, the validity of the normal curve model for the sampling distribution of the mean has been established with a mathematical proof rather than an empirical check.

8.17. As you know from the previous chapter, for IQs, $\mu = 100$ and $\sigma = 15$. What is the probability that a first-grade classroom of 25 students who are chosen randomly from the population will have a mean IQ of 105 or greater? 90 or less?

8.18. Now you are in a position to return to the story of the two students at the beginning of this chapter. Find, for the junior, the probability that a sample of 40 from a population with $\mu = \$77{,}000$ and $\sigma = \$20{,}000$ would produce a mean of \$73,900 or less. Write an interpretation.

Using your answer to problem 8.18, the junior might say to the sophomore,

". . . and so, the mean of \$73,900 doesn't lead to a clear-cut decision about the population. The standard deviation is large, and with $N = 40$, samples can bounce all over the place. Why, if the real mean for our campus is the same as the \$77,000 national average, we would *expect* about one-sixth of all our random samples of 40 students to have means of \$73,900 or less."

"Yeah," said the sophomore, "but that *if* is a big one. What if the campus mean is really \$73,900? Then the sample we got was right on the nose. After all, a sample mean is an *unbiased estimator* of the population parameter."

"I see your point. And I see mine, too. It seems like either of us could be correct. That leaves me uncertain about the real campus parameter."

"Me, too. Let's go get another ice cream cone."

Not all statistical stories end with so much uncertainty (or calories). However, I said that one of the advantages of random sampling is that you can measure the uncertainty. You measured it, and there was a lot. Remember, if you agree to use a sample, you agree to accept some uncertainty about the results.

Constructing a Sampling Distribution When σ Is Not Available

Let's review what you just did. You answered some questions about sample means by relying on a table of the normal curve. Your justification for saying that sample means are distributed normally was the Central Limit Theorem. The Central Limit Theorem always applies when the sample size is adequate and you know σ, both of which were true for the problems you worked. For those problems, you were given σ or calculated it from the population data you had available.

In the world of empirical research, however, you often do not know σ and you don't have population data to calculate it. Because researchers are always inventing new dependent variables, an unknown σ is common. Without σ, the justification for using the normal curve evaporates. What to do if you don't have σ? Can you suggest something?

One solution is to use $\hat{s}$ as an estimate of σ. (Was that your suggestion?) This was the solution used by researchers about a century ago. They knew that $\hat{s}$ was only an estimate and that the larger the sample, the better the estimate. Thus, they chose problems for which they could gather huge samples. (Remember Karl Pearson and Alice Lee's data on father–daughter heights? They had a sample of 1376.) Very large samples produce an $\hat{s}$ that is identical to σ, for all practical purposes.

Other researchers, however, could not gather that much data. One of those we remember today is W. S. Gosset (1876–1937), who worked for Arthur Guinness, Son & Co., a brewery headquartered in Dublin, Ireland. Gosset had majored in chemistry and mathematics at Oxford, and his job at Guinness was to make recommendations to the brewers that were based on scientific experiments. The experiments, of course, used samples.

Gosset was familiar with the normal curve and the strategy of using large samples to accurately estimate σ. Unfortunately, his samples were small. Gosset (and other statisticians) knew that such small-sample $\hat{s}$ values were not accurate estimators of σ and thus the normal curve could not be relied on for accurate probabilities.

Gosset's solution was to work out a new set of distributions that were based on $\hat{s}$. He found that the distribution depended on the sample size, with a different distribution for each N. These distributions make up a family of curves that have come to be called the ***t* distribution.**[6] The t distribution is an important tool for those who analyze data. I will use it for confidence interval problems in this chapter and for four other kinds of problems in later chapters.

***t* distribution**
Theoretical distribution used to determine probabilities when σ is unknown.

The *t* Distribution

The different curves that make up the t distribution are distinguished from one another by their **degrees of freedom.** Degrees of freedom (abbreviated *df*) range from 1 to ∞. Knowing the degrees of freedom for your data tells you which t distribution to use.[7]

degrees of freedom
Concept in mathematical statistics that determines the distribution that is appropriate for sample data.

Determining the correct number of degrees of freedom for a particular problem can become fairly complex. For the problems in this chapter, however, the formula is simple: $df = N - 1$. Thus, if the sample consists of 12 members, $df = 11$. In later chapters I will give you a more thorough explanation of degrees of freedom (and additional formulas).

Figure 8.5 is a picture of three t distributions. Their degrees of freedom are 2, 9, and ∞. You can see that as the degrees of freedom increase, less and less of the curve is in the tails. Note that the t values on the horizontal axis are quite similar to the z scores used with the normal curve.

The *t* Distribution Table

Look at **Table D** (page 392), the t distribution table. The first column shows degrees of freedom, ranging from 1 to ∞. Degrees of freedom are determined by an analysis of the problem that is to be solved. There are three rows across the top of the table; the row you use depends on the kind of problem you are working on. In this chapter, use the top row because the problems are about confidence intervals. Each column is associated with

[6] Gosset spent several months in 1906–07 studying with Karl Pearson in London. It was during this period that the t distribution was developed.

[7] Traditionally, the t distribution is written with a lowercase t. A capital T is used for another distribution (covered in Chapter 15) and for standardized test scores. However, because some early computer programs did not print lowercase letters, t became T on some printouts (and often in text based on that printout). Be alert.

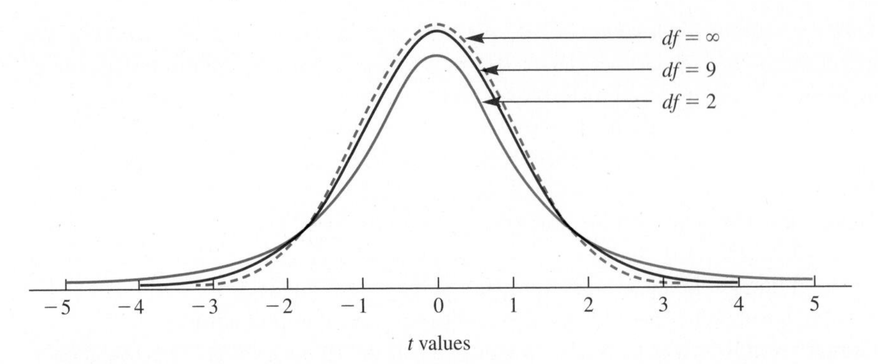

FIGURE 8.5 Three different *t* distributions

probability (which can be converted to a percent). The body of the table contains *t* values. Table D is used most frequently to find the probability associated with a particular *t* value.[8]

Table D differs in several ways from the normal curve table you have been using. In the normal curve table, the *z* scores are in the margin and the probability figures are in the body of the table. In Table D, the *t* values are in the body of the table and the probability figures are in the headings at the top. In the normal curve table, there are hundreds of probability figures, and in Table D there are only six. These six are the ones commonly used by researchers.

Dividing a Distribution into Portions

Look at **Figure 8.6,** which shows separated versions of the three distributions in Figure 8.5. Each distribution is divided into a large middle portion and smaller tail portions. In all three curves in Figure 8.6, the middle portion is 95 percent and the other 5 percent is divided evenly between the two tails.

Notice that the *t* values that separate equal portions of the three curves are different. The fewer the *df*, the larger the *t* value that is required to separate out the extreme 5 percent of the curve.

The *t* values in Figure 8.6 came from Table D. Look at Table D for the 9 *df* row. Now look in the 95% confidence interval column. The *t* value at the intersection is 2.26, the number used in the middle curve of Figure 8.6.

Finally, look at the top curve in Figure 8.6. When $df = \infty$, the *t* scores that separate the middle 95 percent from the extreme 5 percent are -1.96 and $+1.96$. You have seen these numbers before. In fact, ± 1.96 are the *z* scores you used in the preceding chapter to find heart rates that characterize the extreme 5 percent of the population. (See Figure 7.16.) Thus, a *t* distribution with $df = \infty$ is a normal curve. In summary, as *df* increases, the *t* distribution approaches the normal distribution.

Your introduction to the *t* distribution is in place. After working the problems that follow, you will be ready to calculate confidence intervals.

---

[8] For data with a *t* value intermediate between two tabled values, be conservative and use the *t* value associated with the smaller *df*.

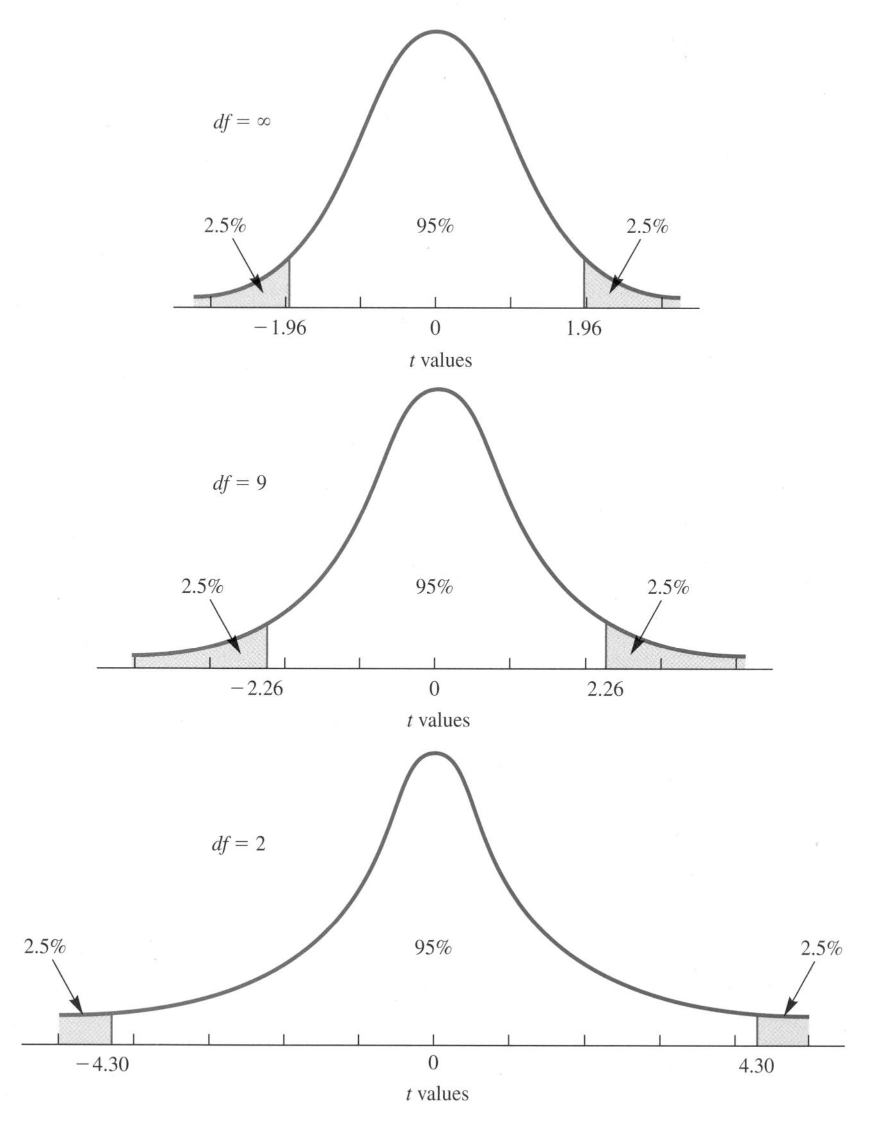

FIGURE 8.6 Three *t* distributions showing the *t* values that enclose 95 percent of the cases

PROBLEMS

8.19. Fill in the blanks in the statements with "*t*" or "normal."

a. There is just one __________ distribution, but there are many __________ distributions.

b. Given a value of +2.00 on the horizontal axis, a __________ distribution has a greater proportion to the right than does a __________ distribution.

c. The __________ distribution has a larger standard deviation than does the __________ distribution.

8.20. To know which t distribution to use, determine __________.

8.21. Who invented the t distribution? For what purpose?

Confidence Interval about a Population Mean

confidence interval
Range of scores that is expected, with specified confidence, to capture a parameter.

Although population means are generally unknowable, samples can be obtained and means and standard errors calculated. A **confidence interval** is an inferential statistic that uses sample statistics to establish two brackets (limits) between which a population parameter is expected (with a particular degree of confidence) to be. Confidence intervals are most often used to capture a population mean, but they can be used for other population parameters as well. Commonly chosen degrees of confidence are 95 percent and 99 percent.

clue to the future

Analyzing data with confidence intervals is increasingly popular among behavioral and medical scientists. Confidence intervals are used again in Chapter 10.

To find the lower and upper limits of a confidence interval about a population mean, use the formulas

$$\text{LL} = \overline{X} - t_{\alpha}(s_{\overline{X}})$$
$$\text{UL} = \overline{X} + t_{\alpha}(s_{\overline{X}})$$

where $\overline{X}$ is the mean of the sample from the population
t_{α} is the value from the t distribution table for a particular α value
$s_{\overline{X}}$ is the standard error of the mean, calculated from a sample

The formula for $s_{\overline{X}}$ is quite similar to the formula for $\sigma_{\overline{X}}$:

$$s_{\overline{X}} = \frac{\hat{s}}{\sqrt{N}}$$

where $\hat{s}$ is the standard deviation of the sample (using $N - 1$).

A confidence interval is useful for evaluating quantitative claims such as those made by manufacturers. For example, some lightbulb manufacturers claim that their 13-watt compact fluorescent bulbs last an average of 10,000 hours. Is this claim true? One way to find the answer is to gather data (although more than a year is required). The following are fictitious data for hours to failure:

$$N = 16$$
$$\Sigma X = 156{,}000$$
$$\Sigma X^2 = 1{,}524{,}600{,}000$$

PROBLEM

***8.22.** For the data on the lives of 13-watt compact fluorescent lightbulbs, calculate the mean ($\overline{X}$), the standard deviation ($\hat{s}$), and the standard error of the mean ($s_{\overline{X}}$).

You can calculate a 95 percent confidence interval about the sample mean in problem 8.22 and use it to evaluate the claim that bulbs last 10,000 hours. With $N = 16$, $df = N - 1 = 15$. The t value in Table D for a 95 percent confidence interval with $df = 15$ is 2.131. Using the answers you found in problem 8.22,

$$\text{LL} = \overline{X} - t_{\alpha}(s_{\overline{X}}) = 9750 - 2.131(122.47) = 9489 \text{ hours}$$
$$\text{UL} = \overline{X} + t_{\alpha}(s_{\overline{X}}) = 9750 + 2.131(122.47) = 10{,}011 \text{ hours}$$

Thus, you are 95 percent confident that the interval of 9489 to 10,011 hours contains the true mean lighting time of the 13-watt bulbs. The advertised claim of 10,000 hours *is* in the interval, but just barely. A reasonable interpretation is to recognize that samples are changeable, variable things and that you do not have good evidence showing that the lives of bulbs are shorter than claimed. But because the interval *almost* does not capture the advertised claim, you might have an incentive to gather more data.

To obtain a 99 percent or a 90 percent confidence interval, use the t values under those headings in Table D.

error detection

The sample mean is always in the exact middle of any confidence interval.

As another example, normal body temperature is commonly given as 98.6°F. In Chapters 2, 3, 4, and 5, you worked with data based on actual measurements (Mackowiak, Wasserman, and Levine, 1992) that showed that average temperature is *less* than 98.6°F. Is it true that we have been wrong all these years? Or, perhaps the study's sample mean is within the expected range of sample means taken from a population with a mean of 98.6°F. A 99 percent confidence interval will help in choosing between these two alternatives. The summary statistics are:

$$N = 40$$
$$\Sigma X = 3928.0$$
$$\Sigma X^2 = 385{,}749.24$$

These summary statistics produce the following values:

$$\overline{X} = 98.2°\text{F}$$
$$\hat{s} = 0.710°\text{F}$$
$$s_{\overline{X}} = 0.112°\text{F}$$
$$df = 39$$

The next task is to find the appropriate t value in Table D. With 39 df, be conservative and use 30 df. Look under the heading for a 99 percent confidence interval. Thus, t_{99} (30 df) = 2.75. The 99 percent confidence limits about the sample mean are:

$$\text{LL} = \overline{X} - t_{\alpha}(s_{\overline{X}}) = 98.2 - 2.75(0.112) = 97.9°\text{F}$$
$$\text{UL} = \overline{X} + t_{\alpha}(s_{\overline{X}}) = 98.2 + 2.75(0.112) = 98.5°\text{F}$$

What conclusion do you reach about the 98.6°F standard? We *have* been wrong all these years. On the basis of actual measurements, you can be 99 percent confident that normal body temperature is between 97.9°F and 98.5°F.[9] 98.6°F is not in the interval.

The 95 percent confidence interval is the basis for *margin of error* statistics in survey reports. If a survey shows that a proposal has the support of 60 percent of the population, plus or minus a margin of error of 3 percent, it means that you can be 95 percent confident that the true level of support is between 57 percent and 63 percent.

In summary, confidence intervals for population means produce an interval statistic (lower and upper limits) that is destined to contain the population mean 95 percent of the time (or 99 or 90). Whether or not a particular confidence interval contains the unknowable μ is uncertain, but you have control over the degree of uncertainty.

Confidence Intervals Illustrated

The concept of confidence intervals has been known to be troublesome. Here is another explanation of confidence intervals, this time with a picture.

Look at **Figure 8.7,** which has these features:

- A population of scores (top curve)[10]
- A sampling distribution of the mean when $N = 25$ (small curve)
- Twenty 95 percent confidence intervals based on random samples ($N = 25$) from the population

On each of the 20 horizontal lines, the endpoints represent the lower and upper limits and the filled circles show the mean. As you can see, nearly all the confidence intervals have "captured" μ. One has not. Find it. Does it make sense to you that 1 out of 20 of the 95 percent confidence intervals does not contain μ?

Adding Confidence Intervals to Graphs

As you know from reading textbooks and articles, a bar graph that shows two or more means that appear different conveys the idea that the means really are different. That is, it never occurs to most of us that the observed difference could be due to the chance differences that go with sampling. However, if confidence intervals are added, they direct our attention to the degree of chance fluctuation that is expected for sample means.

Figure 8.8 is a bar graph of the Chapter 3 puberty data for females and males. In addition to the two means (12.75 years and 14.75 years), 95 percent confidence intervals are indicated by lines (often called *bars*) that extend in either direction from the means. On the left side of Figure 8.8, look at the extent of the confidence interval for females. The sample mean of the males is not within the confidence interval calculated for the females. (Use a straightedge to assure yourself, if necessary.) Thus, you can be more than 95 percent confident that the interval that estimates the population mean of the females does not include the sample mean of the males. In short, these confidence intervals provide assurance that the difference in the two means is due not to differences that go with sampling but to a difference in the population means.

---

[9] C. R. A. Wunderlich, a German investigator in the mid-19th century, was influential in establishing 98.6°F (37°C) as the normal body temperature.

[10] It is not necessary that the population be normal.

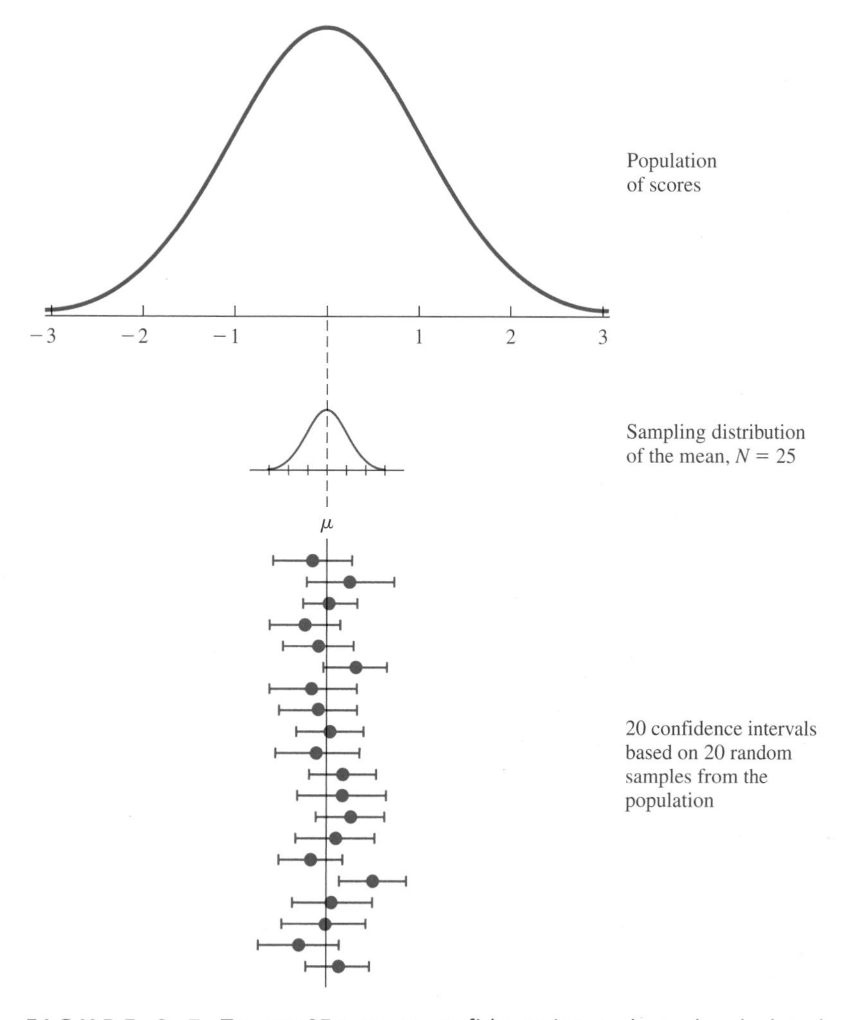

FIGURE 8.7 Twenty 95 percent confidence intervals, each calculated from a random sample of 25 from a normal population. Each sample mean is represented by a dot.

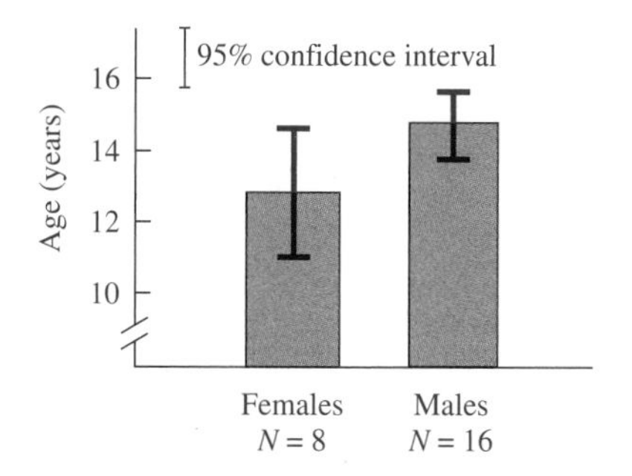

FIGURE 8.8 Puberty data for females and males (means and confidence intervals)

Categories of Inferential Statistics

Two of the major categories of inferential statistics are confidence intervals and hypothesis testing. A confidence interval establishes an interval within which a population parameter is expected to lie (with a certain degree of confidence). Confidence intervals were derived by Jerzy Neyman and introduced in 1934 (Salsburg, 2001). In recent years they have become more fashionable. The APA Publication Manual (2010) recommends them strongly (p. 34).

A second category of inferential statistics, hypothesis testing, allows you to use sample data to test a hypothesis about a population parameter. Popular for analyzing data from experiments since the 1920's, hypothesis testing is a topic in each of the remaining chapters in this book.

PROBLEMS

8.23. A social worker conducted an 8-week assertiveness training workshop. Afterward, the 14 clients took the Door Manifest Assertiveness Test, which has a national mean of 24.0. Use the data that follow to construct a 95 percent confidence interval about the sample mean. Write an interpretation about the effectiveness of the workshop.

24	25	31	25	33	29	21
22	23	32	27	29	30	27

8.24. Airplane manufacturers use only slow-burning materials. For the plastic in coffee pots the maximum burn rate is 4 inches per minute. An engineer at Hawker-Beechcraft randomly sampled from a case of 120 Krups coffee pots and obtained burn rates of 1.20, 1.10, and 1.30 inches per minute. Calculate a 99.9 percent confidence interval and write an interpretation that tells about using Krups coffee makers in airplanes.

8.25. This problem will give you practice in constructing confidence intervals and in discovering how to reduce their size (without sacrificing confidence). Smaller confidence intervals tell you the value of μ more precisely.

a. How wide is the 95 percent confidence interval about a sample mean when $\hat{s} = 4$ and $N = 4$?

b. How much smaller is the 95 percent confidence interval about a sample mean when N is increased fourfold to 16 ($\hat{s}$ remains 4)?

c. Compared to your answer in part **a,** how much smaller is the 95 percent confidence interval about a sample mean when N remains 4 and $\hat{s}$ is reduced to 2?

8.26. In Figure 8.7, the 20 lines that represent confidence intervals vary in length. What aspect of the sample causes this variation?

8.27. Figure 8.7 showed 95 percent confidence intervals for $N = 25$. Imagine a graphic with confidence intervals based on $N = 100$.

a. How many of the 20 intervals are expected to capture μ?

b. Will the confidence intervals be wider or narrower? By how much?

8.28. Figure 8.7 showed 95 percent confidence intervals. Imagine that the lines are 90 percent confidence intervals.

a. How many of the 20 intervals are expected to capture μ?

b. Will the confidence intervals be wider or narrower?

8.29. Use a ruler to estimate the lower and upper limits of the confidence intervals in Figure 8.8.

8.30. In the accompanying figure, look at the curve of performance over trials, which is generally upward except for the dip at trial 2. Is the dip just a chance variation, or is it a reliable, nonchance drop? By indicating the lower and upper confidence limits for each mean, a graph can convey at a glance the reliability that goes with each point. Calculate a 95 percent confidence interval for each mean. Pencil in the lower and upper limits for each point on the graph. Write a sentence explaining the dip at trial 2.

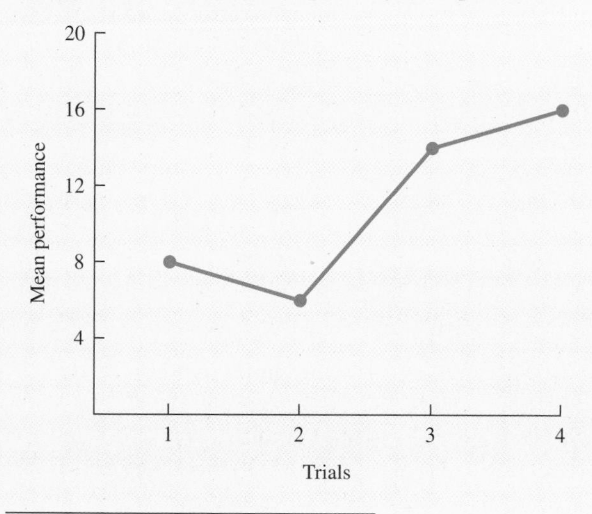

	Trials			
	1	2	3	4
$\overline{X}$	8	6	14	16
$\hat{s}$	5	5	5	5
N	25	25	25	25

8.31. Here's a problem you probably were expecting. Review the objectives at the beginning of the chapter. You *are* using this valuable, memory-consolidating technique, aren't you?

ADDITIONAL HELP FOR CHAPTER 8

Visit *cengage.com/psychology/spatz.* At the Student Companion Site, you'll find multiple-choice tutorial quizzes, flashcards with definitions and workshops. For this chapter there are Statistical Workshops on Sampling Distribution, Standard Error, and Central Limit Theorem.

KEY TERMS

Biased sample (p. 151)
Central Limit Theorem (p. 154)
Confidence interval (p. 164)
Degrees of freedom (p. 161)
Expected value (p. 153)
Random sample (p. 148)
Sampling distribution (p. 152)
Standard error (p. 153)
t distribution (p. 161)

transition passage
to hypothesis testing

YOU ARE IN the midst of material on inferential statistics, which, as you probably recall, is a technique that allows you to use sample data to help you make decisions about populations. In the last chapter, you studied and worked problems on *confidence intervals.* Those confidence intervals allowed you to decide something about the mean of an unmeasured population.

The next chapter, Chapter 9, covers the basics of *null hypothesis statistical testing* (NHST), a more widely used inferential statistics technique. NHST techniques result in a yes/no decision about a population parameter.

In Chapter 9, you test hypotheses about population means and population correlation coefficients. In Chapter 10, you test hypotheses about the difference between the means of two populations and also calculate confidence intervals about the difference between sample means. In both of these chapters, you determine probabilities by using t distributions. Also in both chapters, you calculate from sample data the effect size index d, the descriptive statistic you learned about in Chapter 5.

One of the most important uses of statistics is to analyze data from experiments. Chapter 10 discusses the basics of simple experiments.

CHAPTER

9

Hypothesis Testing and Effect Size: One-Sample Designs

OBJECTIVES FOR CHAPTER 9

After studying the text and working the problems in this chapter, you should be able to:

1. Explain the procedure called null hypothesis statistical testing (NHST)
2. Define the null hypothesis in words
3. Define the three alternative hypotheses in words
4. Define α, *significance level, rejection region,* and *critical value*
5. Use a one-sample *t* test and the *t* distribution to decide if a sample mean came from a population with a hypothesized mean, and write an interpretation
6. Decide if a sample *r* came from a population in which the correlation between the two variables is zero, and write an interpretation
7. Explain what rejecting the null hypothesis means and what retaining the null hypothesis means
8. Distinguish between Type I and Type II errors
9. Interpret the meaning of *p* in the phrase $p < .05$
10. Describe the difference between a one-tailed and a two-tailed statistical test
11. Calculate and interpret an effect size index for a one-sample experiment

null hypothesis statistical testing (NHST)
Process that produces probabilities that are accurate when the null hypothesis is true.

THIS CHAPTER INTRODUCES **null hypothesis statistical testing (NHST).** In the social, behavioral, and biological sciences, NHST is used more frequently than any other inferential statistics technique. Every chapter from this point on assumes that you can use the concepts of null hypothesis statistical testing. I suggest that you resolve now to achieve an understanding of NHST.

Let's begin by looking in on our two students from Chapter 8, the ones who gathered family income data. As we catch up with them, they are hanging out at the Campus Center, sharing a bag of Doritos® tortilla chips, nacho cheese flavor. The junior picks up the almost empty bag, shakes the remaining contents into the outstretched hand of the sophomore, and says,

"Hmmm, the bag says it had 269.3 grams in it."

"Hummph. I doubt it," complained the suspicious sophomore.

"Oh, I don't think they would cheat us," replied the genial junior, "but I'm surprised that they can measure chips that precisely."

"Well, let's find out," said the sophomore. "Let's buy a fresh package, find a scale, and get an answer."

"OK, but maybe we ought to buy more than one package," said the junior, thinking ahead.

While our two investigators go off to buy chips and secure a scale, let's examine what is going on. The Frito-Lay® company claims that their package contains 269.3 grams of tortilla chips. Of course, we know that the claim cannot be true for *every* package because there is always variation in the real world. A reasonable interpretation is that the company claims that their packages *on average* contain 269.3 grams of chips.

Now, let's put our story into the context of null hypothesis statistical testing (NHST). A claim is made that these bags contain 269.3 grams of chips. Data can be gathered to test the claim. Any conclusion about the claim will be based on sample data, so there will be some uncertainty, but as you learned in the previous chapter, uncertainty can be measured and if there isn't much, you can state a conclusion *and* your degree of uncertainty. As will become apparent, the logic used in NHST applies not just to Doritos tortilla chips but to claims and comparisons on topics that range from anthropology and business to zoology and zymurgy.

PROBLEM

9.1. NHST is an acronym that stands for what phrase?

The Logic of Null Hypothesis Statistical Testing (NHST)

I will tell the story of NHST in general terms, but also present the Doritos problem side by side as a specific example. NHST always begins with a claim about a parameter. For claims about a population mean, the symbol is μ_0. In the case of the Frito-Lay company, it claims that the mean weight of bags of Doritos tortilla chips is 269.3 grams. Thus, in formal terms the company claims that $\mu_0 = 269.3$ grams. Of course, the population of weights of the bags has an actual mean, which may or may not be equal to 269.3 grams. The symbol for this unknown population mean is μ_1. With a sample of weights and NHST, you may be able to reach a conclusion about μ_1.

NHST does not result in an exact value for μ_1 or even a confidence interval that captures μ_1. Rather, NHST results in a conclusion about the *relationship* between μ_1

TABLE 9.1 The general case of the relationship between μ_1 and μ_0 and its application to the Doritos example

General case	The Doritos example
1. $\mu_1 = \mu_0$ hypothesis of equality	The average weight of the chips is 269.3 grams, which is what the company claims
2. $\mu_1 \neq \mu_0$ hypothesis of difference	The average weight of the chips is not 269.3 grams

and μ_0. In **Table 9.1,** the possible relationships between μ_1 and μ_0 are shown on the left; their applications to the Doritos example are on the right. Table 9.1 covers *all* of the logical possibilities of the relationship between μ_1 and μ_0. The two means are either equal to each other (1) or they are not equal to each other (2).

NHST is a procedure that allows, if the data permit, strong statements of support for the hypothesis of difference (2). However, NHST *cannot* result in strong support for the hypothesis of equality (1), *regardless* of how the data come out.

The logic that NHST uses is somewhat roundabout. Researchers (much like the suspicious sophomore) usually believe that the hypothesis of difference is correct. To get support for the hypothesis of difference, the hypothesis of equality must be discredited. Fortunately, the hypothesis of equality produces predictions about outcomes. NHST tests these predictions against actual outcomes.

If the hypothesis of equality is true, sample means of 269.3 grams are more likely than larger values or smaller values. Some values are so far from 269.3 that they are very unlikely. Now suppose that the actual observations did produce a sample mean that was far from 269.3 (and thus, very unlikely according to the predictions of the hypothesis of equality). Because of this failure to predict, the hypothesis of equality is discredited.

To capture this logic with a different set of words, let's begin with the actual outcome. If this outcome is unlikely when the hypothesis of equality is true, we can reject the hypothesis of equality. With the hypothesis of equality removed, the only hypothesis left is the hypothesis of difference. This paragraph and the two previous paragraphs deserve marks in the margin. Reread them now and again later.

At this point, it will be helpful to have those data from the two students who bought more bags of chips and weighed them. Let's listen in again.

"I'm glad we went ahead and bought eight bags. The leftovers from this experiment are delicious."

"OK, we've got the eight measurements, but what statistics do we need? The mean, of course. Anything else?"

"Knowing about variability always helps my understanding. Let's calculate a standard deviation."

"OK, let's see, the mean is 270.675 grams and the standard deviation is 0.523 gram. How about that! The mean is *more* than what the company claims," said the surprised sophomore. "Surely they wouldn't deliberately put in *more* than 269.3 grams?"

"I don't know, but look at that tiny standard deviation," said the junior. "It's only about a half a gram, which is less than a 50th of an ounce. They put almost exactly the same amount in every bag!"

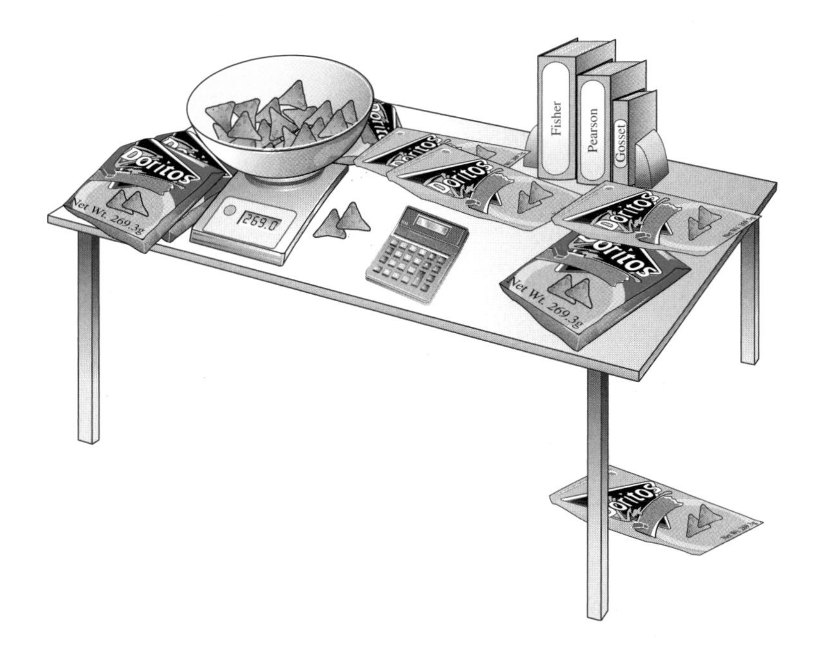

At this point the data have changed the question from whether the company was cheating to whether the company might be deliberately overfilling the bags. The alternative to the *overfilling* hypothesis is that sample means of 270.675 grams or more are common—such means are just examples of the normal fluctuation that goes with production. NHST may be able to resolve this issue by providing the exact probability of obtaining sample means equal to or greater than 270.675 grams from a population with a mean of 269.3 grams (for samples with $N = 8$). What follows is a repeat of the logic of NHST, but this version includes technical terms.

For any null hypothesis statistical test, there is a **null hypothesis** (symbolized $\boldsymbol{H_0}$). The null hypothesis is always a statement about a parameter (or parameters). It always includes an equals mark. For problems in this chapter, the null hypothesis comes from some claim about how things are. In the Doritos problem, the null hypothesis is that $\mu_0 = \mu_1 = 269.3$ grams. Thus, H_0: $\mu_0 = \mu_1 = 269.3$ grams.

null hypothesis (H_0)
Hypothesis about a population or the relationship among populations.

Here's how NHST provides an exact probability. NHST begins by tentatively assuming that the null hypothesis is true. We know that random samples ($N = 8$) from the population produce an array of sample means that center around 269.3 grams. Except for the means that are exactly 269.3, half will be greater and half will be less than 269.3 grams. If 269.3 is subtracted from each sample mean, the result is a distribution of differences, centered around *zero*. This distribution is a sampling distribution of differences.

Once you have a sampling distribution, it can be used to determine the probability of obtaining any particular difference. The particular difference we are interested in is the difference between our sample mean and the null hypothesis mean, which is 1.375 grams. That is, $270.675 - 269.3 = 1.375$ grams.

If the probability of a difference of 1.375 grams is very small when the null hypothesis is true, we have evidence that the null hypothesis is not correct and should be rejected. If the probability is large, we have evidence that is *consistent* with the null hypothesis. Unfortunately, large probabilities are also consistent with hypotheses other than the null hypothesis.[1] Large probabilities do not permit you to adopt or accept the null hypothesis but only to retain it as one among many hypotheses that the data support. Here's an important point: *The probability that NHST produces is the probability of the data actually obtained, if the null hypothesis is true.*

alternative hypothesis (H_1) Hypothesis about population parameters that is accepted if the null hypothesis is rejected.

Central to the logic of NHST is the competition between a hypothesis of equality and a hypothesis of difference. The hypothesis of equality is called, as you already know, the null hypothesis. The hypothesis of difference is called the **alternative hypothesis;** its symbol is $\boldsymbol{H_1}$. If H_0 is rejected, only H_1 remains. Actually, for any NHST problem there are three *possible* alternative hypotheses. In practice, a researcher chooses one of the three H_1's *before the data are collected.* The choice of a specific alternative hypothesis helps determine the conclusions that are possible. The three alternative hypotheses for the Doritos data are:

H_1: $\mu_1 \neq 269.3$ grams. This alternative states that the mean of the population of Doritos weights is not 269.3 grams. It doesn't specify whether the actual mean is greater than or less than 269.3 grams. If you reject H_0 and accept this H_1, you must examine the sample mean to determine whether its population mean is greater than or less than the one specified by H_0.

H_1: $\mu_1 < 269.3$ grams. This alternative hypothesis states that the sample is from a population with a mean less than 269.3 grams.

H_1: $\mu_1 > 269.3$ grams. This alternative hypothesis states that the sample is from a population with a mean greater than 269.3 grams.

For the Doritos problem, the first alternative hypothesis, H_1: $\mu_1 \neq 269.3$ grams, is the best choice because it allows the conclusion that the Doritos bags contain more than *or* less than the company claims. The section on one-tailed and two-tailed tests later in this chapter gives you more information on choosing an alternative hypothesis.

In summary, the null hypothesis (H_0) meets with one of two fates at the hands of the data. It may be rejected or it may be retained. Rejecting the null hypothesis provides strong support for the alternative hypothesis, which is a statement of *greater than* or *less than.* If H_0 is retained, *it is not proved as correct;* it is simply retained as one among many possibilities. A retained H_0 leaves you unable to choose between H_0 and H_1. You do, however, have the data you gathered. They are informative and helpful, even if they do not support a strong NHST conclusion. And, of course, a completed experiment, regardless of its outcome, can lead to a new experiment and then to better understanding.

clue to the future

You just completed a section that explains reasoning that is at the heart of inferential statistics. This reasoning is important in every chapter that follows and for every effort to comprehend and explain statistics in the years to come.

[1] For example, hypotheses that specify that the difference between μ_1 and μ_0 is very small (but not zero) are supported by large probabilities.

PROBLEMS

9.2. NHST techniques produce probability figures such as .20. What event has a probability of .20?

9.3. Is the null hypothesis a statement about a statistic or about a parameter?

9.4. Distinguish between the null hypothesis and the alternative hypothesis, giving the symbol for each.

9.5. List the three alternative hypotheses in the Doritos problem using symbols and numbers.

9.6. In your own words, outline the logic of null hypothesis statistical testing using technical terms and symbols.

9.7. Agree or disagree: The students' sample of Doritos weights is a random sample.

Using the *t* Distribution for Null Hypothesis Statistical Testing

To get a probability figure for an NHST problem, you must have a sampling distribution. Fortunately for the Doritos problem, the sampling distribution that is appropriate is a *t* distribution and you already know about *t* distributions from your previous study. For the task at hand (testing the null hypothesis that $\mu_0 = 269.3$ grams using a sample of eight), a *t* distribution with 7 *df* gives you correct probabilities. You may recall that Table D in Appendix C provides probabilities for 34 *t* distributions. Table D values and their probabilities are generated by a mathematical formula that assumes that the null hypothesis is true.

Figure 9.1 shows a *t* distribution with 7 *df*. It is a picture of the distribution of the mean weights of samples ($N = 8$) from a population with a mean weight of 269.3 grams. The numbers in Figure 9.1 are based on the assumption that the null hypothesis is true.

Examine Figure 9.1, paying careful attention to the three sets of values on the abscissa. The *t* values on the top line and the probability figures at the bottom apply to any problem that uses a *t* distribution with 7 *df*. Sandwiched between are differences between sample means and the hypothesized population mean—differences that are specific to the Doritos data.

In your examination of Figure 9.1, note the following characteristics:

1. The mean is zero.
2. As the size of the difference between the sample mean and the population mean increases, *t* values increase.
3. As differences and *t* values become larger, probabilities become smaller.
4. As always, small probabilities go with small areas at each end of the curve.

Using Figure 9.1, you can make a decision about the null hypothesis for the Doritos data. (Perhaps you have already anticipated the conclusion?) The difference between the mean weight of the sample and the advertised weight is 1.375 grams.

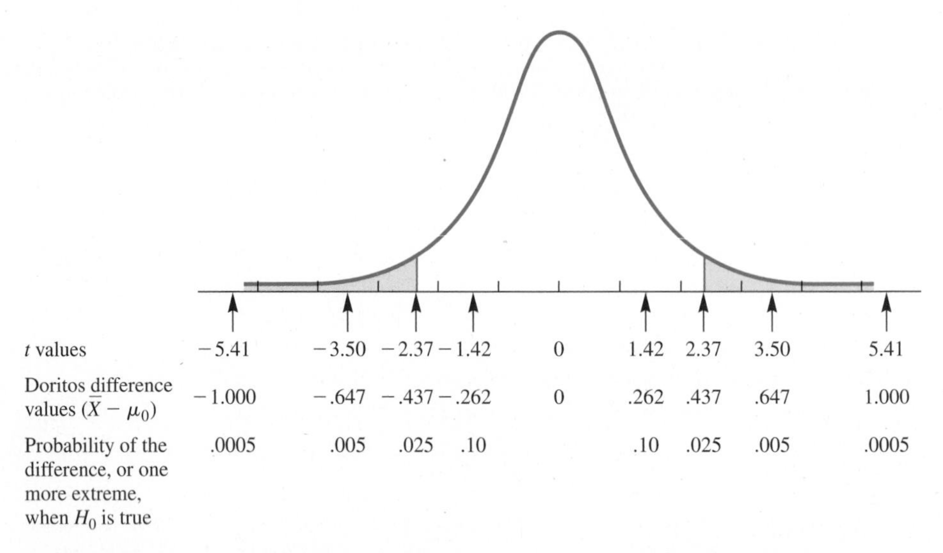

FIGURE 9.1 A *t* distribution. The *t* values and probability figures are those that go with 7 *df*. The difference values apply only to the Doritos data.

This difference is not even shown on the distribution, so it must be extremely rare—*if the null hypothesis is true.* How rare is it?

Let's take the rarest events shown and use them as a reference point. As seen in **Figure 9.1,** a difference as extreme as 1.000 gram or greater occurs with a probability of .0005. In addition, a difference as extreme as −1.000 or less occurs with the same probability, .0005. Adding these two probabilities together gives a probability of .001.[2] Thus, the probability is .001 of getting a difference of ±1.000 gram or greater. The difference found in the data (1.375 grams) is even more rare, so the *p* value of the data actually obtained is less than .001. Thus, if the null hypothesis is true, the probability of getting the difference actually obtained is less than 1 in 1000 ($p < .001$).

The reasonable thing to do is abandon (reject) the null hypothesis. With the null hypothesis gone, only the alternative hypothesis remains: H_1: $\mu_1 \neq 269.3$ grams. Knowing that the actual weight is *not* 269.3 grams, the next step is to ask whether it is greater than or less than 269.3 grams. The sample mean is greater, so a conclusion can be written: "The contents of Doritos bags weigh more than the company claims, $p < .001$."

To summarize, we determined the probability of the data we observed if the null hypothesis is true. In this case, the probability is quite small. Rather than conclude that a very rare event occurred, we concluded that the null hypothesis is false. **Table 9.2** summarizes the NHST technique (left column) and its application to the Doritos example (right column). Please study Table 9.2 now and mark it for review later.

[2] I'll explain this step in the section on one- and two-tailed tests.

TABLE 9.2 A summary of NHST and its application to the Doritos example

General statement of NHST for means	Application of NHST to Doritos example
Recognize two possibilities for the population mean: The null hypothesis, H_0 The alternative hypothesis, H_1	H_0: $\mu_0 = 269.3$ grams H_1: $\mu_1 \neq 269.3$ grams
Assume for the time being that H_0 is true.	Assume for the time being that the mean weight of all bags of chips is 269.3 grams.
Use a sampling distribution that shows the differences between sample means and the null hypothesis mean when H_0 is true.	The *t* distribution is appropriate.
Gather sample data from the population. Calculate a mean, $\overline{X}_1$.	$\overline{X}_{\text{Doritos}} = 270.675$ grams
Calculate the difference between the sample mean and the null hypothesis mean.	$270.675 - 269.3 = 1.375$ grams
Using the sampling distribution, determine the probability of the difference that was actually observed (or one more extreme).	$p < .001$
Small probability: Reject the null hypothesis.	Reject the null hypothesis that $\mu_0 = 269.3$ grams.
If $\overline{X}_1 > \mu_0$, conclude that $\mu_1 > \mu_0$.	The mean weight of Doritos chips in bags is greater than the company's claim of 269.3 grams.
If $\overline{X}_1 < \mu_0$, conclude that $\mu_1 < \mu_0$.	Not applicable
Large probability: Conclude that the data are consistent with the null hypothesis.	Not applicable

A Problem and the Accepted Solution

The hypothesis that the sample of Doritos weights came from a population with a mean of 269.3 grams was so unlikely ($p < .001$) that it was easy to reject the hypothesis. But what if the probability had been .01, or .05, or .25, or .50? The problem that this section addresses is where on the probability continuum you should change the decision from "reject the null hypothesis" to "retain the null hypothesis."

It may already be clear to you that any solution will be an arbitrary one. Breaking a continuum into two parts often leaves you uncomfortable near the break. Nevertheless, those who use statistics have adopted a solution.

Setting Alpha (α) or Establishing a Significance Level

alpha (α)
Probability of a Type I error.

significance level
Probability (α) chosen as the criterion for rejecting the null hypothesis.

The widely accepted solution to the problem of where to break the continuum is to use the .05 mark. The choice of a probability value is called setting **alpha (α)**. It is also called establishing a **significance level.**

In an experiment, the researcher sets α, gathers data, and uses a sampling distribution to find the probability (p) of such data. This p value is correct only when the null hypothesis is true. If $p \leq \alpha$, reject H_0. Thus, if $\alpha = .05$ and p is .05 or less, reject H_0. If $\alpha = .05$ and $p > .05$, retain H_0. Of course, if $p = .03$, or .01, or .001, reject H_0. If the probability is .051 or greater, retain H_0. When H_0 is rejected, the difference is described as **statistically significant.** Sometimes, when the context is clear, this is shortened to simply "significant." When H_0 is retained (the difference is not significant), the abbreviation **NS** is often used.

statistically significant
Difference so large that chance is not a plausible explanation for the difference.

NS
Difference is not significant.

In some research reports, an α level is not specified; only data-based probability values are given. Statistical software programs produce probabilities such as .0004, .008, or .049, which are described as significant. If tables are used to determine probabilities, such differences are identified as $p < .001$, $p < .01$, and $p < .05$. Regardless of how the results are reported, however, researchers view .05 as an important cutoff (Nelson, Rosenthal, and Rosnow, 1986). When .10 or .20 is used as an α level, a justification should be given.

Rejection Region

The **rejection region** is the area of the sampling distribution that includes all the differences that have a probability *equal to* or *less than* α. Thus, any event in the rejection region leads to rejection of the null hypothesis.[3] In Figure 9.1 the rejection region is shaded (for a significance level of .05). Thus, t values more extreme than ± 2.37 lead to rejecting H_0.

rejection region
Area of a sampling distribution that corresponds to test statistic values that lead to rejection of the null hypothesis.

Critical Values

Most statistical tables do not provide you with probability figures for every outcome of a statistical test. What they provide are statistical test values that correspond to commonly chosen α values. These specific test values are called **critical values.**

I will illustrate critical values using the t distribution table (Table D in Appendix C). The t values in that table are called critical values when they are used in hypothesis testing. The α levels that researchers use are listed across the top in rows 2 and 3. Degrees of freedom are in column 1. Remember that if you have a df that is not in the table, it is conventional to use the next *smaller df* or to interpolate a t value that corresponds to the exact df.

critical value
Number from a sampling distribution that determines whether the null hypothesis is rejected.

The t values in the table separate the rejection region from the rest of the sampling distribution (sometimes referred to as the *acceptance region*). Thus, data-produced t values that are equal to or greater than the critical value fall in the rejection region, and the null hypothesis is rejected.

Figure 9.2 is a t distribution with 14 df. It shows that the critical value separates the rejection region from the rest of the distribution. I will indicate critical values with an expression such as $t_{.05}$ (14 df) = 2.145. This expression indicates the sampling distribution that was used (t), α level (.05), degrees of freedom (14), and critical value from the table (2.145, for a two-tailed test).

[3] Another term for rejection region is *critical region*.

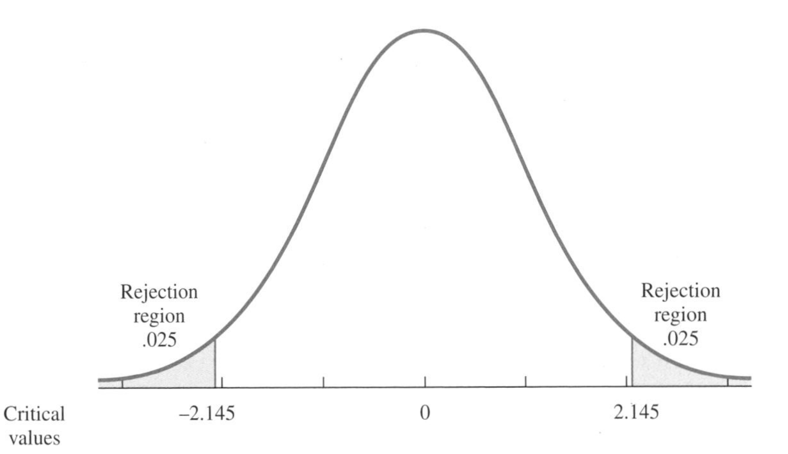

FIGURE 9.2 A t distribution with 14 *df*. For $\alpha = .05$ and a two-tailed test, critical values and the rejection region are illustrated.

PROBLEMS

9.8. True or false? Sampling distributions used for hypothesis testing are based on the assumption that H_0 is false.

9.9. What is a significance level? What is the largest significance level you can use without giving a justification?

9.10. Circle all phrases that go with events that are in the rejection region of a sampling distribution.

p is small

Retain the null hypothesis

Accept H_1

Middle section of the sampling distribution

9.11. Use Table D to find critical values for two-tailed tests for these values.

a. α level of .01; $df = 17$

b. α level of .001; $df = 46$

c. α level of .05; $df = \infty$

The One-Sample *t* Test

Some students suspect that it is not necessary to construct an entire sampling distribution such as the one in Figure 9.1 each time there is a hypothesis to test. You may be such a student, and if so, you are correct. The conventional practice is to analyze the data using the **one-sample *t* test.** The formula for the one-sample *t* test is

one-sample *t* test
Statistical test of the hypothesis that a sample with mean $\overline{X}$ came from a population with mean μ.

$$t = \frac{\overline{X} - \mu_0}{s_{\overline{X}}}; \quad df = N - 1$$

where $\bar{X}$ is the mean of the sample
μ_0 is the hypothesized mean of the population
$s_{\bar{X}}$ is the standard error of the mean

For the chip data, the question to be answered is whether the Frito-Lay company is justified in claiming that bags of Doritos tortilla chips weigh 269.3 grams. The null hypothesis is that the mean weight of the chips is 269.3 grams. The alternative hypothesis is that the mean weight is *not* 269.3 grams. A sample of eight bags produced a mean of 270.675 grams with a standard deviation (using $N - 1$) of 0.523 gram. To test the hypothesis that the sample weights are from a population with a mean of 269.3 grams, you should calculate a t value using the one-sample t test. Thus,

$$t = \frac{\bar{X} - \mu_0}{s_{\bar{X}}} = \frac{\bar{X} - \mu_0}{\hat{s}/\sqrt{N}} = \frac{270.675 - 269.3}{0.523/\sqrt{8}} = 7.44; \quad 7\ df$$

To interpret a t value of 7.44 with 7 *df*, turn to Table D and find the appropriate critical value. For α levels for two-tailed tests, go down the .05 column until you reach 7 *df*. The critical value is 2.365. The t test produced a larger value (7.44), so the difference observed (1.375 grams) is statistically significant at the .05 level. To find out if 7.44 is significant at an even lower α level, move to the right, looking for a number that is greater than 7.44. There isn't one. The largest number is 5.408, which is the critical value for an α level of .001. Because 7.44 is greater than 5.408, you can reject the null hypothesis at the .001 level.[4]

The last step is to write an informative conclusion that tells the story of the variable that was measured. Here's an example: "The mean weight of the eight Doritos bags was 270.675 grams, which is significantly greater than the company's claim of 269.3 grams, $p < .001$."

A good interpretation is more informative than simply, "the null hypothesis was rejected, $p < .05$." A good interpretation always:

1. Uses the terms of the experiment
2. Tells the direction of the difference between the means

The algebraic sign of the t in the t test is ignored when you are finding the critical value for a two-tailed test. Of course, the sign tells you whether the sample mean is greater or less than the null hypothesis mean—an important point in any statement of the results.

I want to address two additional issues about the Doritos conclusion; both have to do with generalizability. The first issue is the nature of the sample: It most certainly was not a random sample of the population in question. Because of this, any conclusion about the population mean is not as secure as it would be for a random sample. How much less secure is it? I don't really know, and inferential statistics provides me with no guidelines on this issue. So, I state both my conclusion (contents weigh more than company claims) and my methods (nonrandom sample, $N = 8$) and then see whether this conclusion will be challenged or supported by others with more or better data. (I'll just have to wait to see who responds to my claim.)

[4] Some statisticians simply determine if the t test value is significant at the .05 level and stop there. They say that looking for smaller α levels amounts to changing the α level after looking at the data.

The other generalizability issue is whether the conclusion applies to other-sized bags of Doritos, to other Frito-Lay products, or to other manufacturers of tortilla chips. These are good questions but, again, inferential statistics does not give any guidelines. You'll have to use aids such as your experience, library research, correspondence with the company, or more data gathering.

Finally, I want to address the question of the usefulness of the one-sample t test. The t test is useful for checking manufacturers' claims (where an advertised claim provides the null hypothesis), comparing a sample to a known norm (such as an IQ score of 100), and comparing behavior against a "no error" standard (as used in studies of lying or visual illusions). In addition to its use in analyzing data, the one-sample t test is probably the simplest example of hypothesis testing (an important consideration when you are being introduced to the topic).

PROBLEM

9.12. What are the two characteristics of a good interpretation?

An Analysis of Possible Mistakes

To some forward-thinking students, the idea of adopting an α level of 5 percent seems preposterous. "You gather the data," they say, "and if the difference is large enough, you reject H_0 at the .05 level. You then state that the sample came from a population with a mean other than the hypothesized one. But in your heart you are uncertain. Hey, perhaps a rare event happened."

This line of reasoning is fairly common. Many thoughtful students take the next step. "For me, I won't use the .05 level. I'll use an α level of 1 in one million. That way I can reduce the uncertainty."

It is true that adopting a .05 α level leaves some room for mistaking a chance difference for a real difference. It is probably clear to you that lowering the α level (to a probability such as .01 or .001) *reduces* the probability of this kind of mistake. Unfortunately, lowering the α level *increases* the probability of a different kind of mistake.

Look at **Table 9.3,** which shows the two ways to make a mistake. Cell 1 shows the situation when the null hypothesis is true and you reject it. Your sample data and hypothesis testing have produced a mistake—a mistake called a **Type I error.** However, if H_0 is true and you retain it, you have made a correct decision (cell 2).

Type I error
Rejection of a null hypothesis when it is true.

Now, suppose that the null hypothesis is actually false (the second column). If, on the basis of your sample data, you reject H_0, you have made a correct decision (cell 3). Finally, if the null hypothesis is false and you retain it (cell 4), you have made a mistake—a mistake called a **Type II error.**

Type II error
Failure to reject a null hypothesis that is false.

Let's attend first to a Type I error. The probability of a Type I error is α, and you control α when you adopt a significance level. If the data produce a p value that is less than α, you reject H_0 and conclude that the difference observed is statistically significant. If you reject H_0 (and H_0 is true), you make a Type I

TABLE 9.3 **Type I and Type II errors***

		True situation in the population	
		H_0 true	H_0 false
Decision made on the basis of sample data	Reject H_0	1. Type I error	3. Correct decision
	Retain H_0	2. Correct decision	4. Type II error

* In this case, *error* means mistake.

error, but the probability of this error is controlled by your choice of α level. Note that rejecting H_0 is wrong only when the null hypothesis is true.

beta (β)
Probability of a Type II error.

Now for a Type II error. A Type II error is possible only if you retain the null hypothesis. If the null hypothesis is false and you retain it, you have made a Type II error. The probability of a Type II error is symbolized by **β (beta).**

Controlling α by setting a significance level is straightforward; determining the value of β is much more complicated. For one thing, a Type II error requires that the sample be from a population that is different from the null hypothesis population. Naturally, the more different the two populations are, the more likely you are to detect it and, thus, the lower β is.

In addition, the values of α and β are inversely related. As α is reduced (from, say, .05 to .01), β goes up. To explain, suppose you insist on a larger difference between means before you say the difference is "not chance." (You reduce α.) In this case, you will be less able to detect a small, real, nonchance difference. (β goes up.) The following description further illustrates the relationship between α and β.

Figure 9.3 shows two populations. The population on the left has a mean $\mu_0 = 10$; the one on the right has a mean $\mu_1 = 14$. An investigator draws a large sample from the population on the right and uses it to test the null hypothesis H_0: $\mu_0 = 10$. In the real world of data, every decision carries some uncertainty, but in this textbook example, the correct decision is clear: Reject the null hypothesis.

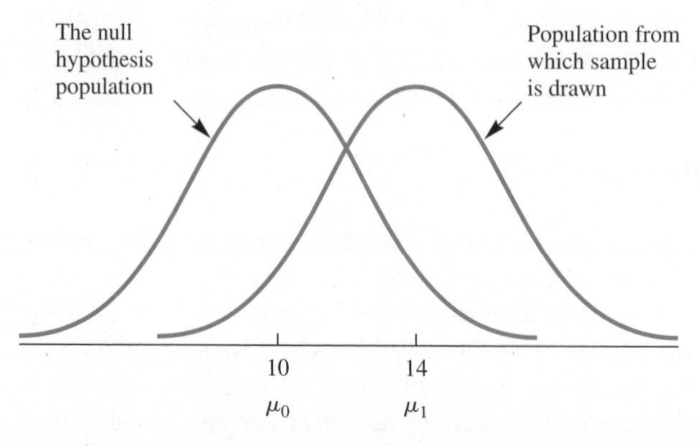

FIGURE 9.3 Frequency distribution of scores when H_0 is false

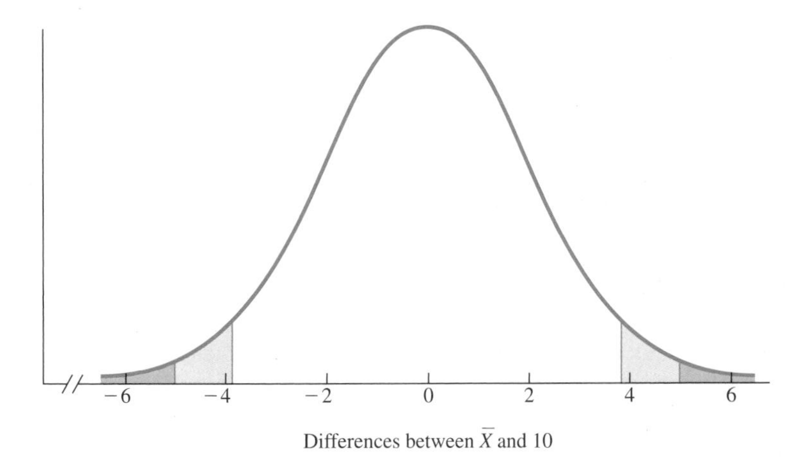

FIGURE 9.4 A sampling distribution from the population on the left in Figure 9.3

The actual decision, however, will be made by evaluating the difference between the sample mean $\overline{X}$ and the mean specified by the null hypothesis, $\mu_0 = 10$.

Now, let's suppose that the sample produced a mean of 14. That is, the sample tells the exact truth about its population. Under these circumstances, will the hypothesis-testing procedure produce a correct decision?

Figure 9.4 shows the sampling distribution for this problem. As you can see, if the 5 percent α level is used (*all* the blue areas), a difference of 4 falls in the rejection region. Thus, if $\alpha = .05$, you will correctly reject the null hypothesis. However, if a 1 percent α level is used (the *dark* shaded areas only), a difference of 4 *does not* fall in the rejection region. Thus, if $\alpha = .01$, you will not reject H_0. This failure to reject the false null hypothesis is a Type II error.

At this point, I can return to the discussion of setting the α level. The suggestion was: Why not reduce the α level to 1 in one million? From the analysis of the potential mistakes, you can answer that when you lower the α level to reduce the chance of a Type I error, you increase β, the probability of a Type II error. More protection from one kind of error increases the liability to another kind of error. Most who use statistics adopt an α level (usually at .05) and let β fall where it may.

To summarize, some uncertainty always goes with a conclusion based on statistical evidence. A famous statistics textbook put it this way: "If you agree to use a sample, you agree to accept some uncertainty about the results."

The Meaning of *p* in *p* < .05

The symbol p is quite common in data-based investigations, regardless of the topic of investigation. The p in $p < .05$ deserves a special section.

Every statistical test such as the t test has its own sampling distribution. A sampling distribution shows the probability of various sample outcomes when the null hypothesis is true. For every statistical test, p is always the probability of the statistical test value when H_0 is true. If a t test gives you $p < .05$, it means that the sample results you actually got or results more extreme occur fewer than 5 times in 100 *when the null hypothesis is true.*

There are a number of interpretations of p that might seem correct, but they are not. To illustrate, here are some cautions about the meaning of p.

- p is not the probability that H_0 is true.
- p is not the probability of a Type I error.
- p is not the probability that the data are due to chance.
- p is not the probability of making a wrong decision.
- the complement of $p(1 - p)$ is not the probability that the alternative hypothesis is true

For any statistical test, p is the probability of the data observed, if the null hypothesis is true. Mistaken ideas about the interpretation of p are fairly common. Kline (2004, Chapter 3) explains five misinterpretations and how widespread they are. He also gives guidelines on how behavioral scientists can reduce their reliance on NHST tests.

Perhaps the best thing for beginners to do is to memorize a definition of p and use it (or close variations) in the interpretations they write. In statistics, p is the probability of the data obtained, if the null hypothesis is true. To be just a little more technical, p is the probability of the statistical test value.

What about the situation when the null hypothesis is false? Statisticians address this situation under the topic of *power.* There is a short discussion of power in the next chapter.

One-Tailed and Two-Tailed Tests

As mentioned earlier, researchers choose, before the data are gathered, a null hypothesis, an α level, and one of three possible alternative hypotheses. The choice of an alternative hypothesis determines the conclusions that are possible after the statistical analysis is finished. Here's the general expression of these alternative hypotheses and the conclusions they allow you:

Two-tailed test: H_1: $\mu_1 \neq \mu_0$. The population means differ, but no direction of the difference is specified. Conclusions that are possible: The sample is from a population with a mean *less than* that of the null hypothesis *or* the sample is from a population with a mean *greater than* that of the null hypothesis.

One-tailed test: H_1: $\mu_1 < \mu_0$. Conclusion that is possible: The sample is from a population with a mean less than that of the null hypothesis.

H_1: $\mu_1 > \mu_0$. Conclusion that is possible: The sample is from a population with a mean greater than that of the null hypothesis.

two-tailed test of significance
Statistical test for a difference in population means that can detect positive and negative differences.

If you want to have two different conclusions available to you—that the sample is from a population with a mean *less than* that of the null hypothesis *or* that the sample is from a population with a mean *greater than* that of the null hypothesis—you must have a rejection region in each tail of the sampling distribution. Such a test is called a **two-tailed test of significance** for reasons that should be obvious from **Figure 9.4** and **Figure 9.2.**

If your only interest is in showing (if the data will permit you) that the sample comes from a population with a mean greater than that of the null hypothesis or from a population with a mean less than that of the null hypothesis, you should conduct a **one-tailed test of significance,** which puts all of the rejection region into one tail of the sampling distribution. **Figure 9.5** illustrates this for a *t* distribution with 30 *df*.

one-tailed test of significance
Statistical test that can detect a positive difference in population means or a negative difference, but not both.

The probability figures for two-tailed tests are in the second row of Table D (p. 392); probability figures for one-tailed tests are in row 3.

For most research situations, a two-tailed test is the appropriate one. The usual goal of a researcher is to discover the way things actually are, and a one-tailed test does not allow you to discover, for example, that the populations are exactly reversed from the way you expect them to be. Researchers always have expectations (but they sometimes get surprised).

In many applied situations, however, a one-tailed test is appropriate. Some new procedure, product, or person, for example, is being compared to an already existing standard. The only interest is in whether the new is better than the old; that is, only a difference in one direction will produce a decision to change. In such a situation, a one-tailed test seems appropriate.

I chose a two-tailed test for the Doritos data because I wanted to be able to conclude that the Frito-Lay company was giving more than it claimed *or* that it was giving less than it claimed. Of course, the data would have their say about the conclusion, but either outcome would be informative.

I'll end this section on one- and two-tailed tests by telling you that statistics instructors sometimes smile or even laugh at the subtitle of this book. The phrase "tales of distributions" brings to mind "tails of the distribution."

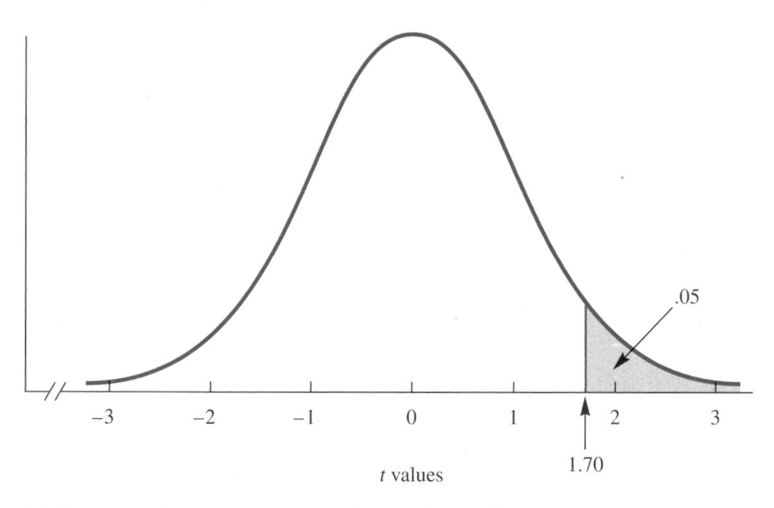

FIGURE 9.5 A one-tailed test of significance, with $\alpha = .05$, $df = 30$. The critical value is 1.70.

PROBLEMS

9.13. Distinguish between α and p.

9.14. What happens when you make a Type II error?

9.15. Suppose the actual situation is that the sample comes from a population that is not the one hypothesized by the null hypothesis. If the significance level is .05, what is the probability of making a Type I error? (Careful on this one.)

9.16. Suppose a researcher chooses to use $\alpha = .01$ rather than .05. What effect does this have on the probability of Type I and Type II errors?

9.17. The following table has blanks for you to fill in. Look up critical values in Table D, decide if H_0 should be rejected or retained, and give the probability figure that characterizes your confidence in that decision. Check the answer for each line when you complete it. (Be careful not to peek at the next answer.) I designed this table so that some common misconceptions will be revealed—just in case you acquired one.

	N	α level	Two-tailed or one-tailed test?	Critical value	t-test value	Reject H_0 or retain H_0?	p value
a.	10	.05	Two	______	2.25	______	______
b.	20	.01	One	______	2.57	______	______
c.	35	.05	Two	______	2.03	______	______
d.	6	.001	Two	______	6.72	______	______
e.	24	.05	One	______	1.72	______	______
f.	56	.02	Two	______	2.41	______	______

9.18. The Minnesota Multiphasic Personality Inventory (MMPI-2) has one scale that measures paranoia (a mistaken belief that others are out to get you). The population mean on the paranoia scale is 50. Suppose 24 police officers filled out the inventory, producing a mean of 54.3. Write interpretations of these t-test values.

a. $t = 2.10$ **b.** $t = 2.05$

9.19. A large mail-order company with ten distribution centers had an average return rate of 5.3 percent. In an effort to reduce this rate, a set of testimonials was included with every shipment. The return rate was reduced to 4.9 percent. Write interpretations of these t-test values.

a. $t = 2.10$ **b.** $t = 2.50$

***9.20.** Head Start is a program for preschool children from educationally disadvantaged environments. Head Start began in 1964, and the data that follow are representative of early research on the effects of the program.

A group of children participated in the program and then enrolled in first grade. They took an IQ test, the national norm of which was 100. Perform a t test on the data. Begin your work by stating the null hypothesis and end by writing a conclusion about the effects of the program.

$\Sigma X = 5250 \qquad \Sigma X^2 = 560{,}854 \qquad N = 50$

9.21. Odometers measure automobile mileage. How close to the truth is the number that is registered? Suppose 12 cars traveled exactly 10 miles (measured by surveying) and the following mileage figures were recorded by the odometers. State the null hypothesis, choose an alternative hypothesis, perform a *t* test, and write a conclusion.

9.8 10.1 10.3 10.2 9.9 10.4 10.0 9.9 10.3 10.0 10.1 10.2

***9.22.** The NEO PI-R is a personality test. The E stands for extraversion, one of the personality characteristics the test measures. The population mean for the extraversion test is 50. Suppose 13 people who were successful in used car sales took the test and produced a mean of 56.10 and a standard deviation of 10.00. Perform a *t* test and write a conclusion.

Effect Size Index

Null hypothesis statistical testing addresses this question: Is the population the sample comes from different from the population specified by the null hypothesis? The question of *how* different the two populations are is answered by an **effect size index.** Many effect size indexes have been developed (see Kirk, 2005), but *d*, which you studied in Chapter 5 (pages 79–82), is probably the most widely used. For the one-sample *t* test, the formula for *d* is

$$d = \frac{\overline{X} - \mu_0}{\hat{s}}$$

where $\overline{X}$ is the mean of the sample
μ_0 is the mean specified by the null hypothesis
$\hat{s}$ is the estimate of the population standard deviation based on a sample

The denominator $\hat{s}$ is an estimate of σ. For data sets where σ is known, use σ. However, for the more common research situation in which σ is not known, use $\hat{s}$.

Interpretation of *d*

Positive values of *d* indicate a sample mean larger than the hypothesized mean; negative values of *d* signify that the sample mean is smaller than μ_0. As you may recall from Chapter 5, Cohen (1969) proposed guidelines that help evaluate the absolute size of *d*. Widely adopted by researchers, they are:

Small effect	$d = 0.20$
Medium effect	$d = 0.50$
Large effect	$d = 0.80$

In problem 9.20, you concluded that the Head Start program has a statistically significant effect on the IQs of preschool children. You found that the mean IQ of children in the program (105) was significantly higher than 100. But how big is this effect? The statistic d provides an answer. Knowing that $\sigma = 15$ for IQ tests, you get

$$d = \frac{\overline{X} - \mu_0}{\sigma} = \frac{105 - 100}{15} = 0.33$$

An effect size index of .33 indicates that the effect of Head Start is midway between a small effect and a medium effect.

Relationship of *p* to *d*

The statistics p and d tell you different things about the population the sample is from. The inferential statistic, p, coupled with NHST logic, provides information about whether the population sampled from is different from the null hypothesis population. The descriptive statistic, d, estimates the degree of separation between the population sampled from and the null hypothesis population. These two bits of information are relatively independent of each other. Knowing that p is large or small doesn't tell you anything about the size of d.

PROBLEMS

9.23. Calculate the effect size index for the students' Doritos data (page 174), and interpret it using the conventions that researchers use.

9.24. In problem 9.22, you found that people who are successful in used car sales have significantly higher extraversion scores than those in the general population. What is the effect size index for those data and is the effect small, medium, or large?

9.25. In Chapters 2 and 8, you worked with data on normal body temperature. You found that the mean body temperature was 98.2°F rather than the conventional 98.6°F. The standard deviation for those data was 0.7°F. Calculate an effect size index and interpret it.

Other Sampling Distributions

You have been learning about the sampling distribution of the mean. There are times, though, when the statistic necessary to answer a researcher's question is not the mean. For example, to find the degree of relationship between two variables, you need a correlation coefficient. To determine whether a treatment causes more variable responses, you need a standard deviation. And, as you know, proportions are commonly used statistics. In each of these cases (and indeed, for any statistic), researchers often use the basic hypothesis-testing procedure you have just learned. Hypothesis testing always involves a sampling distribution of the statistic.

Where do you find sampling distributions for statistics other than the mean? In the rest of this book you will encounter new statistics and new sampling distributions. Tables E–L in Appendix C represent sampling distributions from which probability figures can be obtained. In addition, some statistics have sampling distributions that are t distributions or normal curves. Still other statistics and their sampling distributions are covered in other books.

Along with every sampling distribution comes a standard error. Just as every statistic has its sampling distribution, every statistic has its standard error. For example, the standard error of the median is the standard deviation of the sampling distribution of the median. The standard error of the variance is the standard deviation of the sampling distribution of the variance. Worst of all, the standard error of the standard deviation is the standard deviation of the sampling distribution of the standard deviation. If you follow that sentence, you probably understand the definition of standard error quite well.

The next step in your exploration of sampling distributions will be brief, though useful and (probably) interesting. The *t* distribution will be used to answer a question about correlation coefficients.

Using the *t* Distribution to Test the Significance of a Correlation Coefficient

In Chapter 6, you learned to calculate a Pearson product-moment correlation coefficient, a descriptive statistic that indicates the degree of relationship between two variables in a bivariate distribution. This section is on testing the statistical significance of correlation coefficients. NHST is used to test the null hypothesis that a sample-based *r* came from a population with a parameter coefficient of .00.

The present-day symbol for the Pearson product-moment correlation coefficient of a population is ρ (rho). The null hypothesis that is being tested is

$$H_0: \rho = .00$$

that is, that there is no relationship between the two variables. The alternative hypothesis is that there is a relationship,

$$H_1: \rho \neq .00 \text{ (a two-tailed test)}$$

For $\rho = .00$, the sampling distribution of *r* is a *t* distribution with a mean of .00. The standard error of *r* is $s_r = \sqrt{(1 - r^2)/(N - 2)}$. The test statistic for a sample *r* is a *t* value calculated from the formula

$$t = \frac{r - \rho}{s_r}$$

clue to the future

Note that the formula for the *t* test of a correlation coefficient has the same form as that of the *t* test for a sample mean. This form, which you will see again, shows a difference between a statistic and a parameter, divided by the standard error of the statistic.

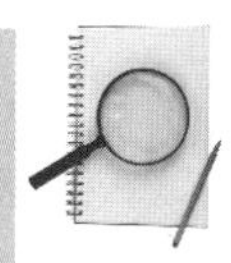

An algebraic manipulation of the formula above produces a formula that is easier to use with a calculator:

$$t = (r)\sqrt{\frac{N-2}{1-r^2}}; \quad df = N - 2, \text{ where } N = \text{number of } pairs$$

This test is the first of several instances in which you will find a t value whose df is *not* 1 less than the number of observations. For testing the null hypothesis H_0: $\rho = .00$, $df = N - 2$. (Having fulfilled my promise to explain one- and two-tailed tests, I hope you will accept my promise to explain degrees of freedom more fully in the next chapter.)

Do you remember your work with correlation coefficients in Chapter 6?[5] Think about a sample $r = -.40$. How likely is a sample $r = -.40$ if the population $\rho = .00$? Does it help to know that the coefficient is based on a sample of 22 pairs? Let's answer these questions by testing the value $r = -.40$ against $\rho = .00$, using a t test. Applying the formula for t, you get

$$t = (r)\sqrt{\frac{N-2}{1-r^2}} = (-.40)\sqrt{\frac{22-2}{1-(-.40)^2}} = -1.95$$

$$df = N - 2 = 22 - 2 = 20$$

Table D shows that, for 20 df, a t value of ± 2.09 (two-tailed test) is required to reject the null hypothesis. The t value obtained for $r = -.40$, where $N = 22$, is less than the tabled t, so the null hypothesis is retained. That is, a coefficient of $-.40$ would be expected by chance alone more than 5 times in 100 from a population in which the true correlation is zero.

In fact, for $N = 22$, the value $r = \pm .43$ is required for significance at the .05 level, and $r = \pm .54$ at the .01 level. As you can see, even medium-sized correlations can be expected by chance alone for samples as small as 22 when the null hypothesis is true. Most researchers strive for N's of 50 or more for correlation problems.

The critical values under "α level for two-tailed test" in Table D are ones that allow you to reject the null hypothesis if the sample r is a large *positive* coefficient or a large *negative* one.

PROBLEM

9.26. Determine whether the Pearson product-moment correlation coefficients are significantly different from .00.

a. $r = .62$; $N = 10$

b. $r = -.19$; $N = 122$

c. $r = .50$; $N = 15$

d. $r = -.34$; $N = 64$

Here is a thought problem. Suppose you have a summer job in an industrial research laboratory testing the statistical significance of hundreds of correlation coefficients based on varying sample sizes. An α level of .05 has been adopted by the management, and your task is to determine whether each coefficient is "significant" or "not significant." How can you construct a table of your own, using Table D and the t formula for testing the significance of a correlation coefficient, that will allow you to label each coefficient without working out a t value for each? Imagine or sketch out your answer.

The table you mentally designed already exists. One version is reproduced in **Table A** in Appendix C. The α values of .10, .05, .02, .01, and .001 are included there. Use Table A in the future to determine whether a Pearson product-moment correlation coefficient is significantly different from .00, but remember that this table is based on the t distribution.

---

[5] If not, take a few minutes to review.

As you might expect, SPSS not only calculates r values, but tests them as well. **Table 6.5** (page 99) shows the correlation between CR SAT verbal and Math SAT scores (.724) and the probability of .724, if the correlation ρ is zero (.042).

A word of caution is in order about testing ρ's other than .00. The sampling distributions for ρ values other than .00 are *not* t distributions. Also, if you want to know whether the *difference* between two sample-based coefficients is statistically significant, you should not use a t distribution because it will not give you correct probabilities. Fortunately, if you have questions that can be answered by testing these kinds of hypotheses, you won't find it difficult to understand the instructions given in intermediate-level texts such as Howell (2010, p. 275).

This chapter has introduced you to two uses of the t distribution:

1. To test a sample mean against a hypothesized population mean
2. To test a sample correlation coefficient against a hypothesized population correlation coefficient of .00

The t distribution has other uses, too, and it has been important in the history of statistics.

Here is a little more background on W. S. Gosset, who invented the t distribution so he could assess probabilities in experiments conducted by the Guinness brewery. Gosset was one of a small group of scientifically trained employees who developed ways for the Guinness company to improve its products.

Gosset wanted to publish his work on the t distribution in *Biometrika,* a journal founded in 1901 by Francis Galton, Karl Pearson, and W. R. F. Weldon. However, at the Guinness company there was a rule that employees could not publish (the rule there, apparently, was publish *and* perish). Because the rule was designed to keep brewing secrets from escaping, there was no particular ferment within the company when Gosset, in 1908, published his new mathematical distribution under the pseudonym "Student."[6] The distribution later came to be known as "Student's t." (No one seems to know why the letter t was chosen. E. S. Pearson surmises that t was simply a "free letter"; that is, no one had yet used t to designate a statistic.) Gosset worked for the Guinness company all his life, so he continued to use the pseudonym "Student" for his publications in mathematical statistics. Gosset was devoted to his company, working hard and rising through the ranks. He was appointed head brewer a few months before his death in 1937.[7]

Why .05?

The story of how science and other disciplines came to adopt .05 as the arbitrary point separating differences attributed to chance from those not attributed to chance is not usually covered in textbooks. The story takes place in England at the turn of the century.

It appears that the earliest explicit rule about α was in a 1910 article by Wood and Stratton, "The Interpretation of Experimental Results," which was published in the *Journal of Agricultural Science.* Thomas B. Wood, the principal editor of this journal, advised researchers to take "30:1 as the lowest odds which can be accepted as giving

[6] However, it may be that the Guinness company was completely unaware until Gosset died that he published as "Student." Salsburg (2001) gives the details of this possibility.

[7] For a delightful account of "That Dear Mr. Gosset," see Salsburg (2001). Also see "Gosset, W. S." in *Dictionary of National Biography, 1931–1940* (London: Oxford University Press, 1949), or see McMullen and Pearson (1939, pp. 205–253).

practical certainty that a difference . . . is significant" (Wood and Stratton, 1910, p. 438). "Practical certainty" meant "enough certainty for a practical farmer." A consideration of the circumstances of Wood's journal provides a plausible explanation of his advice to researchers.

As the 20th century began, farmers in England were doing quite well. For example, Biffen (1905) reported that wheat production averaged 30 bushels per acre compared to 14 in the United States, 10 in Russia, and 7 in Argentina. Politically, of course, England was near the peak of its influence.

In addition, a science of agriculture was beginning to be established. Practical scientists, with an eye on increased production and reduced costs, conducted studies of weight gains in cattle, pigs, and sheep; amounts of dry matter in milk and mangels (a kind of beet); and yields of many grains. In 1905, the *Journal of Agricultural Science* was founded so these results could be shared around the world. Conclusions were reached after calculating probabilities, but the researchers avoided or ignored the problem created by probabilities between .01 and .25.

Wood's recommendation in his "how to interpret experiments" article was to adopt an α level that would provide a great deal of protection against a Type I error. (Odds of 30:1 against the null hypothesis convert to $p = .0323$.) A Type I error in this context would be a recommendation that, when implemented, did not produce improvement. I think that Wood and his colleagues wanted to be *sure* that when they made a recommendation, it would be correct.

A Type II error (failing to recommend an improvement) would not cause much damage; farmers would just continue to do well (though not as well as they might). A Type I error, however, would, at best, result in no agricultural improvement *and* in a loss of credibility and support for the fledgling institution of agriculture science.

To summarize my argument, it made sense for the agricultural scientists to protect themselves and their institution against Type I errors. They did this by not making a recommendation unless the odds were 30:1 against a Type I error.

When tables were published as an aid to researchers, probabilities such as .10, .05, .02, and .01 replaced odds (e.g., Fisher, 1925). So 30:1 ($p = .0323$) gave way to .05. Those tables were used by researchers in many areas besides agriculture, who appear to have adopted .05 as the most important significance level.

PROBLEMS

9.27. In Chapter 6 you encountered data for 11 countries on per capita cigarette consumption and the male death rate from lung cancer 20 years later. The correlation coefficient was .74. After consulting Table A, write a response to the statement, "The correlation is just a fluke; chance is the likely explanation for this correlation."

CP **9.28.** One of my colleagues gathered data on the time students spent on a midterm examination and their grade. The X variable is time and the Y variable is examination grade. Calculate r. Use Table A to determine if it is statistically significant.

$\Sigma X = 903$ $\Sigma X^2 = 25{,}585$ $\Sigma XY = 46{,}885$

$\Sigma Y = 2079$ $\Sigma Y^2 = 107{,}707$ $N = 42$

9.29. In addition to weighing Doritos chips, I weighed six Snickers candy bars manufactured by Mars, Inc. Test the company's advertised claim that the net weight of a bar is 58.7 grams by performing a *t* test on the following data, which are in grams. Find *d*. Write a conclusion about Snickers bars.

59.1	60.0	58.6	60.2	60.8	62.1

9.30. When people choose a number between 1 and 10, the mean is 6. To investigate the effects of subliminal (below threshold) stimuli on behavior, a researcher embedded many low numbers (1 to 3) in a video clip. Each number appeared so briefly that it couldn't consciously be seen. Afterward the participants were asked to "choose a number between 1 and 10." Analyze the data with a *t* test and an effect size index. Write an interpretation.

7	1	6	2	9	3	8	4	7	5	4
8	2	9	8	7	5	4	2	3	6	4
6	8	4	5	6	2	1	4	7	7	8
8	4	8	5	7	6	6				

9.31. Please review the objectives at the beginning of the chapter. Can you do them?

ADDITIONAL HELP FOR CHAPTER 9

Visit *cengage.com/psychology/spatz*. At the Student Companion Site, you'll find multiple-choice tutorial quizzes, flashcards with definitions and workshops. For this chapter, there are Statistical Workshops on Hypothesis Testing and Single-Sample *t* Test.

KEY TERMS

Alpha (α) (p. 179)
Alternative hypothesis (H_1) (p. 176)
Beta (β) (p. 184)
Correlation coefficient (p. 191)
Critical value (p. 180)
Effect size index (d) (p. 189)
NS (p. 180)
Null hypothesis (H_0) (p. 175)
Null hypothesis statistical testing (NHST) (p. 172)
One-sample *t* test (p. 181)
One-tailed test of significance (p. 187)
p (p. 185)
Rejection region (p. 180)
Significance level (p. 179)
Statistically significant (p. 180)
t distribution (p. 177)
Two-tailed test of significance (p. 186)
Type I error (p. 183)
Type II error (p. 183)

CHAPTER 10

Hypothesis Testing, Effect Size, and Confidence Intervals: Two-Sample Designs

OBJECTIVES FOR CHAPTER 10

After studying the text and working the problems in this chapter, you should be able to:

1. Describe the logic of a simple experiment
2. Explain null hypothesis statistical testing (NHST) for two samples
3. Explain some of the reasoning for determining degrees of freedom
4. Distinguish between independent-samples designs and paired-samples designs
5. Calculate *t*-test values for both independent-samples designs and paired-samples designs and write interpretations
6. Distinguish between statistically significant results and important results
7. Calculate and interpret an effect size index for two-sample designs
8. Calculate confidence intervals for both independent-samples designs and paired-samples designs and write interpretations
9. List and explain assumptions required for accurate probabilities from the *t* distribution
10. Explain how random assignment affects cause-and-effect conclusions
11. Define power and explain the factors that affect power

IN CHAPTER 9 YOU learned to use null hypothesis statistical testing (NHST) to answer questions about a population when data from one sample are available. In this chapter, the same hypothesis-testing reasoning is used, but you have data from two samples.

Two-sample NHST is commonly used to analyze data from experiments. One of the greatest benefits of studying statistics is that it helps you understand experiments and the experimental method. The experimental method is probably our most powerful method of investigating natural phenomena. Besides being powerful, experiments can be interesting. They can answer such questions as:

- If you were lost at sea would you rather have a pigeon or a person searching for you?
- Does the removal of 20 percent of the cortex of the brain have an effect on the memory of tasks learned before the operation?
- Can you *reduce* people's ability to solve a problem by educating them?

In this chapter I discuss the simplest kind of experiment and then show you how the logic of null hypothesis statistical testing, which you studied in the preceding chapter, can be expanded to answer questions like those above.

A Short Lesson on How to Design an Experiment

The basic ideas of a simple two-group experiment are not very complicated.

> *The logic of an experiment:* Start with two equivalent groups. Treat them exactly alike except for one thing. Measure both groups. Attribute any statistically significant difference between the two to the one way in which they were treated differently.

This summary of an experiment is described more fully in **Table 10.1.** The question that the experiment in Table 10.1 sets out to answer is: What is the effect of Treatment A on a person's ability to perform task Q? In formal statistical terms, the question is: For task Q scores, is the mean of the population of those who have had Treatment A different from the mean of the population of those who have not had Treatment A?

To conduct this experiment, a group of participants is identified and two samples are assembled. These samples should be (approximately) equivalent. Treatment A is then administered to one group (generically called the **experimental group**) but not to the other group (generically called the **control group**). Except for Treatment A, both groups are treated exactly the same way; that is, extraneous variables are held constant or balanced out for the two groups. Both groups perform task Q, and the mean score for each group is calculated.

The two sample means will almost surely differ. The question is whether the difference is due to Treatment A or is just the usual chance difference that would be expected of two samples from the same population. This question can be answered by applying the logic of NHST.

This generalized example has an independent variable with two levels (Treatment A and no Treatment A) and one dependent variable (scores on task Q). The word **treatment** is recognized by all experimentalists; it refers to different levels of the independent variable. The experiment in Table 10.1 has two treatments.

experimental group
Group that receives treatment in an experiment and whose dependent-variable scores are compared to those of a control group.

control group
No-treatment group to which other groups are compared.

treatment
One value (or level) of the independent variable.

In many experiments, it is obvious that there are *two* populations of participants to begin with—for example, a population of men and a population of women. The question, however, is whether they are equal on the dependent variable.[1]

[1] Two samples from one population are the same statistically as two samples from two identical populations.

TABLE 10.1 Summary of a Simple Experiment

Key words	Tasks for the researcher	
Population	A population of participants	
	↙	↘
Assignment	Randomly assign one-half of *available* participants to the experimental group.	Randomly assign one-half of *available* participants to the control group.
	↓	↓
Independent variable	Give Treatment A to participants in experimental group.	Withhold Treatment A from participants in control group.
	↓	↓
Dependent variable	Measure participants' behavior on task Q.	Measure participants' behavior on task Q.
	↓	↓
Descriptive statistics	Calculate mean score on task Q, $\overline{X}_e$.	Calculate mean score on task Q, $\overline{X}_c$.
	↘	↙
Descriptive statistics	Calculate effect size index, *d*.	
	↓	
Inferential statistics	Compare $\overline{X}_e$ and $\overline{X}_c$ using a hypothesis-testing statistic. Reject or retain the null hypothesis.	
	↓	
Interpretation	Write a conclusion about the effect of Treatment A on task Q scores.	

In some experimental designs, participants are randomly assigned to treatments by the researcher. In other designs, the researcher uses a group of participants who have already been "treated" (for example, being males or being children of alcoholic parents). In either of these designs, the methods of inferential statistics are the same, although the interpretation of the first kind of experiment is usually less open to attack.[2]

The experimental procedure is versatile. Experiments have been used to decide a wide variety of issues such as how much sugar to use in a cake recipe, what kind of persuasion is effective, whether a drug is useful in treating cancer, and the effect of alcoholic parents on the personality of their children.

The *raison d'être*[3] of experiments is to be able to tell a story about the universal effect of one variable on a second variable. All the work with samples is just a way to

[2] I am bringing up an issue that is beyond the scope of this statistics book. Courses with titles such as "Research Methods" and "Experimental Design" address the interpretation of experiments in more detail.

[3] *Raison d'être* means "reason for existence." (Ask someone who can pronounce French phrases to teach you this one. If you already know how to pronounce the phrase, use it in conversation to identify yourself as someone to ask.)

get evidence to support the story, but samples, of course, have chance built into them. NHST takes chance factors into account when the story is told.

NHST: The Two-Sample Example

Once the data are gathered, analysis follows. The NHST technique is a popular one among researchers, although it has been criticized (see Dillon, 1999; Erceg-Hurn and Mirosevich, 2008; Nickerson, 2000; Spatz, 2000.) NHST with two samples is quite similar to NHST with one sample, which you studied in Chapter 9. In fact, the logic is identical. (See pages 173–176 for a review.) Here is the logic of NHST for a two-sample experiment.

In a well-designed, well-executed experiment, all imaginable results are included in the statement—Either Treatment A has an effect or it does not have an effect. Begin by making a tentative assumption that Treatment A does *not* have an effect. Gather data. Using a sampling distribution based on the assumption that Treatment A has no effect, find the probability of the data obtained. If the probability is low, abandon your tentative assumption and draw a strong conclusion: Treatment A has an effect.[4] If the probability is not low, your analysis does not permit you to make strong statements regarding Treatment A.

The outline that follows combines the logic of hypothesis testing with the language of an experiment:

1. Begin with two logical possibilities, the null hypothesis, H_0, and an alternative hypothesis, H_1. H_0 is a hypothesis of equality about parameters. H_1 is a hypothesis of difference, again about parameters.

   H_0: Treatment A *does not* have an effect; that is, the mean of the population of scores of those who receive Treatment A is equal to the mean of the population of scores of those who do not receive Treatment A. The difference between population means is zero. In statistical language:

   $$H_0\colon \mu_{\text{A}} = \mu_{\text{no A}} \qquad \text{or} \qquad H_0\colon \mu_{\text{A}} - \mu_{\text{no A}} = 0$$

   H_1: Treatment A *does* have an effect; that is, the mean of the population of scores of those who receive Treatment A is *not* equal to the mean of the population of scores of those who do not receive Treatment A. The alternative hypothesis, which should be chosen before the data are gathered, can be two-tailed or one-tailed. The alternative hypothesis consists of *one* of the three H_1's that follow.

Two-tailed alternative: $H_1\colon \mu_{\text{A}} \neq \mu_{\text{no A}}$

The two-tailed alternative hypothesis says that Treatment A has an effect, but it does not indicate whether the treatment improves or disrupts performance on task Q. A two-tailed test allows either conclusion.

One-tailed alternatives: $H_1\colon \mu_{\text{A}} > \mu_{\text{no A}}$

$H_1\colon \mu_{\text{A}} < \mu_{\text{no A}}$

---

[4] Always tell whether the treatment raises scores or lowers scores.

The first one-tailed alternative hypothesis allows you to conclude that Treatment A increases scores. However, no outcome of the experiment can lead to the conclusion that Treatment A produces lower scores than no treatment. The second alternative hypothesis permits a conclusion that Treatment A reduces scores, but not that it increases scores.

2. Tentatively assume that Treatment A has no effect (that is, assume H_0). If H_0 is true, the two samples will be alike except for the usual chance variations in samples.
3. Decide on an α level. (Usually, $\alpha = .05$.)
4. Choose an appropriate statistical test. For a two-sample experiment, a *t* test is appropriate. Calculate a *t*-test value.
5. Compare the calculated *t*-test value to the critical value for α. (Use Table D in Appendix C.)
6. If the data-based *t*-test value is greater than the critical value from the table, reject H_0. If the data-based *t*-test value is less than the critical value, retain H_0.
7. Calculate an effect size index (d).
8. Write a conclusion that is supported by the data analysis. Your conclusion should describe how any differences in the dependent variable means are related to the two levels of the independent variable.

As you know from the previous chapter, sample means from a population are distributed as a *t* distribution. It is also the case that the *differences* between sample means, each drawn from the same population, are distributed as a *t* distribution.[5]

Researchers sometimes are interested in statistics other than the mean, and they often have more than two samples in an experiment. These situations call for sampling distributions that will be covered later in this book—sampling distributions such as *F*, chi square, *U*, and the normal distribution. NHST can be used for these situations as well. The only changes in the preceding list are in steps 4, 5, and 6, starting with "Choose an appropriate test." I hope that you will return to this section for review when you encounter new sampling distributions. If this seems like a good idea, mark this page.

PROBLEMS

10.1. In your own words, outline a simple experiment.

10.2. What procedure is used in the experiment described in Table 10.1 to ensure that the two samples are equivalent before treatment?

10.3. For each experiment that follows, identify the independent variable and its levels. Identify the dependent variable. State the null hypothesis in words.

a. A psychologist compared the annual income of people with high Satisfaction With Life Scale scores to the income of those with low SWLS scores.

b. A sociologist compared the numbers of years served in prison by those convicted of robbery and those convicted of embezzlement.

c. A physical therapist measured flexibility after 1 week of treatment and again after 6 weeks of treatment for 40 patients.

[5] Differences are distributed as *t*, *if* the assumptions discussed near the end of this chapter are true.

10.4. In your own words, outline the logic of NHST for a two-group experiment.
10.5. What is the purpose of an experiment, according to your textbook?

Degrees of Freedom

In your use of the *t* distribution so far, you found **degrees of freedom** using rule-of-thumb techniques: $N - 1$ for a one-sample *t* test, and $N - 2$ when you determine whether a correlation coefficient *r* is significantly different from .00. Now it is time to explain degrees of freedom more thoroughly.

The *freedom* in *degrees of freedom* refers to the freedom of a number to have any possible value. If you are asked to pick two numbers and there are no restrictions, both numbers are free to vary (take any value) and you have 2 degrees of freedom. If a restriction, such as $\Sigma X = 0$, is imposed, then 1 degree of freedom is lost because of that restriction; that is, when you now pick the two numbers, only one of them is free to vary. As an example, if you choose 3 for the first number, the second number *must be* −3. Because of the restriction that $\Sigma X = 0$, the second number is not free to vary. In a similar way, if you are to pick five numbers with a restriction that $\Sigma X = 0$, you have 4 degrees of freedom. Once four numbers are chosen (say, −5, 3, 16, and 8), the last number (−22) is determined.

The restriction that $\Sigma X = 0$ may seem to you to be an "out-of-the-blue" example and unrelated to your earlier work in statistics, but some of the statistics you calculated have such a restriction built in. For example, when you found , as required in the formula for the one-sample *t* test, you used some algebraic version of

$$s_{\bar{X}} = \frac{\hat{s}}{\sqrt{N}} = \frac{\sqrt{\dfrac{\Sigma(X - \bar{X})^2}{N - 1}}}{\sqrt{N}}$$

The built-in restriction is that $\Sigma(X - \bar{X})$ is always zero and, in order to meet that requirement, one of the *X*'s is determined. All *X*'s are free to vary except one, and the degrees of freedom for $s_{\bar{X}}$ is $N - 1$. Thus, for the problem of using the *t* distribution to determine whether a sample came from a population with a mean μ_0, $df = N - 1$. Walker (1940) summarized this reasoning by stating: "A universal rule holds: The number of degrees of freedom is always equal to the number of observations minus the number of necessary relations obtaining among these observations." A necessary relationship for $s_{\bar{X}}$ is that $\Sigma(X - \bar{X}) = 0$.

Another approach to explaining degrees of freedom is to emphasize the parameters that are being estimated by statistics. The rule for this approach is that *df* is equal to the number of observations minus the number of parameters that are estimated with a sample statistic. In the case of $s_{\bar{X}}$, 1 *df* is subtracted because $\bar{X}$ is used as an estimate of μ.

Now, let's turn to your use of the *t* distribution to determine whether a correlation coefficient *r* is significantly different from .00. The test statistic was calculated using the formula

$$t = (r)\sqrt{\frac{N - 2}{1 - r^2}}$$

The r in the formula is a *linear* correlation coefficient, which is based on a linear regression line. The formula for the regression line is

$$\hat{Y} = a + bX$$

There are two parameters in the regression formula, a and b. Each of these parameters costs one degree of freedom, so the df for testing the significance of correlation coefficients is $N - 2$.

These explanations of df for the one-sample t test and for testing the significance of r will prepare you for the reasoning I will give for determining df when you analyze two-group experiments.

Paired-Samples Designs and Independent-Samples Designs

paired-samples design
Experimental design in which scores from each group are logically matched.

independent-samples design
Experimental design with samples whose dependent-variable scores cannot logically be paired.

***t* test**
Test of a null hypothesis that uses *t* distribution probabilities.

An experiment with two groups can be either a **paired-samples design** or an **independent-samples design.** For either design, the appropriate statistical test is a ***t* test.** However, the two different designs require different formulas for calculating the t test, so you must decide what kind of design you have before you analyze the data.

In a paired-samples (or paired-scores) design, each dependent-variable score in one treatment is matched or paired with a particular dependent-variable score in the other treatment. This pairing is based on some logical reason for matching up two scores and not on the size of the scores. In an independent-samples design, there is no reason to pair up the scores in the two groups.

You cannot tell the difference between the two designs just by knowing the independent variable and the dependent variable. And, after the data are analyzed, you cannot tell the difference from the t-test value or from the interpretation of the experiment. To tell the difference, you must know whether scores in one group are paired with scores in a second group.

clue to the future

Most of the rest of this chapter will be organized around independent-samples and paired-samples designs. In Chapters 11 and 13 the procedures you will learn are appropriate for only independent samples. The design in Chapter 12 is a paired-samples design with more than two levels of the independent variable. Three-fourths of Chapter 15 is also organized around these two designs.

Paired-Samples Design

The paired-samples design is a favorite of researchers if their materials permit.[6] The logical pairing required for this design can be created three ways: natural pairs,

[6] Other terms for this design are *correlated samples, related samples, dependent samples, matched groups, within subject, repeated measures,* and *split plots.*

matched pairs, and repeated measures. Fortunately, the arithmetic of calculating a t-test value is the same for all three.

Natural pairs In a **natural pairs** investigation, the researcher does not assign the participants to one group or the other; the pairing occurs naturally, prior to the investigation. **Table 10.2** identifies one way in which natural pairs may occur in family relationships—fathers and sons. In such an investigation, you might ask whether fathers are shorter than their sons (or more religious, or more racially prejudiced, or whatever). Notice, though, that it is easy to decide that these are paired-samples data: There is a logical pairing of the scores by family. Pairing is based on membership in the same family.

natural pairs
Paired-samples design in which pairing occurs without intervention by the researcher.

Matched pairs In some situations, the researcher can pair up participants on the basis of similar (or identical) scores on a pretest that is related to the dependent variable. Then one member of each pair is randomly assigned to one of the treatments. This ensures that the two groups are fairly equivalent on the dependent variable at the beginning of the experiment.

For example, in an experiment on the effect of hypnosis on problem solving, college students might initially be paired on the basis of SAT or ACT scores, which are known predictors of problem-solving ability. Participants are then separated into two groups. One solves problems while hypnotized and the other works without being hypnotized. If the two groups differ in problem solving, you have some assurance that the difference is not due to cognitive ability because the groups were matched on SAT or ACT scores.

Another variation of **matched pairs** is the split-litter technique used with animal subjects. A pair from a litter is selected, and one individual is put into each group. In this way, the genetics and prenatal conditions are matched for the two groups. The same technique has been used in human experiments with twins or siblings. Gosset's barley experiments used this design; Gosset started with two similar subjects (adjacent plots of ground) and assigned them at random to one of two treatments.

matched pairs
Paired-samples design in which individuals are paired by the researcher before the experiment.

A third example of the matched-pairs technique occurs when a pair is formed *during* the experiment. Two subjects are "yoked" so that what happens to one, happens to the other (except for the independent and dependent variables). For example, if the

TABLE 10.2 Illustration of a paired-samples design

Father	Height (in.) X	Son	Height (in.) Y
Michael Smith	74	Mike, Jr.	74
Christopher Johnson	72	Chris, Jr.	72
Matthew Williams	70	Matt, Jr.	70
Joshua Jones	68	Josh, Jr.	68
Daniel Brown	66	Dan, Jr.	66
David Davis	64	Dave, Jr.	64

procedure allows an animal in the experimental group to get a varying amount of food (or exercise or punishment), then that same amount of food (or exercise or punishment) is administered to the "yoked" control subject.

The difference between the matched-pairs design and a natural-pairs design is that, with the matched pairs, the investigator can randomly assign one member of the pair to a treatment. In the natural-pairs design, the investigator has no control over assignment. Although the statistics are the same, the natural-pairs design is usually open to more interpretations than the matched-pairs design.

repeated measures
Experimental design in which each subject contributes to more than one treatment.

Repeated measures A third kind of paired-samples design is called a **repeated-measures** design because more than one measure is taken on each participant. This design may take the form of a before-and-after experiment. A pretest is given, some treatment is administered, and a posttest is given. The mean of the scores on the posttest is compared with the mean of the scores on the pretest to determine the effectiveness of the treatment. Clearly, two scores should be paired: the pretest and the posttest scores of each participant. In such an experiment, each person is said to serve as his or her own control.

For paired-samples designs, researchers typically label one level of the independent variable X and the other Y. The first score under treatment X is paired with the first score under treatment Y because the two scores are natural pairs, matched pairs, or repeated measures on the same subject.

Independent-Samples Design

The experiment outlined in Table 10.1 is an independent-samples design.[7] In this design, researchers typically label the two levels of the independent variable X_1 and X_2. The distinguishing characteristic of an independent-samples design is that there is no reason to pair a score in one treatment (X_1) with a particular score in the other treatment (X_2).

In creating an independent-samples design, researchers often begin with a pool of participants and then randomly assign individuals to the groups. Thus, there is no reason to suppose that the first score under X_1 should be paired with the first score under X_2. Random assignment can be used with paired-samples designs, too, but in a different way. In a paired-samples design, one member of a *pair* is assigned randomly.

The basic difference between the two designs is that with a paired-samples design there is a logical reason to pair up scores from the two groups; with the independent-samples design there is not. Thus, in a paired-samples design, the two scores in the first row of numbers belong there because they come from two sources that are logically related.

[7] Other terms used for this design are *between subjects, unpaired, randomized,* and *uncorrelated.*

PROBLEMS

10.6. Describe the two ways to state the rule governing degrees of freedom.

10.7. Identify each of the following as an independent-samples design or a paired-samples design. For each, identify the independent variable and its levels and the dependent variable. Work all six problems before checking your answers.

a. An investigator gathered many case histories of situations in which identical twins were raised apart—one in a "good" environment and one in a "bad" environment. The group raised in the "good" environment was compared with that raised in the "bad" environment on attitude toward education.

***b.** A researcher counted the number of aggressive encounters among children who were playing with worn and broken toys. Next, the children watched other children playing with new, shiny toys. Finally, the first group of children resumed playing with the worn and broken toys, and the researcher again counted the number of aggressive encounters. (For a classic experiment on this topic, see Barker, Dembo, and Lewin, 1941.)

c. Patients with seasonal affective disorder (a type of depression) spent 2 hours a day under bright artificial light. Before treatment the mean depression score was 20.0, and at the end of one week of treatment the mean depression score was 6.4.

d. One group was deprived of REM (rapid eye movement) sleep and the other was not. At the sleep lab, participants were paired and randomly assigned to one of the two conditions. When those in the deprivation group began to show REM, they were awakened, thus depriving them of REM sleep. The other member of the pair was then awakened for an equal length of time during non-REM sleep. The next day all participants filled out a mood questionnaire.

e. One group was deprived of REM sleep and the other was not. At the sleep lab, participants were randomly assigned to one of the two groups. When those in the deprivation group began to show REM, they were awakened, thus depriving them of REM sleep. The other participants were awakened during non-REM sleep. The next day all participants filled out a mood questionnaire.

f. Thirty-two freshmen applied for a sophomore honors course. Only 16 could be accepted, so the instructor flipped a coin for each applicant, with the result that 16 were selected. At graduation, the instructor compared the mean grade point average of those who had taken the course to the GPA of those who had not to see if the sophomore honors course had an effect on GPA.

The *t* Test for Independent-Samples Designs

When a *t* test is used to decide whether two populations have the same mean, the null hypothesis is

$$H_0\colon \mu_1 = \mu_2$$

where the subscripts 1 and 2 are assigned arbitrarily to the two populations. If this null hypothesis is true, any difference between two sample means is due to chance. The task is to establish an α level, calculate a t-test value, and compare that value with a critical value of t in Table D. If the t value calculated from the data is greater than the critical value (that is, less probable than α), reject H_0 and conclude that the two samples came from populations with different means. If the data-based t value is not as large as the critical value, retain H_0. I expect that this sounds familiar to you. For an independent-samples design, the formula for the t test is

$$t = \frac{\overline{X}_1 - \overline{X}_2}{s_{\overline{X}_1 - \overline{X}_2}}$$

standard error of a difference
Standard deviation of a sampling distribution of differences between means.

The term $s_{\overline{X}_1 - \overline{X}_2}$ is the **standard error of a difference,** and **Table 10.3** shows formulas for calculating it. Use the formula at the top of the table when the two samples have an unequal number of scores. In the situation $N_1 = N_2$, the formula simplifies to those shown in the lower portion of Table 10.3. These formulas are called the *pooled error* by some statistical software.

The previous t-test formula is the "working formula" for the more general case in which the numerator is $(\overline{X}_1 - \overline{X}_2) - (\mu_1 - \mu_2)$. For the examples in this book, the hypothesized value of $\mu_1 - \mu_2$ is zero, which reduces the general case to the working formula shown above. Thus, the t test, like many other statistical tests, consists of a difference between a statistic and a parameter divided by the standard error of the statistic.

The formula for degrees of freedom for independent samples is $df = N_1 + N_2 - 2$. Here is the reasoning. For each sample, the number of degrees of freedom is $N - 1$ because,

TABLE 10.3 Formulas for $s_{\overline{X}_1 - \overline{X}_2}$, the standard error of a difference, for independent-samples t tests

If $N_1 \neq N_2$:

$$s_{\overline{X}_1 - \overline{X}_2} = \sqrt{\left(\frac{\Sigma X_1^2 - \frac{(\Sigma X_1)^2}{N_1} + \Sigma X_2^2 - \frac{(\Sigma X_2)^2}{N_2}}{N_1 + N_2 - 2}\right)\left(\frac{1}{N_1} + \frac{1}{N_2}\right)}$$

If $N_1 = N_2$:

$$s_{\overline{X}_1 - \overline{X}_2} = \sqrt{s_{\overline{X}_1}^2 + s_{\overline{X}_2}^2}$$

$$= \sqrt{\left(\frac{\hat{s}_1}{\sqrt{N_1}}\right)^2 + \left(\frac{\hat{s}_2}{\sqrt{N_2}}\right)^2}$$

$$= \sqrt{\frac{\Sigma X_1^2 - \frac{(\Sigma X_1)^2}{N_1} + \Sigma X_2^2 - \frac{(\Sigma X_2)^2}{N_2}}{N_1(N_2 - 1)}}$$

for each sample, a mean has been calculated with the restriction that $\Sigma(X - \bar{X}) = 0$. Thus, the total degrees of freedom is $(N_1 - 1) + (N_2 - 1) = N_1 + N_2 - 2$.

Let's use the independent-samples design to analyze data that illustrate a classic social psychology phenomenon. The question is whether you work harder when you are part of a group or when you are alone. In the experimental setup, blindfolded participants were instructed to "pull as hard as you can on this rope." Some participants thought they were part of a group and others thought they were pulling alone. In fact, all were pulling alone, and their effort (in kilograms) was recorded. (See Ingham et al., 1974, for a similar experiment.)

The null hypothesis is that group participation has no effect on effort. A two-tailed test is appropriate because we want to be able to detect any effect that group participation has, either to enhance *or* inhibit performance.

The scores of the participants and the t test are shown in **Table 10.4.** Because the N's are unequal for the two samples, the longer formula for the standard error must be

TABLE 10.4 Effort in kilograms for participants in the social psychology experiment

	Group (X_1)	Alone (X_2)
	34	39
	52	57
	26	68
	47	74
	42	49
	37	57
	40	
ΣX	278	344
ΣX^2	11,478	20,520
N	7	6
$\bar{X}$	39.71	57.33

Applying the t test gives

$$t = \frac{\bar{X}_1 - \bar{X}_2}{s_{\bar{X}_1 - \bar{X}_2}} = \frac{\bar{X}_1 - \bar{X}_2}{\sqrt{\left(\dfrac{\Sigma X_1^2 - \dfrac{(\Sigma X_1)^2}{N_1} + \Sigma X_2^2 - \dfrac{(\Sigma X_2)^2}{N_2}}{N_1 + N_2 - 2}\right)\left(\dfrac{1}{N_1} + \dfrac{1}{N_2}\right)}}$$

$$= \frac{39.71 - 57.33}{\sqrt{\left(\dfrac{11{,}478 - \dfrac{(278)^2}{7} + 20{,}520 - \dfrac{(344)^2}{6}}{7 + 6 - 2}\right)\left(\dfrac{1}{7} + \dfrac{1}{6}\right)}}$$

$$= \frac{-17.62}{\sqrt{\left(\dfrac{437.43 + 797.33}{11}\right)(0.31)}} = \frac{-17.62}{5.90} = -2.99$$

$$df = N_1 + N_2 - 2 = 7 + 6 - 2 = 11$$

used. The t value for these data is -2.99, a negative value. For a two-tailed test, always use the absolute value of t. Thus, $|-2.99| = 2.99$.

To evaluate a data-based t value of 2.99 with 11 *df*, you need a critical value from Table D. Begin by finding the row with 11 *df*. The critical value in the column for a two-tailed test with $\alpha = .05$ is 2.201; that is, $t_{.05}(11) = 2.201$. Thus, the null hypothesis can be rejected. Because 2.99 is also greater than the critical value at the .02 level (2.718), the results would usually be reported as "significant at the .02 level."

An SPSS analysis of the data in Table 10.4 is shown in **Table 10.5.** Descriptive statistics for the two conditions in the experiment are in the upper panel; most of the output from the independent-samples t test is in the lower panel. In the lower panel, the first line, the one with 11 *df*, corresponds to the t test that I've provided the formulas for. The t value of -2.989 has a probability of .012.

The final step is to interpret the results. Stop for a moment and compose your interpretation of what this experiment shows. My version follows.

> The experiment shows that participants exert less effort when working with a group than when working alone. Participants exerted significantly less effort pulling a rope when they thought they were part of a group ($\overline{X} = 39.71$ kg) than they did when they thought they were pulling alone ($\overline{X} = 57.33$ kg), $p < .02$. The phenomenon that people slack off when they are part of a group, compared to working alone, is called *social loafing*.

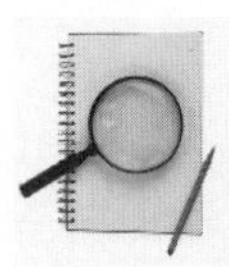

clue to the future

The end result of statistical software programs is a p value. Such p values make critical values obsolete. If $p \leq \alpha$, reject H_0. If $p > \alpha$, retain H_0. Tables and their critical values are not needed.

TABLE 10.5 SPSS output for independent-samples *t* test of the data in Table 10.4

Group Statistics

	Perception	N	Mean	Std. Deviation	Std. Error Mean
Effort	Group	7	39.7143	8.53843	3.22722
	Alone	6	57.3333	12.62801	5.15536

Independent Samples Test

		t test for Equality of Means			95% Confidence Interval of the Difference	
		t	df	Sig. (2-tailed)	Lower	Upper
Effort	Equal variances assumed	-2.989	11	.012	-30.59262	-4.64548
	Equal variances not assumed	-2.897	8.588	.019	-31.47926	-3.75884

PROBLEMS

***10.8.** Imagine being lost at sea, bobbing around in an orange life jacket. The Coast Guard sends out two kinds of observers to spot you, fellow humans and pigeons. All observers are successful; the search time for each is given in the table in minutes. Decide on a one- or two-tailed test, analyze the data, and write a conclusion.

Fellow humans	Pigeons
45	31
63	24
39	20

***10.9.** Karl Lashley (1890–1958) studied how the brain stores memories. In one experiment, rats learned a simple maze. Afterward, they were anesthetized and for half the rats, 20 percent of the cortex of the brain was removed. For the other half, no brain tissue was removed (a sham operation). The rats recovered and learned a maze. The number of errors was recorded. Name the independent variable, its levels, and the dependent variable. Analyze the data in the table and write a conclusion about the effect of a 20 percent loss of cortex on retention of memory for a simple maze.

	Percent of cortex removed	
	0	20
ΣX	208	252
ΣX^2	1706	2212
N	40	40

10.10. A mail-order firm was considering a new software package that provided information on products (availability, history, cost, and so forth). To test the software, the next 17 new employees were randomly assigned to use the old or the new software. After training, each employee completed a search report on 40 products. Analyze the data, which are in minutes, and write a conclusion about the new software package.

New package	Old package
4.3	6.3
4.7	5.8
6.4	7.2
5.2	8.1
3.9	6.3
5.8	7.5
5.2	6.0
5.5	5.6
4.9	

***10.11.** Two sisters attend different universities. Each is sure that she has to compete with brighter students than her sister. They decide to settle their disagreement by comparing ACT admission scores of freshmen at the two schools. Suppose they bring you the following data and ask for an analysis. The *N*'s represent all freshmen for one year.

	The U.	State U.
$\bar{X}$	23.4	23.5
$\hat{s}$	3	3
N	8000	8000

The sisters agree to a significance level of .05. Begin by deciding whether to use a one-tailed or a two-tailed test. Calculate a *t* test and interpret your results. Be sure to carry several decimal places in your work on this problem.

The *t* Test for Paired-Samples Designs

The paired-samples *t* test has a familiar theme: a difference between means $(\bar{X} - \bar{Y})$ divided by the standard error of a difference $(s_{\bar{D}})$. The working formula for this *t* test is

$$t = \frac{\bar{X} - \bar{Y}}{s_{\bar{D}}} = \frac{\bar{D}}{s_{\bar{D}}}$$

I will explain each of these elements and then conclude with a discussion of degrees of freedom. I begin with $s_{\bar{D}}$, the standard error of a difference, which I will explain two ways. First, the definitional formula is

$$s_{\bar{D}} = \sqrt{s_{\bar{X}}^2 + s_{\bar{Y}}^2 - 2r_{XY}(s_{\bar{X}})(s_{\bar{Y}})}$$

Compare this standard error of a difference to that for an independent-samples *t* test for equal *N*'s (lower portion of Table 10.3). The difference in the two procedures is the term $-2r_{XY}(s_{\bar{X}})(s_{\bar{Y}})$. As you can see, when $r_{XY} = 0$, this term becomes zero, and the standard error becomes the same as for independent samples.

Now, notice what happens to the formula when $r > 0$: The standard error of the difference is reduced. Reducing the size of the standard error increases the size of the *t*-test value. Whether this reduction increases the likelihood of rejecting the null hypothesis depends on the size of the reduction because, as will be explained, there are fewer degrees of freedom in a paired-samples design than in an independent-samples design.

Understanding the formula $s_{\bar{D}} = \sqrt{s_{\bar{X}}^2 + s_{\bar{Y}}^2 - 2r_{XY}(s_{\bar{X}})(s_{\bar{Y}})}$ should help you better understand the difference between the two designs in this chapter. Unfortunately, this formula requires a number of separate calculations. A second formula, the *direct difference method,* is algebraically equivalent and much simpler.

As you know, in a paired-samples t test, the scores for one group (X) are aligned with their partners in the other group (Y). To calculate $s_{\bar{D}}$ using the direct difference method, find the difference between each pair of scores, calculate the standard deviation of the difference scores, and divide the standard deviation by the square root of the number of pairs. The formula is

$$s_{\bar{D}} = \frac{\hat{s}_D}{\sqrt{N}} = \frac{\sqrt{\dfrac{\Sigma D^2 - \dfrac{(\Sigma D)^2}{N}}{N - 1}}}{\sqrt{N}}$$

where $D = X - Y$
N = number of *pairs* of scores
$\hat{s}_D$ = standard deviation of the difference scores

In my examples I use this direct difference method, but for one of the problems, you'll need the definitional formula.

The numerator of the paired-samples t test is the difference between the means of the two samples. This value can be found by subtracting one mean from the other $(\bar{X} - \bar{Y})$ or by averaging the D values $(\bar{D})$. The null hypothesis for a paired-samples t test is $H_0\text{: } \mu_X - \mu_Y = 0$. Thus, the general case of the null hypothesis of the numerator is $(\bar{X} - \bar{Y}) - (\mu_X - \mu_Y)$.

With the elements explained, the working formula for a paired-samples t test is

$$t = \frac{\bar{X} - \bar{Y}}{s_{\bar{D}}} = \frac{\bar{D}}{\dfrac{\hat{s}_D}{\sqrt{N}}}$$

where $df = N - 1$, and N = number of pairs.

The degrees of freedom in a paired-samples t test is the number of pairs minus 1. Although each pair has two values, once one value is determined, the other is expected to be a similar value (not free to vary independently). In addition, another degree of freedom is subtracted when $s_{\bar{D}}$ is calculated. This loss is similar to the loss of 1 *df* when $s_{\bar{X}}$ is calculated.

Here is an example of a paired-samples design and a t-test analysis. Suppose you are interested in the effects of interracial contact on racial attitudes. You have a fairly reliable test of racial attitudes in which high scores indicate more positive attitudes. You administer the test one Monday morning to a multiracial group of fourteen 12-year-old girls who do not know one another but who have signed up for a weeklong community day camp. The campers then spend the next week taking nature walks, playing ball, eating lunch, swimming, making things, and doing the kinds of things that camp directors dream up to keep 12-year-old girls busy. On Saturday morning the girls are again given the racial attitude test. Thus, the data consist of 14 pairs of before-and-after scores. The null hypothesis is that the mean of the population of "after" scores is equal to the mean of the population of "before" scores or, in terms of the specific experiment, that a week of interracial contact has no effect on racial attitudes.

Suppose researchers obtain the data in **Table 10.6.** Using the sums of the D and D^2 columns in Table 10.6, you can find $\hat{s}_D$:

$$\hat{s}_D = \sqrt{\frac{\Sigma D^2 - \frac{(\Sigma D)^2}{N}}{N-1}} = \sqrt{\frac{897 - \frac{(-81)^2}{14}}{13}} = \sqrt{32.951} = 5.740$$

$$s_{\bar{D}} = \frac{\hat{s}_D}{\sqrt{N}} = \frac{5.740}{\sqrt{14}} = 1.534$$

Also, $\Sigma D = -81$, and $N = 14$. Thus, $\bar{D} = \Sigma D/N = -81/14 = -5.78$. Now you can find the t-test value:

$$t = \frac{\bar{X} - \bar{Y}}{s_{\bar{D}}} = \frac{28.36 - 34.14}{1.534} = \frac{-5.78}{1.534} = -3.77$$

$$df = N - 1 = 14 - 1 = 13$$

Because $t_{.01}(13\ df) = 3.012$, a t value of 3.77 is significant beyond the .01 level; that is, $p < .01$. Because the "after" mean is greater than the "before" mean, conclude that racial attitudes were significantly more positive after camp than before.

An SPSS analysis of the data in Table 10.6 is in **Table 10.7.** The upper panel shows descriptive statistics of the racial attitudes of the girls before camp and after camp. The lower panel shows numbers you saw as you worked through the paired-samples t test of the difference; it confirms the t-test value of -3.77.

TABLE 10.6 Hypothetical data from a racial attitudes study

	Racial attitude scores			
Name*	Before day camp X	After day camp Y	D	D^2
Jessica	34	38	−4	16
Ashley	22	19	3	9
Brittany	25	36	−11	121
Amanda	31	40	−9	81
Samantha	27	36	−9	81
Sarah	32	31	1	1
Stephanie	38	43	−5	25
Jennifer	37	36	1	1
Elizabeth	30	30	0	0
Lauren	26	31	−5	25
Megan	16	34	−18	324
Emily	24	31	−7	49
Nicole	26	36	−10	100
Kayla	29	37	−8	64
Sum	397	478	−81	897
Mean	28.36	34.14		

* The names are, in order, the 14 most common for baby girls in the United States in 1990 (*www.ssa.gov/cgi-bin/popularnames.cgi*).

TABLE 10.7 Output of an SPSS paired-samples *t* test of the data in Table 10.6

Paired Samples Statistics

	Mean	N	Std. Deviation	Std. Error Mean
Before Camp	28.3571	14	5.94341	1.58844
After Camp	34.1429	14	5.72252	1.52941

Paired Samples Test

	Paired Differences							
				95% Confidence Interval of the Difference				
	Mean	Std. Deviation	Std. Error Mean	Lower	Upper	*t*	*df*	Sig. (2-tailed)
Before Camp After Camp	−5.78571	5.74026	1.53415	−9.10004	−2.47139	−3.771	13	.002

error detection

Deciding whether a study is a paired-samples or independent-samples design is difficult for many beginners. Here are two hints. If the two groups have different *N*'s, the design is independent samples. If *N*'s are equal, look at the top row of numbers and ask if they are on the top row together for a reason (such as natural pairs, matched pairs, or repeated measures). If yes, paired-samples design; if no, independent-samples design.

PROBLEMS

10.12. Give a formula and definition for each symbol.

a. $\hat{s}_D$

b. D

c. $s_{\overline{D}}$

d. $\overline{Y}$

10.13. The accompanying table shows data based on the aggression and toys experiment described in problem 10.7b. Reread the problem, analyze the scores, and write a conclusion.

Before	After
16	18
10	11
17	19
4	6
9	10
12	14

***10.14.** When I first began statistical consulting, a lawyer asked me to analyze salary data for employees at a large rehabilitation center to determine if there was evidence of sex discrimination. At the center, men's salaries were about 25 percent higher than women's. To rule out explanations such as "more educated" and "more experienced," a subsample of employees with bachelor's degrees was separated out. From this group, men and women with equal experience were paired, resulting in an N of 16 pairs. Analyze the data. In your conclusion (which you can direct to the judge), explain whether education, experience, or chance could account for the observed difference. (*Note:* You will have to think to set up this problem.)

	Women, X	Men, Y
ΣX or ΣY	\$204,516	\$251,732
ΣX^2 or ΣY^2	2,697,647,000	4,194,750,000
ΣXY	3,314,659,000	

***10.15.** Which do you respond faster to, a visual signal or an auditory signal? The reaction times (RT) of seven students were measured for both signals. Decide whether the study is an independent- or paired-samples design, analyze the data, and write a conclusion.

Auditory RT (seconds)	Visual RT (seconds)
0.16	0.18
0.19	0.20
0.14	0.20
0.14	0.17
0.13	0.14
0.18	0.17
0.20	0.21

10.16. To study the effects of primacy and recency, two groups were chosen randomly from a large sociology class. Both groups read a two-page description of a person at work, which included a paragraph telling how the person had been particularly helpful to a new employee. For half of the participants the "helping" paragraph was near the beginning (primacy), and for the other half the "helping" paragraph was near the end (recency). Afterward, each participant wrote a paragraph describing the worker's leisure activities. The dependent variable was the number of positive adjectives in the paragraph (words such as *good, exciting, cheerful*). Based on these data, write a conclusion about primacy and recency.

Primacy	Recency
0	1
6	6
10	8
8	4
5	4

10.17. Asthma is a reemerging health problem among children. In one experiment, a group whose attacks began before age 3 was compared with a group whose attacks began later (Mrazek, Schuman, and Klinnert, 1998). The dependent variable was the child's score on the Behavioral Screening Questionnaire (BSQ), which was filled out by the mother. High BSQ scores indicate more problems. Analyze the difference between means and write a conclusion.

	Onset of asthma	
	Before age 3	After age 6
$\bar{X}$	60	36
$\hat{s}$	12	6
N	45	45

Significant Results and Important Results

When I was a rookie instructor, I did some research with an undergraduate student, David Cervone, who was interested in hypnosis. We used two rooms in the library to conduct the experiment. When the results were analyzed, two of the groups were significantly different. One day, a librarian asked how the experiment had come out.

"Oh, we got some significant results," I said.

"I imagine David thought they were more significant than you did," was the reply.

At first I was confused. Why would David think they were more significant than I would? Point oh-five was point oh-five. Then I realized that the librarian and I were using the word *significant* in two quite different ways. I was using *significant* in the statistical sense: The difference was a *reliable* one that would be expected to occur again if the study was run again. The librarian meant *important.* Of course, *important* is important, so let's pursue the librarian's meaning and ask about the importance of a difference.

First of all, if a difference is not statistically significant, its importance is questionable because such differences might reasonably be attributed to chance. If a difference is NS, you have no assurance that a second experiment will produce a similar difference.

If a difference is statistically significant, you have to go beyond NHST to decide if the difference is important. Arthur Irion (1976) captured this limitation of NHST (and the two meanings of the word *significant*) when he reported that his statistical test "reveals that there is significance among the differences although, of course, it doesn't reveal what significance the differences have."

In two problems that you worked, you found differences that were statistically significant. In problem 10.11 the difference in college ACT admissions scores between two schools was one-tenth of a point, which doesn't qualify as important. In contrast, in problem 10.17 the big difference in children's behavioral problems that depended on age of asthma onset seems like an important difference.

Thus, NHST tests a difference only for statistical significance. The next two sections will describe other statistics that help you decide about importance.

Effect Size Index

The effect size index, d, may help you decide whether a significant difference is important. You probably recall from Chapter 5 that the mathematical formula for d is

$$d = \frac{\mu_1 - \mu_2}{\sigma}$$

You may also recall that for the problems in Chapter 5 you were given σ. In Chapter 9 you calculated $\hat{s}$ from data to find d. In this chapter, you use $\hat{s}$ again. The formula for $\hat{s}$, however, depends on whether the design is independent samples or paired samples.

Effect Size Index for Independent Samples

For independent-samples designs, the working formula for d is

$$d = \frac{\bar{X}_1 - \bar{X}_2}{\hat{s}}$$

The calculation of $\hat{s}$ depends on whether or not the sample sizes are equal. When $N_1 = N_2$,

$$\hat{s} = \sqrt{\frac{N_1}{2}}(s_{\bar{X}_1-\bar{X}_2})$$

where N_1 is the sample size for *one group*.

In the case $N_1 \neq N_2$, you cannot find $\hat{s}$ directly from $s_{\bar{X}_1-\bar{X}_2}$. To find $\hat{s}$ when $N_1 \neq N_2$, use the formula

$$\hat{s} = \sqrt{\frac{\hat{s}_1^{\,2}(df_1) + \hat{s}_2^{\,2}(df_2)}{df_1 + df_2}}$$

where df_1 and df_2 are the degrees of freedom $(N - 1)$ for each of the two samples
$\hat{s}_1$ and $\hat{s}_2$ are the sample standard deviations.

The formula for $s_{\bar{X}_1-\bar{X}_2}$ for $N_1 \neq N_2$ (Table 10.3) does not include separate elements for $\hat{s}_1$ and $\hat{s}_2$. Thus, you have to calculate $\hat{s}_1$ and $\hat{s}_2$ from the raw data for each sample.

Effect Size Index for Paired Samples

For paired samples, the working formula for d is

$$d = \frac{\bar{X} - \bar{Y}}{\hat{s}_D}$$

The formula for $\hat{s}_D$ for paired samples is

$$\hat{s}_D = \sqrt{N}(s_{\bar{D}})$$

where N is the number of pairs of participants.

If you like to play algebra, you can use the formulas for the paired-samples case to prove an equivalent rule for d, namely

$$d = \frac{t}{\sqrt{N}}$$

which is a formula that often involves less work.

Interpretation of *d*

The conventions proposed by Cohen (1969) are commonly used to interpret d:

Small effect	$d = 0.20$
Medium effect	$d = 0.50$
Large effect	$d = 0.80$

In studies that compare an experimental group to a control group, it is conventional to subtract the control group mean from the experimental group mean. Thus, if the treatment increases scores, d is positive; if the treatment reduces scores, d is negative.

To illustrate the interpretation of d, I calculated an effect size index for two problems that you have already worked. For both of these, the t test revealed a significant difference.

When you compared the ACT scores at two state universities (problem 10.11), you found that the mean for State U. freshmen (23.5) was significantly higher than the mean for freshmen at The U. (23.4). The standard error of the difference for this $N_1 = N_2$ study was 0.0474. Thus,

$$d = \frac{\bar{X}_1 - \bar{X}_2}{\hat{s}} = \frac{23.5 - 23.4}{\sqrt{\frac{8000}{2}}(0.0474)} = \frac{0.1}{3.00} = 0.03$$

An effect size index of 0.03 is very, very small. Thus, although the difference between the two schools is statistically significant, the size of this difference seems to be of no consequence.

In problem 10.14 you found that there was a significant difference between the salaries of men and women. An effect size index will reveal whether this difference can be considered small, medium, or large.

$$d = \frac{\bar{X} - \bar{Y}}{\hat{s}_D} = \frac{12{,}782.25 - 15{,}733.25}{\sqrt{16}(718.44)} = \frac{-2951}{2873.76} = -1.03$$

An effect size index of 1.03 qualifies as large. Thus, the statistical analysis of these data, which includes a *t* test and an effect size index, supports the conclusion that sex discrimination was taking place and that the effect was large. In fact, on the basis of these and other considerations, the defendants agreed to an out-of-court settlement.

Establishing a Confidence Interval About a Mean Difference

Another way to measure the size of the effect that an independent variable has is to establish a confidence interval. You probably recall from Chapter 8 that a confidence interval establishes a range of values that captures a parameter with a certain degree of confidence. In this section the parameter that is being captured is the *difference* between population means, $\mu_1 - \mu_2$. Researchers usually choose 95 percent or 99 percent as the amount of confidence they want.

Confidence Intervals for Independent Samples

To find the lower and upper limits of the confidence interval about a mean difference for data from an independent-samples design, use these formulas:

$$\text{LL} = (\bar{X}_1 - \bar{X}_2) - t_\alpha(s_{\bar{X}_1-\bar{X}_2})$$

$$\text{UL} = (\bar{X}_1 - \bar{X}_2) + t_\alpha(s_{\bar{X}_1-\bar{X}_2})$$

Of course, $\bar{X}_1$ and $\bar{X}_2$ come from the two samples. The standard error of the difference, $s_{\bar{X}_1-\bar{X}_2}$, is calculated using the appropriate formula from Table 10.3. The value for t_α comes from Table D. Look again at Table D. The top row of the table gives you commonly used confidence interval percents.

As an example, let's return to problem 10.8, the time required by humans and pigeons to find a person lost at sea. The data produce a mean of 49 minutes for the three humans and 25 minutes for the three pigeons. The standard error of the difference, $s_{\bar{X}_1-\bar{X}_2}$, is 7.895. For this example, let's find the lower and upper limits of a 95 percent confidence interval about the true difference between humans and pigeons. The t_α (4 *df*) value from Table D for 95 percent confidence is 2.776. You get

$$\text{LL} = (\bar{X}_1 - \bar{X}_2) - t_\alpha(s_{\bar{X}_1-\bar{X}_2}) = (49 - 25) - 2.776(7.895) = 2.08 \text{ minutes}$$
$$\text{UL} = (\bar{X}_1 - \bar{X}_2) + t_\alpha(s_{\bar{X}_1-\bar{X}_2}) = (49 - 25) + 2.776(7.895) = 45.92 \text{ minutes}$$

The interpretation of this confidence interval is that we can expect, with 95 percent confidence, that the true difference between the people and the pigeons in the time required to find a person lost at sea is between 2.08 and 45.92 minutes.

Not only does a confidence interval about a mean difference tell you the size of the difference but *it also tests the null hypothesis* in the process. The null hypothesis is

H_0: $\mu_1 - \mu_2 = 0$. A 95 percent confidence interval gives you a lower limit for $\mu_1 - \mu_2$ and an upper limit for $\mu_1 - \mu_2$. If zero is not between the two limits, you can be 95 percent confident that the difference between μ_1 and μ_2 is not zero. This is equivalent to rejecting the null hypothesis at the .05 level. Because the interval for search time, 2.08 to 45.92 minutes, does not contain zero, you can reject the hypothesis that humans and pigeons take equal amounts of time to spot a person lost at sea. This is the conclusion you reached when you first worked problem 10.8.

Thus, one of the arguments for confidence intervals is that they do everything that hypothesis testing does, plus they give you an estimate of just how much difference there is between the two groups. The APA publication manual gives this endorsement: "Because confidence intervals ... can often be directly used to infer significance levels, they are, in general, the best reporting strategy" (APA, 2010, p. 35). When used as error bars on graphs, confidence intervals are especially informative (Cumming and Finch, 2005).

Confidence Interval for Paired Samples

To find the confidence interval about a mean difference for paired-samples data, use these formulas:

$$\text{LL} = (\bar{X} - \bar{Y}) - t_\alpha(s_{\bar{D}})$$
$$\text{UL} = (\bar{X} - \bar{Y}) + t_\alpha(s_{\bar{D}})$$

Remember that with paired-samples data, *df* is $N - 1$, where N is the number of pairs of data.

Both the effect size index and confidence intervals about a mean difference are statistics that may help you decide about the importance of a difference that is statistically significant. If you are a little hazy in your understanding of confidence intervals about mean differences, it might help to reread the material in Chapter 8 that introduced confidence intervals.

SPSS routinely displays a confidence interval when you have it calculate a t test. Both Table 10.5 and Table 10.7 show a 95 percent confidence interval about the difference in sample means.

error detection

For confidence intervals for either independent or paired samples, use a t value from Table D, not a t-test value calculated from the data.

PROBLEMS

10.18. In an experiment teaching Spanish, the total physical response method produced higher test scores than the lecture–discussion method. A report of the experiment included the phrase "$p < .01$." What event does the p refer to?

10.19. Write sentences that distinguish between a statistically significant difference and an important difference.

10.20. Problem 10.9 revealed that there was no significant loss in the memory of rats that had 20 percent of their cortex removed. Calculate the effect size index and write an interpretation that incorporates both the effect size index and the t test. (*Hint:* My interpretation notes that the nonsignificant t test was based on 80 rats, which is a large sample.)

10.21. In the example problem for the paired-samples t test (page 211), you found that a week of camp activities significantly improved racial attitudes. Calculate and interpret an effect size index for those data.

10.22. How does sleep affect memory? To find out, eight students learned a list of 10 nonsense words. For the next 4 hours, half of them slept and half engaged in daytime activities. Then each was tested on the list. The next day everyone learned a new list. Those who had slept did daytime activities and those who were awake the day before slept for 4 hours. Each recalled the new list. The table shows the number of words each person recalled under the two conditions. Establish a 95 percent confidence interval about the difference between the two means, and write an interpretation that includes a statement about the fate of the null hypothesis when $\alpha = .05$. (Based on a 1924 study by Jenkins and Dallenbach.)

Asleep	Awake
4	2
3	2
5	4
3	0
2	4
4	3
3	1
6	4

10.23. Twenty depressed patients were randomly assigned to two groups of ten. Each group spent 2 hours each day working and visiting at the clinic. One group worked in bright artificial light; the other did not. At the end of the study, the mean depression score for those who received light was 6; for those who did not, the mean score was 11. (High scores mean more depression.) The standard error of the difference was 1.20. Establish a 99 percent confidence interval about the mean difference and write an interpretation.

Reaching Correct Conclusions

The last step in any statistical analysis is to reach a conclusion (and to express it clearly). A good conclusion describes the relationship of the dependent variable to the independent variable, often utilizing p values, d magnitudes, or confidence interval limits. But can you be sure that your conclusion is correct? Of course, with practice you can be sure that your expressed conclusion is correct for the probabilities that result.

But are the probabilities correct? Are there considerations that should temper your description of the relationship of the dependent variable to the independent variable? The two subsections that follow give cautions about probabilities and the description of the relationship between the dependent variable and the independent variable.

Nature of the Populations the Samples Come From

For an independent-samples *t* test, the two populations the samples come from should:

1. Be normally distributed
2. Have variances that are equal

The reason the populations should have these characteristics is that the probability figures in Table D were derived mathematically from populations that were normally distributed and had equal variances. What if you are uncertain about the characteristics of the populations in your own study? Are you likely to reach the wrong conclusion by using Table D?

There is no simple answer to this important question. Several studies suggest that the answer is no because the *t* test is a robust test. (*Robust* means that a test gives fairly accurate probabilities even when the populations do not have the assumed characteristics.) In actual practice, researchers routinely use *t* tests unless the data are clearly not normally distributed or the variances are not equal.

Random Assignment

Extraneous variables are bad news when it comes time to draw conclusions. If a study has an uncontrolled extraneous variable, you cannot write a simple cause-and-effect conclusion relating the dependent variable to the independent variable.

The most common way for researchers to control for extraneous variables is to randomly assign available participants to treatment groups. Using random assignment creates two groups that are approximately equal on all extraneous variables. When random assignment is used, strong conclusions are warranted. Phrases such as *caused*, *produced*, and *responsible* for mean that changes in the independent variable created differences in scores on the dependent variable.

Unfortunately, for many interesting investigations, random assignment is just not possible. Studies of gender, socioeconomic status, and age are examples. When random assignment isn't possible, researchers typically use additional control groups and logic to reduce the number of extraneous variables.

When random assignment is not used, phrases such as *related to, associated with,* and *correlated* are used to describe the relationship between the dependent variable and the independent variable.

This section introduced three considerations that affect your confidence when you draw conclusions from an independent-samples t test. The first two (normal distribution and equal variances) are referred to as "assumptions of the independent-samples *t* test" and are related to the value of *p* obtained from the data. The general topic "assumptions of the test" will be addressed again in later chapters. The other consideration (random assignment) in related to the strength of the conclusion you can make about the effect of the independent variable on the dependent variable.

To prepare you for the next section, remember that the probabilities the t distribution gives are for differences that occur when the null hypothesis is true. When those probabilities are correct and you use an α level of .05, you are assured that if you reject the null hypothesis, the probability that you've made a Type I error is .05 or less. Thus, you have good protection against making a mistake when the null hypothesis is true. However, what about when the null hypothesis is false?

Statistical Power

The theme of this section is the same as that of the previous one: how to arrive at correct conclusions. Look again at Table 9.3 (page 184). Table 9.3 shows that you will be correct if you retain H_0 when the null hypothesis is true and that you will be correct if you reject H_0 when the null hypothesis is false. Of course, rejecting H_0 is what a researcher usually wants. When H_0 is rejected, there is a clear-cut conclusion: The two populations are different. Statistical power is about making the correct decision when H_0 is false.

When the null hypothesis is false, the probability that you will either make a correct decision or make a Type II error is 1.00 (see Table 9.3). As you may recall from Chapter 9, the probability of a Type II error is symbolized β (beta). Thus, $1 - \beta$ is the probability of a correct decision. For statisticians, **power** $= 1 - \beta$. Here are two equivalent word versions of this formula:

power
Power $= 1 - \beta$; the probability of rejecting a false null hypothesis.

- Power is the probability of not making a Type II error.
- Power is the probability of a correct decision when the null hypothesis is false.

Naturally, researchers want to maximize power. The three factors that affect the power of a statistical analysis are effect size, standard error, and α.

1. *Effect size.* The larger the effect size, the more likely that you will reject H_0. For example, the greater the difference between the mean of the experimental group population and the mean of the control group population, the more likely that the samples will lead you to correctly conclude that the populations are different. Of course, determining effect size *before* the data are gathered is difficult, and experience helps.

2. *Standard error of a difference.* Look at the formulas for t on pages 206 and 210. You can see that as $s_{\bar{X}_1 - \bar{X}_2}$ and $s_{\bar{D}}$ get smaller, t gets larger and you are more likely to reject H_0. Here are two ways you can reduce the size of the standard error:

   a. *Sample size.* The larger the sample, the smaller the standard error of a difference. Figure 8.3 shows this to be true for the standard error of the mean; it is also true for the standard error of a difference. How big should a sample be? Cohen (1992) provides a helpful table (as well as a discussion of the other factors that affect power). Many times, of course, sample size is dictated by practical considerations—time, money, or availability of participants.

   b. *Sample variability.* Reducing the variability in the sample data will produce a smaller standard error. You can reduce variability by using reliable measuring instruments, recording data correctly, being consistent, and, in short, reducing the "noise" or random error in your experiment.

3. *Alpha*. The larger α is, the more likely you are to reject H_0. As you know, the conventional limit is .05. Values above .05 begin to raise doubts in readers' minds, although there has been a healthy recognition in recent years of the arbitrary nature of the .05 rule. Researchers whose results come out with $p = .08$ (or near there) are very likely to remain convinced that a real difference exists and proceed to gather more data, expecting that the additional data will establish a statistically significant difference.

Actually incorporating these three factors into a power analysis is a topic covered in Chapter 17, which is on the Student Companion Website that accompanies this book, and in intermediate textbooks.[8] Researchers use a power analysis before gathering data to help decide how large N should be. In addition, a power analysis after the data are gathered and analyzed can help determine the value of an experiment in which the null hypothesis was retained.

I will close this section on power by asking you to imagine that you are a researcher directing a project that could make a Big Difference. (Because you are imagining this, the difference can be in anything you would like to imagine—the health of millions, the destiny of nations, your bank account, whatever.) Now suppose that the success or failure of the project hinges on one final statistical test. One of your assistants comes to you with the question, "How much power do you want for this last test?"

"All I can get," you answer.

If you examine the list of factors that influence power, you will find that there is only one item that you have some control over and that is the standard error of a difference. Of the factors that affect the size of the standard error of a difference, the one that most researchers can best control is N. So, allocate plenty of power to important statistical tests—use large N's.

PROBLEMS

10.24. Your confidence that the probabilities are accurate for an independent-samples t test depends on certain assumptions. List them.

10.25. Using words, define statistical power.

10.26. List the four topics that determine the power of an experiment. [If you can recall them without looking at the text, you know that you have been reading actively (and effectively).]

10.27. What is the effect of set (previous experience) on problem solving? In a classic experiment by Birch and Rabinowitz (1951), the problem was to tie together two strings (A and B in the illustration) that were suspended from the ceiling. The solution to the problem is to tie a weight to string A, swing it out, and then catch it on the return. The researcher supplied the weight (an electrical light switch) under one of two conditions. In one condition, participants had wired the switch into an electrical circuit and used it to turn on a light. In the other condition, participants had no experience with the switch. Does education (wiring experience) have an effect on the time required to solve the problem? State the null hypothesis, analyze the data in the accompanying table with a t test and an effect size index, and write a conclusion.

[8] See Howell (2010, Chap. 8), Howell (2008, Chap. 15) or Aron, Aron, and Coups (2009, Chap. 6).

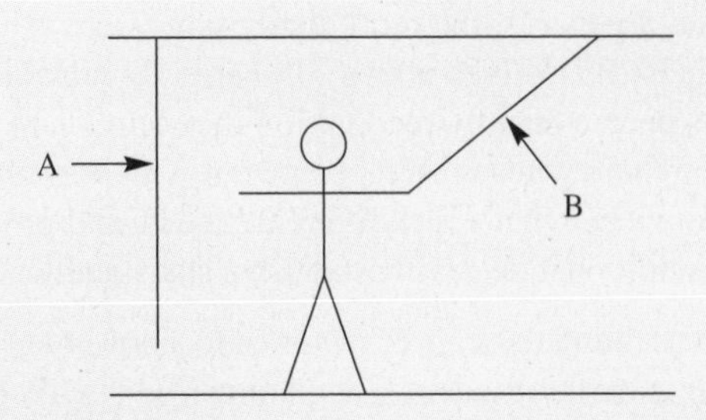

	Wired the switch into circuit	No previous experience with switch
$\overline{X}$	7.40 minutes	5.05 minutes
$\hat{s}$	2.13 minutes	2.31 minutes
N	20	20

10.28. How does cell phone use affect driving performance? Strayer, Drews, and Johnston (2003) had undergraduates drive a high-fidelity automobile simulator as their only task and while engaged in a hands-free cell phone conversation. Participants applied the brake when the brake lights of the vehicle in front of them came on. The numbers that follow are reaction times in seconds. Analyze the data and write a conclusion about the effect of cell phone conversations on driving performance. Carry four decimals in your calculations.

No cell phone	Cell phone
1.03	1.12
0.97	1.31
0.82	0.96
1.09	1.08
1.15	0.99
0.72	1.04
0.95	1.16
0.87	1.22

10.29. A French teacher tested the claim that "concrete, immediate experience enhances vocabulary learning." Driving from the college into the city, he described in French the terrain, signs, distances, and so forth, which he recorded with a digital voice recorder. ("L'auto est sur le pont. Carrefour prochain est dangereux.") For the next part of his investigation, his French II students were ranked from 1 to 10 in competence. The 10 were divided in such a way that numbers 1, 4, 5, 8, and 9 listened to the record while riding to the city (concrete, immediate experience) and numbers 2, 3, 6, 7, and 10 listened to the record in the language laboratory. The next day each student took a 25-item vocabulary test; the number of errors was recorded. Think about the design, analyze the data, and write a conclusion about immediate, concrete experience.

#1	7	#2	7	#3	15	#4	7	#5	12
#6	22	#7	24	#8	12	#9	21	#10	32

10.30. When we speak, we convey not only information but emotional tone as well. Ambady et al. (2002) investigated the relationship between the emotional tone used by surgeons as they talked to patients and the surgeon's history of being sued for malpractice. Audiotapes of the doctor's conversations were filtered to remove high-frequency sounds, thus making the words unintelligible but leaving expressive features such as intonation, pitch, and rhythm. Judges rated the sounds for the emotional tone of dominance on a scale of 1 to 7, with 7 being extremely dominant. The surgeons in this study had never been sued or had been sued twice. Analyze the scores and write an interpretation.

Malpractice suits	
0	2
3	4
4	3
2	2
6	7
3	5
1	5
3	4
2	7
	5

10.31. Skim over the objectives at the beginning of the chapter as a way to consolidate what you have learned.

ADDITIONAL HELP FOR CHAPTER 10

Visit *cengage.com/psychology/spatz*. At the Student Companion Site, you'll find multiple-choice tutorial quizzes, flashcards with definitions and workshops. For this chapter there are Statistical Workshops on Independent versus Repeated *t* Tests and Statistical Power.

KEY TERMS

alternative hypothesis (p. 199)
assumptions (p. 221)
confidence interval (p. 218)
control group (p. 197)
degrees of freedom (p. 201)
effect size index (d) (p. 216)
experimental group (p. 197)
important results (p. 215)
independent-samples design (p. 202)
matched pairs (p. 203)
natural pairs (p. 203)
NHST (p. 199)
null hypothesis (p. 199)
one-tailed tests (p. 199)
paired-samples design (p. 202)
power (p. 222)
repeated measures (p. 204)
standard error of a difference (p. 206)
t test (p. 202)
treatment (p. 197)
two-tailed test (p. 199)

What Would You Recommend? Chapters 7–10

It is time again for a set of *What would you recommend?* problems. No calculations are required. Reviewing Chapters 7 through 10 is permitted and even encouraged.

Read an item. Think about the techniques you studied in the Chapters 7 to 9. (Perhaps you even have a list?) Choose one or more statistics that would help you write an answer to the question that is posed or a conclusion about the topic. In my answers, I'll give some explanation for my choices.

a. Do babies smile to establish communication with others or to express their own emotion? During 10-minute sessions in which the number of smiles was recorded, there was an audience for 5 minutes (a requirement for communication) and no audience for the other 5 minutes.
b. Imagine a social worker with a score of 73 on a personality inventory that measures gregariousness. The scores on this inventory are distributed normally with a population mean of 50 and a standard deviation of 10. How can you determine the percent of the population that is less gregarious than this social worker?
c. A psychologist conducted a workshop to help people improve their social skills. After the workshop, the participants filled out a personality inventory that measured gregariousness. (See problem b.) The mean gregariousness score of the participants was 55. What statistical test can help determine whether the workshop produced participants whose mean gregariousness score was greater than the population mean?
d. A sample of high school students takes the SAT test each year. The mean score for the students in a particular state is readily available. What statistical technique will give you a set of brackets that "captures" a population mean?
e. One interpretation of the statistical technique in problem d is that the brackets capture the *state's* population mean. This interpretation is not correct. Why not?
f. In a sample of 50 undergraduates, there was a correlation of −.39 between disordered eating and self-esteem. How could you determine the likelihood of such a correlation coefficient if there is actually no relationship between the two variables?
g. Colleges and universities in the United States are sometimes divided into six categories: public universities, 4-year public institutions, 2-year public institutions, private universities, 4-year private institutions, and 2-year private institutions. In any given year, each category has a certain number of students. Which of the six categories does your institution belong to? How can you find the probability that a student, chosen at random, is from your category?
h. Half the school-age children in an experiment were placed on a diet high in sugar for 3 weeks and then switched to a low-sugar diet. The other half of the participants began on the low-sugar diet and then changed to the high-sugar diet. All children were assessed on cognitive and behavioral measures. [See Wolraich et al. (1994), who found no significant differences.]

transition passage
to more complex designs

THE TRADITIONAL STATISTICAL analysis of a two-group experiment involves an effect size index and a *t* test or a confidence interval. The statistical analysis of experiments with more than two groups requires a more complex technique.

This more complex technique is called the analysis of variance (ANOVA for short). ANOVA is based on the very versatile concept of assigning the variability in the data to various sources. (This often means that ANOVA identifies the variables that produce changes in the scores.) Among social and behavioral scientists, it is used more frequently than any other inferential statistical technique. ANOVA was invented by Sir Ronald A. Fisher, an English biologist and statistician.

So the transition this time is from a *t* test and its sampling distribution, the *t* distribution, to ANOVA and its sampling distribution, the *F* distribution. In Chapter 11 you will use ANOVA to compare means from independent-samples designs that involve more than two treatments. Chapter 12 shows you how to use ANOVA to compare means from a repeated-measures design that has more than two treatments. In Chapter 13 the ANOVA technique is used to analyze experiments with two independent variables, both of which may have two or more levels of treatment.

The analysis of variance is a widely used statistical technique, and Chapters 11–13 are devoted to an introduction to its elementary forms. Intermediate and advanced books explain more sophisticated (and complicated) analysis-of-variance designs. [Howell (2010) is particularly accessible.]

CHAPTER

11

Analysis of Variance: One-Way Classification

OBJECTIVES FOR CHAPTER 11

After studying the text and working the problems in this chapter, you should be able to:

1. Identify the independent and dependent variables in a one-way ANOVA
2. Explain the rationale of ANOVA
3. Define *F* and explain its relationship to *t* and the normal distribution
4. Compute sums of squares, degrees of freedom, mean squares, and *F* for an ANOVA
5. Construct a summary table of ANOVA results
6. Interpret the *F* value for a particular experiment and explain what the experiment shows
7. Distinguish between *a priori* and *post hoc* tests
8. Use the Tukey Honestly Significant Difference (HSD) test to make all pairwise comparisons
9. List and explain the assumptions of ANOVA
10. Calculate and interpret *d* and *f*, effect size indexes

analysis of variance
An inferential statistics technique for comparing means, comparing variances, and assessing interactions.

one-way ANOVA
Statistical test of the hypothesis that two or more population means in an independent-samples design are equal.

IN THIS CHAPTER you will work with the simplest **analysis of variance** design, a design called **one-way ANOVA**. (ANOVA stands for analysis of variance. Remember, a *variance* is just a squared standard deviation.) The *one* in one-way means that the experiment has one independent variable. One-way ANOVA is used to find out if there are any differences among three or more population means.[1] Experiments with more than two treatment levels are common in all disciplines that use statistics. Here are some examples, each of which is covered in this chapter:

1. People with phobias received treatment using one of four methods. Afterward, the group means were compared to

[1] As will become apparent, ANOVA can also be used when there are just two groups.

determine which method of therapy was most effective in reducing fear responses.
2. Four different groups learned a new task. A different schedule of reinforcement was used for each group. Afterward, persistence in doing the task was measured.
3. Suicide rates were compared for countries that represented low, medium, and high degrees of modernization.

Except for the fact that these experiments have more than two groups, they are like those you analyzed in Chapter 10. There is one independent and one dependent variable. The null hypothesis is that the population means are the same for all groups. The subjects in each group are independent of the subjects in the other groups. The only difference is that the independent variable has more than two levels. To analyze the differences among three or more means, use a one-way ANOVA.[2]

In example 1, the independent variable is method of treatment, and it has four levels. The dependent variable is fear responses. The null hypothesis is that the four methods are equally effective in reducing fear responses—that is,
$H_0: \mu_1 = \mu_2 = \mu_3 = \mu_4$.

PROBLEM

***11.1.** For examples 2 and 3 in the text, identify the independent variable, the number of levels of the independent variable, the dependent variable, and the null hypothesis.

Although this chapter does not have "hypothesis testing" or "effect size" in the title (as did the two previous chapters on *t* tests), ANOVA is an NHST technique, and effect size indexes will be calculated.

A common first reaction to the task of determining if there is a difference among three or more population means is to compute *t* tests on all possible pairs of sample means. For three populations, three *t* tests are required ($\overline{X}_1$ vs. $\overline{X}_2$, $\overline{X}_1$ vs. $\overline{X}_3$, and $\overline{X}_2$ vs. $\overline{X}_3$). For four populations, six tests are needed.[3] This multiple *t*-test approach *will not work* (and not just because doing lots of *t* tests is tedious). Here's why.

Suppose you have 15 samples that all come from the same population. (Because there is just one population, the null hypothesis is clearly true.) These 15 sample means will vary from one another as a result of chance factors. Now suppose you perform every possible *t* test (all 105 of them), retaining or rejecting each null hypothesis at the .05 level. How many times would you reject the null hypothesis?

The answer is about five. When the null hypothesis is true (as in this example) and α is .05, 100 *t* tests will produce about five Type I errors.

Now, let's move from this theoretical analysis to the reporting of the experiment. Suppose you conducted a 15-group experiment, computed 105 *t* tests, and found five significant differences. If you then pulled out those five and said they were reliable

[2] The ANOVA technique for analyzing paired-samples designs is covered in Chapter 12.
[3] The formula for the number of combinations of *n* things taken two at a time is $[n(n - 1)]/2$.

differences (that is, differences not due to chance), people who understand the preceding two paragraphs would realize that you don't understand statistics (and would recommend that your manuscript not be published). You can protect yourself from such a disaster if you use some other statistical technique—one that keeps the overall risk of a Type I error at an acceptable level (such as .05 or .01).

Sir Ronald A. Fisher (1890–1962) developed such a technique. Fisher did brilliant work in genetics, but it has been overshadowed by his fundamental work in statistics. In genetics, it was Fisher who showed that Mendelian genetics is compatible with Darwinian evolution. (For a while after 1900, genetics and evolution were in opposing camps.) And, among other contributions, Fisher showed how a recessive gene can become established in a population.

In statistics, Fisher invented ANOVA, the topic of this chapter and the next two. He also discovered the exact sampling distribution of r (1915), developed a general theory of estimation, and wrote *the* book on statistics. (*Statistical Methods for Research Workers* was first published in 1925 and went through 14 editions and several translations by 1973.)

Before getting into biology and statistics in such a big way, Fisher worked for an investment company for 2 years and taught in a private school for 4 years. For biographical sketches, see Field (2005b), Edwards (2001), Yates (1981), Hald (1998, pp. 734–739), or Salsburg (2001).

Rationale of ANOVA

The question that ANOVA addresses is whether the populations that the samples come from have the same μ. If the answer is no, then at least one of the populations has a different μ. I've grouped the rationale of ANOVA into two subsections. The first is a description of null hypothesis statistical testing (NHST) as applied in ANOVA. The second describes the F distribution, which supplies probabilities, much as the t distribution provided probabilities for a two-sample experiment.

Null Hypothesis Statistical Testing (NHST)

1. The populations you are comparing (the different treatments) may be exactly the same *or* one or more of them may have a different mean. These two descriptions cover all the logical possibilities and are, of course, the null hypothesis (H_0) and the alternative hypothesis (H_1).
2. Tentatively assume that the null hypothesis is correct. If the populations are all the same, any differences among sample means will be the result of chance.
3. Choose a sampling distribution that shows the probability of various differences among sample means when H_0 is true. For more than two sample means, use the sampling distribution that Fisher invented, which is now called the F distribution.
4. Obtain data from the populations you are interested in. Perform calculations on the data using the procedures of ANOVA until you have an F value.
5. Compare your calculated F value to the critical value of F in the sampling distribution (F distribution). From the comparison you can determine the probability of obtaining the data you did, *if the null hypothesis is true.*

6. Come to a conclusion about H_0. If the probability is less than α, reject H_0. With H_0 eliminated, you are left with H_1. If the probability is greater than α, retain H_0, leaving you with both H_0 and H_1 as possibilities.
7. Tell the story of what the data show. If you reject H_0 when there are three or more treatments, a conclusion about the relationships of the specific groups requires further data analysis. If you retain H_0, you have your data, but no strong conclusion about H_0 and H_1.

I'm sure that the ideas in this list are more than somewhat familiar to you. In fact, I suspect that your only uncertainties are about step 3, the *F* distribution, and step 4, calculating an *F* value. The next several pages will explain the *F* distribution and then what to do to get an *F* value from the data.

The *F* Distribution

The ***F* distribution** used in ANOVA is yet another sampling distribution. Like the *t* distribution and the normal distribution, it gives the probability of various outcomes when all samples come from identical populations (the null hypothesis is true). The concepts of ANOVA and the *F* distribution are not simple. Understanding them will take careful study now and review later. You should mark this section and reread it later.

***F* distribution**
Theoretical sampling distribution of *F* values.

F is a ratio of two numbers. The numerator is a variance and the denominator is a variance. When the null hypothesis is true, the variances are equal. To put this in symbol form:

$$F = \frac{\sigma^2}{\sigma^2}$$

When the two variances are equal, the value of *F* is 1. Of course, the two σ^2's are parameters, and parameters are never available in sample-based research. Each σ^2 must be estimated with a statistic, $\hat{s}^2$.

Fisher's insight was to calculate $\hat{s}^2$ for the numerator with a formula that is accurate (on the average) when the null hypothesis is true but is too large (on the average) when the null hypothesis is false. The denominator, however, is calculated with a formula that is accurate (on the average) regardless of whether the null hypothesis is true or false.

Thus, when the null hypothesis is true, Fisher's ANOVA produces an *F* value that is approximately 1. When the null hypothesis is false, ANOVA produces an *F* that is greater than 1. The sampling distribution of *F* allows you to determine whether a particular *F* value is so large that the null hypothesis should be rejected. I'll explain more about the *F* ratio using both algebra and graphs.

An algebraic explanation of the *F* ratio Begin with the idea that each level (treatment) of the independent variable produced a group of scores. Assume that the null hypothesis is true.

Numerator

1. For each treatment level, calculate a sample mean.

2. Using the treatment means as data, calculate a standard deviation. This standard deviation of treatment means is an acquaintance of yours, the standard error of the mean, $s_{\bar{X}}$.
3. Remember that $s_{\bar{X}} = \hat{s}/\sqrt{N}$. Squaring both sides gives $s_{\bar{X}}{}^2 = \hat{s}^2/N$. Multiplying both sides by N and rearranging give $\hat{s}^2 = Ns_{\bar{X}}{}^2$.
4. $\hat{s}^2$ is, of course, an estimator of σ^2.

Thus, starting with a null hypothesis that is true and a set of treatment means, you can obtain an estimate of the variance of the populations that the samples come from.

Notice that the accuracy of this estimate of σ^2 depends on the assumption that the samples are all drawn from the same population. If one or more samples come from a population with a larger or smaller mean, the treatment means will vary more, and $\hat{s}^2$ will overestimate σ^2.

Denominator

The formula for $\hat{s}^2$ in the denominator is based on the fact that each sample variance is an independent estimate of the population variance.[4] An average of these independent estimators produces a more reliable estimate of σ^2. This denominator of the F ratio has a special name, the **error term**. (It is also referred to as the *error estimate* and as the *residual error*.) One important point about the error term is that it is an unbiased estimate of σ^2 even if the null hypothesis is false.[5]

error term
Variance due to factors not controlled in the experiment.

To summarize the algebra of the F ratio, one student described it as "the variance of the means divided by the mean of the variances." If the treatment means all come from the same population, their variance will be small. If they come from different populations, their variance will be larger. The standard of comparison is the population variance itself.

Finally, here are two additional terms. They are widely used because they are descriptive of the process that goes into the numerator and the denominator of the F ratio. Because the estimate of σ^2 in the numerator is based on variation among the treatment means, the numerator is referred to as the *between-treatments estimate* or *between-groups estimate*. At least two groups are needed to find a between-treatments estimate. The estimate of σ^2 in the denominator (the error term) is referred to as the *within-treatments estimate* or *within-groups estimate* because any one group by itself can produce a statistic that is an estimate of the population variance. That is, the estimate can be obtained from within one group.

A graphic explanation of the *F* ratio The three graphs that follow (Figures 11.1, 11.2, and 11.3) also help explain the F ratio. Each figure shows three treatment populations (A, B, and C). Each population produces a sample and its sample mean ($\bar{X}_A$, $\bar{X}_B$, or $\bar{X}_C$). In Figure 11.1 and Figure 11.3, the null hypothesis is true; in Figure 11.2, the null hypothesis is false.

Figure 11.1 shows three treatment populations that are identical. Each population produced a sample whose mean is projected onto a dependent-variable line at the

[4] As will be explained later, one of the assumptions of ANOVA is that the populations the samples are from have variances that are equal.

[5] Just a reminder: *Error* in statistics means random variation.

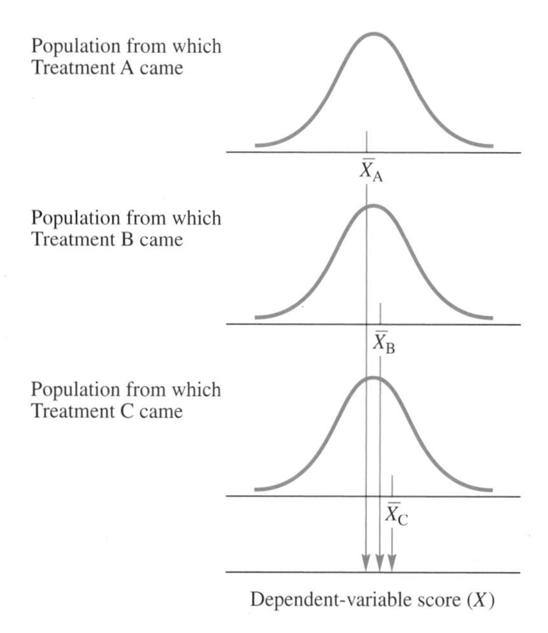

FIGURE 11.1 H_0 is true. All three treatment groups are from identical populations. Not surprisingly, samples from those populations produce means that are similar. (Note that the points are close together on the X axis.)

bottom of the figure. Note that the three means are fairly close together. Thus, a variance calculated from these three means will be small. This is a between-treatments estimate, the numerator of the F ratio.

In **Figure 11.2,** the null hypothesis is false. The mean of Population C is greater than that of Populations A and B. Look at the projection of the sample means onto the dependent-variable line. A variance calculated from these sample means will be larger than the one for the means in Figure 11.1. By studying Figures 11.1 and 11.2, you can convince yourself that the between-treatments estimate of the variance (the numerator of the F ratio) is larger when the null hypothesis is false.

Figure 11.3 is like Figure 11.1 in that the null hypothesis is true, but in Figure 11.3, the population variances are larger. Because of the larger population variance, the sample means are likely to be more variable (as shown by the larger spaces between the sample mean projections at the bottom).

So, a large between-treatments estimate may be due to a false null hypothesis (Figure 11.2) or to population variances that are large (Figure 11.3). To distinguish between the two situations, divide by the population variance (the error term). In Figure 11.2, the small population variance produces an F ratio greater than 1.00. In Figure 11.3, the larger population variance produces an F ratio of about 1.00. Thus, the F ratio is large when H_0 is false and about 1.00 when H_0 is true.

Table F One of Fisher's central tasks while developing ANOVA was to determine the sampling distribution of the ratio that today we call F. Part of the problem was that there is a sampling distribution of between-treatments variances (the numerator) *and* a

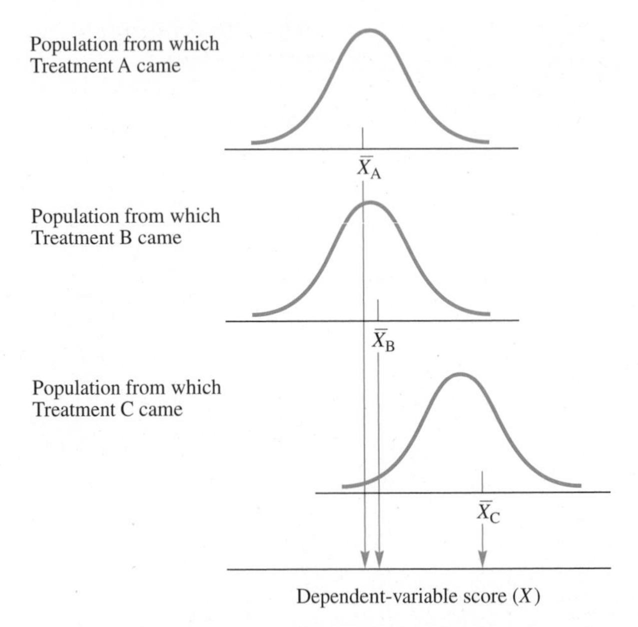

FIGURE 11.2 H_0 is false. Two treatment groups are from identical populations. A third group is from a population with a larger mean. The sample means show more variability than those in Figure 11.1.

sampling distribution of error terms (the denominator). As a result, every different between-treatment variance can be paired with a whole array of different error-term values. The result is a matrix of possibilities. As it turns out, both numerator and denominator values are identified by their degrees of freedom. Thus, there is an F value that goes with 3 *df* in the numerator and 20 *df* in the denominator, and a different F for 3 *df* and 21 *df* (and a different one for 2 *df* and 21 *df*, and so forth).

In 1934, George W. Snedecor of Iowa State University compiled these ratios into tables that provided critical values for $\alpha = .05$ and $\alpha = .01$. He named the ratio F in honor of Fisher (who had been a visiting professor at Iowa in 1931). **Table F** in Appendix C (to be explained later) gives you the critical values of F that you need for ANOVA problems.

Figure 11.4 shows two of the many F curves. As you can see on the abscissa, their F values range from 0 to greater than 4. For the solid curve ($df = 10, 20$), 5 percent of the curve has F values equal to or greater than 2.35. Notice that both curves are skewed, which is typical of F curves.

I'd like for you to be aware of the relationships among the normal distribution, the t distribution, and the F distribution. There is just one normal distribution, but t and F are each a family of curves, characterized by degrees of freedom. As you already know (page 162), a t distribution with $df = \infty$ is a normal curve. The F distributions that have 1 *df* in the numerator are directly related to the t distributions in Table C of Appendix C. The relationship is that $F = t^2$. And to conclude, an F distribution with ∞ *df* in the numerator and ∞ *df* in the denominator is a normal distribution.

Fisher and Gosset (who derived the t distribution) were friends for years, each respectful of the other's work [Fisher's daughter reported their correspondence

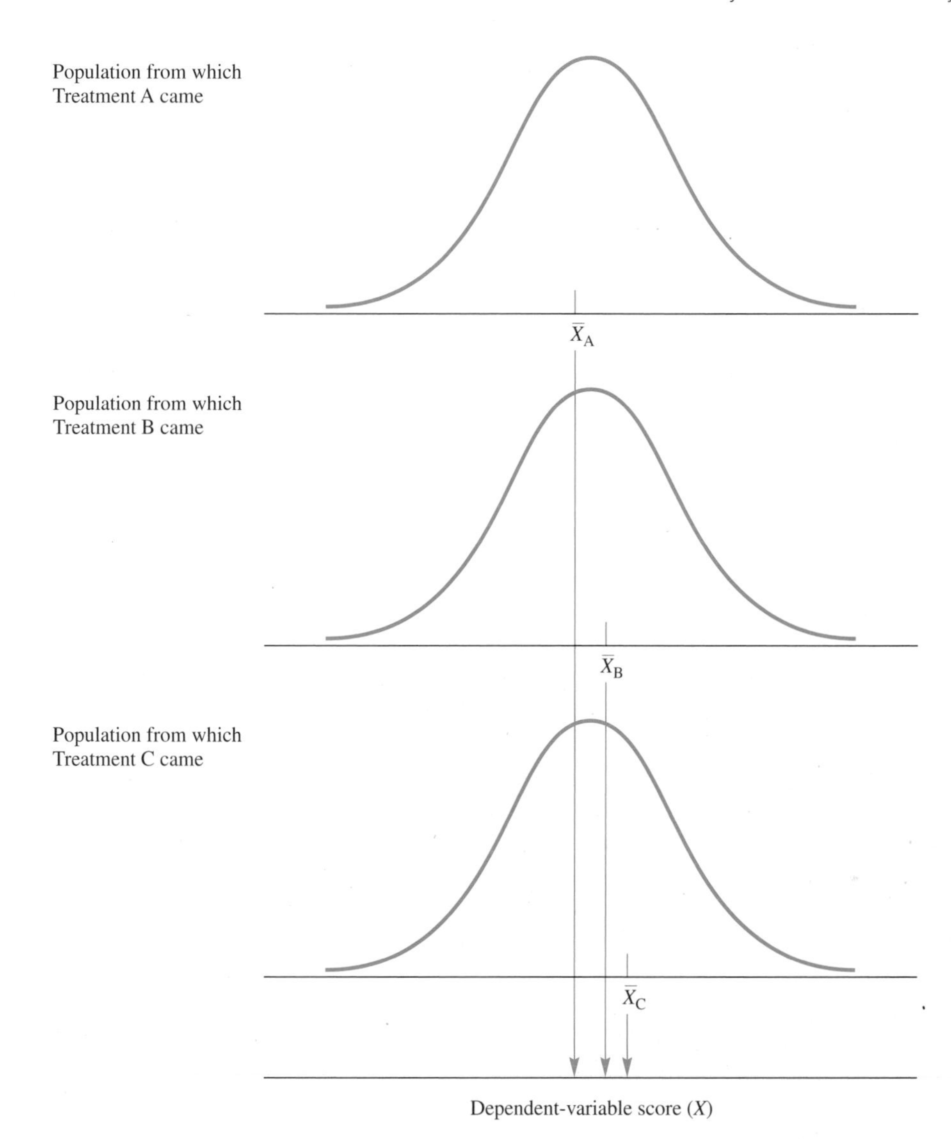

FIGURE 11.3 H_0 is true. All three treatment groups are from identical populations. Because the populations have larger variances than those in Figure 11.1, the sample means show more variability; that is, the arrow points are spaced farther apart

(Box, 1981)]. Fisher recognized that the Student's *t* test was really a ratio of two measures of variability and that such a concept was applicable to experiments with more than two groups.[6]

---

[6] That is,

$$t = \frac{\overline{X}_1 - \overline{X}_2}{s_{\overline{X}_1 - \overline{X}_2}}$$

$$= \frac{\text{A range}}{\text{Standard error of a difference between means}} = \frac{\text{A measure of variability between treatments}}{\text{A measure of variability within treatments}}$$

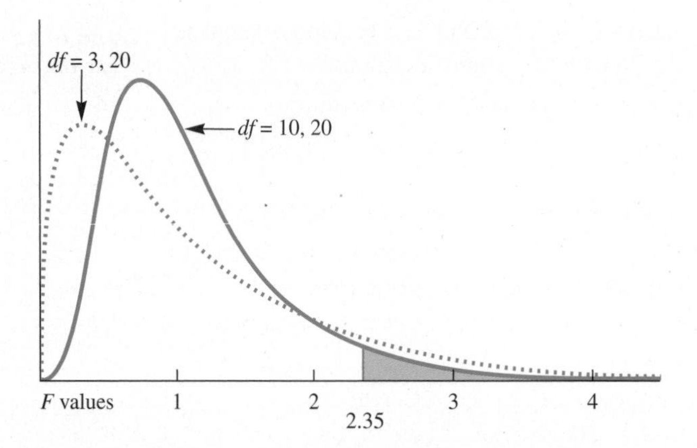

FIGURE 11.4 Two F distributions, showing their positively skewed nature

PROBLEMS

11.2. Suppose three sample means came from a population with $\mu = 100$ and a fourth from a population with a smaller mean, $\mu = 50$. Will the between-treatments estimate of variability be larger or smaller than the between-treatments estimate of variability of four means drawn from a population with $\mu = 100$?

11.3. Define F using parameters.

11.4. Describe the difference between the between-treatments estimate of variance and the within-treatments estimate.

11.5. In your own words, explain the rationale of ANOVA. Identify the two components that combine to produce an F value. Explain what each of these components measures, both when the null hypothesis is true and when it is false.

11.6. **a.** Suppose an ANOVA for example 1 at the beginning of this chapter produced a very small F value. Write an interpretation, using the terms of the example.

b. Suppose an ANOVA for example 3 at the beginning of this chapter produced a very large F value. Write an interpretation, using the terms of the example.

More New Terms

Sum of squares In the "Clue to the Future" on page 67 you read that parts of the formula for the standard deviation would be important in future chapters. That future is now. The numerator of the basic formula for the standard deviation, $\Sigma(X - \overline{X})^2$, is the

sum of squares (abbreviated *SS*). So, $SS = \Sigma(X - \overline{X})^2$. A more complete name for sum of squares is "sum of the squared deviations."

Mean square **Mean square (*MS*)** is the ANOVA term for a variance, $\hat{s}^2$. The mean square is a sum of squares divided by its degrees of freedom.

Grand mean The **grand mean (GM)** is the mean of all scores; it ignores the fact that the scores come from different groups (treatments).

tot A subscript *tot* (total) after a symbol means that the symbol stands for all such numbers in the experiment; ΣX_{tot} is the sum of all *X* scores.

t The subscript *t* after a symbol means that the symbol applies to a treatment group; for example, $\Sigma(\Sigma X_t)^2$ tells you to sum the scores in each group, square each sum, and then sum these squared values. N_t is the number of scores in one treatment group.

K *K* is the number of treatments in the experiment. *K* is the same as the number of levels of the independent variable.

F test An ***F* test** is the outcome of an analysis of variance.

sum of squares (*SS*)
Sum of the squared deviations from the mean.

mean square (*MS*)
The variance; a sum of squares divided by its degrees of freedom.

grand mean (GM)
The mean of all scores, regardless of treatment.

***F* test**
Test of the statistical significance of differences among means, or variances, or of an interaction.

clue to the future

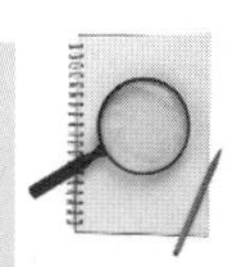

In this chapter and the next two, you will work problems with several sets of numbers. Unless you are using a computer, you will need the sum, the sum of squares, and other values for each set. Most calculators produce both the sum and the sum of squares with just one entry of each score. If you know how (or take the time to learn) to exploit your calculator's capabilities, you can spend less time on statistics problems and make fewer errors as well.

Sums of Squares

Analysis of variance is called that because the variability of all the scores in an experiment is assigned to two or more sources. In the case of simple analysis of variance, just two sources contribute all the variability seen among the scores. One source is the variability between treatments, and the other source is the variability within the scores of each treatment. The sum of these two sources is equal to the total variability.

Now that you have studied the rationale of ANOVA, it is time to do the calculations. **Table 11.1** shows fictional data from an experiment in which there were three treatments for patients with psychological disorders. The independent variable was drug therapy. Three levels of the independent variable were used—Drug A, Drug B, and Drug C. The dependent variable was the number of psychotic episodes each patient had during the therapy period.

In this experiment (and in all one-way ANOVAs), the null hypothesis is that the populations are identical. In terms of treatment means in the drug experiment, H_0: $\mu_A = \mu_B = \mu_C$. If the null hypothesis is true, the differences among the numbers of psychotic episodes for the three treatment groups is due to sampling fluctuation and not to the drugs administered.

TABLE 11.1 Computation of sums of squares of fictional data from a drug study

	Drug A		Drug B		Drug C	
	X_A	X_A^2	X_B	X_B^2	X_C	X_C^2
	9	81	9	81	4	16
	8	64	7	49	3	9
	7	49	6	36	1	1
	5	25	5	25	1	1
Σ	29	219	27	191	9	27
$\bar{X}$	7.25		6.75		2.25	

$$\Sigma X_{tot} = 29 + 27 + 9 = 65$$

$$\Sigma X_{tot}^2 = 219 + 191 + 27 = 437$$

$$SS_{tot} = \Sigma X_{tot}^2 - \frac{(\Sigma X_{tot})^2}{N_{tot}} = 437 - \frac{65^2}{12} = 84.917$$

$$SS_{drugs} = \Sigma\left[\frac{(\Sigma X_t)^2}{N_t}\right] - \frac{(\Sigma X_{tot})^2}{N_{tot}}$$

$$= \frac{29^2}{4} + \frac{27^2}{4} + \frac{9^2}{4} - \frac{65^2}{12}$$

$$= 210.25 + 182.25 + 20.25 - 352.083 = 60.667$$

$$SS_{error} = \Sigma\left[\Sigma X_t^2 - \frac{(\Sigma X_t)^2}{N_t}\right] = \left(219 - \frac{29^2}{4}\right) + \left(191 - \frac{27^2}{4}\right) + \left(27 - \frac{9^2}{4}\right)$$

$$= 8.750 + 8.750 + 6.750 = 24.250$$

Check: $SS_{treat} + SS_{error} = SS_{tot}$; $60.667 + 24.250 = 84.917$

In this particular experiment, the researcher believed that Drug C would be better than Drug A or B, which were in common use. In this case, *better* means fewer psychotic episodes. A researcher's belief is often referred to as the *research hypothesis*. This belief is not ANOVA's alternative hypothesis, H_1. The alternative hypothesis in ANOVA is that one or more populations are different from the others. No *greater than* or *less than* is specified.

The first step in any data analysis is to calculate descriptive statistics. No doubt the researcher would be pleased to see in **Table 11.1** that the mean number of psychotic episodes for Drug C is less than those for Drugs A and B. An ANOVA can determine if the reduction is statistically significant, and an effect size index tells how big the reduction is. The first step in an ANOVA is to calculate the *sum of squares*, which is illustrated in Table 11.1.

First, please focus on the *total* variability as measured by the total sum of squares (SS_{tot}). As you will see, you are already familiar with SS_{tot}. To find SS_{tot}, subtract the grand mean from each score, square these deviation scores, and sum them:

$$SS_{tot} = \Sigma(X - \bar{X}_{GM})^2$$

To compute SS_{tot}, use the algebraically equivalent raw-score formula:

$$SS_{tot} = \Sigma X_{tot}^2 - \frac{(\Sigma X_{tot})^2}{N_{tot}}$$

which you may recognize as the numerator of the raw-score formula for $\hat{s}$. To calculate SS_{tot}, square each score and sum the squared values to obtain ΣX_{tot}^2. Next, the scores are summed and the sum squared. That squared value is divided by the total number of scores to obtain $(\Sigma X_{tot})^2/N_{tot}$. Subtraction of $(\Sigma X_{tot})^2/N_{tot}$ from ΣX_{tot}^2 yields the total sum of squares. For the data in Table 11.1,

$$SS_{tot} = \Sigma X_{tot}^2 - \frac{(\Sigma X_{tot})^2}{N_{tot}} = 437 - \frac{65^2}{12} = 84.917$$

Thus, the total variability of all the scores in Table 11.1 is 84.917 when measured by the sum of squares. This total comes from two sources: the between-treatments sum of squares and the error sum of squares. Each of these can be computed separately.

The *between-treatments sum of squares* (SS_{treat}) is the variability of the group means from the grand mean of the experiment, weighted by the size of the group:

$$SS_{treat} = \Sigma[N_t(\bar{X}_t - \bar{X}_{GM})^2]$$

SS_{treat} is more easily computed by the raw-score formula:

$$SS_{treat} = \Sigma\left[\frac{(\Sigma X_t)^2}{N_t}\right] - \frac{(\Sigma X_{tot})^2}{N_{tot}}$$

This formula tells you to sum the scores for each treatment group and then square the sum. Each squared sum is then divided by the number of scores in its group. These values (one for each group) are then summed, giving you $\Sigma[(\Sigma X_t)^2/N_t]$. From this sum is subtracted the value $(\Sigma X_{tot})^2/N_{tot}$, a value that was obtained previously in the computation of SS_{tot}.

Although the general case of an experiment is described in general terms such as *treatment* and SS_{treat}, a specific experiment is described in the words of the experiment. For the experiment in Table 11.1, that word is *drugs*. Thus,

$$SS_{drugs} = \Sigma\left[\frac{(\Sigma X_t)^2}{N_t}\right] - \frac{(\Sigma X_{tot})^2}{N_{tot}}$$

$$= \frac{29^2}{4} + \frac{27^2}{4} + \frac{9^2}{4} - \frac{65^2}{12} = 60.667$$

The other source of variability is the *error sum of squares* (SS_{error}), which is the sum of the variability within each of the groups. SS_{error} is defined as

$$SS_{error} = \Sigma(X_1 - \bar{X}_1)^2 + \Sigma(X_2 - \bar{X}_2)^2 + \cdots + \Sigma(X_K - \bar{X}_K)^2$$

or the sum of the squared deviations of each score from the mean of its treatment group added to the sum of the squared deviations from all other groups for the experiment. As with the other *SS* values, one arrangement of the arithmetic is easiest. For SS_{error}, it is

$$SS_{error} = \Sigma\left[\Sigma X_t^{2} - \frac{(\Sigma X_t)^2}{N_t}\right]$$

This formula tells you to square each score in a treatment group and sum them (ΣX_t^2). Subtract from this a value that you obtain by summing the scores, squaring the sum, and dividing by the number of scores in the group: $(\Sigma X_t)^2/N_t$. For each group, a value is calculated, and these values are summed to get SS_{error}. For the data in Table 11.1,

$$SS_{error} = \Sigma\left[\Sigma X_t^{2} - \frac{(\Sigma X_t)^2}{N_t}\right]$$

$$= \left(219 - \frac{29^2}{4}\right) + \left(191 - \frac{27^2}{4}\right) + \left(27 - \frac{9^2}{4}\right) = 24.250$$

Please pause and look at what is happening in calculating the sums of squares, which is the first step in any ANOVA problem. The total variation (SS_{tot}) is found by squaring *every* number (score) and adding them up. From this total, you subtract $(\Sigma X_{tot})^2/N_{tot}$. In a similar way, the variation that is due to treatments is found by squaring the sum of each treatment and dividing by N_t, adding these up, and then subtracting the same component you subtracted before, $(\Sigma X_{tot})^2/N_{tot}$. This factor,

$$\frac{(\Sigma X_{tot})^2}{N_{tot}}$$

is sometimes referred to as the *correction factor* and is found in all ANOVA analyses. Finally, the error sum of squares is found by squaring each number *within* a treatment group and subtracting that group's correction factor. Summing the differences from all the groups gives you SS_{error}.

Another way of expressing these ideas is to say that SS_{error} is the variation *within* the groups and SS_{treat} is the variation *between* the groups. In a one-way ANOVA, the total variability (as measured by SS_{tot}) consists of these two components, SS_{treat} and SS_{error}. So,

$$SS_{treat} + SS_{error} = SS_{tot}$$

Thus, 60.667 + 24.250 = 84.917 for the drug experiment. This relationship is shown graphically in **Figure 11.5** as a flowchart and as a pie chart.

error detection

1. Sums of squares are always positive numbers or zero.
2. Always use the check $SS_{tot} = SS_{treat} + SS_{error}$. This check, however, will not catch errors made in summing the scores (ΣX) or in summing the squared scores $(\Sigma X)^2$.

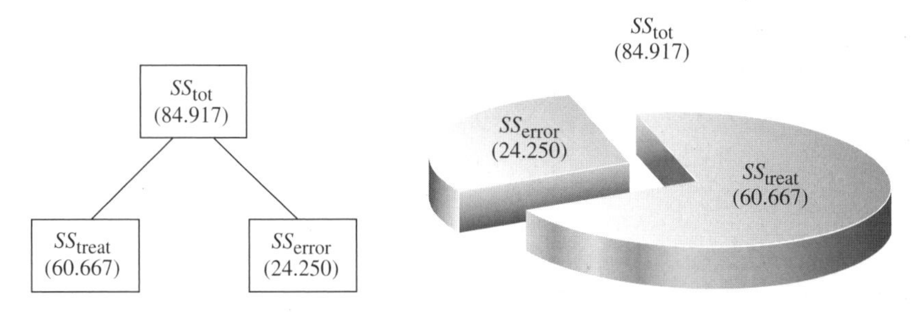

FIGURE 11.5 $SS_{treat} + SS_{error} = SS_{tot}$

PROBLEMS

***11.7.** Here are three groups of numbers. This problem and the ones that use this data set are designed to illustrate some relationships and to give you practice calculating ANOVA values. To begin, show that $SS_{treat} + SS_{error} = SS_{tot}$.

X_1	X_2	X_3
5	6	12
4	4	8
3	2	4

***11.8.** Emile Durkheim (1858–1917), a founder of sociology, believed that modernization produced social problems such as suicide. He compiled data from several European countries and published *Suicide* (1897/1951), a book that helped establish quantification in social science disciplines. The data in this problem were created so that the conclusion you reach will mimic that of Durkheim (and present-day sociologists).

On variables such as electricity consumption, newspaper circulation, and gross national product, 13 countries were classified as to their degree of modernization (low, medium, or high). The numbers in the table are annual suicide rates per 100,000 persons. For the data in the table, name the independent variable, the number of levels it has, and the dependent variable. Compute SS_{tot}, SS_{mod}, and SS_{error}.

Degree of modernization		
Low	Medium	High
4	17	20
8	10	22
7	9	19
5	12	9
		14

***11.9.** The best way to cure phobias is known. Bandura, Blanchard, and Ritter (1969) conducted an experiment that provided individuals with one of four treatments for their intense fear of snakes. One group worked with a model—a person who handled a 4-foot king snake and encouraged others to imitate her. One group watched a film of adults and children who enjoyed progressively closer contact with a king snake. A third group received desensitization therapy, and a fourth group served as a control, receiving no treatment. The numbers of snake-approach responses after treatment are listed in the table. Name the independent variable and its levels. Name the dependent variable. Compute SS_{tot}, SS_{treat}, and SS_{error}.

Model	Film	Desensitization	Control
29	22	21	13
27	18	17	12
27	17	16	9
21	15	14	6

Mean Squares and Degrees of Freedom

After you calculate *SS*, the next step in an ANOVA is to find the mean squares. A mean square is simply a sum of squares divided by its degrees of freedom. It is an estimate of the population variance, σ^2, when the null hypothesis is true.

Each sum of squares has a particular number of degrees of freedom associated with it. In a one-way classification, the *df* are df_{tot}, df_{treat}, and df_{error}. The relationship among degrees of freedom is the same as that among sums of squares:

$$df_{tot} = df_{treat} + df_{error}$$

The formula for df_{tot} is $N_{tot} - 1$. The df_{treat} is the number of treatment groups minus one $(K - 1)$. The df_{error} is the sum of the degrees of freedom for each group $[(N_1 - 1) + (N_2 - 1) + \cdots + (N_K - 1)]$. If there are equal numbers of scores in the K groups, the formula for df_{error} reduces to $K(N_t - 1)$. A little algebra will reduce this still further:

$$df_{error} = K(N_t - 1) = KN_t - K$$

However, $KN_t = N_{tot}$, so $df_{error} = N_{tot} - K$. This formula for df_{error} works whether or not the numbers in all groups are the same. In summary,

$$df_{tot} = N_{tot} - 1$$
$$df_{treat} = K - 1$$
$$df_{error} = N_{tot} - \mathrm{K}$$

error detection

$df_{tot} = df_{treat} + df_{error}$. Degrees of freedom are always positive numbers.

Mean squares are found using the following formulas:[7]

$$MS_{treat} = \frac{SS_{treat}}{df_{treat}} \quad \text{where} \quad df_{treat} = K - 1$$

$$MS_{error} = \frac{SS_{error}}{df_{error}} \quad \text{where} \quad df_{error} = N_{tot} - \text{K}$$

For the data in Table 11.1,

$$df_{drugs} = K - 1 = 3 - 1 = 2$$
$$df_{error} = N_{tot} - K = 12 - 3 = 9$$
$$\textit{Check}: df_{treat} + df_{error} = df_{tot}; \quad 2 + 9 = 11$$

Now, returning to the drug problem and calculating mean squares,

$$MS_{drugs} = \frac{SS_{drugs}}{df_{drugs}} = \frac{60.667}{2} = 30.333$$

$$MS_{error} = \frac{SS_{error}}{df_{error}} = \frac{24.250}{9} = 2.694$$

Notice that although $SS_{treat} + SS_{error} = SS_{tot}$ and $df_{treat} + df_{error} = df_{tot}$, mean squares are *not* additive; that is, $MS_{treat} + MS_{error} \neq MS_{tot}$.

Calculation and Interpretation of *F* Values Using the *F* Distribution

The next two steps in an ANOVA are to calculate an F value and to interpret it using the F distribution.

The *F* Test: Calculation

You learned earlier that F is a ratio of two estimates of the population variance. MS_{treat} is an estimate based on the variability between groups. MS_{error} is an estimate based on the variances within the groups. The F test is the ratio:

$$F = \frac{MS_{treat}}{MS_{error}}$$

[7] MS_{tot} is not used in ANOVA; only MS_{treat} and MS_{error} are calculated.

Every F value has 2 degrees of freedom associated with it. The first is the *df* associated with MS_{treat} (the numerator) and the second is the *df* associated with MS_{error} (the denominator). For the data in Table 11.1,

$$F = \frac{MS_{drugs}}{MS_{error}} = \frac{30.333}{2.694} = 11.26; \qquad df = 2, 9$$

F Value: Interpretation

The next question is: What is the probability of obtaining $F = 11.26$ if all three samples come from populations with the same mean? If that probability is less than α, then the null hypothesis should be rejected.

Turn now to **Table F** in Appendix C, which gives the critical values of F when $\alpha = .05$ and when $\alpha = .01$. Across the top of the table are degrees of freedom associated with the numerator (MS_{treat}). For the data in Table 11.1, $df_{drugs} = 2$, so 2 is the column you want. The rows of the table show degrees of freedom associated with the denominator (MS_{error}). In this case, $df_{error} = 9$, so look for 9 along the side. The tabled value for 2 and 9 *df* is 4.26 at the .05 level (lightface type) and 8.02 at the .01 level (boldface type). If the F value from the data is as great as or greater than the tabled value, reject the null hypothesis. If the F value from the data is not as great as the tabled value, the null hypothesis must be retained.

Because $11.26 > 8.02$, you can reject the null hypothesis at the .01 level. The three samples do not have a common population mean. At least one of the drugs produced scores (psychotic episodes) that were different from the rest.

At this point, your interpretation must stop. An ANOVA does not tell you which of the population means is different from the others. Such an interpretation requires an additional statistical analysis, which is covered later in this chapter.

It is customary to present all of the calculated ANOVA statistics in a *summary table.* **Table 11.2** is an example. Look at the right side of the table under p (for probability). The notation "$p < .01$" is shorthand for "the probability is less than 1 in 100 of obtaining treatment means as different as the ones actually obtained, if, in fact, the samples all came from identical populations."

Sometimes Table F does not contain an F value for the *df* in your problem. For example, an F with 2, 35 *df* or 4, 90 *df* is not tabled. When this happens, be conservative; use the F value that is given for *fewer df* than you have. Thus, the

TABLE 11.2 Summary table of ANOVA for data in Table 11.1

Source	*SS*	*df*	*MS*	*F*	*p*
Drugs	60.667	2	30.333	11.26	$<.01$
Error	24.250	9	2.694		
Total	84.917	11			

$F_{.01}(2, 9\ df) = 8.02$

proper F values for those two examples would be based on 2, 34 *df* and 4, 80 *df*, respectively.[8]

Schedules of Reinforcement— A Lesson in Persistence

Persistence is shown when you keep on trying even though the rewards are scarce or nonexistent. It is something that seems to vary a great deal from person to person, from task to task, and from time to time. What causes this variation? What has happened to lead to such differences in people, tasks, and times?

Persistence can often be explained if you know how frequently reinforcement (rewards) occurred in the past. I will illustrate with data produced by pigeons, but the principles illustrated hold for other forms of life, including students and professors.

The data in **Table 11.3** are typical results from a study of *schedules of reinforcement.* A hungry pigeon is taught to peck at a disk on the wall. A peck produces food according to a schedule the experimenter set up. For some pigeons every peck produces food—a continuous reinforcement schedule (crf). For other birds every other peck produces food—a fixed-ratio schedule of 2:1 (FR2). A third group of pigeons get food after every fourth peck—an FR4 schedule. Finally, for a fourth group, eight pecks are required to produce food—an FR8 schedule. After all groups receive 100 reinforcements, no more food is given (extinction begins). Under such conditions pigeons will continue to peck for a while and then stop. The dependent variable is the number of minutes a bird continues to peck (persist) after the food stops. As you can see in Table 11.3, three groups had five birds and one had seven.

Table 11.3 shows the raw data and the steps required to get the three *SS* figures and the three *df* figures. Work through **Table 11.3** now. When you finish, examine the summary table (**Table 11.4**), which continues the analysis by giving the mean squares, F test, and the probability of such an F.

The SPSS output of a one-way ANOVA on the data in Table 11.3 is a summary table that looks like Table 11.4, except that the treatment name *Schedules* is replaced with a generic *Between Groups* and the error term is labeled *Within Groups*. In addition, SPSS always produces a specific probability rather than a figure such as $<.01$.

When an experiment produces an F with a probability value less than .05, a conclusion is called for. Here it is: "The schedule of reinforcement used during learning significantly changes the persistence of responding during extinction. It is unlikely that the four samples have a common population mean."

---

[8] The F distribution may be used to test hypotheses about variances as well as hypotheses about means. To determine the probability that two sample variances came from the same population (or from populations with equal variances), form a ratio with the larger sample variance in the numerator. The resulting F value can be interpreted with Table F. The proper *df* are $N_1 - 1$ and $N_2 - 1$ for the numerator and denominator, respectively. For more information, see Kirk (2008, p. 362).

You might feel unsatisfied with this conclusion. You might ask which groups differ from the others, or which groups are causing that "significant change." Good questions. You will find answers in the next section, but first, here are a few problems to reinforce what you learned in this section.

TABLE 11.3 Partial analysis of the data for the number of minutes to extinction after four schedules of reinforcement during learning

	Schedule of reinforcement during learning			
	crf	FR2	FR4	FR8
	3	5	8	10
	5	7	12	14
	6	9	13	15
	2	8	11	13
	5	11	10	11
		10		
		6		
ΣX	21	56	54	63
ΣX^2	99	476	598	811
$\bar{X}$	4.2	8.0	10.8	12.6
N	5	7	5	5

$$\Sigma X_{tot} = 21 + 56 + 54 + 63 = 194$$

$$\Sigma X^2_{tot} = 99 + 476 + 598 + 811 = 1984$$

$$SS_{tot} = \Sigma X^2_{tot} - \frac{(\Sigma X_{tot})^2}{N_{tot}} = 1984 - \frac{(194)^2}{22} = 1984 - 1710.727 = 273.273$$

$$SS_{schedules} = \Sigma\left[\frac{(\Sigma X_t)^2}{N_t}\right] - \frac{(\Sigma X_{tot})^2}{N_{tot}} = \frac{(21)^2}{5} + \frac{(56)^2}{7} + \frac{(54)^2}{5} + \frac{(63)^2}{5} - \frac{(194)^2}{22} = 202.473$$

$$SS_{error} = \Sigma\left[\Sigma X_t^2 - \frac{(\Sigma X_t)^2}{N_t}\right]$$

$$= \left(99 - \frac{(21)^2}{5}\right) + \left(476 - \frac{(56)^2}{7}\right) + \left(598 - \frac{(54)^2}{5}\right) + \left(811 - \frac{(63)^2}{5}\right)$$

$$= 70.800$$

$$df_{tot} = N_{tot} - 1 = 22 - 1 = 21$$

$$df_{schedules} = K - 1 = 4 - 1 = 3$$

$$df_{error} = N_{tot} - K = 22 - 4 = 18$$

TABLE 11.4 Summary table of ANOVA analysis of the schedules of reinforcement study

Source	*SS*	*df*	*MS*	*F*	*p*
Schedules	202.473	3	67.491	17.16	<.01
Error	70.800	18	3.933		
Total	273.273	21			

$F_{.05}(3, 18) = 3.16$ $F_{.01}(3, 18) = 5.09$

PROBLEMS

11.10. Suppose your analysis produced $F = 2.56$ with 5, 75 *df*. How many treatment groups are in your experiment? What are the critical values for the .05 and .01 levels? What conclusion should you reach?

11.11. Here are two more questions about those raw numbers you found sums of squares for in problem 11.7.

a. Perform an F test and compose a summary table. Look up the critical value of F and make a decision about the null hypothesis. Write a sentence explaining what the analysis has shown.

b. Perform an F test using only the data from group 2 and group 3. Perform a t test on these same two groups.

***11.12.** Perform an F test on the modernization and suicide data in problem 11.8 and write a conclusion.

***11.13.** Perform an F test on the data from Bandura and colleagues' phobia treatment study (problem 11.9). Write a conclusion.

Comparisons Among Means

A significant F by itself tells you only that it is not likely that the samples have a common population mean. Usually you want to know more, but exactly what you want to know often depends on the particular problem you are working on. For example, in the schedules of reinforcement study, many would ask which schedules are significantly different from others (all pairwise comparisons). Another question might be whether there are sets of schedules that do not differ among themselves but do differ from other sets. In the drug study, the focus of interest is the new experimental drug. Is it better than either of the standard drugs or is it better than the average of the standards? This is just a *sampling* of the kinds of questions that can be asked.

Questions such as these are answered by analyzing the data using specialized statistical tests. As a group, these tests are referred to as **multiple-comparisons tests**. Unfortunately, no one multiple-comparisons method is satisfactory for all questions. In fact, not even three or four methods will handle all questions satisfactorily.

multiple-comparisons test Tests for statistical significance between treatment means or combination of means.

The reason there are so many methods is that specific questions are best answered by tailor-made tests. Over the years, statisticians have developed many tests for specific situations, and each test has something to recommend it. For an excellent chapter on the tests and their rationale, see Howell (2010, "Multiple Comparisons Among Treatment Means").

With this background in place, here is the plan for this elementary statistics book. I will discuss the two major *categories* of tests subsequent to ANOVA (*a priori* and *post hoc*) and then cover one *post hoc* test (the Tukey HSD test).

A Priori and *Post Hoc* Tests

***A priori* tests** require that you *plan* a limited number of tests before gathering the data. *A priori* means "based on reasoning, not on immediate observation." Thus, for *a priori* tests, you must choose the groups to compare when you design the experiment.

a priori test
Multiple-comparisons test that must be planned before examination of the data.

post hoc test
Multiple-comparisons test that is appropriate after examination of the data.

***Post hoc* tests** can be chosen after the data are gathered and you have examined the means. Any and all differences for which a story can be told are fair game for *post hoc* tests. *Post hoc* means "after this" or "after the fact."

The necessity for two kinds of tests is evident if you think again about the 15 samples that were drawn from the same population. Before the samples are drawn, what is your expectation that two randomly chosen means will be significantly different if tested with an independent-samples t test? About .05 is a good guess. Now, what if the data are drawn, the means compiled, and you then pick out the largest and the smallest means and test them with a t test? What is your expectation that these two will be significantly different? Much higher than .05, I would suppose. Thus, if you want to make comparisons after the data are gathered (and keep α at .05), then larger differences should be required before a difference is declared "not due to chance."

The preceding paragraphs should give you a start on understanding that there is a good reason for having more than one statistical test to answer questions subsequent to ANOVA. For this text, however, I have confined us to just one test, a *post hoc* test named the Tukey Honestly Significant Difference test.

The Tukey Honestly Significant Difference Test

Tukey Honestly Significant Difference (HSD) Test
Significance test for all possible pairs of treatments in a multitreatment experiment.

For many research problems, the principal question is: Are any of the treatments different from the others? John W. Tukey (pronounced "too-key") designed a test to answer this question when N is the same for all samples. A **Tukey Honestly Significant Difference (HSD) test** pairs each sample mean with every other mean, a procedure called *pairwise comparisons.* For each pair, the Tukey HSD test tells you if one sample mean is significantly larger than the other.

When N_t is the same for all samples, the formula for HSD is

$$\text{HSD} = \frac{\overline{X}_1 - \overline{X}_2}{s_{\overline{X}}}$$

$$\text{where } s_{\overline{X}} = \sqrt{\frac{MS_{\text{error}}}{N_t}}$$

N_t = the number in each treatment; the number that an $\overline{X}$ is based on

The critical value against which HSD is compared is found in Table G in Appendix C. Turn to **Table G**. Critical values for $\text{HSD}_{.05}$ are on the first page; those for $\text{HSD}_{.01}$ are on the second. For either page, enter the row that gives the *df* for the MS_{error} from the ANOVA. Go over to the column that gives K for the experiment.

If the HSD you calculated from the data is equal to or greater than the number at the intersection in Table G, reject the null hypothesis and conclude that the two means are significantly different. Finally, describe the direction of the difference using the names of the independent and dependent variables.

To illustrate Tukey HSD tests, I will make comparisons of the three groups in the drug experiment (Tables 11.1 and 11.2). With three groups, there are three pairwise comparisons, A and B, B and C, and A and C. The three tests are:

$$\text{HSD} = \frac{\bar{X}_A - \bar{X}_B}{s_{\bar{X}}} = \frac{7.25 - 6.75}{\sqrt{2.694/4}} = \frac{0.50}{0.821} = 0.61$$

$$\text{HSD} = \frac{\bar{X}_B - \bar{X}_C}{s_{\bar{X}}} = \frac{6.75 - 2.25}{\sqrt{2.694/4}} = \frac{4.50}{0.821} = 5.48$$

$$\text{HSD} = \frac{\bar{X}_A - \bar{X}_C}{s_{\bar{X}}} = \frac{7.25 - 2.25}{\sqrt{2.694/4}} = \frac{5.00}{0.821} = 6.09$$

In Table G, the critical values for experiments with three treatments and $df_{\text{error}} = 9$ are $\text{HSD}_{.05} = 3.95$ and $\text{HSD}_{.01} = 5.43$. At this point, all the numbers are calculated and tables consulted. It's time for thinking and interpretation.

A good beginning is to replace the generic levels of the independent variable, A, B, and C, with information specific to the experiment. Drug C was a new drug being compared with two existing standards, Drug A and Drug B. Interpretation: The mean number of psychotic episodes during the therapy period was only 2.25 for patients taking a new medicine, Drug C. This number is significantly less ($p < .01$) than that recorded for patients taking either of two standard drugs (Drug A = 7.25 episodes and Drug B = 6.75 episodes). Drug A and Drug B are not significantly different, $p > .05$.

SPSS calculates HSD tests when Tukey is selected from among the *post hoc* comparisons available with a one-way ANOVA. **Table 11.5** shows the output for the drug study data in Table 11.1. Each comparison of a pair is printed twice, and 95 percent confidence intervals are provided for each mean difference. Note that the mean differences of 4.50 and 5.00 are significant ($p = .009$ and $p = .005$) and that the mean difference of 0.50 is not ($p = .904$).

TABLE 11.5 SPSS output of Tukey HSD tests for the drug study

Multiple Comparisons

Dependent Variable: Episodes
Tukey HSD

(I) Drug	(J) Drug	Mean Difference (I-J)	Std. Error	Sig.	95% Confidence Interval Lower Bound	95% Confidence Interval Upper Bound
A	B	.50000	1.16070	.904	-2.7407	3.7407
	C	5.00000*	1.16070	.005	1.7593	8.2407
B	A	-.50000	1.16070	.904	-3.7407	2.7407
	C	4.50000*	1.16070	.009	1.2593	7.7407
C	A	-5.00000*	1.16070	.005	-8.2407	-1.7593
	B	-4.50000*	1.16070	.009	-7.7407	-1.2593

*The mean difference is significant at the .05 level.

Tukey HSD when N_t's are not equal The Tukey HSD test was developed for studies in which the N_t's are equal for all groups. Sometimes, despite the best of intentions, unequal N_t's occur. A modification in the denominator of HSD produces a statistic that can be tested with the values in Table G. The modification, recommended by Kramer (1956) and shown to be accurate (R. A. Smith, 1971), is

$$s_{\bar{X}} = \sqrt{\frac{MS_{\text{error}}}{2}\left(\frac{1}{N_1} + \frac{1}{N_2}\right)}$$

SPSS calculations of Tukey HSD values use this formula when N's are unequal.

Tukey HSD and ANOVA

In the analysis of an experiment, it has been common practice to calculate an ANOVA and then, if the F value is significant, conduct Tukey HSD tests. Mathematical statisticians (see Zwick, 1993) have shown, however, that this practice fails to detect real differences (the procedure is too conservative). So, if the *only* thing you want to know from your data is whether any *pairs* of treatments are significantly different, then you may calculate the Tukey HSD test without first showing that the ANOVA produced a significant F. Other multiple comparisons tests besides HSD, however, do require that an initial ANOVA produce a significant F. So, for the beginning statistics student, I recommend that you learn both ANOVA and Tukey HSD and apply them in that order.

PROBLEMS

11.14. Distinguish between the two categories of multiple-comparisons tests: *a priori* and *post hoc.*

11.15. For Durkheim's modernization and suicide data (problems 11.8 and 11.12), make all possible pairwise comparisons and write a conclusion telling what the data show.

11.16. For the schedules of reinforcement data, make all possible pairwise comparisons and write a conclusion about reinforcement schedules and persistence.

***11.17.** For the data on treatment of phobias (problems 11.9 and 11.13), compare the best treatment with the second best, and the poorest treatment with the control group. Write a conclusion about the study.

Assumptions of the Analysis of Variance and Random Assignment

Reaching conclusions by analyzing data with an ANOVA is fairly straightforward. But are there any behind-the-scenes requirements? Of course.

Two of the requirements are about the populations the samples come from. If these "requirements" (assumptions) are met, you are assured that the probability figures of

.05 and .01 are correct for the F values in the body of Table F. If the third "requirement" is met, cause-and-effect conclusions are supported.

Assumptions of ANOVA

1. *Normality.* The dependent variable is assumed to be normally distributed in the populations from which samples are drawn. In some areas of research, populations are known to be skewed, and researchers in those fields may decide that ANOVA is not appropriate for their data analysis. However, unless there is a reason to suspect that populations depart severely from normality, the inferences made from the F test will probably not be affected. ANOVA is robust. (It gives correct probabilities even when the populations are not exactly normal.) Where there is suspicion of a severe departure from normality, however, use the nonparametric method explained in Chapter 15.

2. *Homogeneity of variance.* The variances of the dependent-variable scores for each of the populations are assumed to be equal. Figures 11.1, 11.2, and 11.3, which were used to illustrate the rationale of ANOVA, show populations with equal variances. Several methods of testing this assumption are presented in advanced texts, such as Winer, Brown, and Michels (1991), Kirk (1995), and Howell (2010). When variances are grossly unequal, the nonparametric method discussed in Chapter 15 may be called for.

Random Assignment

Random assignment of participants to conditions distributes the effects of extraneous variables equally over all levels of the independent variable. For experiments where random assignment is used, strong statements of cause-and-effect are possible. When random assignment is not used, cause-and-effect conclusions should be avoided or, if tentatively advanced, supported by additional control groups, previous research, or logical considerations.

I hope these requirements seem familiar to you. They are the same as those you learned for the t distribution in Chapter 10. This makes sense; t is the special case of F when there are two samples.

Effect Size Indexes

The important question, Are the populations different? can be answered with an NHST F test. The equally important question, Are the differences large, medium, or small? requires an *effect size index.*

The Index *d*

Values for d, the effect size index covered in Chapters 5, 9, and 10, can be calculated for any two pairs of treatments in a one-way ANOVA design. The formula is

$$d = \frac{\bar{X}_1 - \bar{X}_2}{\hat{s}_{error}}$$

where $\hat{s}_{error} = \sqrt{MS_{error}}$

For the schedules of reinforcement example in Table 11.3, how much of an effect does changing from an FR4 to an FR8 schedule have on persistence? Calculating *d* for the two groups,

$$d = \frac{\overline{X}_{FR4} - \overline{X}_{FR8}}{\hat{s}_{error}} = \frac{10.8 - 12.6}{\sqrt{3.933}} = -0.91$$

Based on the convention that *d* values of 0.20, 0.50, and 0.80 correspond to small, medium, and large effect sizes, conclude that changing a schedule of reinforcement from FR4 to FR8 greatly increases persistence.

The Index *f*

Many other indexes of effect size have been developed to use on ANOVA data. Most provide a measure of the overall magnitude that the levels of the independent variable have on the dependent variable. Examples include ω^2 (omega squared), η^2 (eta squared), and *f* (lowercase). ω^2 and η^2 are explained in Howell (2008, 2010), and Field (2005a). As for *f*,

$$f = \frac{\sigma_{treat}}{\sigma_{error}}$$

To find *f*, *estimators* of σ_{treat} and σ_{error} are required. The estimator of σ_{treat} is symbolized by $\hat{s}_{treat}$ and is found using the formula[9]

$$\hat{s}_{treat} = \sqrt{\frac{K - 1}{N_{tot}}(MS_{treat} - MS_{error})}$$

The estimator of σ_{error} is symbolized by $\hat{s}_{error}$, and as you know, it is the square root of the error variance:

$$\hat{s}_{error} = \sqrt{MS_{error}}$$

Thus,

$$f = \frac{\hat{s}_{treat}}{\hat{s}_{error}}$$

Using the drug data in Table 11.1, the calculation of *f* is

$$f = \frac{\sqrt{\frac{K - 1}{N_{tot}}(MS_{treat} - MS_{error})}}{\sqrt{MS_{error}}} = \frac{\sqrt{\frac{3 - 1}{12}(30.333 - 2.694)}}{\sqrt{2.694}} = 1.31$$

[9] This formula is appropriate for an ANOVA with equal numbers of participants in each treatment group (Kirk, 1995, p. 181). For unequal *N*'s, see Cohen (1988, pp. 359–362).

The next step is to interpret an f value of 1.31. Jacob Cohen, who promoted the awareness of effect size and also contributed formulas, provided guidance for judging f values as small, medium, and large (Cohen, 1969, 1992):

Small	$f = 0.10$
Medium	$f = 0.25$
Large	$f = 0.40$

Thus, an f value of 1.31 indicates that the effect size in the drug study was just huge. (This is not surprising because textbook authors often manufacture fictional data so that an example will produce a large effect.) As another example of a huge effect size index, the value of f for the distributions pictured in Figure 11.2 is about 1.2.

The statistic f can be helpful when ANOVA produces an F that is not statistically significant. Facing a nonsignificant difference, should you pour more resources into your study or call it quits and look for other variables to investigate? If f is small, then perhaps any differences you might find are so small as to be unimportant and a new line of research is appropriate. On the other hand, if f is large, additional resources may reveal differences that the first study failed to detect.

The following set of problems provides an opportunity for you to apply all the ANOVA techniques you have studied.

PROBLEMS

11.18. ANOVA is based on two assumptions about the populations the samples are from. List them

11.19. What is the advantage of random assignment of participants to treatments in an experiment?

11.20. For the data from Bandura and colleagues' phobia treatment experiment (problems 11.9, 11.13, and 11.17), find f and write a sentence of interpretation.

11.21. Hermann Rorschach, a Swiss psychiatrist, began to get an inkling of an idea in 1911. By 1921 he had developed, tested, and published his test, which measures motivation and personality. One of the several ways that the test is scored is to count the number of responses the test-taker gives (R scores). The data in the table are representative of those given by people with schizophrenia, depression, and those with no diagnosable disorder. Analyze the data with an F test and a set of Tukey HSDs. Determine d for the schizophrenia–depression comparison. Write a conclusion about Dr. Rorschach's test.

Disorder		
Schizophrenia	Depression	None
17	16	23
20	11	18
22	13	31
19	10	14
21	12	22
14	10	28
25		20
		22

11.22. Many health professionals recommend that women practice monthly breast self-examination (BSE) to detect breast cancer in its early stages (when treatment is much more effective). Several methods are used to teach BSE. T. S. Spatz (1991) compared the method recommended by the American Cancer Society (ACS) to the 4MAT method, a method based on different learning styles that different people have. She also included an untrained control group. The dependent variable was the amount of information retained three months after training. Analyze the data in the table with an ANOVA, HSD tests, and *f.* Write a conclusion.

	ACS	4MAT	Control
ΣX	178	362	62
ΣX^2	2619	7652	1192
N	20	20	20

11.23. Common sense says that a valuable goal is worth suffering for. Is that the way it works? In a classic experiment in social psychology (Aronson and Mills, 1959), college women had to "qualify" to be in an experiment. The qualification activity caused severe embarrassment, mild embarrassment, or no embarrassment. After qualifying, the women listened to a recorded discussion, which was, according to the experimenters, "one of the most worthless and uninteresting discussions imaginable." The dependent variable was the women's rating of the discussion. Analyze the ratings using the techniques in this chapter (high scores = favorable ratings). Write an interpretation.

Degree of embarrassment		
Severe	Mild	None
18	18	17
23	12	15
14	14	9
20	15	12
25	10	13

11.24. Look over the objectives at the beginning of the chapter. Can you do them?

ADDITIONAL HELP FOR CHAPTER 11

Visit *cengage.com/psychology/spatz*. At the Student Companion Site, you'll find multiple-choice tutorial quizzes, flashcards with definitions and workshops. For this chapter there is a Statistical Workshop on One-Way ANOVA.

KEY TERMS

A priori test (p. 247–248)
Alternative hypothesis (p. 230)
Analysis of variance (p. 228)
Assumptions of ANOVA (p. 251)
Between-treatments estimate (p. 232)
Degrees of freedom (p. 242)
Effect size index *d* (p. 251)
Effect size index *f* (p. 252)
Error term (p. 232)
F distribution (p. 231)
F test (p. 237)
Grand mean (p. 237)
Mean square (*MS*) (p. 237)
Multiple-comparisons test (p. 247)
Null hypothesis (p. 229, 230)
NHST (p. 230)
One-way ANOVA (p. 228)
Post hoc test (p. 248)
Sum of squares (*SS*) (p. 237)
Summary table (p. 244)
Tukey HSD test (p. 248)
Within-treatment estimate (p. 232)

CHAPTER 12

Analysis of Variance: One-Factor Repeated Measures

OBJECTIVES FOR CHAPTER 12

After studying the text and working the problems in this chapter, you should be able to:

1. Describe the characteristics of a one-factor repeated-measures analysis of variance (ANOVA)
2. For a one-factor repeated-measures ANOVA, identify the sources of variance and calculate their sums of squares, degrees of freedom, mean squares, and the *F* value
3. Interpret the *F* value from a one-factor repeated-measures ANOVA
4. Use a Tukey Honestly Significant Difference test to make pairwise comparisons
5. Distinguish between Type I and Type II errors
6. List and explain advantages and cautions that come with a one-factor repeated-measures ANOVA

Let's begin with a review of the designs in the two previous chapters and then fit this chapter into that context. In Chapter 10 you studied *t* tests. Both the *independent-samples t test* and the *paired-samples t test* are null hypothesis statistical testing (NHST) techniques that are used to analyze experiments with two levels of one independent variable. The difference is that with the *paired-samples t test* you have a reason to pair the scores from one level of the independent variable with the scores from the second level—a reason other than that the scores are similar. The reason might be that the participants who made the scores are natural pairs, or matched pairs, or that the scores are made by the same person (or rat or fruit fly).

In Chapter 11 you studied analysis of variance (ANOVA), a NHST technique that helps you evaluate data from experiments that have more than two levels of one independent variable. The technique in Chapter 11, one-way ANOVA, is used when the treatment levels are independent of each other. It is the multilevel counterpart of the independent-samples *t* test.

repeated-measures ANOVA Statistical technique for designs with repeated measures of subjects or groups of matched subjects.

The multilevel counterpart of the paired-samples *t* test is a **repeated-measures ANOVA**, a NHST technique that is the topic of this chapter. Like the paired-samples design, the scores of a repeated-measures ANOVA are grouped together for a reason. For an experiment with three levels of the independent variable, scores might be grouped together because they are natural triplets, or the three participants had similar scores on a pretest, or the three scores were all made by the same participant. The term *repeated measures* refers to all three of these ways to group scores together.

factor Independent variable.

Here are two points about ANOVA terminology. First, the word **factor** is often substituted for *independent variable*. A one-factor design has one independent variable and a two-factor design has two independent variables. Thus, the title of this chapter indicates that it is about a technique used to analyze data from multilevel experiments (ANOVA) with one independent variable ("One Factor") and whose scores can be grouped together for some logical reason ("Repeated Measures").

The second point is that statistics terminology in one discipline is different from terminology in other disciplines. This is because terms that are quite descriptive in one field do not get adopted by researchers in other fields. In agriculture, the name of the repeated-measures design is the *split-plot* design. When several plots are available for planting, each is split into subplots equal to the number of treatments. Thus, a shady plot has all the treatments within it (as does a sunny plot and a hillside plot). In this way, the effects of shade, sun, and slope occur equally in all treatments. Besides *repeated-measures* and *split-plot*, the design in this chapter is also referred to as a *randomized blocks* design and a *subjects x treatments* design.

A Data Set

To begin your study of repeated-measures ANOVA, look at the data in **Table 12.1.** There are three levels of the factor (named simply X at this point). The mean score for each treatment is shown at the bottom of its column. Using the skills you learned in Chapter 11, estimate whether an F ratio for these data will be significant. Remember that $F = MS_{treat}/MS_{error}$. Go ahead, stop reading and make your estimate. (If you also write down how you arrived at your estimates, you can compare them to the analysis that follows.)

TABLE 12.1 A repeated-measures data set

	Levels of factor X		
Subjects	X_1	X_2	X_3
S_1	57	60	64
S_2	71	72	74
S_3	75	76	78
S_4	93	92	96
$\overline{X}$	74	75	78

Here's one line of thinking that might be expected from a person who remembers the lessons of Chapter 11:

> Well, OK, let's start with the means of the three treatments. They are pretty close together: 74, 75, and 78. Not much variability there. How about *within* each treatment? Uh-huh, scores range from 57 to 93 in the first column, so there is a good bit of variability there. Same story in the second and the third columns, too. There's lots of variability within the groups, so MS_{error} will be large. Nope, I don't think the F ratio will be very big. My guess is that F won't be significant—that the means are not significantly different.

Most of us who teach statistics would say that this reasoning shows a good understanding of what goes into the F value in a one-way ANOVA. Such a description would put an approving smile on our faces. With a repeated-measures ANOVA, however, there is an *additional* consideration.

The additional consideration for the data in Table 12.1 is that *each subject*[1] contributed a score to *each level* of the independent variable. To explain, look at the second column, X_2, which has scores of 60, 72, 76, and 92. The numbers are quite variable. But notice that some of this variability is *predictable* from the X_1 scores. That is, low scores (such as 60) are associated with S_1 (who had a low score of 57 on X_1). Notice, too, that in column X_3, S_1 had a score of 64, the lowest in the X_3 column. Thus, knowing that a score was made by S_1, you can predict that the score will be low. Please do a similar analysis on the scores for the person labeled S_4.

In a repeated-measures ANOVA, then, some of the variability of the scores within a level of the factor is predictable if you know which subject contributed the score. If you could *remove* the variability that goes with the differences between the subjects, you could *reduce* the variability within a level of the factor.

One way to reduce the variability between subjects is to use subjects who are alike. Although this would work well, getting subjects—whether people, agricultural plots, or production units—who are all alike is almost impossible. Fortunately, a repeated-measures ANOVA provides a statistical alternative that accomplishes the same goal.

One-Factor Repeated-Measures ANOVA: The Rationale

A one-factor repeated-measures ANOVA is a null hypothesis statistical testing technique. Like all NHST techniques, there is a null hypothesis, and for most research situations the null hypothesis is that the populations the samples are from are identical. Thus, for a three-treatment study, H_0: $\mu_1 = \mu_2 = \mu_3$.

Samples are collected from each population. This produces means of $\bar{X}_1$, $\bar{X}_2$, and $\bar{X}_3$, which typically vary from each other. A repeated-measures ANOVA allows you to calculate and then discard the variability among the means that comes from the differences between the subjects. The remaining variability in the data set is then partitioned into two components: one due to the differences between treatments (between-treatment variance) and another due to the inherent variability in the

[1] "Participant" is usually preferable to "subject" when referring to people whose scores are analyzed in a behavioral science or medical study. Statistical methods, however, are used by all manner of disciplines; Dependent variable scores come from agricultural plots, production units, and *Rattus norvegicus* as well as from *Homo sapiens*. Thus, statistics books often use the more general term *subjects*.

measurements (the error variance). These two variances are put into a ratio; the between-treatments variance is divided by the error variance. This ratio is an F value. This F ratio is a more sensitive measure of any differences that may exist among the treatments because the variability produced by differences among the subjects has been removed.

An Example Problem

The numbers in Table 12.1 can be used to illustrate the procedures for calculating a one-factor repeated-measures ANOVA. However, to include the important step of telling what the analysis shows, you must know what the independent and dependent variables are. The numbers in the table (the dependent variable) are scores on a multiple-choice test. The independent variable is the instruction given to those taking the multiple-choice test. The three different instructions were:

1. If you become doubtful about an item, always return to your first choice.
2. Write notes and comments about the items on the test itself.
3. No instructions.

Thus, the design compares two strategies and a control group.[2] Do you use either of these strategies? Do you have an opinion about the value of either of these strategies? Perhaps you would be willing to express your opinion by deciding which of the columns in Table 12.1 represent which of the two strategies and which column represents the control condition (taking the test without instructions). (If you want to guess, do it now, because I'm going to label the columns in the next paragraph.)

The numbers in the X_2 column represent typical multiple-choice test performances (control condition). The mean is 75. The X_1 column shows scores when test-takers follow strategy 1 (go with your initial choice). The scores in column X_3 represent performance when test-takers make notes on the test itself (though these notes are not graded in any way). Does either strategy have a significant effect on scores? A repeated-measures ANOVA will give you an answer.

Terminology for N

The three different N's in a one-factor repeated-measures ANOVA are distinguished by three different subscripts. Two of these terms you have used before; the new one is N_k. Definitions:

N_{tot} = total number of scores. In Table 12.1, $N_{\text{tot}} = 12$.

N_t = number of scores that receive one treatment. In Table 12.1, $N_t = 4$.

N_k = number of treatments. In Table 12.1, $N_k = 3$.

Sums of Squares

The first set of calculations in the analysis of test-taking strategies is of sums of squares. They are shown in the lower portion of **Table 12.2.** The upper portion of

[2] By the way, should you be one of those people who find yourself taking multiple-choice tests from time to time, understanding this problem will reveal a strategy for improving your score on such exams.

TABLE 12.2 A repeated-measures ANOVA that compares two test-taking strategies and a control condition (same data set as Table 12.1)

	Strategies			
Subjects	Use first choice	Control	Write notes	ΣX_k
S_1	57	60	64	181
S_2	71	72	74	217
S_3	75	76	78	229
S_4	93	92	96	281
ΣX_t	296	300	312	908
$\bar{X}$	74	75	78	

$$\Sigma X_{\text{tot}} = 296 + 300 + 312 = 908$$

$$\Sigma X^2_{\text{tot}} = 57^2 + 60^2 + \cdots + 92^2 + 96^2 = 70{,}460$$

$$SS_{\text{tot}} = \Sigma X^2_{\text{tot}} - \frac{(\Sigma X_{\text{tot}})^2}{N_{\text{tot}}} = 70{,}460 - 68{,}705.333 = 1754.667$$

$$SS_{\text{subjects}} = \Sigma\left[\frac{(\Sigma X_k)^2}{N_k}\right] - \frac{(\Sigma X_{\text{tot}})^2}{N_{\text{tot}}} = \frac{181^2}{3} + \frac{217^2}{3} + \frac{229^2}{3} + \frac{281^2}{3} - \frac{908^2}{12}$$

$$= 70{,}417.333 - 68{,}705.333 = 1712.000$$

$$SS_{\text{strategies}} = \Sigma\left[\frac{(\Sigma X_t)^2}{N_t}\right] - \frac{(\Sigma X_{\text{tot}})^2}{N_{\text{tot}}} = \frac{296^2}{4} + \frac{300^2}{4} + \frac{312^2}{4} - \frac{908^2}{12}$$

$$= 68{,}740.000 - 68{,}705.333 = 34.667$$

$$SS_{\text{error}} = SS_{\text{tot}} - SS_{\text{subjects}} - SS_{\text{strategies}}$$

$$= 1754.667 - 1712.000 - 34.667 = 8.000$$

$$df_{\text{tot}} = N_{\text{tot}} - 1 = 12 - 1 = 11$$

$$df_{\text{subjects}} = N_t - 1 = 4 - 1 = 3$$

$$df_{\text{strategies}} = N_k - 1 = 3 - 1 = 2$$

$$df_{\text{error}} = (N_t - 1)(N_k - 1) = (3)(2) = 6$$

Table 12.2 is identical to Table 12.1, except that labels and sums have been added. I did not include columns of X^2 values like those in previous chapters. Line 2 in the sum of squares portion of the table shows the calculation of ΣX^2_{tot}. The formula for SS_{tot} is the same as that for all ANOVA problems:

$$SS_{\text{tot}} = \Sigma X^2_{\text{tot}} - \frac{(\Sigma X_{\text{tot}})^2}{N_{\text{tot}}}$$

For the data in Table 12.2, $SS_{\text{tot}} = 1754.667$.

SS_{subjects} is the variability due to differences between subjects. As you know, people differ in their abilities when it comes to multiple-choice exams. These differences are reflected in the different row totals. The sum of one subject's scores (a row total) is symbolized ΣX_k. This sum depends on the number of treatments, which is symbolized N_k. The general formula for $SS_{subjects}$ is

$$SS_{subjects} = \Sigma\left[\frac{(\Sigma X_k)^2}{N_k}\right] - \frac{(\Sigma X_{tot})^2}{N_{tot}}$$

Work through the calculations in Table 12.2 that lead to the conclusion $SS_{subjects} = 1712.000$. Perhaps you have already noted that 1712 is a big portion of the total sum of squares.

SS_{treat} is the variability that is due to differences among the strategies the subjects used; that is, SS_{treat} is the variability due to the independent variable. The sum of the scores for one treatment level is ΣX_t. This sum depends on the number of subjects in that treatment, N_t. The general formula for SS_{treat} is

$$SS_{treat} = \Sigma\left[\frac{(\Sigma X_t)^2}{N_t}\right] - \frac{(\Sigma X_{tot})^2}{N_{tot}}$$

For the analysis in Table 12.2, the generic term *treatments* has been replaced with the term specific to these data: *strategies*. As you can see, $SS_{strategies} = 34.667$.

SS_{error} is the variability that remains in the data when the effects of the other identified sources have been removed. To find SS_{error}, calculate

$$SS_{error} = SS_{tot} - SS_{subjects} - SS_{treat}$$

For these data, $SS_{error} = 8.000$.

PROBLEMS

12.1. You just worked through three *SS* values for a repeated-measures ANOVA. But is it a repeated-measures design? (I gave no description of the procedures.) Describe a way to form groups to create a repeated-measures design.

12.2. Figure 11.5 shows how SS_{tot} is partitioned into its components in a one-way ANOVA. Construct a graphic that shows the partition of the components of sums of squares in a one-factor repeated-measures ANOVA.

Degrees of Freedom, Mean Squares, and *F*

After you calculate sums of squares, the next step is to determine their degrees of freedom. Degrees of freedom for one-way repeated-measures ANOVA follow the general rule: number of observations minus 1.

In general:	***For the strategy study:***
$df_{tot} = N_{tot} - 1$	$df_{tot} = 12 - 1 = 11$
$df_{subjects} = N_t - 1$	$df_{subjects} = 4 - 1 = 3$
$df_{treat} = N_k - 1$	$df_{strategies} = 3 - 1 = 2$
$df_{error} = (N_t - 1)(N_k - 1)$	$df_{error} = (3)(2) = 6$

As is always the case for ANOVAs, mean squares are found by dividing a sum of squares by its *df*. With mean squares in hand, you can form any appropriate *F* ratios.

TABLE 12.3 ANOVA summary table for the multiple-choice strategy study

Source	*SS*	*df*	*MS*	*F*	*p*
Subjects	1712.000	3			
Strategies	34.667	2	17.333	13.00	<.01
Error	8.000	6	1.333		
Total	1754.667	11			

$F_{.05}(2, 6\ df) = 5.14 \quad F_{.01}(2, 6\ df) = 10.92$

Because the usual purpose of this design is to provide a more sensitive test of the treatment variable, $F_{\text{strategies}}$ is the only F to calculate. Thus, there is no need to calculate MS_{subjects}.

$$MS_{\text{strategies}} = \frac{SS_{\text{strategies}}}{df_{\text{strategies}}} = \frac{34.667}{2} = 17.333$$

$$MS_{\text{error}} = \frac{SS_{\text{error}}}{df_{\text{error}}} = \frac{8.000}{6} = 1.333$$

To find F, divide $MS_{\text{strategies}}$ by the error term:

$$F = \frac{MS_{\text{strategies}}}{MS_{\text{error}}} = \frac{17.333}{1.333} = 13.003$$

An ANOVA summary table is the conventional way to present the results of an analysis of variance. **Table 12.3** shows the analysis that assessed strategies for taking multiple-choice exams.

Interpretation of *F*

Recall that the null hypothesis for a three-treatment, repeated-measures ANOVA is H_0: $\mu_1 = \mu_2 = \mu_3$. As you can see in Table 12.3, the tabled values for F with 2 and 6 *df* are 5.14 and 10.92 at the .05 and .01 α levels, respectively. Thus, for an obtained F value of 13.00, reject the null hypothesis and conclude that the three strategies for taking multiple-choices tests do not have a common population mean. Such a conclusion, of course, doesn't answer the obvious question: Should I use or avoid either of these strategies?

With this question in mind, look at the means of each of the strategies in Table 12.2. What pairwise comparison is the most interesting to you?

Tukey HSD Tests

The one pairwise comparison that I'll do for you is the comparison between the Write notes strategy and the Control group. The formula for a Tukey HSD is the same as the one you used in Chapter 11:

$$\text{HSD} = \frac{\bar{X}_1 - \bar{X}_2}{s_{\bar{X}}}$$

where $s_{\overline{X}} = \sqrt{\frac{MS_{\text{error}}}{N_t}}$

N_t = the number of scores in each treatment; the number that an $\overline{X}$ is based on.

Applying the formula to the data for the Write notes strategy and the Control group, you get

$$\text{HSD}_{\text{write v. cont}} = \frac{78 - 75}{\sqrt{1.333/4}} = \frac{3}{0.577} = 5.20; \quad p < .05$$

The critical value of HSD at the .05 level for three groups and a $df_{\text{error}} = 6$ is 4.34. Because the data-produced HSD is greater than the critical value of HSD, conclude that making notes on a multiple-choice exam produces a significantly better score.[3]

OK, now it is time for you to put into practice what you have been learning.

PROBLEMS

12.3. Many studies show that interspersing rest between practice sessions (spaced practice) produces better performance than a no-rest condition (massed practice). Most studies use short rests, such as minutes or hours. Bahrick et al. (1993), however, studied the effect of weeks of rest. Thirteen practice sessions of learning the English equivalents of foreign words were interspersed with rest periods of 2, 4, or 8 weeks. The scores (percent of recall) for all four participants in this study are in the accompanying table. Analyze the data with a one-factor repeated-measures ANOVA, construct a summary table, perform Tukey HSD tests, and write an interpretation.

	Rest interval (weeks)		
Subject	2	4	8
1	40	50	58
2	58	56	65
3	44	70	69
4	57	61	74

12.4. Using the data in Table 12.2 and a Tukey HSD, compare the scores of those who returned to their first choice to those of the control group.[4]

Type I and Type II Errors

In problem 12.3 you reached the conclusion that an 8-week rest between practice sessions is better than a 2-week rest. Could this conclusion be in error? Of course it

[3] For confirmation, see McKeachie, Pollie, and Speisman (1955).

[4] This comparison is based on data presented by Benjamin, Cavell, and Shallenberger (1984).

could be, although it isn't very likely. But suppose the conclusion is erroneous. What kind of a statistical error is it: a Type I error or a Type II error?

In problem 12.4 you compared the strategy of *return to your initial choice* to the *control* condition. You did not reject the null hypothesis. If this conclusion is wrong, then you've made an error. Is it a Type I error or a Type II error?

Perhaps these two review questions have accomplished their purpose—to provide you with a spaced practice trial in your efforts to learn about Type I and Type II errors. As a reminder, a Type I error is rejecting the null hypothesis when it is true. A Type II error is failing to reject a false null hypothesis. (To review, see pages 183–185 and 222–223.)

Some Behind-the-Scenes Information about Repeated-Measures ANOVA

Now that you can analyze data from a one-factor repeated-measures ANOVA, you are ready to appreciate something about what is really going on when you perform this statistical test. I'll begin by addressing the *fundamental* issue in any ANOVA: partitioning the variance.

Think about SS_{tot}. You may recall that the definitional formula for SS_{tot} is $\Sigma(X - \overline{X}_{GM})^2$. When you calculate SS_{tot}, you get a number that measures how variable the scores are. If the scores are all close together, SS_{tot} is small. (Think this through. Do you agree?) If the scores are quite different, SS_{tot} is large. (Agree?)

Regardless of the amount of total variance, an ANOVA divides it into separate, independent parts. Each part is associated with a different source.

Look again at the raw scores in Table 12.2. Some of the variability in those scores is associated with different subjects. You can see this in the *row totals*—there is a lot of variability from S_1 to S_2 to S_3 to S_4; that is, there is variability *between the subjects.*

Now, focus on the variability *within the subjects.* That is, look at the variability within S_1. S_1's scores vary from 57 to 60 to 64. S_2's scores vary from 71 to 72 to 74. Where does this variability within a subject come from? Clearly, treatments are implicated because, as treatments vary from the first to the second to the third column, scores increase.

But the change in treatments doesn't account for all the variation. That is, for S_1 the change in treatments produced a 3-point or 4-point increase in the scores, but for S_2 the change in treatments produced only a 1-point or 2-point increase, going from the first to the third treatment. Thus, the change from the first treatment to the third is not consistent; although an increase usually comes with the next treatment, the change isn't the same for each subject. This means that some of the variability within a subject is accounted for by the treatments, but not all. The variability that remains in the data in Table 12.2 is used to calculate an error term. It is sometimes called the **residual variance.**

residual variance
Variability due to unknown or uncontrolled variables; error term.

Advantages of Repeated-Measures ANOVA

With repeated-measures ANOVA there is a reason the scores in each row belong together. That reason might be that the subjects were matched before the experiment began, that the subjects were "natural triplets" or "natural quadruplets," or that the same subject contributed all the scores in one row.

When the same subjects are used for all levels of the independent variable, there is a saving of time and effort. To illustrate, it takes time to recruit participants, give explanations, and provide debriefing. So, if your study has three levels of the independent variable, and each person can participate in all three conditions, you can probably save two-thirds of the time it takes to recruit, explain, and debrief. Clearly, a repeated-measures design can be more efficient.

What about statistical power? You may recall that the more powerful the statistical test, the more likely it will reject a false null hypothesis (see pages 222–223). Any procedure that reduces the size of the error term (MS_{error}) will produce a larger F ratio. When the between-subjects variability is removed from the analysis, the error term is reduced, making a repeated-measures design more powerful. To summarize, two of the advantages of a repeated-measures ANOVA are *efficiency* and *power.*

Cautions about Repeated-Measures ANOVA

The previous section, titled "Advantages of . . . ," leads good readers to respond internally with "OK, *disadvantages* is next." Well, this section gives you three *cautions* that go with repeated-measures ANOVAs. To address the first, work the problem that follows.

PROBLEM

12.5. One of the following experiments does not lend itself to a repeated-measures design using the same participants in every treatment. Figure out which one and explain why.

a. You want to design an experiment to find out the physiological effects of meditation. Your plan is to measure oxygen consumption 1 hour before meditation, during meditation, and 1 hour after a meditation session.

b. A friend, interested in the treatment of depression, wants to compare drug therapy, interpersonal therapy, cognitive-behavioral therapy, and a placebo drug using Beck Depression Inventory scores as a dependent variable.

The first caution about repeated-measures ANOVA is that it may not be useful if one level of the independent variable continues to affect the participant's response in the next treatment condition (a *carryover* effect). In problem 12.5a, there is no problem with the experiment because there is no reason to expect that measuring oxygen consumption before meditation will carry over and affect consumption at a later time. However, in problem 12.5b, the effect of therapy is long-lasting and will affect depression scores at a later time. (To use a repeated-measures ANOVA on problem 12.5b, some method of matching patients could be used.)

Having cautioned you about carryover effects, I must mention that *sometimes* you want and expect carryover effects. For example, to evaluate a workshop on computerizing medical records, you might measure compliance and accuracy by giving the participants a pretest, a test after the workshop, and a follow-up test 6 months later. The important thing is carryover effects, and a repeated-measures ANOVA is the way to measure these effects.

The second caution is that the levels of the independent variable must be chosen by the researcher (a fixed-effects model) rather than being selected randomly from a number of possible levels. This requirement is commonly met in behavioral science experiments. For the random-effects model, see Howell (2010, Chapter 14) or Kirk (1995, Chapter 7).

The third caution that goes with repeated-measures ANOVA is one that goes with all NHST techniques—the caution about the assumptions of the test. In deriving formulas for statistical tests, mathematical statisticians begin by assuming that certain characteristics of the population are true. If the populations the samples come from have these characteristics, the probability figures that the tests produce are exact. If the populations don't have these characteristics, the accuracy of the *p* values is not assured. For this chapter's repeated-measures ANOVA, the only familiar assumption is that the populations are normally distributed. Other assumptions are covered in advanced statistics courses.

PROBLEMS

12.6. What are three cautions about using the repeated-measures ANOVA in this chapter?

12.7. What is a carryover effect?

***12.8.** Many college students are familiar with meditation exercises. One question about meditation that has been answered is: Are there physiological effects? The data that follow are representative of answers to this question. These data on oxygen consumption (in cubic centimeters per minute) are based on Wallace and Benson's article in *Scientific American* (1972). Analyze the data. Include HSD tests and an interpretation that describes the effect of meditation on oxygen consumption.

	Oxygen consumption (cc/min)		
Subjects	Before meditation	During meditation	After meditation
S_1	175	125	180
S_2	320	290	315
S_3	250	210	250
S_4	270	215	260
S_5	220	190	240

12.9. In problem 12.8 you drew three conclusions based on HSD tests. Each could be wrong. For each conclusion, identify what type of error is possible.

CP **12.10.** Most of the problems in this book are already set up for you in tabular form. Researchers, however, usually begin with fairly unorganized raw scores in front of them. For this problem, study the data until you can arrange it into a usable form.

The data are based on the report of Hollon, Thrase, and Markowitz (2002), who compared three kinds of therapy for depression with a control group. I have given you characteristics that allow you to match up a group of four subjects for each row. The dependent variable is Beck Depression Inventory (BDI) scores recorded at the end of treatment. The lower the score, the less the depression. Begin by identifying the independent variable. Study the data until it becomes clear how to set up the analysis. Analyze the data and write a conclusion.

Female, age 20	BDI = 10	Drug therapy
Male, age 30	BDI = 9	Interpersonal therapy
Female, age 45	BDI = 16	Drug therapy
Female, age 20	BDI = 11	Interpersonal therapy
Female, age 20	BDI = 8	Cognitive-behavioral therapy
Female, age 20	BDI = 18	Placebo
Female, age 45	BDI = 10	Cognitive-behavioral therapy
Male, age 30	BDI = 10	Drug therapy
Male, age 30	BDI = 15	Cognitive-behavioral therapy
Female, age 45	BDI = 24	Placebo
Female, age 45	BDI = 16	Interpersonal therapy
Male, age 30	BDI = 21	Placebo

12.11. Look over the objectives at the beginning of the chapter. Can you do them?

ADDITIONAL HELP FOR CHAPTER 12

Visit *cengage.com/psychology/spatz*. At the Student Companion Site, you'll find multiple-choice tutorial quizzes and flashcards with definitions.

KEY TERMS

Assumptions (p. 266)
Carryover effects (p. 265)
Degrees of freedom (p. 261)
Error sum of squares (p. 261)
Factor (p. 257)
F value (p. 262)
Mean squares (p. 261)
Partitioning the variance (p. 264)
Repeated-measures ANOVA (p. 257)
Residual variance (p. 264)
Subjects sum of squares (p. 260)
Treatment sum of squares (p. 261)
Tukey HSD (p. 262)
Type I and Type II errors (p. 263)

CHAPTER

Analysis of Variance: Factorial Design

13

OBJECTIVES FOR CHAPTER 13

After studying the text and working the problems in this chapter, you should be able to:

1. Define the terms *factorial design, factor, cell, main effect*, and *interaction*
2. Name the sources of variance in a factorial design
3. Compute sums of squares, degrees of freedom, mean squares, and *F* values in a factorial design
4. Determine whether or not *F* values are significant
5. Interpret an interaction
6. Interpret main effects when the interaction is not significant
7. Make pairwise comparisons between means with Tukey HSD tests
8. List the assumptions required for a factorial design ANOVA

Three chapters in this book (Chapters 11 through 13) are about the analysis of variance (ANOVA). No other NHST technique gets as much space from me or as much time from you. This emphasis mirrors ANOVA's importance and widespread use. When I surveyed 50 consecutive empirical articles in the most recent issues of *Psychological Science*, I found that 56 percent reported using ANOVA. Of those that used ANOVA, 93 percent used a factorial ANOVA. Among behavioral scientists, ANOVA is the most widely used inferential statistics technique and factorial ANOVA is its most popular version.

factor
Independent variable.

factorial ANOVA
Experimental design with two or more independent variables.

Chapters 11 and 12 covered the concepts, procedures and interpretation of ANOVA designs for experiments that have one **factor.** This chapter explains a design for experiments that have two or more factors, a design called **factorial ANOVA.** An attractive feature of factorial ANOVAs is that they not only give you a NHST test of each separate factor but they also tell you about the interaction between the factors.

TABLE 13.1 Illustration of a two-treatment design that can be analyzed with a *t* test

Factor *A*	
A_1	A_2
Scores on the dependent variable	Scores on the dependent variable
$\bar{X}_{A_1}$	$\bar{X}_{A_2}$

Using factorial terminology, here is a review of designs covered in Chapters 10 through 12 to help prepare you for factorial ANOVA and its bonus information, the interaction.

In Chapter 10 you learned to analyze data from a two-group experiment, schematically shown in **Table 13.1.** There is *one independent variable,* Factor *A*, with data from two levels, A_1 and A_2. For the *t* test, the question is whether the population that $\bar{X}_{A_1}$ is from is different from the population that $\bar{X}_{A_2}$ is from. You learned to determine this for both independent and paired-samples designs.

In Chapters 11 and 12 you learned to analyze data from a two-or-more-treatment experiment, schematically shown in **Table 13.2** for a four-treatment design. There is *one independent variable,* Factor *A*, with data from four levels, A_1, A_2, A_3, and A_4. The question is whether the four treatment means could have come from populations with identical means. The ANOVA technique covered in Chapter 11 is for independent-samples designs. The technique for samples that are matched in some fashion (repeated measures) is explained in Chapter 12.

Factorial Design

A factorial design has two or more factors, each of which has two or more levels. In a factorial design, every level of a factor occurs with every level of the other factor(s). The technique I explain in this chapter is for independent samples.[1] **Table 13.3** illustrates this design. One factor (Factor *A*) has three levels (A_1, A_2, and A_3) and the other factor (Factor *B*) has two levels (B_1 and B_2).

TABLE 13.2 Illustration of a four-treatment design that can be analyzed with an *F* test

Factor A			
A_1	A_2	A_3	A_4
Scores on the dependent variable	Scores on the dependent variable	Scores on the dependent variable	Scores on the dependent variable
$\bar{X}_{A_1}$	$\bar{X}_{A_2}$	$\bar{X}_{A_3}$	$\bar{X}_{A_4}$

[1] Advanced textbooks such as Howell (2010) and Kirk (1995) discuss the analysis of factorial repeated-measures designs and three-or-more-factor designs.

TABLE 13.3 Illustration of a 2 × 3 factorial design

		Factor A			
		A_1	A_2	A_3	
Factor B	B_1	Cell A_1B_1 Scores on the dependent variable	Cell A_2B_1 Scores on the dependent variable	Cell A_3B_1 Scores on the dependent variable	$\overline{X}_{B_1}$
	B_2	Cell A_1B_2 Scores on the dependent variable	Cell A_2B_2 Scores on the dependent variable	Cell A_3B_2 Scores on the dependent variable	$\overline{X}_{B_2}$
		$\overline{X}_{A_1}$	$\overline{X}_{A_2}$	$\overline{X}_{A_3}$	

Table 13.4 shows the NHST tests that are explained in Chapters 10 through 13 organized according to whether the tests are for independent samples or related samples. *Related samples* is a general term that includes pairing, matching, and repeated measures. Studying **Table 13.4** now will help you put this chapter's topic into the matrix of what you already know.

Factorial Design Notation

Factorial designs are identified with a shorthand notation such as "2 × 3" (two by three) or "3 × 5." The general term is $R \times C$ (Rows × Columns). The first number is the number of levels of one factor; the second number is the number of levels of the other factor. Thus, the design in **Table 13.3** is a 2 × 3 design. Assignment of a factor to a row or column is arbitrary; Table 13.3 could also be rearranged into a 3 × 2 table.

cell
Scores that receive the same combination of treatments.

Table 13.3 has six cells. The participants in each **cell** are treated differently. Participants in the upper left cell are given treatment A_1

TABLE 13.4 Catalog of statistical tests arranged by kind of design

Independent variables and number of levels	Design of experiment: Independent samples	Design of experiment: Related samples
One IV, 2 levels	Independent-samples t test (Chapter 10)	Paired-samples t test (Chapter 10)
One IV, 2 or more levels	One-way ANOVA (Chapter 11)	One-factor repeated-measures ANOVA (Chapter 12)
Two IVs, each with 2 or more levels	Factorial ANOVA (Chapter 13)	

and treatment B_1; that cell is identified as Cell A_1B_1. Participants in the lower right cell are given treatment A_3 and treatment B_2, and that cell is called Cell A_3B_2.

Factorial Design Information

A factorial ANOVA gives you two kinds of information. First, it gives you the same information that you would get from two separate one-factor ANOVAs. Look again at **Table 13.3.** A factorial ANOVA can determine the probability of obtaining means such as $\bar{X}_{A_1}$, $\bar{X}_{A_2}$, and $\bar{X}_{A_3}$ from identical populations. The same factorial ANOVA can determine the probability of $\bar{X}_{B_1}$ and $\bar{X}_{B_2}$, if the null hypothesis is true. These comparisons, which are like one-factor ANOVAs, are called **main effects.**

main effect
Significance test of the deviations of the mean levels of one independent variable from the grand mean.

In addition, a factorial ANOVA helps you decide whether the two factors *interact* with each other to affect scores on the dependent variable. An **interaction** between two factors means that *the effect that one factor has on the dependent variable depends on which level of the other factor is being administered.*

interaction
When the effect of one independent variable on the dependent variable depends on the level of another independent variable.

The experiment in Table 13.3 would have an interaction if the difference between Level B_1 scores and Level B_2 scores depended on whether you looked at the A_1 column, the A_2 column, or the A_3 column. To expand on this, each of the three columns represents an experiment that compares B_1 to B_2. If there is *no* interaction, then the difference between B_1 scores and B_2 scores will be the same for all three experiments. If there *is* an interaction, however, the difference between B_1 and B_2 will *depend* on whether you are examining the experiment at A_1, at A_2, or at A_3.

Factorial Design Examples

Perhaps a couple of factorial design examples will help. The first has an interaction; the second doesn't. Suppose several students are sitting in a dormitory lounge one Monday discussing the big party they all went to on Saturday. There is a lot of disagreement; some loved the party, while others just shrug. Now, imagine that everyone goes quantitative and rates the party on a scale from 1 to 10. The resulting numbers are quite variable. Why are they so variable?

One tactic for explaining variability is to give reasons that help explain the variability. For example, some people were at the party with one special person and others were there with several friends. Perhaps that helps explain some of the variability. Also, there were two kinds of activities at the party, talking and playing games. Perhaps some of the variability in ratings depends on the activity people engaged in. With two variables identified (number of companions and activities) and the ratings of the party (the dependent variable), you have the makings of a factorial ANOVA.

In **Table 13.5,** locate the two factors, companions and activities. Each factor has two levels, so this is a 2 × 2 factorial design. The numbers in the cells are the mean ratings of the party by those in that cell.

The cell means in Table 13.5 show an interaction between the two factors. Those who went to the party with one companion rated it highly if they talked but poorly if they played games. For those who went with several companions, the effect was just the opposite; talking resulted in low ratings and games resulted in high ratings.

TABLE 13.5 Illustration of a factorial design with an interaction (The numbers are mean ratings of a party.)

		Activity	
		Talk	Games
Companions	One	8	3
	Several	3	8

To return to the definition of an interaction, the effect of the number of companions depends on which column you look at. In the *talk* column, being with one companion results in a higher rating than being with several, but in the *games* column, being with one companion results in a lower rating than being with several.

As a second example, let's take a different Monday group who attended the same party where the activities were talking and playing games. In this second group, there were two personality types, shy and outgoing.

Table 13.6 shows the mean party ratings for this second group. The shy people gave the party a low rating if they talked and a slightly higher rating if they played games. Outgoing people show the same pattern as you look from talk to games—ratings go up slightly. The data in Table 13.6 show no interaction because the effect that personality type has on ratings does not depend on the type of activity. Although there is no interaction, there is an effect of personality—outgoing people rated the party higher than shy people.

PROBLEMS

13.1. Use $R \times C$ notation to identify the following factorial designs:

***a.** Men and women worked complex problems at either 7:00 A.M. or 7:00 P.M.

b. Three methods of teaching Chinese were used with 4-year-olds and 8-year-olds.

c. Four strains of mice were infected with three types of virus.

TABLE 13.6 Illustration of a factorial design with no interaction (The numbers are mean ratings of a party.)

		Activity	
		Talk	Games
Personality	Shy	2	3
	Outgoing	7	8

d. (This one is a little different.) Four varieties of peas were grown with three amounts of fertilizer in soil of low, medium, or high acidity.

13.2. Group together the designs that have the same number of factors.
a. Paired t test
b. Independent t test
c. One-way ANOVA with four treatments
d. A 2×4 factorial ANOVA
e. Repeated-measures ANOVA with three conditions

13.3. Group together the designs that have an independent variable with two levels.
a. Paired t test
b. Independent t test
c. One-way ANOVA with four treatments
d. A 2×4 factorial ANOVA
e. Repeated-measures ANOVA with three conditions

13.4. In your own words, define an interaction.

13.5. Make up an outcome for problem 13.1a that has an interaction.

Main Effects and Interaction

A two-way factorial ANOVA produces three statistical tests: two main effects and one interaction. That is, a factorial ANOVA gives you a test for Factor *A* and a separate test for Factor *B*. The test for the interaction (symbolized *AB*) tells you whether the scores for the different levels of Factor *A* depend on the levels of Factor *B*.

An Example That Does Not Have an Interaction

Table 13.7 shows a 2×3 factorial ANOVA. The numbers in the cells are cell means. I'll use Table 13.7 to illustrate how to examine a table of cell means and how to look for the two main effects and the interaction.

TABLE 13.7 **Illustration of a 2 × 3 factorial design with no interaction (The number in each cell is the mean of the scores in the cell. *N* is the same for each cell.)**

		Factor *A*			
		A_1	A_2	A_3	Factor *B* means
Factor *B*	B_1	10	20	60	30
	B_2	50	60	100	70
	Factor *A* means	30	40	80	Grand mean 50

Main effects Begin by looking at the margin means for Factor B ($B_1 = 30$, $B_2 = 70$ in Table 13.7). The B main effect gives the probability of obtaining means of 30 and 70 if both samples came from identical populations. Using this probability you can decide whether the null hypothesis H_0: $\mu_{B_1} = \mu_{B_2}$ should be rejected or not.

The A main effect gives you the probability of getting means of A_1, A_2, and A_3 (30, 40, and 80) if $\mu_{A_1} = \mu_{A_2} = \mu_{A_3}$. Again, this probability can result in a rejection or a retention of the null hypothesis.

Notice that a comparison of B_1 to B_2 satisfies the requirements for an experiment (page 199) in that, except for being treated with either B_1 or B_2, the groups are alike. That is, Group B_1 and Group B_2 both have equal numbers of participants who received A_1 (and also A_2 and A_3). The effect of receiving A_1 occurs as much in the B_1 group as it does in the B_2 group; B_1 and B_2 differ only in their levels of Factor B. This same line of reasoning when comparing A_1, A_2, and A_3 also satisfies the requirements for an experiment.

Interaction To check for an interaction, begin by looking at the two cell means in the A_1 column of **Table 13.7** (10 and 50). Changing from B_1 to B_2 increases the mean score by 40 points. Now look at level A_2. There is an increase of 40 points. The same increase is also found at level A_3. Thus, the effect of changing levels of Factor B *does not* depend on Factor A; the 40-point effect is the same, regardless of the level of A.

In a similar way, if you start with the B_1 row, you see that there is a mean increase of 10 points and then 40 points as you move from A_1 to A_2 to A_3. This same change is found in the B_2 row (an increase of 10 points and then an increase of 40 points). Thus, the effect of changing levels of Factor A *does not* depend on Factor B; the change is the same regardless of which level of B you look at.

An Example That Has an Interaction

Table 13.8 shows a 2×3 factorial design in which there *is* an interaction between the two independent variables. The main effect of Factor A is seen in the margin means along the bottom. The average effect of a change from A_1 to A_2 to A_3 is to *reduce* the mean score by 10 (55 to 45 to 35). But look at the cells. For B_1, the effect of changing from A_1 to A_2 to A_3 is to *increase* the mean score by 10 points. For B_2, the effect is to

TABLE 13.8 Illustration of a 2 × 3 factorial design with an interaction between factors (The number in each cell is the mean of the scores within that cell. *N* is the same for each cell.)

		Factor A			
		A_1	A_2	A_3	Factor B means
Factor B	B_1	10	20	30	20
	B_2	100	70	40	70
	Factor A means	55	45	35	Grand mean 45

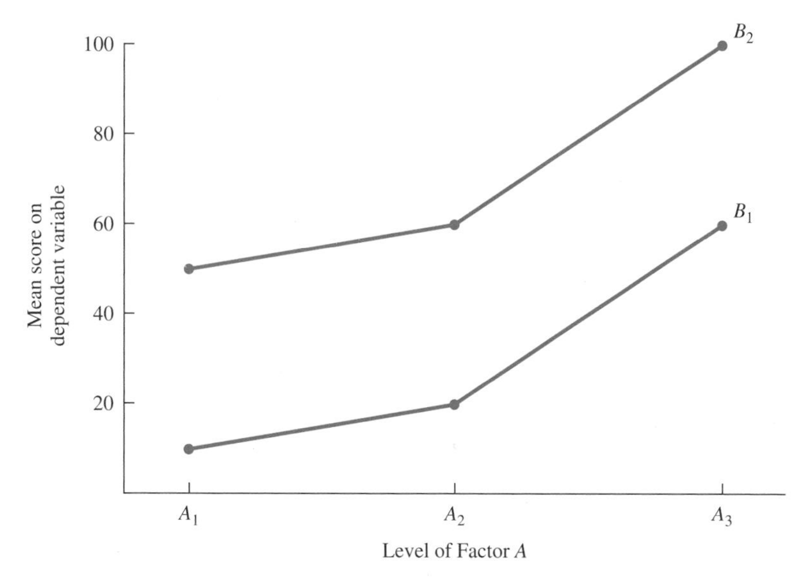

FIGURE 13.1 Line graph of data in Table 13.7. Parallel lines indicate there is no interaction.

decrease the score by 30 points. These data have an interaction because the effect of one factor *depends* on which level of the other factor you administer.

Now, let's do this same interaction analysis in **Table 13.8** by examining the two levels of factor B. In condition A_1, what is the effect of changing from B_1 to B_2? The effect is to increase scores by 90. How about for A_2? In condition A_2, the mean increase is 50 points. And for A_3, the mean increase from B_1 to B_2 is only 10 points. The effect of Factor B depends on the level of Factor A you are examining.[2]

To examine Table 13.8 for main effects, look at the margin means. A main effect for Factor B tests the hypothesis that the mean of the population of B_1 scores is identical to the mean of the population of B_2 scores. A main effect for Factor A gives the probability that the three means (55, 45, and 35) came from populations with a common mean.[3]

Graphs of Factorial ANOVA Designs

Because graphs are so helpful, the cell means of factorial ANOVA data are often presented as either a line graph or a bar graph. In both cases, the dependent variable is on the Y axis and one of the independent variables is on the X axis. The other independent variable is identified in the legend.

Figure 13.1 shows a line graph of the cell means in Table 13.7. Factor A is plotted on the X axis. Factor B is plotted as two separate curves, B_1 and B_2. Note that the two

[2] The concept of interaction is not easy to grasp. Usually, a presentation of the same concept in different words helps. After you finish the chapter, two good sources for additional explanations are Howell (2008, pp. 427–429) and an encyclopedia entry, Aiken and West (2005).

[3] A factorial ANOVA actually compares each margin mean to the grand mean, but for interpretation purposes I've described the process as a comparison of margin means.

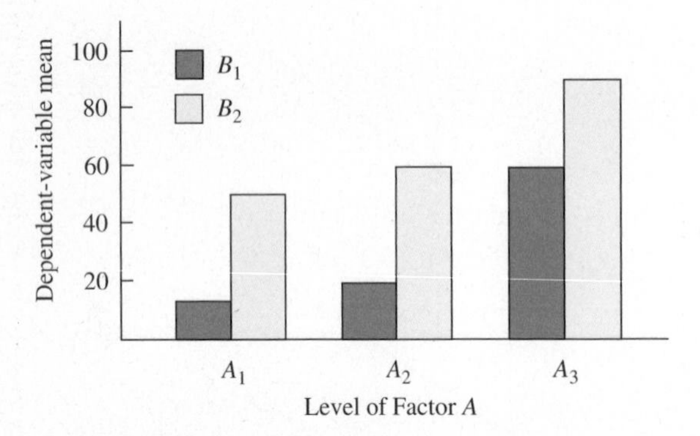

FIGURE 13.2 Bar graph of data in Table 13.7. There is no interaction.

curves are parallel, which is a characteristic of factorial data with no interaction. Of course, you cannot determine the statistical significance of an interaction for sure just by looking at a graph of the means, because means are subject to sampling variability. A statistical test of the interaction is required.

In a similar way, you cannot determine the significance of main effects just by looking at a graph, but an examination encourages helpful estimates. For example, in Figure 13.1 the B_1 curve is well below the B_2 curve, which suggests that the main effect for Factor *B* may be significant. Turning to Factor *A*, there are three levels to compare. The A_1 mean is the mean of the two points, A_1B_1 and A_1B_2. Thus, the A_1 mean is about 30. In a similar way, you can estimate the A_2 mean and the A_3 mean. These three means, 30, 40, and 80, seem quite different, which suggests that Factor *A* might prove to be statistically significant.

Figure 13.2 is a bar graph of the same data (Table 13.7). With bar graphs, comparing the *pattern* of the heights of the bars allows you to estimate the significance of the interaction and the main effects. For the interaction, the pattern of the stair steps in the three Factor *A* cells is identical, so the bar graph indicates there is *no* interaction. To assess the main effect of Factor *B*, judge the average height of the B_1 bars and the average height of the B_2 bars. The B_1 bars are shorter, so we can estimate that Factor *B* may be significant. Comparing the three levels of Factor *A* is about as difficult with a bar graph as it is with a line graph; the mean of the two A_1 conditions must be compared to the mean of the A_2 conditions and to the A_3 conditions. Please make those comparisons yourself.

Figure 13.3 shows two graphs of the data in Table 13.8 in which there *is* an interaction. The line graph in the left panel has curves that are not parallel, which indicates an interaction. Factor *B* appears to be significant, because the B_1 curve is always below the B_2 curve. Factor *A* appears to be not significant because the three means appear to be about the same, although there is some decrease from A_1 to A_2 to A_3.

Examining the bar graph in the right panel of Figure 13.3, you can see that the stair-step pattern of the three B_1 measures (going up) is *not* the pattern in the three B_2 measures (going down). Different patterns indicate there may be a significant interaction. Looking at main effects, the B_1 bars on the left are shorter than the B_2 bars on the right; Factor *B* appears to be significant. The difficult-to-assess Factor *A* with its three levels shows three means that are similar.

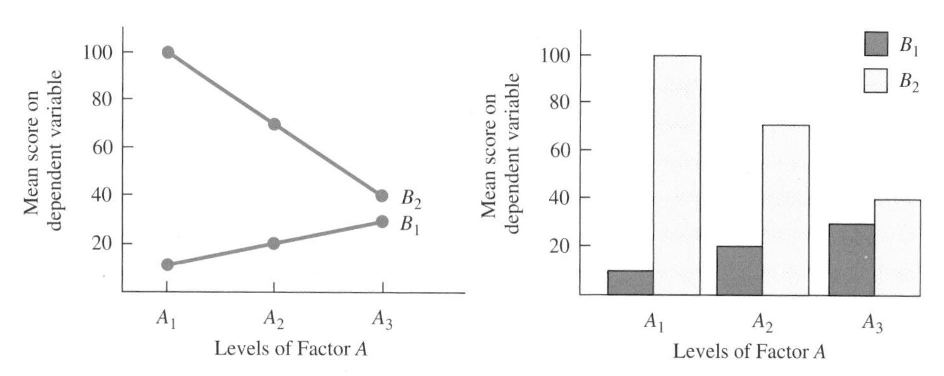

FIGURE 13.3 Line graph and bar graph of cell means in Table 13.8. Both graphs show the interaction.

For additional graphing examples, let's return to the ratings of the party described previously in the chapter. For the first group of partygoers, there was an interaction between the two factors, *companions* and *activities*. **Figure 13.4** shows a line graph and a bar graph of the cell means in Table 13.5. The line graph shows decidedly unparallel lines; they form an *X*, which is a strong indication of an interaction. The main effect of companions is not significant; both means are 5.5. Likewise, the main effect of activities isn't significant; both means are 5.5. For the bar graph in the right panel, an interaction is clearly indicated because the pattern of steps on the left is reversed on the right.

For the second group of partygoers (Table 13.6), the factors were *personality* and *activities*. The cell means are graphed in **Figure 13.5.** The line graph has two parallel lines, which indicate there is no interaction. The main effect of personality may be significant because the two means are so different (2.5 and 7.5). The main effect of activities does not appear to be significant; the means are similar (4.5 and 5.5). The bar graph indicates there is no interaction because the pattern of steps on the left is repeated on the right. As for main effects, personality makes a big difference, but the average of *talk* and *games* produces similar means.

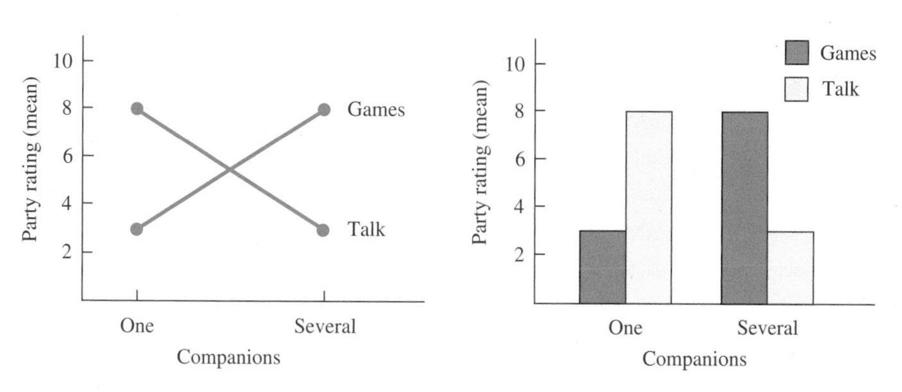

FIGURE 13.4 Line graph and bar graph of party ratings in Table 13.5 (shows interaction).

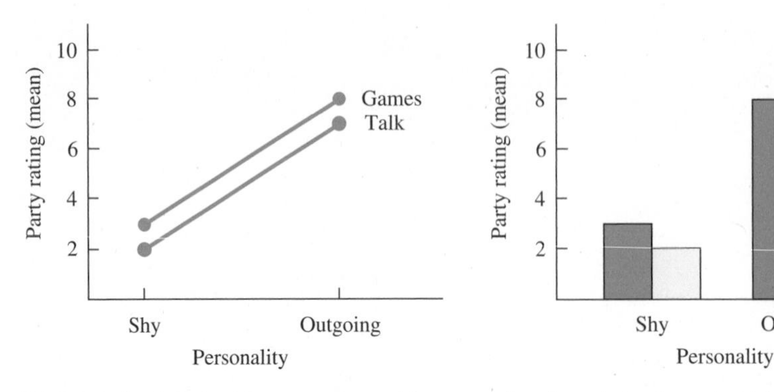

FIGURE 13.5 Line graph and bar graph of party ratings in Table 13.6 (no interaction).

clue to the future

When the interaction is statistically significant, the interpretation of the main effects may not be as straightforward as I have indicated. Be alert for this issue later in the chapter.

PROBLEMS

13.6. What is a main effect?

13.7. Descriptions of four studies follow (i–iv). High scores indicate better, or positive, outcomes. For each study, provide answers to a–e.

a. Graph the cell means, using a line graph or bar graph as indicated.

b. Decide whether an interaction appears to be present. State that an interaction does or does not appear to be present. Interactions, like main effects, are subject to sampling variations.

c. Explain how the two factors interact or do not interact.

d. For problems with no interaction, decide whether either main effect appears to be significant.

e. Explain the main effects in those problems that do not show an interaction.

i. The dependent variable for this ANOVA is attitude scores. Attitudes were expressed toward card games and computer games by the young and the old. Graph as a bar graph.

		Factor *A* (age)	
		A_1 (young)	A_2 (old)
Factor *B* (games)	B_1 (cards)	50	75
	B_2 (computer)	75	50

ii. This is a study of the effect of the dose of a new drug on three types of schizophrenia. The dependent variable is scores on a competency test. Use a line graph.

		A (diagnosis)		
		A_1 (disorganized)	A_2 (catatonic)	A_3 (paranoid)
B (dose)	B_1 (small)	5	10	70
	B_2 (medium)	20	25	85
	B_3 (large)	35	40	100

iii. The dependent variable is the score on a test that measures self-consciousness. Use a bar graph.

		A (grade)		
		A_1 (5th grade)	A_2 (9th grade)	A_3 (12th grade)
B (height)	B_1 (short)	10	30	20
	B_2 (tall)	5	60	30

iv. The dependent variable is attitude toward lowering the taxes on profits from investments. Graph as a line graph.

		A (socioeconomic status)		
		A_1 (low)	A_2 (middle)	A_3 (high)
B (gender)	B_1 (men)	10	35	60
	B_2 (women)	15	35	55

A Simple Example of a Factorial Design

As you read the following story, identify the dependent variable and the two independent variables (factors).

The semester was over. A dozen students who had all been in the same class were talking about what a bear of a test the comprehensive final exam had been. Competition surfaced.

"Of course, I was ready. Those of us who major in natural science disciplines just study more than you humanities types."

"Poot, poot, and balderdash," exclaimed one of the humanities types, who was a sophomore.

A third student said, "Well, there are some of both types in this group; let's just gather some data. How many hours did each of you study for that exam?"

"Wait a minute," exclaimed one of the younger students. "I'm a natural science type, but this is my first term in college. I didn't realize how much time I would need to cover the readings. I don't want my score to pull down my group's tally."

"Hmmm," mused the data-oriented student. "Let's see. Look, some of you are in your first term and the rest of us have had some college experience. We'll just take experience into account. Maybe the dozen of us will divide up evenly."

TABLE 13.9 Scores, cell means, and summary statistics for the study time data

		Previous college experience (Factor *A*)				
		Some (A_1)		None (A_2)		
		X	X^2	X	X^2	
Major (Factor *B*)	Humanities (B_1)	12	144	8	64	
		11	121	7	49	
		10	100	7	49	
		$\Sigma X_{cell} = 33$		$\Sigma X_{cell} = 22$		$\Sigma X_{B_1} = 55$
		$\Sigma X^2_{cell} = 365$		$\Sigma X^2_{cell} = 162$		$\Sigma X^2_{B_1} = 527$
		$\bar{X}_{cell} = 11.00$		$\bar{X}_{cell} = 7.3333$		$\bar{X}_{B_1} = 9.1667$
	Natural science (B_2)	12	144	10	100	
		10	100	9	81	
		9	81	8	64	
		$\Sigma X_{cell} = 31$		$\Sigma X_{cell} = 27$		$\Sigma X_{B_2} = 58$
		$\Sigma X^2_{cell} = 325$		$\Sigma X^2_{cell} = 245$		$\Sigma X^2_{B_2} = 570$
		$\bar{X}_{cell} = 10.3333$		$\bar{X}_{cell} = 9.00$		$\bar{X}_{B_2} = 9.6667$
		$\Sigma X_{A_1} = 64$		$\Sigma X_{A_2} = 49$		$\Sigma X_{tot} = 113$
		$\Sigma X^2_{A_1} = 690$		$\Sigma X^2_{A_2} = 407$		$\Sigma X^2_{tot} = 1097$
		$\bar{X}_{A_1} = 10.6667$		$\bar{X}_{A_2} = 8.1667$		$\bar{X}_{GM} = 9.4167$

This story establishes the conditions for a 2 × 2 factorial ANOVA. One of the factors is area of interest; the two levels are natural science and humanities. The other factor is previous experience in college; the two levels are none and some. The dependent variable is hours of study for a comprehensive final examination. In explaining this example, I will add the specific terms of the experiment to the *AB* terminology I've used so far. Interpretations should use the terms of the experiment.

It probably won't surprise you to find out that when I created a data set to go with this story, the dozen did divide up evenly—there were three students in each of the four cells. The data are shown in **Table 13.9,** along with ΣX, means, and ΣX^2, which are used in all factorial ANOVA problems.

Developing a preliminary understanding of the data in even more helpful with factorial designs than with simpler designs. Cell means, margin means, and graphs are a big help.

PROBLEM

13.8. Perform a preliminary analysis on the data in Table 13.9. Graph the cell means using the technique that you prefer. What is your opinion about whether the interaction is significant? Give your opinion about each of the main effects.

TABLE 13.10 Sums of squares and computation check for the study time data

$$SS_{\text{tot}} = \Sigma X^2_{\text{tot}} - \frac{(\Sigma X_{\text{tot}})^2}{N_{\text{tot}}} = 1097 - \frac{(113)^2}{12} = 1097 - 1064.0833 = 32.9167$$

$$SS_{\text{cells}} = \Sigma\left[\frac{(\Sigma X_{\text{cell}})^2}{N_{\text{cell}}}\right] - \frac{(\Sigma X_{\text{tot}})^2}{N_{\text{tot}}} = \frac{(33)^2}{3} + \frac{(22)^2}{3} + \frac{(31)^2}{3} + \frac{(27)^2}{3} - \frac{(113)^2}{12}$$
$$= 363 + 161.3333 + 320.3333 + 243 - 1064.0833 = 23.5833$$

$$SS_{\text{experience}} = \frac{(\Sigma X_{A_1})^2}{N_{A_1}} + \frac{(\Sigma X_{A_2})^2}{N_{A_2}} - \frac{(\Sigma X_{\text{tot}})^2}{N_{\text{tot}}} = \frac{(64)^2}{6} + \frac{(49)^2}{6} - \frac{(113)^2}{12}$$
$$= 682.6667 + 400.1667 - 1064.0833 = 18.7501$$

$$SS_{\text{major}} = \frac{(\Sigma X_{B_1})^2}{N_{B_1}} + \frac{(\Sigma X_{B_2})^2}{N_{B_2}} - \frac{(\Sigma X_{\text{tot}})^2}{N_{\text{tot}}} = \frac{(55)^2}{6} + \frac{(58)^2}{6} - \frac{(113)^2}{12}$$
$$= 504.1667 + 560.6667 - 1064.0833 = 0.7501$$

$$SS_{AB} = N_{\text{cell}}[(\overline{X}_{A_1B_1} - \overline{X}_{A_1} - \overline{X}_{B_1} + \overline{X}_{\text{GM}})^2 + (\overline{X}_{A_2B_1} - \overline{X}_{A_2} - \overline{X}_{B_1} + \overline{X}_{\text{GM}})^2$$
$$+ (\overline{X}_{A_1B_2} - \overline{X}_{A_1} - \overline{X}_{B_2} + \overline{X}_{\text{GM}})^2 + (\overline{X}_{A_2B_2} - \overline{X}_{A_2} - \overline{X}_{B_2} + \overline{X}_{\text{GM}})^2]$$
$$= 3[(11.00 - 10.6667 - 9.1667 + 9.4167)^2 + (7.3333 - 8.1667 - 9.1667 + 9.4167)^2$$
$$+ (10.3333 - 10.6667 - 9.6667 + 9.4167)^2 + (9.00 - 8.1667 - 9.6667 + 9.4167)^2]$$
$$= 4.0836$$

Check: $SS_{\text{cells}} - SS_A - SS_B = SS_{AB}$; $23.5833 - 18.7501 - 0.7501 = 4.0831$

$$SS_{\text{error}} = \Sigma\left[\Sigma X^2_{\text{cell}} - \frac{(\Sigma X_{\text{cell}})^2}{N_{\text{cell}}}\right]$$
$$= \left(365 - \frac{(33)^2}{3}\right) + \left(162 - \frac{(22)^2}{3}\right) + \left(325 - \frac{(31)^2}{3}\right) + \left(245 - \frac{(27)^2}{3}\right)$$
$$= 2 + 0.6667 + 4.6667 + 2 = 9.3333$$

Check: $SS_{\text{cells}} + SS_{\text{error}} = SS_{\text{tot}}$; $23.5833 + 9.3333 = 32.9167$

Sources of Variance and Sums of Squares

Remember that in Chapter 11 you identified three sources of variance in the one-way analysis of variance: (1) the total variance, (2) the between-treatments variance, and (3) the error variance. In a factorial design with two factors, the same sources of variance can be identified. However, the between-treatments variance, which is called the between-cells variance in factorial ANOVA, is further partitioned into three components. These are the two *main effects* and the *interaction.* That is, of the variability among the four cell means in Table 13.9, some can be attributed to the A main effect, some to the B main effect, and the rest to the AB interaction. Calculation of the sums of squares is shown in **Table 13.10.** I'll explain each calculation in the text that follows.

Total sum of squares Calculating the total sum of squares will be easy for you because it is the same as SS_{tot} in a one-way ANOVA and a one-factor repeated-measures ANOVA. Defined as $\Sigma(X - \overline{X}_{GM})^2$, SS_{tot} is the sum of the squared deviations of each score in the experiment from the grand mean of the experiment. To compute SS_{tot}, use the formula

$$SS_{tot} = \Sigma X_{tot}^2 - \frac{(\Sigma X_{tot})^2}{N_{tot}}$$

For the study time data (Table 13.10),

$$SS_{tot} = 1097 - \frac{(113)^2}{12} = 32.9167$$

Between-cells sums of squares To find the main effects and interaction, first find the between-cells sum of squares and then partition it into its component parts. SS_{cells} is defined as $\Sigma[N_{cell}(\overline{X}_{cell} - \overline{X}_{GM})^2]$. A "cell" in a factorial ANOVA is a group of participants treated alike; for example, humanities majors with no previous college experience constitute a cell.

The computational formula for SS_{cells} is similar to that for a one-way analysis:

$$SS_{cells} = \Sigma\left[\frac{(\Sigma X_{cell})^2}{N_{cell}}\right] - \frac{(\Sigma X_{tot})^2}{N_{tot}}$$

For the study time data (Table 13.10),

$$SS_{cells} = \frac{(33)^2}{3} + \frac{(22)^2}{3} + \frac{(31)^2}{3} + \frac{(27)^2}{3} - \frac{(113)^2}{12} = 23.5833$$

After SS_{cells} is obtained, it is partitioned into its three components: the A main effect, the B main effect, and the AB interaction.

Main effects sum of squares The sum of squares for each main effect is somewhat like a one-way ANOVA. The sum of squares for Factor A ignores the existence of Factor B and considers the deviations of the Factor A means from the grand mean. Thus,

$$SS_A = N_{A_1}(\overline{X}_{A_1} - \overline{X}_{GM})^2 + N_{A_2}(\overline{X}_{A_2} - \overline{X}_{GM})^2$$

where N_{A_1} is the total number of scores in the A_1 cells. For Factor B,

$$SS_B = N_{B_1}(\overline{X}_{B_1} - \overline{X}_{GM})^2 + N_{B_2}(\overline{X}_{B_2} - \overline{X}_{GM})^2$$

Computational formulas for the main effects are like formulas for SS_{treat} in a one-way design:

$$SS_A = \frac{(\Sigma X_{A_1})^2}{N_{A_1}} + \frac{(\Sigma X_{A_2})^2}{N_{A_2}} - \frac{(\Sigma X_{tot})^2}{N_{tot}}$$

And for the study time data (Table 13.10),

$$SS_{\text{experience}} = \frac{(64)^2}{6} + \frac{(49)^2}{6} - \frac{(113)^2}{12} = 18.7501$$

The computational formula for the B main effect simply substitutes B for A in the previous formula. Thus,

$$SS_B = \frac{(\Sigma X_{B_1})^2}{N_{B_1}} + \frac{(\Sigma X_{B_2})^2}{N_{B_2}} - \frac{(\Sigma X_{\text{tot}})^2}{N_{\text{tot}}}$$

For the study time data (Table 13.10),

$$SS_{\text{major}} = \frac{(55)^2}{6} + \frac{(58)^2}{6} - \frac{(113)^2}{12} = 0.7501$$

error detection

SS_{cells}, SS_A, and SS_B are calculated by adding a series of terms and then subtracting

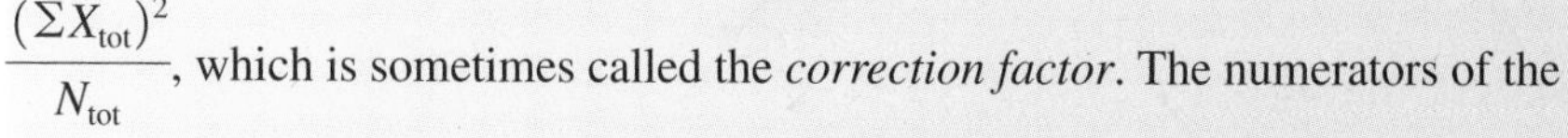

$\frac{(\Sigma X_{\text{tot}})^2}{N_{\text{tot}}}$, which is sometimes called the *correction factor*. The numerators of the terms that are added are equal to $(\Sigma X_{\text{tot}})^2$. The denominators of the terms that are added are equal to N_{tot}.

Interaction sum of squares To find the sum of squares for the interaction, use this formula:

$$SS_{AB} = N_{\text{cell}}[(\overline{X}_{A_1B_1} - \overline{X}_{A_1} - \overline{X}_{B_1} + \overline{X}_{\text{GM}})^2 + (\overline{X}_{A_2B_1} - \overline{X}_{A_2} - \overline{X}_{B_1} + \overline{X}_{\text{GM}})^2$$
$$+ (\overline{X}_{A_1B_2} - \overline{X}_{A_1} - \overline{X}_{B_2} + \overline{X}_{\text{GM}})^2 + (\overline{X}_{A_2B_2} - \overline{X}_{A_2} - \overline{X}_{B_2} + \overline{X}_{\text{GM}})^2]$$

For the study time data (Table 13.10),

$$\begin{aligned} SS_{AB} = {} & 3[(11.00 - 10.6667 - 9.1667 + 9.4167)^2 \\ & + (7.3333 - 8.1667 - 9.1667 + 9.4167)^2 \\ & + (10.3333 - 10.6667 - 9.6667 + 9.4167)^2 \\ & + (9.00 - 8.1667 - 9.6667 + 9.4167)^2] \\ = {} & 4.0836 \end{aligned}$$

Because SS_{cells} contains only the components SS_A, SS_B, and the interaction SS_{AB}, you can also obtain SS_{AB} by subtraction:

$$SS_{AB} = SS_{\text{cells}} - SS_A - SS_B$$

For the study time data (Table 13.10),

$$SS_{AB} = 23.5833 - 18.7501 - 0.7501 = 4.0831$$

Error sum of squares The error variability is due to the fact that participants treated alike differ from one another on the dependent variable. Because all were treated the same, this difference must be due to uncontrolled variables or random variation. SS_{error} for a 2 × 2 design is defined as

$$SS_{error} = \Sigma(X_{A_1B_1} - \overline{X}_{A_1B_1})^2 + \Sigma(X_{A_2B_1} - \overline{X}_{A_2B_1})^2 + \Sigma(X_{A_1B_2} - \overline{X}_{A_1B_2})^2 + \Sigma(X_{A_2B_2} - \overline{X}_{A_2B_2})^2$$

In words, SS_{error} is the sum of the sums of squares for each *cell* in the experiment. The computational formula is

$$SS_{error} = \Sigma\left[\Sigma X^2_{cell} - \frac{(\Sigma X_{cell})^2}{N_{cell}}\right]$$

For the study time data (Table 13.10),

$$SS_{error} = \left(365 - \frac{(33)^2}{3}\right) + \left(162 - \frac{(22)^2}{3}\right) + \left(325 - \frac{(31)^2}{3}\right) + \left(245 - \frac{(27)^2}{3}\right) = 9.3333$$

As in a one-way ANOVA, the total variability in a factorial ANOVA is composed of two components: SS_{cells} and SS_{error}. Thus,

$$23.5833 + 9.3333 = 32.9167$$

As you read earlier, SS_{cells} can be partitioned among SS_A, SS_B, and SS_{AB} in a factorial ANOVA. Thus,

$$23.5833 = 18.7501 + 0.7501 + 4.0831$$

These relationships among sums of squares in a factorial ANOVA are shown graphically in **Figure 13.6.**

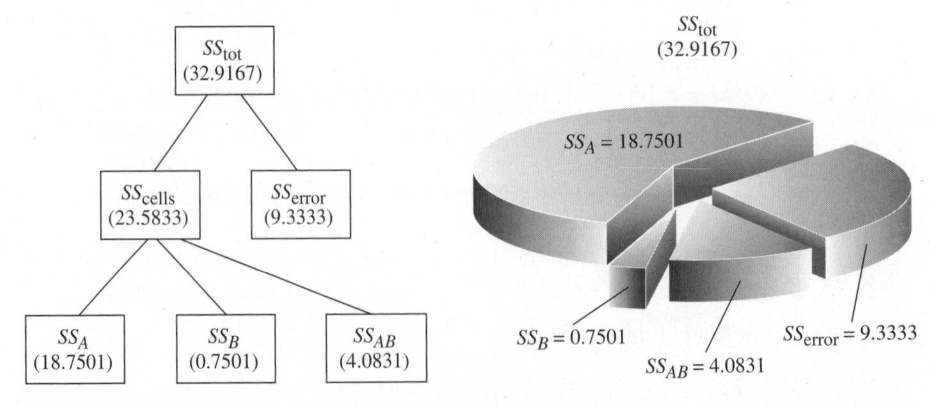

FIGURE 13.6 The partition of sums of squares in a factorial ANOVA problem

Did you notice that the direct calculation of SS_{AB} produced 4.0836 but that the value obtained by subtraction was 4.0831? The difference is due to rounding. By the time mean squares and an F test are calculated (in the next section), the difference disappears.

I'll interrupt the analysis of the study time data now so that you may practice what you have learned about the sums of squares in a factorial ANOVA.

error detection

One computational check for a factorial ANOVA is similar to that for the one-way classification: $SS_{cells} + SS_{error} = SS_{tot}$. A more complete version is $SS_A + SS_B + SS_{AB} + SS_{error} = SS_{tot}$. As before, this check will not catch errors in calculating ΣX or ΣX^2.

PROBLEMS

***13.9.** You know that women live longer than men. What about right-handed and left-handed people? Coren and Halpern's review (1991) included data on both gender and handedness, as well as age at death. An analysis of the age-at-death numbers in this problem produces conclusions like those reached by Coren and Halpern.

		Gender	
		Women	Men
Handedness	Left-handed	76 74 69	67 61 58
	Right-handed	82 78 74	76 72 68

a. Identify the design using $R \times C$ notation.

b. Name the independent variables, their levels, and the dependent variable.

c. Calculate SS_{tot}, SS_{cells}, SS_{gender}, $SS_{handedness}$, SS_{AB}, and SS_{error}.

***13.10.** Clinical depression occurs in about 15 percent of the population. Many treatments are available. The data that follow are based on Hollon, Thrase, and Markowitz's study (2002) comparing: (1) *psychodynamic therapy,* which uses free association and dream analysis to explore unconscious conflicts from childhood, (2) *interpersonal therapy,* which progresses through a three-stage treatment that alters the patient's response to recent life events, and (3) *cognitive-behavioral therapy,* which focuses on changing the client's thought and behavior patterns. The second factor in this data set is gender. The numbers are improvement scores for the 36 individuals receiving therapy.

a. Identify the design using $R \times C$ notation.
b. Identify the independent variables, their levels, and the dependent variable.
c. Calculate SS_{tot}, SS_{cells}, $SS_{therapy}$, SS_{gender}, SS_{AB}, and SS_{error}.

	Therapies		
	Psychodynamic	Interpersonal	Cognitive-behavioral
Women	22	41	33
	42	57	67
	30	75	41
	49	68	59
	15	48	49
	34	59	51
Men	37	48	36
	20	52	56
	56	41	44
	39	67	72
	48	33	52
	28	59	64

***13.11.** Many experiments have investigated the concept of *state-dependent memory.* Participants learn a task in a particular state—say, under the influence of a drug or not under the influence of the drug. They recall what they have learned in the same state or in the other state. The dependent variable is recall score (memory). The data in the table are designed to illustrate the phenomenon of state-dependent memory. Calculate SS_{tot}, SS_{cells}, SS_{learn}, SS_{recall}, SS_{AB}, and SS_{error}.

		Learn with	
		Drug	No drug
Recall with	Drug	$\Sigma X = 43$ $\Sigma X^2 = 400$ $N = 5$	$\Sigma X = 20$ $\Sigma X^2 = 100$ $N = 5$
	No drug	$\Sigma X = 25$ $\Sigma X^2 = 155$ $N = 5$	$\Sigma X = 57$ $\Sigma X^2 = 750$ $N = 5$

Degrees of Freedom, Mean Squares, and *F* Tests

Now that you are skilled at calculating sums of squares, you can proceed with the analysis of the study time data. Mean squares, as before, are sums of squares divided by their appropriate degrees of freedom. Formulas for degrees of freedom are:

In general:	***For the study time data:***
$df_{\text{tot}} = N_{\text{tot}} - 1$	$df_{\text{tot}} = 12 - 1 = 11$
$df_A = A - 1$	$df_{\text{experience}} = 2 - 1 = 1$
$df_B = B - 1$	$df_{\text{major}} = 2 - 1 = 1$
$df_{AB} = (A - 1)(B - 1)$	$df_{AB} = (1)(1) = 1$
$df_{\text{error}} = N_{\text{tot}} - (A)(B)$	$df_{\text{error}} = 12 - (2)(2) = 8$

In these equations, A and B stand for the number of levels of Factor A and Factor B, respectively.

error detection

Always use these checks:

$df_A + df_B + df_{AB} + df_{\text{error}} = df_{\text{tot}}$

$df_A + df_B + df_{AB} = df_{\text{cells}}$

A mean square is always the sum of squares divided by its degrees of freedom. Mean squares for the study time data are

$$MS_{\text{experience}} = \frac{SS_{\text{experience}}}{df_{\text{experience}}} = \frac{18.7501}{1} = 18.7501$$

$$MS_{\text{major}} = \frac{SS_{\text{major}}}{df_{\text{major}}} = \frac{0.7501}{1} = 0.7501$$

$$MS_{AB} = \frac{SS_{AB}}{df_{AB}} = \frac{4.0831}{1} = 4.0831$$

$$MS_{\text{error}} = \frac{SS_{\text{error}}}{df_{\text{error}}} = \frac{9.3333}{8} = 1.1667$$

The F values are computed by dividing mean squares by the error mean square:

$$F_{\text{experience}} = \frac{MS_{\text{experience}}}{MS_{\text{error}}} = \frac{18.7501}{1.1667} = 16.07$$

$$F_{\text{major}} = \frac{MS_{\text{major}}}{MS_{\text{error}}} = \frac{0.7501}{1.1667} = 0.64$$

$$F_{AB} = \frac{MS_{AB}}{MS_{\text{error}}} = \frac{4.0831}{1.1667} = 3.50$$

Results of a factorial ANOVA are usually presented in a summary table. Examine **Table 13.11,** the summary table for the study time data.

You now have three F values from the data. Each has 1 degree of freedom in the numerator and 8 degrees of freedom in the denominator. To find the probabilities of obtaining such F values if the null hypothesis is true, use Table F in Appendix C. Table F yields the following critical values: $F_{.05}(1, 8) = 5.32$ and $F_{.01}(1, 8) = 11.26$.

TABLE 13.11 **ANOVA summary table for the study time data**

Source	*SS*	*df*	*MS*	*F*	*p*
Experience (*A*)	18.7501	1	18.7501	16.07	<.01
Major (*B*)	0.7501	1	0.7501	0.64	>.05
AB	4.0831	1	4.0831	3.50	>.05
Error	9.3333	8	1.1667		
Total	32.9167	11			

The next step is to compare the obtained *F*'s to the critical value *F*'s and then tell what your analysis reveals about the relationships among the variables. With factorial ANOVAs, you should first interpret the interaction *F* and then proceed to the *F*'s for the main effects. (As you continue this chapter, be alert for the explanation of *why* you should use this order.)

The interaction F_{AB} (3.50) is less than 5.32, so the interaction between experience and major is not significant. Although the effect of experience was to increase study time by 50 percent for humanities majors and only about 15 percent for natural science majors, this interaction difference was not statistically significant. When *the interaction is not significant,* interpret each main effect as if it came from a one-way ANOVA.

The main effect for experience ($F_{experience}$) is significant beyond the .01 level ($16.07 > 11.26$). The null hypothesis $\mu_{some} = \mu_{none}$ can be rejected. By examining the margin means (8.17 and 10.67), you can conclude that experienced students study significantly more than those without experience. F_{major} was less than 1, and *F* values less than 1 are never significant. Thus, these data do not provide evidence that students in either major study more than the others. For the problems that follow, complete the factorial ANOVAs that you began earlier.

PROBLEMS

***13.12.** For the longevity data in problem 13.9, plot the cell means with a bar graph. Compute *df*, *MS*, and *F* values. Arrange these in a summary table that footnotes appropriate critical values. Tell what the analysis shows.

***13.13.** For the data on the effectiveness of therapies and gender in problem 13.10, plot the cell means using a line graph. Compute *df*, *MS*, and *F* values. Arrange these in a summary table and note the appropriate critical values. Tell what the analysis shows.

***13.14.** For the data on state-dependent memory in problem 13.11, plot the cell means using a bar graph. Compute *df*, *MS*, and *F* values. Arrange these in a summary table and note the appropriate critical values.

Analysis of a 2 × 3 Design

Here's a question for you. Read the question, take 10 seconds, and compose an answer. *Who is taller, boys or girls?*

A halfway good answer is "It depends." A completely good answer gives the other variable that the answer depends on; for example, "It depends on the *age* of the boys and girls."

TABLE 13.12 Data and summary statistics for height example

Age (Factor B)	Male (A_1) X_{A_1}	Male (A_1) $X^2_{A_1}$	Female (A_2) X_{A_2}	Female (A_2) $X^2_{A_2}$	Summary values
Age 6 (B_1)	44	1936	44	1936	
	45	2025	44	1936	
	46	2116	46	2116	$\Sigma X_{B_1} = 362$
	47	2209	46	2116	$\Sigma X^2_{B_1} = 16{,}390$
Sum	182	8286	180	8104	$\bar{X}_{B_1} = 45.25$
Mean	45.50		45.00		
Age 12 (B_2)	56	3136	58	3364	
	57	3249	60	3600	
	59	3481	61	3721	$\Sigma X_{B_2} = 474$
	60	3600	63	3969	$\Sigma X^2_{B_2} = 28{,}120$
Sum	232	13,466	242	14,654	$\bar{X}_{B_2} = 59.25$
Mean	58.00		60.50		
Age 18 (B_3)	66	4356	63	3969	
	69	4761	65	4225	
	72	5184	67	4489	$\Sigma X_{B_3} = 544$
	73	5329	69	4761	$\Sigma X^2_{B_3} = 37{,}074$
Sum	280	19,630	264	17,444	$\bar{X}_{B_3} = 68.00$
Mean	70.00		66.00		
Summary values	$\Sigma X_{A_1} = 694$; $\Sigma X^2_{A_1} = 41{,}382$; $\bar{X}_{A_1} = 57.83$		$\Sigma X_{A_2} = 686$; $\Sigma X^2_{A_2} = 40{,}202$; $\bar{X}_{A_2} = 57.17$		$\Sigma X_{\text{tot}} = 1380$; $\Sigma X^2_{\text{tot}} = 81{,}584$; $\bar{X}_{\text{GM}} = 57.50$

Gender (Factor A)

Factorial ANOVA designs are often used when a researcher thinks that the answer to a question about the effect of variable A is "It depends on variable B." As you may have already anticipated, the way to express this idea statistically is to talk about significant interactions.

The analysis that follows is for a 2×3 factorial design. The factor with two levels is gender: males and females. The second factor is age: 6, 12, and 18 years old. The dependent variable is height in inches.

I've constructed data that mirror fairly closely the actual situation for Americans. The data and the summary statistics are given in **Table 13.12.** Orient yourself to this table by noting the column headings and the row headings. Examine the six cell means and the five margin means. Note that $N = 4$ for each cell.

Next, examine **Figure 13.7,** which shows a line graph and a bar graph of the cell means. Make a preliminary guess about the significance of the interaction and the two main effects.

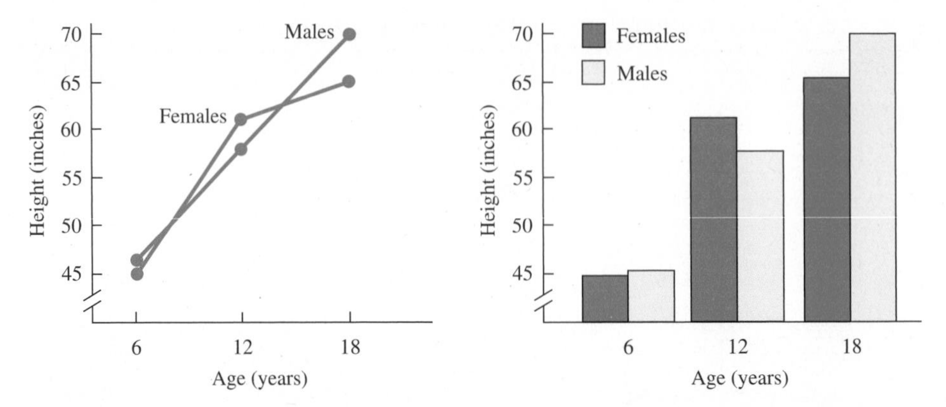

FIGURE 13.7 Line graph and bar graph of height data. An interaction is present.

Table 13.13 shows the calculation of the components of a factorial ANOVA. The headings in this table are familiar to you: sums of squares, degrees of freedom, mean squares, and *F* values. Work your way through each line in **Table 13.13** now.

The results of a factorial ANOVA are presented as a summary table. **Table 13.14** is the summary table for the height example. First, check the interaction *F* value. It is statistically significant.

Table 13.15 is an SPSS summary table from the factorial analysis of the height data. The lines labeled *Gender, Age,* and *Gender*Age* correspond to the factorial ANOVA analyses covered in this chapter.

Interpreting a Significant Interaction

Telling the story of a significant interaction in words is facilitated by referring to the graph of the cell means. Look at **Figure 13.7.** Boys and girls are about the same height at age 6, but during the next 6 years the girls grow faster, and they are taller at age 12. In the following 6 years, however, the boys grow faster, and at age 18 the boys are taller than the girls are. The question of whether boys or girls are taller *depends* on what age is being considered.

So, one way to interpret a significant interaction is to examine a graph of the cell means and describe the changes in the variables. Though a good start, this technique is just a beginning. A complete interpretation of interactions requires techniques that are beyond the scope of a textbook for elementary statistics. (To begin a quest for more understanding, see Rosnow and Rosenthal, 2005, for an explanation of residuals.)

Interpreting Main Effects When the Interaction Is Significant

When the interaction is *not* significant, the interpretation of a main effect is the same as the interpretation of the independent variable for a one-way ANOVA. However, when

TABLE 13.13 Calculation of the components of the 2 × 3 factorial ANOVA of height

Sums of squares

$$SS_{\text{tot}} = \Sigma X^2_{\text{tot}} - \frac{(\Sigma X_{\text{tot}})^2}{N_{\text{tot}}} = 81{,}584 - \frac{(1380)^2}{24} = 2234.000$$

$$SS_{\text{cells}} = \Sigma\left[\frac{(\Sigma X_{\text{cell}})^2}{N_{\text{cell}}}\right] - \frac{(\Sigma X_{\text{tot}})^2}{N_{\text{tot}}}$$

$$= \frac{(182)^2}{4} + \frac{(180)^2}{4} + \frac{(232)^2}{4} + \frac{(242)^2}{4} + \frac{(280)^2}{4} + \frac{(264)^2}{4} - \frac{(1380)^2}{24} = 2152.000$$

$$SS_{\text{gender}(A)} = \frac{(\Sigma X_{A_1})^2}{N_{A_1}} + \frac{(\Sigma X_{A_2})^2}{N_{A_2}} - \frac{(\Sigma X_{\text{tot}})^2}{N_{\text{tot}}} = \frac{(694)^2}{12} + \frac{(686)^2}{12} - \frac{(1380)^2}{24} = 2.667$$

$$SS_{\text{age}(B)} = \frac{(\Sigma X_{B_1})^2}{N_{B_1}} + \frac{(\Sigma X_{B_2})^2}{N_{B_2}} + \frac{(\Sigma X_{B_3})^2}{N_{B_3}} - \frac{(\Sigma X_{\text{tot}})^2}{N_{\text{tot}}} = \frac{(362)^2}{8} + \frac{(474)^2}{8} + \frac{(544)^2}{8} - \frac{(1380)^2}{24} = 2107.000$$

$$SS_{AB} = N_{\text{cell}}[(\bar{X}_{A_1B_1} - \bar{X}_{A_1} - \bar{X}_{B_1} + \bar{X}_{\text{GM}})^2 + (\bar{X}_{A_2B_1} - \bar{X}_{A_2} - \bar{X}_{B_1} + \bar{X}_{\text{GM}})^2$$
$$+ (\bar{X}_{A_1B_2} - \bar{X}_{A_1} - \bar{X}_{B_2} + \bar{X}_{\text{GM}})^2 + \cdots + (\bar{X}_{A_2B_3} - \bar{X}_{A_2} - \bar{X}_{B_3} + \bar{X}_{\text{GM}})^2]$$
$$= 4[(45.50 - 57.83 - 45.25 + 57.50)^2 + (45.00 - 57.17 - 45.25 + 57.50)^2$$
$$+ (58.00 - 57.83 - 59.25 + 57.50)^2 + (60.50 - 57.17 - 59.25 + 57.50)^2$$
$$+ (70.00 - 57.83 - 68.00 + 57.50)^2 + (66.00 - 57.17 - 68.00 + 57.50)^2]$$
$$= 4[10.583] = 42.333$$

Check: $SS_{\text{cells}} - SS_A - SS_B = SS_{AB}$; $2152.000 - 2.667 - 2107.000 = 42.333$

$$SS_{\text{error}} = \Sigma\left[\Sigma X^2_{\text{cell}} - \frac{(\Sigma X_{\text{cell}})^2}{N_{\text{cell}}}\right]$$

$$= \left(8286 - \frac{(182)^2}{4}\right) + \left(8104 - \frac{(180)^2}{4}\right) + \left(13{,}466 - \frac{(232)^2}{4}\right)$$
$$+ \left(14{,}654 - \frac{(242)^2}{4}\right) + \left(19{,}630 - \frac{(280)^2}{4}\right) + \left(17{,}444 - \frac{(264)^2}{4}\right)$$
$$= 82.000$$

Check: $2152.000 + 82.000 = 2234.000$

(continued)

the interaction *is* significant, the interpretation of main effects becomes complicated. To illustrate the problem, consider the main effect of gender in **Table 13.14.** The simple interpretation of $F = 0.59$ is that gender does not have a significant effect on height. But clearly, that isn't right! Gender *is* important, as the interpretation of the interaction revealed.

TABLE 13.13 (Continued)

Degrees of freedom

$$df_A = A - 1 = 2 - 1 = 1$$
$$df_B = B - 1 = 3 - 1 = 2$$
$$df_{AB} = (A - 1)(B - 1) = (1)(2) = 2$$
$$df_{error} = N_{tot} - (A)(B) = 24 - (2)(3) = 18$$
$$df_{tot} = N_{tot} - 1 = 24 - 1 = 23$$

F values

$$F_{gender} = \frac{MS_{gender}}{MS_{error}} = \frac{2.667}{4.556} = 0.59$$
$$F_{age} = \frac{MS_{age}}{MS_{error}} = \frac{1053.500}{4.556} = 231.23$$
$$F_{AB} = \frac{MS_{AB}}{MS_{error}} = \frac{21.167}{4.556} = 4.65$$

Mean squares

$$MS_{gender} = \frac{SS_{gender}}{df_{gender}} = \frac{2.667}{1} = 2.667$$
$$MS_{age} = \frac{SS_{age}}{df_{age}} = \frac{2107.000}{2} = 1053.500$$
$$MS_{AB} = \frac{SS_{AB}}{df_{AB}} = \frac{42.333}{2} = 21.167$$
$$MS_{error} = \frac{SS_{error}}{df_{error}} = \frac{82.000}{18} = 4.556$$

In a factorial ANOVA, the analysis of a main effect *ignores* the other independent variable. This causes no problems if there is no interaction, but it can lead to a faulty interpretation when the interaction is significant.

So, what should you do about interpreting a factorial ANOVA that has a significant interaction? First, interpret the interaction based on a graph of the cell means. Then for the main effects, you might (1) seek the advice of your instructor, (2) acknowledge that you don't yet know the techniques needed for a thorough interpretation, or (3) study more advanced statistics textbooks (Kirk, 1995, pp. 383–389, or Howell, 2010, pp. 421–426).

Sometimes, however, a simple interpretation of a main effect is not misleading. The height data provide an example. The interpretation of the significant age effect is that age has an effect on height. In Figure 13.7 this effect is seen in an increase from age 6 to age 12 and another increase from age 12 to age 18. In situations like this, a conclusion such as "The main effect of age was significant, $p < .01$" is appropriate. For more explanation of this approach to interpreting main effects even though the interaction is significant, see Howell (2008, pp. 427–429).

TABLE 13.14 ANOVA summary table for the height example

Source	*SS*	*df*	*MS*	*F*	*p*
Gender (*A*)	2.667	1	2.667	0.59	>.05
Age (*B*)	2107.000	2	1053.500	231.23	<.01
AB	42.333	2	21.167	4.65	<.05
Error	82.000	18	4.556		
Total	2234.000	23			

$F_{.05}(2, 18\ df) = 3.55$ $F_{.01}(2, 18\ df) = 6.01$

TABLE 13.15 SPSS summary table for the 2 × 3 factorial analysis of height data

Tests of Between-Subjects Effects
Dependent Variable: Height

Source	Type III Sum of Squares	*df*	*Mean Square*	*F*	Sig.
Corrected Model	2152.000[a]	5	430.400	94.478	.000
Intercept	79350.000	1	79350.000	17418.293	.000
Gender	2.667	1	2.667	.585	.454
Age	2107.000	2	1053.500	231.256	.000
Gender*Age	42.333	2	21.167	4.646	.024
Error	82.000	18	4.556		
Total	81584.000	24			
Corrected Total	2234.000	23			

[a] *R* Squared = .963 (Adjusted *R* Squared = .953)

PROBLEMS

13.15. These problems are designed to help you learn to *interpret* the results of a factorial experiment. For each problem (**a**) identify the independent variables, their levels, and the dependent variable; (**b**) fill in the rest of the summary table; and (**c**) interpret the results.

i. A clinical psychologist investigates the relationship between humor and aggression. The participant's task is to think up as many captions as possible for four cartoons in 8 minutes. The psychologist simply counts the number of captions produced. In the psychologist's latest experiment, a participant is either insulted or treated in a neutral way by an experimenter. Next, the participant responds to the cartoons at the request of either the same experimenter or a different experimenter. The cell means and the summary table are given.

		Participant was	
		Insulted	Treated neutrally
Experimenter was	Same	31	18
	Different	19	17

Source	*df*	*MS*	*F*	*p*
A (treatments)	1	675.00		
B (experimenters)	1	507.00		
AB	1	363.00		
Error	44	74.31		

ii. An educational psychologist studied response bias. For example, a response bias occurs if the grade an English teacher puts on a theme written by a 10th-grader is influenced by the student's name or by the occupation of his father. For this particular experiment, the psychologist used a "high-prestige" name (David) and a "low-prestige" name (Elmer). In addition, she made up two biographical sketches that differed only in the occupation of the student's father (research chemist or unemployed). The participants in this experiment read a biographical sketch and then graded a theme by that person using a scale of 50 to 100. The same theme was given to each participant.

		Name	
		David	Elmer
Occupation of father	Research chemist	86	81
	Unemployed	80	88

Source	*df*	*MS*	*F*	*p*
A (names)	1	22.50		
B (occupations)	1	2.50		
AB	1	422.50		
Error	36	51.52		

iii. What teaching techniques can instructors use that may help students learn? One is to ask questions during lectures. A second is to conduct demonstrations during class. Of course, both could be used or neither used. An instructor who taught four sections of statistics covered the same material in each class using one of the four conditions. A comprehension test followed.

		Questions	
		Used	Not used
Demonstrations	Used	89	81
	Not used	78	73

Source	*df*	*MS*	*F*	*p*
A (questions)	1	211.25		
B (demonstrations)	1	451.25		
AB	1	11.25		
Error	76	46.03		

12.16. Write an interpretation of your analysis of the data from the state-dependent memory experiment (Problems 13.11 and 13.14).

Comparing Levels within a Factor—Tukey HSD Tests

If the interaction in a factorial ANOVA is not significant, you can compare levels within a factor. Each main effect in a factorial ANOVA is like a one-way ANOVA; the question that remains after testing a main effect is: Are any of the levels of this factor significantly different from any others? You will be pleased to learn that the solution you learned for one-way ANOVA also applies to factorial ANOVA.

Of course, for some main effects you do not need a test subsequent to ANOVA; the F test on that main effect gives you the answer. This is true for any main effect with only *two levels*. Just like a t test, a significant F tells you that the two means are from different populations. Thus, if you are analyzing a 2×2 factorial ANOVA, you do not need a test subsequent to ANOVA for *either* factor.

When there are more than two levels of one factor in a factorial design, a Tukey HSD test can be used to test for significant differences between pairs of levels. There is one restriction, however, which is that the Tukey HSD test is appropriate *only if the interaction is **not** significant.*[4]

The HSD formula for a factorial ANOVA is the same as that for a one-way ANOVA:

$$\text{HSD} = \frac{\overline{X}_1 - \overline{X}_2}{s_{\overline{X}}}$$

$$\text{where } s_{\overline{X}} = \sqrt{\frac{MS_{\text{error}}}{N_t}}$$

Two of the terms need a little explanation. The means in the numerator are *margin* means. HSD tests the difference between two levels of a factor, so all the scores for one level are used to calculate that mean. The term N_t is the number of scores used to calculate a mean in the numerator (a margin N).

I'll illustrate HSD for factorial designs with the improvement scores in Problems 13.10 and 13.13. The two factors were gender (two levels) and the kind of therapy used to treat depression (three levels). The interaction was not significant, so Tukey HSD tests on kind of therapy are appropriate. To refresh your recall, look at the data in Problem 13.10 (pages 285–286).

Begin by calculating the three differences between treatment means:

Psychodynamic and interpersonal	$35 - 54 = -19$
Psychodynamic and cognitive-behavioral	$35 - 52 = -17$
Interpersonal and cognitive-behavioral	$54 - 52 = 2$

Because none of the three therapies is a "control" condition, the sign of the difference is not important. The intermediate difference is -17, so it should be tested first.

$$\text{HSD} = \frac{\overline{X}_P - \overline{X}_C}{s_{\overline{X}}} = \frac{35 - 52}{\sqrt{\frac{172}{12}}} = \frac{-17}{3.786} = -4.49$$

[4] A nonsignificant interaction is a condition required by many *a priori* and *post hoc* tests.

From Table G in Appendix C, $HSD_{.01}$ is 4.46 for $K = 3$ and $df_{error} = 30$. Because $4.49 > 4.46$, you can conclude that cognitive-behavioral therapy produced significantly greater improvement scores than psychodynamic therapy, $p < .01$.

The difference in psychodynamic and interpersonal therapy (19) is even greater than the difference between psychodynamic and cognitive-behavioral therapy (17). Thus, conclude that the interpersonal therapy scores were significantly greater than those for psychodynamic therapy, $p < .01$.

The other difference is between interpersonal and cognitive-behavioral therapy:

$$HSD = \frac{\overline{X}_I - \overline{X}_C}{s_{\overline{X}}} = \frac{54 - 52}{\sqrt{\frac{172}{12}}} = \frac{2}{3.786} = 0.53$$

Values of HSD that are less than 1.00 are never significant (just like F). Thus, there is no strong evidence that interpersonal therapy is better than cognitive-behavioral therapy.

Note that the value of $s_{\overline{X}}$, 3.786, is the same in both HSD problems. For factorial ANOVA problems such as those in this text (cells all have the same N), $s_{\overline{X}}$ always has the same value for every comparison.

Effect Size Indexes for Factorial ANOVA

The two effect size indexes that I discussed in Chapter 11, d and f, can be used with factorial designs *if the interaction is not significant.* I'll illustrate with the depression therapy study, in which the interaction was not significant.

Calculations for d are made using column totals (for Factor A comparisons) or row totals (for Factor B comparisons). This disregard of the factor not under consideration is the same disregard found in the preceding section on calculating Tukey HSD values for factorial ANOVAs. How much difference is there between psychodynamic and cognitive-behavioral therapy? The effect size index d provides an answer:

$$d = \frac{\overline{X}_1 - \overline{X}_2}{\hat{s}_{error}} = \frac{\overline{X}_P - \overline{X}_C}{\sqrt{MS_{error}}} = \frac{-17}{12.599} = -1.35$$

A d value of 1.35 is much larger than 0.80, the d value that indicates a large effect (see page 80). Cognitive-behavioral therapy produced considerably greater improvement scores than did psychodynamic therapy.

The effect size index f for the therapy variable is

$$f = \frac{\sqrt{\frac{K-1}{N_{tot}}(MS_{therapy} - MS_{error})}}{\sqrt{MS_{error}}} = \frac{\sqrt{\frac{3-1}{36}(1308.00 - 158.73)}}{\sqrt{158.73}} = 0.63$$

A value of 0.63 is well above the value of 0.40 that qualifies as a large effect size index (see page 253).

Restrictions and Limitations

I have emphasized throughout this book the limitations that go with each statistical test you learn. For the factorial analysis of variance presented in this chapter, the restrictions include the three that you learned for one-way ANOVA. If you cannot recall those three, reread the section near the end of Chapter 11, "Assumptions of the Analysis of Variance and Random Assignment." In addition to the basic assumptions of ANOVA, the factorial ANOVA presented in this chapter requires the following:

1. The number of scores in each cell must be equal. For techniques dealing with unequal *N*'s, see Kirk (1995) or Howell (2010).
2. The cells must be independent. One way to accomplish this is to randomly assign participants to only one of the cells. This restriction means that these formulas should not be used with any type of paired-samples or repeated-measures design.
3. The levels of both factors are chosen by the experimenter. The alternative is that the levels of one or both factors are chosen at random from several possible levels of the factor. The techniques of this chapter are used when the levels are *fixed* by the experimenter and not chosen randomly. For a discussion of fixed and random models of ANOVA, see Howell (2010) or Winer, Brown, and Michels (1991).

For a general discussion of the advantages and disadvantages of factorial ANOVA, see Kirk (1995).

PROBLEMS

13.17. To use the techniques in this chapter for factorial ANOVA and have confidence about the *p* values, the data must meet six requirements. List them.

13.18. Two social psychologists asked freshman and senior college students to write an essay supporting recent police action on campus. (The students were known to be against the police action.) The students were given \$20, \$10, \$5, or \$1 for their essay. Later, each student's attitude toward the police was measured on a scale of 1 to 20. Perform an ANOVA, fill out a summary table, and write an interpretation of the analysis at this point. (See Brehm and Cohen, 1962, for a similar study of Yale students.) If appropriate, calculate HSDs for all pairwise comparisons. Write an overall interpretation of the results of the study.

			Reward			
			\$20	\$10	\$5	\$1
Students	Freshmen	ΣX	73	83	99	110
		ΣX^2	750	940	1290	1520
		N	8	8	8	8
	Seniors	ΣX	70	81	95	107
		ΣX^2	820	910	1370	1610
		N	8	8	8	8

13.19. The conditions that make for happy people have been researched by psychologists for a long time. The data that follow are based on a review of the literature by Diener and colleagues (1999). Participants were asked their marital status and how often they engaged in religious behavior. They also indicated how happy they were on a scale of 1 to 10. Analyze the data with a factorial ANOVA and, if appropriate, Tukey HSD tests.

		Frequency of religious behavior		
		Never	Occasionally	Often
Marital status	Married	6 2 4	3 7 5	7 8 9
	Unmarried	4 2 3	3 1 5	3 7 5

13.20. Review the objectives at the beginning of the chapter.

ADDITIONAL HELP FOR CHAPTER 13

Visit cengage.com/psychology/spatz. At the Student Companion Site, you'll find multiple-choice tutorial quizzes, flashcards with definitions and workshops. For this chapter there is a Statistical Workshop on Two-Way ANOVA.

KEY TERMS

Assumptions (p. 297)
Bar graph (p. 276)
Between-cells sums of squares (p. 282)
Cell (p. 270)
Degrees of freedom (p. 287)
Effect size index *d* (p. 296)
Effect size index *f* (p. 296)
Error sum of squares (p. 284)
F tests (p. 287)
Factor (p. 268)
Factorial ANOVA (p. 268)
Interaction (p. 271)
Line graph (p. 275)
Main effect (p. 271)
Mean squares (p. 287)
Related samples (p. 270)
Sums of squares (p. 281)
Tukey HSD (p. 295)

transition passage
to nonparametric statistics

SO FAR IN your practice of null hypothesis statistical testing, you have used three different sampling distributions. The normal curve is appropriate when you know σ (Chapter 8). When you don't know σ, you can use the *t* distribution or the *F* distribution (Chapters 8–13) if the populations you are sampling from have certain characteristics such as being normally distributed and having equal variances. The *t* test and *F* test are called *parametric tests* because they make assumptions about parameters such as σ^2.

In the next two chapters, you will learn about statistical tests that require neither knowledge of σ nor that the data have the characteristics needed for *t* and *F* tests. These tests are called *nonparametric tests* because they don't assume that populations are normally distributed or have equal variances. These nonparametric NHST tests have sampling distributions that will be new to you. They do, however, have the same purpose as those you have been working with—providing you with the probability of obtaining the observed sample results, *if the null hypothesis is true.*

In Chapter 14, "Chi Square Tests," you will learn to analyze frequency count data. These data result when observations are classified into categories and the frequencies in each category are counted. In Chapter 15, "More Nonparametric Tests," you will learn four techniques for analyzing scores that are ranks or can be reduced to ranks.

The techniques in Chapters 14 and 15 are sometimes described as "less powerful." This means that *if* the populations you are sampling from have the characteristics required for *t* and *F* tests, then a *t* or an *F* test is more likely than a nonparametric test to reject H_0 if it should be rejected. To put this same idea another way, *t* and *F* tests have a smaller probability of a Type II error if the population scores have the characteristics the tests require.

CHAPTER 14

Chi Square Tests

OBJECTIVES FOR CHAPTER 14

After studying the text and working the problems in this chapter, you should be able to:

1. Identify the kind of data that require a chi square test for hypothesis testing
2. Distinguish between problems that require tests of independence and those that require goodness-of-fit tests
3. For tests of independence: State the null hypothesis and calculate and interpret chi square values
4. For goodness-of-fit tests: State the null hypothesis and calculate and interpret chi square values
5. Calculate a chi square value from 2 × 2 tables using the shortcut method
6. Calculate an effect size index for 2 × 2 chi square problems
7. Discuss issues associated with chi square tests based on small samples

ALEX'S PLAN FOR his senior year project was coming into focus—conducting experiments on risk taking. Being somewhat shy and reserved, Alex admired fellow students who seemed brave and adventurous; a year-long project gathering data on risk taking might just cure him, he thought. What he had in mind was for participants to actually engage in risky behavior such as rock climbing, eating unfamiliar food, and initiating conversations with strangers. His advisor suggested that his first step should be to find out if students at his college would volunteer for such experiments.

After gathering questionnaire data from 112 students, Alex reported to his advisor that more than half were willing to participate. His advisor wanted to see the data, because "exploring data usually produces good ideas." Alex responded that 62 of the 112 indicated they would volunteer for experiments such as Alex listed.

"No, no," said his advisor, "that's not what I mean. Separate the data into categories and make up some tables. Then, let's get together and look at them."

One of the tables was:

	Willing to participate	
	Yes	No
Women	32	38
Men	30	12
Total	62	50
Percentage	55.4%	44.6%

"See," said Alex pointing to the bottom line, 55.4 percent is more than half,"

"What I see," said his advisor "is that you may have to worry about a gender bias in your project. Women seem less willing than men to participate in your activities. Of the 70 women, just 32 said yes, but 30 of the 42 men said yes."

"Uh, hang on a minute," said Alex, stalling for time as he tried to process the numbers. He pulled out his calculator and punched in some numbers.

"OK, I see what you mean. Only 40 percent of the women said yes, but two-thirds of the men agreed. But still, couldn't that difference be just the usual variation we often find in samples?"

"Could be," said his advisor, "or maybe there really is a gender difference in willingness to sign up for experiments that involve risky behavior."[1]

Do the two positions at the end of the story sound familiar? This issue seems like it could be resolved with a NHST test. The null hypothesis is that women and men are the same in their willingness to volunteer for experiments that involve risky behavior. What is needed is a statistical test and its sampling distribution. Together, the test and the distribution produce the probability of the data observed, if the null hypothesis is true. With a probability in hand, you can reject or retain the null hypothesis.

In the case of Alex's data, the statistical test needed is a chi square test and the sampling distribution is the chi square distribution. The probability is that of sample proportions of .40 and .67 coming from identical populations, if the null hypothesis is true.[2]

Chi square (pronounced "ki," as in *kind,* "square," and symbolized χ^2) is a sampling distribution that gives probabilities about frequencies. Frequencies like those in the table above are distributed approximately as chi square, so the chi square distribution provides the probabilities needed for decision making about the difference between the responses of women and men.

The characteristic that distinguishes chi square from the techniques in previous chapters is the *kind of data*. For *t* tests and ANOVA, the data consist of a set of scores, such as attitudes, time, errors, heights, and so forth. Each subject has one quantitative score.

[1] The data in this scenario are based on Byrnes, Miller, and Schafer (1999), Gender Differences in Risk Tasking: A Meta-analysis.

[2] Casting data in a chi square problem as proportions helps you interpret the results. However, this book does not cover direct tests of proportions using chi square. To test proportions see Kirk, 2008, pp. 485–488

With chi square, however, the data are *frequency counts* in categories. Each subject is observed and placed in one category. The frequencies of observations in categories are counted and the chi square test is calculated from the frequency counts. What a chi square analysis does is compare the observed frequencies of a category to frequencies that would be expected if the null hypothesis is true.

error detection

To determine if chi square is appropriate, look at the data for one subject. If the subject has a quantitative score, chi square is not appropriate. If the subject is classified into one category among several, chi square may be appropriate.

Karl Pearson (1857–1936), of Pearson product-moment correlation coefficient fame, published the first article on chi square in 1900. Pearson wanted a way to measure the "fit" between data generated by a theory and data obtained from observations. Until chi square was dreamed up, theory testers presented theoretical predictions and empirical data side by side, followed by a declaration such as "good fit" or "poor fit."

The most popular theories at the end of the 19th century predicted that data would be distributed as a normal curve. Many data gatherers had adopted Quetelet's position that measurements of almost any social phenomenon would be normally distributed if the number of cases was large enough. Pearson (and others) thought this was not true, and they proposed other curves. By inventing chi square, Pearson provided everyone with a quantitative method of *choosing* the curve of best fit.

Chi square turned out to be very versatile, being applicable to many problems besides curve-fitting ones. As a test statistic, it is used by psychologists, sociologists, health professionals, biologists, educators, political scientists, economists, foresters, and others. In addition to its use as a test statistic, the chi square *distribution* has come to occupy an important place in theoretical statistics. As further evidence of the importance of Pearson's chi square, it was selected as one of the 20 most important discoveries of the 20th century (Hacking, 1984).

Pearson, too, turned out to be very versatile, contributing data and theory to both biology and statistics. He is credited with naming the standard deviation and seven other concepts in this textbook (David, 1995). He (and Galton and Weldon) founded the journal *Biometrika* to publicize and promote the marriage of biology and mathematics.[3] In addition to his scientific efforts, Pearson was an advocate of women's rights and the father of an eminent statistician, Egon Pearson.

As a final note, Pearson's overall goal was not to be a biologist or a statistician. His goal was a better life for the human race. An important step in accomplishing this was to "develop a methodology for the exploration of life" (Walker, 1968, pp. 499–500).

---

[3] In 1901 when *Biometrika* began, the Royal Society, the principal scientific society in Great Britain, accepted articles on biology and articles on mathematics, but it would not permit papers that combined the two. (See Galton, 1901, for a thinly disguised complaint against the establishment.)

The Chi Square Distribution and the Chi Square Test

The **chi square distribution** is a theoretical distribution, just as *t* and *F* are. Like them, as the number of degrees of freedom increases, the shape of the distribution changes. **Figure 14.1** shows how the shape of the chi square distribution changes as degrees of freedom change from 1 to 5 to 10. Note that χ^2, like the *F* distribution, is a positively skewed curve.

chi square distribution
Theoretical sampling distribution of chi square values.

Critical values for χ^2 are given in Table E in Appendix C for α levels of .10, .05, .02, .01, and .001. Look at **Table E.**

The design of the χ^2 table is similar to that of the *t* table: α levels are listed across the top and each row shows a different *df* value. Notice, however, that with chi square, as degrees of freedom increase (looking down columns), larger and larger values of χ^2 are required to reject the null hypothesis. *This is just the opposite of what occurs in the t and F distributions.* This will make sense if you examine the three curves in **Figure 14.1.**

Once again, I will symbolize a critical value by giving the statistic (χ^2), the α level, *df*, and the critical value. Here's an example (that you will see again): $\chi^2_{.05}(1\ df) = 3.84$.

To calculate a **chi square test** value to compare to a critical value from Table E, use the formula

chi square test
NHST technique that compares observed frequencies to expected frequencies.

$$\chi^2 = \Sigma\left[\frac{(O - E)^2}{E}\right]$$

where O = observed frequency
E = expected frequency

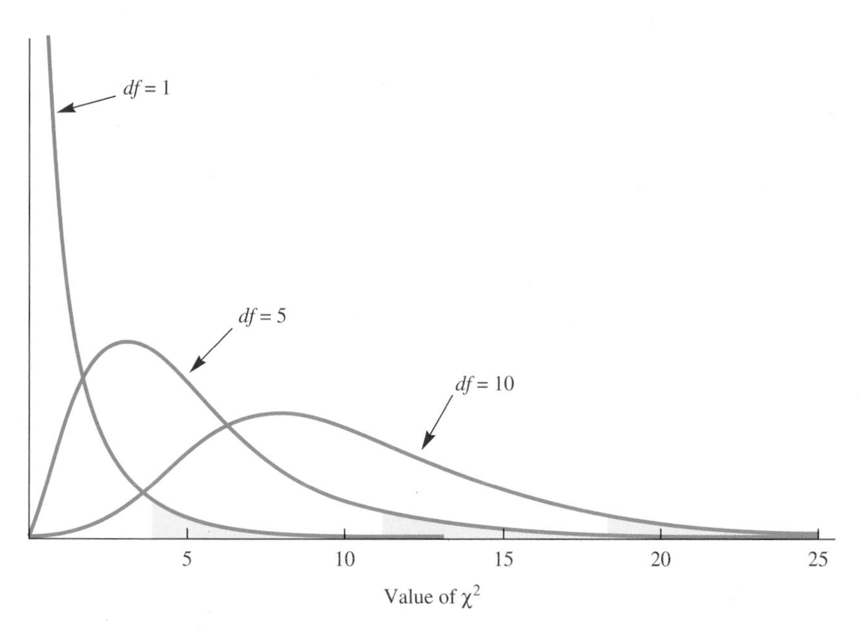

FIGURE 14.1 Chi square distribution for three different degrees of freedom. Rejection regions at the .05 level are shaded.

observed frequency
Count of actual events in a category.

expected frequency
Theoretical frequency derived from the null hypothesis.

Obtaining the **observed frequency** is simple enough—count the events in each category. Finding the **expected frequency** is a bit more complex. Two methods can be used to determine the expected frequency. One method is used if the problem is a "test of independence" and another method is used if the problem is one of "goodness of fit."

Chi Square as a Test of Independence

Probably the most common use of chi square is to test the independence of two variables. The null hypothesis for a chi square test of independence is that the two variables are *independent*—that there is no relationship between the two. If the null hypothesis is rejected, you can conclude that the two variables are *related* and then tell how they are related. Another word that is an opposite of independent is *contingent*. Tables arranged in the fashion of the one on page 301 are often referred to as contingency tables.

The data gathered by Alex had two variables, willingness to volunteer and gender. Are the two variables independent or are they related? The null hypothesis is that gender and willingness are independent—that knowing a person's gender gives no clue to his or her willingness to volunteer for risky experiments (and that knowing a person's willingness gives no clue about gender). Rejecting the null hypothesis supports the alternative hypothesis, which is that the two variables are contingent—that knowing a person's gender helps predict the person's willingness to volunteer.

Expected Values

A chi square test of independence requires observed values and expected values. **Table 14.1** shows the observed values that you saw before. In addition, it includes the expected values (in parentheses). Expected frequencies are those that are expected if the null hypothesis is true. Please pay careful attention to the logic behind the calculation of expected frequencies.

Let's start with an explanation of the expected frequency in the upper left corner, 38.75, the expected number of women who are willing to participate in risky experiments. Of all the 112 subjects, 70 (row total) were women, so if you chose a subject at random, the probability that the person would be a woman is 70/112 = .6250.

In a similar way, of the 112 subjects, 62 (column total) were willing to volunteer. Thus, the probability that a randomly chosen person would be willing to volunteer is 62/112 = .5536.

TABLE 14.1 **Hypothetical data on willingness to volunteer for risky experiments (Expected frequencies are in parentheses)**

	Willing to volunteer		
	Yes	No	Σ
Women	32 (38.75)	38 (31.25)	70
Men	30 (23.25)	12 (18.75)	42
Σ	62	50	112

Next, if you ask the probability that a person chosen at random is both a woman and a person who is willing to participate, the answer is found by multiplying together the probability of the two separate events.[4] Thus (.6250)(.5536) = .3460.

Finally, the expected frequency of such people is the probability of such a person multiplied by the total number of people. Thus (.3460)(112) = 38.75. Notice what happens to the arithmetic when the steps described above are combined:

$$\left(\frac{70}{112}\right)\left(\frac{62}{112}\right)(112) = \frac{(70)(62)}{112} = 38.75$$

Thus, the formula for the expected value of a cell is its row total, multiplied by its column total, divided by N.

In a similar way, the expected frequency of men who are willing to participate is the probability of a man times the probability of being willing participate times the number of subjects. For an overall view, here is the arithmetic for calculating expected values for all four cells in Table 14.1:

$$\frac{(70)(62)}{112} = 38.75 \qquad \frac{(70)(50)}{112} = 31.25$$

$$\frac{(42)(62)}{112} = 23.25 \qquad \frac{(42)(50)}{112} = 18.75$$

With the expected values (E) and observed values (O) in hand, you have what you need to calculate $(O-E)^2/E$ for each category. **Table 14.2** shows a convenient way to arrange your calculations in a step-by-step fashion. The result is a χ^2 value of 7.02.

error detection

For every chi square test, whether for independence or goodness of fit, the sum of the expected frequencies must equal the sum of the observed frequencies ($\Sigma E = \Sigma O$). Look at the bottom row of Table 14.2. The total of the observed frequencies (112) is equal to the total of the expected frequencies (112.00). Sometimes rounding may lead to slight discrepancies between the two totals.

TABLE 14.2 Calculation of χ^2 for the data in Table 14.1

O	E	$O-E$	$(O-E)^2$	$\frac{(O-E)^2}{E}$
32	38.75	−6.75	45.56	1.176
38	31.25	6.75	45.56	1.458
30	23.25	−6.75	45.56	1.960
12	18.75	6.75	45.56	2.430
Σ112	112.00			$\chi^2 = 7.02$

[4] This is like determining the chances of obtaining two heads in two tosses of a coin. For each toss, the probability of a head is $\frac{1}{2}$. The probability of two heads in two tosses is found by multiplying the two probabilities: $(\frac{1}{2})(\frac{1}{2}) = \frac{1}{4} = .25$. See the section "A Binomial Distribution" in Chapter 7 for a review of this topic.

Degrees of Freedom

Every χ^2 value is accompanied by its *df*. To determine the *df* for any $R \times C$ (rows by columns) table such as Table 14.1, use the formula $(R - 1)(C - 1)$. In this case, $(2 - 1)(2 - 1) = 1$.

Here is the reasoning behind degrees of freedom for tests of independence. Consider the simplest case, a 2×2 table with the four margin totals fixed. How many of the four data cells are free to vary? The answer is one. Once a number is selected for any one of the cells, the numbers in the other three cells must have particular values in order to keep the margin totals the same. You can check this out for yourself by constructing a version of Table 14.1 that has empty cells and margin totals of (reading clockwise) 70, 42, 50, and 62. You are free to choose any number for any of the four cells, but once it is chosen, the rest of the cell numbers are determined. The general rule again is $df = (R - 1)(C - 1)$, where R and C refer to the number of rows and columns.

Interpretation

To determine the significance of $\chi^2 = 7.02$ with 1 *df*, look at Table E in Appendix C. In the first row, you will find that if $\chi^2 = 6.64$, the null hypothesis may be rejected at the .01 level of significance. Because the χ^2 value exceeds this, you can conclude that the attitudes toward taking risks are influenced by gender; that is, gender and attitudes toward risk taking in the population are not independent, but related. By examining the proportions (.40 and .67), you can conclude that men are significantly more likely to indicate they are willing to volunteer to participate in riskyexperiments.

Table 14.3 shows the result of a chi square test on the data in Table 14.1 using an SPSS analysis that follows the path *Descriptive Statistics* and *Crosstabs*. The Pearson Chi-Square in the top row (7.023 with 1 *df*) corresponds to the analysis you just studied. The probability of such a chi square value is .008, or less than .01.

TABLE 14.3 SPSS Crosstabs chi square analysis of the data in Table 14.1

Chi-Square Tests

	Value	df	Asymp. Sig. (2-sided)	Exact Sig. (2-sided)	Exact Sig. (1-sided)
Pearson Chi-Square	7.023[b]	1	.008		
Continuity Correction[a]	6.022	1	.014		
Likelihood Ratio	7.196	1	.007		
Fisher's Exact Test				.011	.007
Linear-by-Linear Association	6.961	1	.008		
N of Valid Cases	112				

[a]Computed only for a 2 × 2 table
[b]0 cells (.0%) have expected count less than 5. The minimum expected count is 18.75

Shortcut for Any 2 × 2 Table

If you have a calculator and are analyzing a 2 × 2 table, the following shortcut will save you time. With this shortcut, you do not have to calculate the expected frequencies, which reduces calculating time by as much as half. Here is the general case of a 2 × 2 table:

			Row totals
	A	B	$A + B$
	C	D	$C + D$
Column totals	$A + C$	$B + D$	N

To calculate χ^2 from a 2 × 2 table, use the formula

$$\chi^2 = \frac{N(AD - BC)^2}{(A + B)(C + D)(A + C)(B + D)}$$

The term in the numerator, $AD - BC$, is the difference between the cross-products. The denominator is the product of the four margin totals.

To illustrate the equivalence of the shortcut method and the method explained in the previous section, I will calculate χ^2 for the data in Table 14.1. Translating cell letters into cell totals produces $A = 32$, $B = 38$, $C = 30$, $D = 12$, and $N = 112$. Applying these figures to the formula gives,

$$\chi^2 = \frac{(112)[(32)(12) - (38)(30)]^2}{(70)(42)(62)(50)} = 7.02$$

Both methods yield $\chi^2 = 7.02$.

A common error in interpreting a χ^2 problem is to tell less than you might. This is especially the case when the shortcut method is used. For example, the statement "There is a relationship between gender and willingness to volunteer for risky experiments, $\chi^2(1) = 7.02$; $p < .01$" leaves unsaid the *direction* of the relationship. In contrast, "Men are more willing than women to volunteer for risky experiments, $\chi^2(1)$ 7.02; $p < .01$" tells the reader not only that there is a relationship but also what the relationship is. So, don't stop with a vague "There is a relationship." Tell what the relationship is, which you can determine by exploring proportions.

Effect Size Index for 2 × 2 Chi Square Data

Although a χ^2 test allows you to conclude that two variables are related, it doesn't tell you the degree of relationship. To know the degree of relationship, you need an effect size index.

phi (ϕ)
Effect size index for a 2 × 2 chi square test of independence.

ϕ (written **phi** and pronounced "fee" by statisticians) is an effect size index for chi square that works for 2 × 2 tables but not for larger tables. ϕ is calculated with the formula

$$\phi = \sqrt{\frac{\chi^2}{N}}$$

where χ^2 is the χ^2 value from a 2 × 2 table
N is the total number of observations

The interpretation of ϕ is much like that of a correlation coefficient. ϕ gives you the degree of relationship between the two variables in a chi square analysis. Thus, a value near 0 means that there is no relationship and a value near 1 means that there is an almost perfect relationship between the two variables.

Here are the guidelines for evaluating ϕ coefficients:

Small effect	$\phi = 0.10$
Medium effect	$\phi = 0.30$
Large effect	$\phi = 0.50$

For the gender-risky experiment data,

$$\phi = \sqrt{\frac{\chi^2}{N}} = \sqrt{\frac{7.02}{112}} = \sqrt{0.0627} = 0.25$$

Thus, for these data (and those of Byrnes, Miller and Schafer, 1999), men are more willing than women to volunteer for risky experiments and this difference produces an effect size value that approaches a medium-sized effect.

PROBLEMS

14.1. In the late 1930s and early 1940s an important sociology experiment took place in the Boston area. At that time 650 boys (median age = $10\frac{1}{2}$ years) participated in the Cambridge–Somerville Youth Study (named for the two economically depressed communities where the boys lived). The participants were *randomly* assigned to either a delinquency-prevention program or a control group. Boys in the delinquency-prevention program had a counselor and experienced several years of opportunities for enrichment. At the end of the study, police records were examined for evidence of delinquency among all 650 boys. Analyze the data in the table and write a conclusion.

	Received program	Control
Police record	114	101
No police record	211	224

14.2. A psychology student was interested in the effect of group size on the likelihood of joining an informal group. On one part of the campus, he had a group of *two* people looking intently up into a tree. On a distant part of the campus, a group of *five* stood looking up into a tree. Single passersby were classified as joiners or nonjoiners depending on whether they looked up for 5 seconds or longer or made some comment to the group. The data in the accompanying table were obtained. Use chi square techniques to determine whether group size had an effect on joining. (See Milgram, 1969.) Use the calculation method you did not choose for problem 14.1. Write a conclusion.

	Group size	
	2	5
Joiners	9	26
Nonjoiners	31	34

14.3. You have probably heard that salmon return to the stream in which they hatched. According to the story, after years of maturing in the ocean, the salmon arrive at the mouth of a river and swim upstream, choosing at each fork the stream that leads to the pool in which they hatched. Arthur D. Hasler's classic research investigated this homing instinct [reported in Hasler (1966) and Hasler and Scholz (1983)]. Here are two sets of data that he gathered.

a. These data help answer the question of whether salmon really do make consistent choices at the forks of a stream in their return upstream. Forty-six salmon were captured from the Issaquah Creek (just east of Seattle, Washington) and another 27 from its East Fork. All salmon were marked and released below the confluence of these two streams. All of the 46 captured in the Issaquah were recaptured there. Of the 27 originally captured in the East Fork, 19 were recaptured from the East Fork and 8 from the Issaquah. Use χ^2 to determine if salmon make consistent choices at the confluence of two streams. Calculate ϕ and write an explanation of the results.

b. Hasler believed that the salmon were making the choice at each fork on the basis of olfactory (smell) cues. He thought that young salmon become imprinted on the particular mix of dissolved mineral and vegetable molecules in their home streams. As adults, they simply make choices at forks on the basis of where the smell of home is coming from. To test this, he captured 70 salmon from the two streams, plugged their nasal openings, and released the salmon below the confluence of the two streams. The fish were recaptured above the fork in one stream or the other. Analyze using chi square techniques and write a sentence about Hasler's hypothesis.

		Recapture site	
		Issaquah	East Fork
Capture site	Issaquah	39	12
	East Fork	16	3

14.4. Here are some data that I analyzed for an attorney. Of the 4200 white applicants at a large manufacturing facility, 390 were hired. Of the 850 black applicants, 18 were hired. Analyze the data and write a conclusion. *Note:* You will have to work on the data some before you set up the table.

Chi Square as a Test for Goodness of Fit

A chi square **goodness-of-fit test** allows an evaluation of a theory and its ability to predict outcomes. The formula is the same as that for a test of independence. Thus, both chi square tests require observed values and expected values. The expected values for a goodness-of-fit test, however, come from a hypothesis, theory, or model, rather than from calculations on the data itself as in tests of independence. With a goodness-of-fit test, you can determine whether or not there is a good fit between the theory and the data.

goodness-of-fit test
Chi square test that compares observed frequencies to frequencies predicted by a theory.

In a chi square goodness-of-fit test, *the null hypothesis is that the actual data fit the expected data.* A rejected H_0 means that the data do *not* fit the model—that is, the model is inadequate. A retained H_0 means that *the data are not at odds with the model.* A retained H_0 does not prove that the model, theory, or hypothesis is true, because other models may also predict such results. A retained H_0 does, however, lend support for the model.

Where do hypotheses, theories, and models come from? At a very basic level, they come from our nature as human beings. Humans are always trying to understand things, and this often results in a guess at how things are. When these guesses are developed, supported, and discussed (and perhaps, published), they earn the more sophisticated label of hypothesis, theory, or model. Some hypotheses, theories, and models make quantitative predictions. Such predictions lead to expected frequencies. By comparing these predicted frequencies to observed frequencies, a chi square analysis provides a test of the theory.

The chi square goodness-of-fit test is used frequently in population genetics, where Mendelian laws predict offspring ratios such as 3:1, or 1:2:1, or 9:3:3:1. For example, the law (model) might predict that crossing two pea plants will result in three times as many seeds in the smooth category as in the wrinkled category (3:1 ratio). If you perform the crosses, you might get 316 smooth seeds and 84 wrinkled seeds, for a total of 400. How well do these actual frequencies fit the expected frequencies of 300 and 100? Chi square gives you a probability figure to help you decide if the data are consistent with the theory.

Although the mechanics of χ^2 require you to manipulate raw frequency counts (such as 316 and 84), you can best understand this test by thinking of proportions. Thus, 316 out of 400 is a proportion of .79 and 84 out of 400 is a proportions of .21. How likely are such proportions if the population proportions are .75 and .25? Chi square provides you with the probability of obtaining the observed proportions if the theory is true. As a result, you have a quantitative way to decide "good fit" or "poor fit."

In the genetics example, the expected frequencies were predicted by a theory. Sometimes, in a goodness-of-fit test, the expected frequencies are predicted by chance. In such a case, a rejected H_0 means that something besides chance is at work. When you interpret the analysis, you should identify what that "something" is. Here is an example.

Suppose you were so interested in sex stereotypes and job discrimination that you conducted the following experiment, which is modeled after that of Mischel (1974). You make up four one-page resumes and four fictitious names—two female, two male. The names and resumes are randomly combined, and each participant is asked to read the resumes and "hire" one of the four "applicants" for a management trainee position. The null hypothesis tested is a "no discrimination" hypothesis. The hypothesis says that gender is *not* being used as a basis for hiring and therefore equal numbers of men and women will be hired. Thus, if the data cause the null hypothesis to be rejected, the hypothesis of "no discrimination" may be rejected.

Suppose you have 120 participants in your study, and they "hire" 75 men and 45 women. Is there statistical evidence that sex stereotypes are leading to discrimination? That is, given the hypothesized result of 60 men and 60 women, how good a fit are the observed data of 75 men and 45 women?

Applying the χ^2 formula, you get

$$\chi^2 = \Sigma\left[\frac{(O - E)^2}{E}\right] = \frac{(75 - 60)^2}{60} + \frac{(45 - 60)^2}{60} = 3.75 + 3.75 = 7.50$$

The number of degrees of freedom for this problem is the number of categories minus 1. There are two categories here—hired men and hired women. Thus, $2 - 1 = 1$ *df*. Looking in Table E, you find in row 1 that if $\chi^2 \geq 6.64$, the null hypothesis may be rejected at the .01 level. Thus, the model may be rejected.

The last step in data analysis (and perhaps the most important one) is to write a conclusion. Begin by returning to the descriptive statistics of the original data. Because 62.5 percent of the hires were men and 37.5 percent were women, you can conclude that discrimination in favor of men was demonstrated.

The term *degrees of freedom* implies that there are some restrictions. For both tests of independence and goodness of fit, one restriction is always that the sum of the expected events must be equal to the sum of the observed events; that is, $\Sigma E = \Sigma O$. If you manufacture a set of expected frequencies from a model, their sum must be equal to the sum of the observed frequencies. In our sex discrimination example, $\Sigma O = 75 + 45 = 120$ and therefore, ΣE must be 120, which it is $(60 + 60)$.

There is no effect size index for a chi square goodness-of-fit test. A little reflection shows why. An effect size index is a measure of how large a difference is. In a goodness-of-fit test, the difference is between the data observed and data predicted by a theory. A chi square p value of less than .05 means that the theory's predictions are wrong. The question "How wrong?" (an effect size index kind of question) doesn't seem necessary. Wrong is wrong.[5]

PROBLEMS

14.5. Using the data on wrinkled and smooth pea seeds on pages 310–311, test how well the data fit a 3:1 hypothesis.

14.6. John B. Watson (1878–1958), the behaviorist, thought there were three basic, inherited emotions: fear, rage, and love (Watson, 1924). He suspected that the wide variety of emotions experienced by adults had been learned. Proving his suspicion would be difficult because "unfortunately there are no facilities in maternity wards for keeping mother and child under close observation for years." So Watson attacked the apparently simpler problem of showing that the emotions of fear, rage, and love could be distinguished in infants. (He recognized the difficulty of proving that these were the *only* basic emotions, and he made no such claim.) Fear could be elicited by dropping the child onto a soft feather pillow (but not by the dark, dogs, white rats, or a pigeon fluttering its wings in the baby's face). Rage could be elicited by holding the child's arms tightly at its sides, and love by "tickling, shaking, gentle rocking, and patting," among other things.

Here is an experiment reconstructed from Watson's conclusions. A child was stimulated so as to elicit fear, rage, or love. An observer then looked at the child and judged the emotion the child was experiencing. Each judgment was scored as correct or incorrect—correct meaning that the judgment (say, love) corresponded to the stimulus (say, patting). Sixty observers made one judgment each, with the results shown in the accompanying table. To find the expected frequencies, think about the chance of being correct or incorrect when there are three possible outcomes. Analyze the data with χ^2 and write a conclusion.

Correct	Incorrect
32	28

Chi Square with More Than 1 Degree of Freedom

The χ^2 values you have found so far have been evaluated by comparing them to a chi square distribution with 1 *df.* In this section, the problems require chi square distributions with more than 1 *df.* Some are tests of independence and some are goodness-of-fit problems.

[5]In away, smaller and smaller p values indicate larger and larger differences between the data and the theory, but, unfortunately, p values are influenced by other factors as well.

Here's a question for you. Suppose that for 6 days next summer you get a free vacation, all expenses paid. Which location would you choose for your vacation?

_____ City

_____ Mountains

_____ Seashore

I asked 78 students that question and found, not surprisingly, that different people chose different locations. The results I got from my sample follow:

Choice			
City	Mountains	Seashore	Total
9	27	42	78

A chi square analysis can test the hypothesis that all three of these locations are preferred equally. In this case, the null hypothesis is that among those in the population, all three locations are equally likely as a first choice. Using sample data and a chi square test, you can reject or retain this hypothesis.

Think for a moment about the expected value for each cell in this problem. If each of the three locations is equally likely, then the expected value for a location is $\frac{1}{3}$ times the number of respondents; that is $(\frac{1}{3})(78) = 26$.

The arrangement of the arithmetic for this χ^2 problem is a simple extension of what you have already learned. Look at the analysis in **Table 14.4.**

What is the *df* for the χ^2 value in Table 14.4? The margin total of 78 is fixed, so only two expected cell frequencies are free to vary. Once two are determined, the third is restricted to whatever value will make the total 78. Thus, there are $3 - 1 = 2$ *df*. For this design, the number of *df* is the number of categories minus 1. From Table E, for $\alpha = .001$, χ^2 with 2 $df = 13.82$. Therefore, reject the hypothesis of independence and conclude that the students were responding in a nonchance manner to the questionnaire.

One characteristic of χ^2 is its *additive* nature. Each $(O - E)^2/E$ value is a measure of deviation of the data from the model, and the final χ^2 is simply the sum of these measures. Because of this, you can examine the $(O - E)^2/E$ values and see which deviations are contributing the most to χ^2. For the vacation preference data, it is a greater number of seashore choices and fewer city choices that make the final χ^2 significant. The mountain choices are about what would be predicted by the "equal likelihood" model.

TABLE 14.4 Calculation of χ^2 for the vacation location preference data

Locations	O	E	$O - E$	$(O - E)^2$	$\frac{(O - E)^2}{E}$
City	9	26	−17	289	11.115
Mountains	27	26	1	1	0.038
Seashore	42	26	16	256	9.846
					$\chi^2 = 20.999$

TABLE 14.5 Recommended family size by those in high school, college, and business

	Number of children			
Subjects	0–1	2–3	4 or more	Σ
High school	26	57	9	92
College	61	38	18	117
Business	17	35	14	66
Σ	104	130	41	275

Let's move now to a more complicated example. Suppose you wanted to gather data on family planning from a variety of people. Your sample consists of high school students, college students, and businesspeople. Your family-planning question is, "A couple just starting a family this year should plan to have how many children?"

_____ 0 or 1

_____ 2 or 3

_____ 4 or more

As participants answer your question, you classify them as high school, college, or business types. Suppose you obtained the 275 responses that are shown in **Table 14.5.**

For this problem, you do not have a theory or model to tell you what the theoretical frequencies should be. The question is whether there is a relationship between attitudes toward family size and group affiliation. A chi square test of independence may answer this question. The null hypothesis is that there is *no* relationship—that is, that recommended family size and group affiliation are independent.

To get the expected frequency for each cell, assume independence (H_0) and apply the reasoning explained earlier about multiplying probabilities:

$$\frac{(92)(104)}{275} = 34.793 \qquad \frac{(92)(130)}{275} = 43.491 \qquad \frac{(92)(41)}{275} = 13.716$$

$$\frac{(117)(104)}{275} = 44.247 \qquad \frac{(117)(130)}{275} = 55.309 \qquad \frac{(117)(41)}{275} = 17.444$$

$$\frac{(66)(104)}{275} = 24.960 \qquad \frac{(66)(130)}{275} = 31.200 \qquad \frac{(66)(41)}{275} = 9.840$$

These expected frequencies are incorporated into **Table 14.6,** which shows the calculation of the χ^2 value.

The *df* for a contingency table such as Table 14.5 is obtained from the formula

$$df = (R - 1)(C - 1)$$

Thus, $df = (R - 1)(C - 1) = (3 - 1)(3 - 1) = 4$.

TABLE 14.6 **Calculation of χ^2 for the family-planning data**

O	E	$O - E$	$(O - E)^2$	$\frac{(O - E)^2}{E}$
26	34.793	−8.793	77.317	2.222
57	43.491	13.509	182.493	4.196
9	13.716	−4.716	22.241	1.622
61	44.247	16.753	280.663	6.343
38	55.309	−17.309	299.602	5.417
18	17.444	0.556	0.309	0.018
17	24.960	−7.960	63.362	2.539
35	31.200	3.800	14.440	0.463
14	9.840	4.160	17.306	1.759
Σ 275	275.000			$\chi^2 = 24.579$

For $df = 4$, $\chi^2 = 18.46$ is required to reject H_0 at the .001 level. The obtained χ^2 exceeds this value, so reject H_0 and conclude that attitudes toward family size are related to group affiliation.[6]

By examining the right-hand column in **Table 14.6,** you can see that 6.343 and 5.417 constitute a large portion of the final χ^2 value. By working backward on those rows, you will discover that college students chose the 0–1 category more often than expected and the 2–3 category less often than expected. The interpretation, then, is that college students think that families should be smaller than high school students and businesspeople do.

There is an effect size index for chi square problems that have more than 1 degree of freedom. It was proposed by Harald Cramér and has been variously referred to as Cramer's ϕ, ϕ_c, and Cramer V. The formula and explanations are available in Aron, Aron, and Coups (2009, pp. 555–556.) and Howell (2010, p. 165).

PROBLEMS

***14.7.** Professor Stickler always grades "on the curve." "Yes, I see to it that my grades are distributed as 7 percent A's, 24 percent B's, 38 percent C's, 24 percent D's, and 7 percent flunks," he explained to a young colleague. The colleague, convinced that the professor is really a softer touch than he sounds, looked at Professor Stickler's grade distribution for the past year. He found frequencies of 20, 74, 120, 88, and 38, respectively, for the five categories, A through F. Perform a χ^2 test for the colleague. What do *you* conclude about Professor Stickler?

[6] The statistical analysis of these data is straightforward; the null hypothesis is rejected. However, a flaw in the experimental design has two variables confounded, leaving the interpretation clouded. Because the three groups differ in both age and education, we are unsure whether the differences in attitude are related to only age, to only education, or to both of these variables. A better design would have groups that differed only on the age or only on the education variable.

14.8. Is problem 14.7 a problem of goodness of fit or of independence?

***14.9.** "Snake eyes," a local gambler, let a group of sophomores know they would be welcome in a dice game. After 3 hours the sophomores went home broke. However, one sharp-minded, sober lad had recorded the results on each throw. He decided to see if the results of the evening fit an "unbiased dice" model. Conduct the test and write a sentence summary of your conclusions.

Number of spots	Frequency
6	195
5	200
4	220
3	215
2	190
1	180

14.10. Identify problem 14.9 as a test of independence or of goodness of fit.

***14.11.** Remember the controversy described in Chapter 1 over the authorship of 12 of *The Federalist* papers? The question was whether they were written by Alexander Hamilton or James Madison. Mosteller and Wallace (1989) selected 48 1000-word passages known to have been written by Hamilton and 50 1000-word passages known to have been written by Madison. In each passage they counted the frequency of certain words. The results for the word *by* are shown in the table. Is *by* used with significantly different frequency by the two writers? Explain how these results help in deciding about the 12 disputed papers.

Rate per 1000 words	Hamilton	Madison
0–6	21	5
7–12	27	31
13–18	0	14

14.12. Is problem 14.11 one of independence or of goodness of fit?

Small Expected Frequencies

The theoretical chi square distribution, which comes from a mathematical formula, is a continuous function that can have any positive numerical value. Chi square test statistics calculated from frequencies do not work this way. They change in discrete steps and when the expected frequencies are very small, the discrete steps are quite large. It is clear to mathematical statisticians that as the expected frequencies approach zero, the theoretical chi square distribution becomes a less and less reliable way to estimate probabilities. In particular, the fear is that such chi square analyses will reject the null hypothesis more often than warranted.

The question for researchers who analyze their data with chi square has always been: Those expected frequencies—how small is too small? For years the usual recommendation was that if an expected frequency was less than 5, a chi square analysis was suspect. A number of studies, however, led to a revision of the answer.

When $df = 1$

Several studies used a computer to draw thousands of random samples from known populations for which the null hypothesis was true, a technique called the *Monte Carlo* method. Each sample was analyzed with a chi square, and the proportion of rejections of the null hypothesis (Type I errors) was calculated. (If this proportion was approximately equal to α, the chi square analysis would be clearly appropriate.) The general trend of these studies has been to show that the theoretical chi square distribution gives accurate conclusions even when the expected frequencies are considerably less than 5. **Yates's correction** used to be recommended for 2×2 tables that had one expected frequency less than 5. The effect of the correction was to reduce the size of the obtained chi square. According to current thinking, Yates's correction resulted in many Type II errors. Therefore, I do not recommend Yates's correction for the kinds of problems found in most research. (See Camilli and Hopkins, 1978, for an analysis of a 2×2 test of independence.)

Yates's correction
Correction for a 2×2 chi square test that has one small expected frequency.

When $df > 1$

When $df > 1$, the same uncertainty exists if one or more expected frequencies is small. Is the chi square test still advisable? Bradley and colleagues (1979) addressed some of the cases in which $df > 1$. Again, they used a computer to draw thousands of samples and analyzed each sample with a chi square test. The proportion of cases in which the null hypothesis was mistakenly rejected when $\alpha = .05$ was calculated. Of course, if the test was working correctly, the proportion rejected should be .05. Nearly all the proportions were in the range of .03 to .06. Exceptions occurred when sample size was small ($N = 20$). The conclusion was that the chi square test gives fairly accurate probabilities if sample size is greater than 20.

An Important Consideration

Now, let's back away from these trees we've been examining and visualize the forest that you are trying to understand. The big question, as always, is, What is the nature of the population? The question is answered by studying a sample, which may lead you to reject or to retain the null hypothesis. Either decision could be in error. The concern of this section so far has been whether the chi square test is keeping the percentage of mistaken rejections (Type I errors) near an acceptable level (.05).

Because the big question is, What is the nature of the population? you should keep in mind the other error you could make: retaining the null hypothesis when it should be rejected (a Type II error). How likely is a Type II error when expected frequencies are small? Overall (1980) has addressed this question and his answer is "Very."

Overall's warning is that if the effect of the variable being studied is not huge, then small expected frequencies have a high probability of dooming you to make the mistake of failing to discover a real difference. You may recall that this issue was discussed in Chapter 10 under the topic of power.

How can this mistake be avoided? Use large N's. The larger the N, the more power you have to detect any real differences. In addition, the larger the N, the more confident you are that the actual probability of a Type I error is your adopted α level.

Combining Categories

Combining categories is a technique that allows you to ensure that your chi square analysis will be more accurate. (It reduces the probability of a Type II error and keeps the probability of a Type I error close to α.) Remember the example in the previous section in which I had you gathering data on recommended family size? Let's return to the idea but with a slightly different design.

Suppose you had just two categories or participants, high school students and college students. However, suppose you expanded their options of number of children so that participants could recommend 0, 1, 2, 3, 4, or 5 or more. For this hypothetical example, the high school students' frequency counts are 13, 12, 18, 5, 2, and 2 and for college students, 12, 13, 3, 6, 3, and 2. Arranging the frequency counts into a contingency table and adding the expected frequencies by calculating them from the margin totals produces Table 14.7.

Examine the expected frequencies (in parentheses) in **Table 14.7.** Several are quite small, well under 5, a value that causes worry among some statisticians. When expected frequencies are so low, the probability figure from a chi square test may be suspect. What to do?

A common resolution to concern about small expected frequencies is to combine categories to create a smaller contingency table. Combining cateogies produces larger expected frequencies. If the categories of *3, 4,* and *5 or more* in Table 14.7 are combined, the result is **Table 14.8.** All of the expected frequencies in Table 14.8 are

TABLE 14.7 Family size recommended by high school and college students (hypothetical data)

	Number of Children						
	0	1	2	3	4	5 or more	Σ
High school	13 (14.29)	12 (14.29)	18 (12.00)	5 (6.29)	2 (2.86)	2 (2.29)	52
College	12 (10.71)	13 (10.71)	3 (9.00)	6 (4.71)	3 (2.14)	2 (1.71)	39
Σ	25	25	21	11	5	4	91

TABLE 14.8 Family size recommended by high school and college students when categories are combined

Subjects	0	1	2	3 or more	Σ
High school	13 (14.29)	12 (14.29)	18 (12.00)	9 (11.43)	52
College	12 (10.71)	13 (10.71)	3 (9.00)	11 (8.57)	39
Σ	25	25	21	20	91

TABLE 14.9 **Calculation of χ^2 for data in Table 14.8**

O	E	$O - E$	$(O - E)^2$	$\frac{(O - E)^2}{E}$
13	14.29	−1.29	1.66	0.116
12	14.29	−2.29	5.24	0.367
18	12.00	6.00	36.00	3.000
9	11.43	2.43	5.91	0.517
12	10.71	1.29	1.66	0.155
13	10.71	2.29	5.24	0.490
3	9.00	−6.00	36.00	4.000
11	8.57	2.43	5.91	0.689
$\Sigma = 91$	91.00			$\chi^2 = 9.33$

well above a worrisome 5, which produces more confidence in the probability figure from a chi square analysis.

Table 14.9 shows the calculation of a chi square statistic for the frequency counts in Table 14.8. For those data, $\chi^2 = 9.33$. Using the formula $(R - 1)(C - 1)$ for degrees of freedom gives $(4 - 1)(2 - 1) = 3$ *df*. By consulting Table E in Appendix C, you can see that a $\chi^2 = 7.82$ is required to reject the null hypothesis at the .05 level (and 11.34 at the .01 level). The null hypothesis can be rejected. The overall interpretation is that high school students and college students give significantly different recommendations when asked to specify how many children a couple starting a family this year should have. Because of the additive nature of chi square, you can examine Table 14.9 and give a more precise interpretation. The two largest components of the chi square value of 9.33 in Table 14.9 are the 3.00 and 4.00 associated with those who recommended that a family have two children. Upon further examination of where the 3.00 and 4.00 come from, you can conclude that high school students recommend two children to a greater degree than college students.

There are other situations in which small expected frequencies may be combined. For example, it is common in opinion surveys to have categories of "strongly agree," "agree," "no opinion," "disagree," and "strongly disagree." If there are few respondents who check the "strongly" categories, those who do might be combined with their "agree" or "disagree" neighbors—producing a table of three cells instead of five. The basic rule on combinations is that they must "make sense." It wouldn't make sense to combine the "strongly agree" and the "no opinion" categories in this example.

When You May Use Chi Square

1. Use chi square when the subjects in your study are identified by category rather than by a quantitative score. With chi square, the raw data are the number of subjects in each category.
2. To use chi square, each observation must be **independent** of other observations. (Independence here means independence of the observations and not independence of the variables.) In practical

independent
The occurrence of one event does not affect the outcome of a second event.

TABLE 14.10 **Comparison of chi square tests of independence and goodness of fit**

Features	Tests of independence	Goodness-of-fit tests
Purpose	Tests independence of two variables	Tests adequacy of a theory
Expected values	Based on assumption of independence	Derived from the theory
Null hypothesis (H_0)	Variables are independent	Data fit the theory
If H_0 rejected	Variables are related	Theory is inadequate
If H_0 retained	Supports independence	Supports theory
Independent observations	Required	Required

terms, independence means that each subject appears only once in the table (no repeated measures on one subject) and that the response of one subject is not influenced by the response of another subject.

3. Chi square conclusions apply to populations of which the samples are representative. Random sampling is one way to ensure representativeness.

error detection

One check for independence is that N (the total number of observations in the analysis) must equal the number of subjects in the study.

Table 14.10 compares the characteristics of chi square tests of independence and goodness-of-fit tests. Study it.

PROBLEMS

14.13. An early hypothesis about schizophrenia (a form of severe mental illness) was that it has a simple genetic cause. In accordance with the theory, one-fourth (a 1:3 ratio) of the offspring of a selected group of parents would be expected to be diagnosed as schizophrenic. Suppose that of 140 offspring, 19.3 percent were schizophrenic. Use this information to test the goodness of fit of a 1:3 model. Remember that a χ^2 analysis is performed on frequency counts.

14.14. How do goodness-of-fit tests and the tests of independence differ in obtaining expected values?

14.15. You may recall from Chapter 9 that I bought Doritos tortilla chips and Snickers candy bars to have data for problems. For this chapter, I bought regular M&Ms. In a 1.69-ounce package I got 9 blues, 11 oranges, 17 greens, 8 yellows, 5 browns, and 7 reds. At the time I bought the package, Mars, Inc., the manufacturer, claimed that the population breakdown is 24 percent blue, 20 percent orange, 16 percent green, 13.3 percent yellow, 13.3 percent brown, and 13.3 percent red. Use a chi square test to evaluate Mars's claim.

14.16. Suppose a friend of yours decides to gather data on sex stereotypes and job discrimination. He makes up ten resumes and ten fictitious names, five male and five female. He asks each of his 24 subjects to "hire" five of the candidates rather than just one. The data show that 65 men and 55 women are hired. He asks you for advice on χ^2. What would you say to your friend?

14.17. For a political science class, students gathered data for a supporters profile of the two candidates for senator, Hill and Dale. One student stationed herself at a busy intersection and categorized the cars turning left according to whether or not they signaled the turn and whose bumper sticker was displayed. After 2 hours, she left with the frequencies shown in the accompanying table. Test these data with a χ^2 test. Write a paragraph of explanation for the supporters profile that can be understood by students who haven't studied statistics. Begin the paragraph with appropriate details about how the data were gathered.

Signaled turn	Sticker	Frequency
No	Hill	11
No	Dale	2
Yes	Hill	57
Yes	Dale	31

14.18. A friend, who knows that you are almost finished with this book, comes to you for advice on a statistical analysis. He wants to know whether there is a significant difference between men and women. Examine the data that follow. What do you tell your friend?

	Men	Women
Mean score	123	206
N	18	23

14.19. Our political science student from problem 14.17 has launched a new experiment—to determine whether affluence is related to voter choice. This time she looks for yard signs for the two candidates and then classifies the houses as brick or one-story frame. Her reasoning is that the one-story frame homes represent less affluent voters in her city. Houses that do not fit the categories are ignored. After 3 hours of driving and three-fourths of a tank of gas, the frequency counts are as shown in the accompanying table. Analyze the data using your chi square techniques. Write a statement of the procedures and results, which can be used by the political science class in the voter profile.

Type of house	Yard signs	Frequency
Brick	Hill	17
Brick	Dale	88
Frame	Hill	59
Frame	Dale	51

14.20. When using χ^2, what is the proper *df* for each of the following tables?

a. 1×4

b. 4×5

c. 2×4

d. 6×3

14.21. Please reread the chapter objectives. Can you do each one?

ADDITIONAL HELP FOR CHAPTER 14

Visit *cengage.com/psychology/spatz*. At the Student Companion Site, you'll find multiple-choice tutorial quizzes and flashcards with definitions. For this chapter there is a Statistics Workshop on Chi-Square.

KEY TERMS

Additive (p. 313)
Chi square distribution (p. 303)
Chi square test (χ^2) (p. 303)
Contingency (p. 304)
Degrees of freedom (p. 306, 311, 318)
Effect size index (ϕ) (p. 308)
Expected frequency (p. 304)
Expected values (p. 304, 310)
Frequency counts (p. 302)
Goodness-of-fit test (p. 310)
Independent (p. 319)
Observed frequency (p. 304)
Phi (ϕ) (p. 308)
Small expected frequencies (p. 316)
Test of independence (p. 304)
Yates's correction (p. 317)

CHAPTER 15

More Nonparametric Tests

OBJECTIVES FOR CHAPTER 15

After studying the text and working the problems in this chapter, you should be able to:

1. Describe the rationale of nonparametric statistical tests
2. Distinguish among designs that require a Mann–Whitney U test, a Wilcoxon matched-pairs signed-ranks T test, a Wilcoxon–Wilcox multiple-comparisons test, and a Spearman r_s correlation coefficient
3. Calculate a Mann–Whitney U test for small or large samples and interpret the results
4. Calculate a Wilcoxon matched-pairs signed-ranks T test for small or large samples and interpret the results
5. Calculate a Wilcoxon–Wilcox multiple-comparisons test and interpret the results
6. Calculate a Spearman r_s correlation coefficient and test the hypothesis that the coefficient came from a population with a correlation of zero

TWO CHILD PSYCHOLOGISTS were talking shop over coffee one morning. (Much research begins with just such sessions.) The topic was intensive early training in athletics. Both were convinced that such training made the child less sociable as an adult, but one psychologist went even further. "I think that really intensive training of young kids is ultimately detrimental to their performance in the sport. Why, I'll bet that, among the top ten men's singles tennis players, those with intensive early training are not in the highest ranks."

"Well, I certainly wouldn't go that far," said the second psychologist. "I think all that early intensive training is quite helpful."

"Good. In fact, great. We disagree and we may be able to decide who is right. Let's get the ground rules straight. For tennis players, how early is early and what is intensive?"

"Oh, I'd say early is starting by age 7 and intensive is playing 5 days a week for 2 or more hours."[1]

"That seems reasonable. Now, let's see, our population is 'excellent tennis players' and these top ten will serve as our representative sample."

"Yes, indeed. What we would have among the top ten players would be two groups to compare. One had intensive early training, and the other didn't. The dependent variable is the player's rank. What we need is some statistical test that will tell us whether the difference in average ranks of the two groups is statistically significant."[2]

"Right. Now, a t test won't give us an accurate probability figure because t tests assume that the population of dependent variable scores is normally distributed. A distribution of ranks is rectangular with each score having a frequency of 1."

"I think there is a nonparametric test that is proper for ranks."

nonparametric tests Statistical techniques that do not require assumptions about the sampled populations.

My story is designed to highlight the reason that **nonparametric tests** exist: Some kinds of data do not meet the assumptions[3] that parametric tests such as t tests and ANOVA are based on. Nonparametric statistical tests (sometimes called *distribution-free tests*) provide correct values for the probability of a Type I error regardless of the nature of the populations the samples come from. As implied in the story, the tests in this chapter use ranks as the dependent variable.

The Rationale of Nonparametric Tests

Many of the reasoning steps in nonparametric statistics are already familiar to you—they are NHST steps. By this time in your study of statistics, you should be able to write two or three paragraphs about the null hypothesis, the alternative hypothesis, gathering data, using a sampling distribution to find the probability of such results when the null hypothesis is true, making a decision about the null hypothesis, and telling a story that the data support. (Can you do this?)

The only part of the hypothesis-testing rationale that is unique to the tests in this chapter is that the sampling distributions are derived from ranks rather than quantitative scores. Here's an explanation of how a sampling distribution based on ranks might be constructed.

Suppose you drew two samples of equal size (for example, $N_1 = N_2 = 10$) from the same population.[4] You then arranged all the scores from both samples into one overall ranking, from 1 to 20. Because the samples are from the same population, the sum of the ranks of one group should be equal to the sum of the ranks of the second group. In this case, the expected sum for each group is 105. (With a little figuring, you can prove this for yourself now, or you can wait for an explanation later in the chapter.)

---

[1] Because the phrase *intensive early training* can mean different things to different people, the second psychologist has provided an operational definition. An operational definition is a definition that specifies how the concept can be measured.

[2] This is how the experts convert vague questionings into comprehensible ideas that can be communicated to others—they identify the independent and the dependent variables.

[3] Two common assumptions are that the population distributions are normal and have equal variances.

[4] As always, drawing two samples from one population is statistically the same as starting with two identical populations and drawing a random sample from each.

Although the expected sum is 105, actual sampling from the population would also produce sums greater than 105 and less than 105. After repeated sampling, all the sums could be arranged into a sampling distribution, which would allow you to determine the likelihood of any sum (if both samples come from the same population).

If the two sample sizes are unequal, the same logic will work. A sampling distribution could be constructed that shows the expected variation in sums of ranks for one of the two groups.

With this rationale in mind, you are ready to learn four new techniques. The first three are NHST methods that produce the probability of the data observed, if the null hypothesis is true. The fourth technique is a correlation coefficient for ranked data, symbolized r_s. You will learn to calculate r_s and then test the hypothesis that a sample r_s came from a population with a correlation coefficient of .00.

The four nonparametric techniques in this chapter and their functions are listed in Table 15.1. In earlier chapters you studied parametric tests that have similar functions, which are listed on the right side of the table. Study **Table 15.1** carefully now.

There are many other nonparametric statistical methods besides the four that you learn in this chapter. Sprent and Smeeton's superb handbook *Applied Nonparametric Statistical Methods* (2007) covers many of these methods. Sprent and Smeeton's book describes applications in fields such as road safety, space research, trade, and medicine, as well as in traditional academic disciplines.

Comparison of Nonparametric to Parametric Tests

In what ways are the nonparametric tests in this chapter similar to and in what ways are they different from parametric tests (such as t tests and ANOVA)? Here are the similarities:

- The hypothesis-testing logic is the same for both nonparametric and parametric tests. Both tests yield the probability of the observed data, *when the null hypothesis is true.* As you will see, though, the null hypotheses of the two kinds of tests are different.

TABLE 15.1 The function of some nonparametric and parametric tests

Function	Nonparametric test	Parametric test
Tests for a significant difference between two independent samples	Mann–Whitney U test	Independent-samples t test
Tests for a significant difference between two paired samples	Wilcoxon matched-pairs signed-ranks T test	Paired-samples t test
Tests for significant differences among all possible pairs of independent samples	Wilcoxon–Wilcox multiple-comparisons test	One-way ANOVA and Tukey HSD tests
Describes the degree of correlation between two variables	Spearman r_s correlation coefficient	Pearson product-moment correlation coefficient, r

- Both nonparametric and parametric tests require you to assign participants randomly to subgroups (or to sample randomly from the populations) if you want to make unambiguous cause-and-effect conclusions.

As for the differences:

- Nonparametric tests do not require the assumptions about the populations that parametric tests require. For example, parametric tests such as *t* tests and ANOVA produce accurate probabilities when the populations are normally distributed and have equal variances. Nonparametric tests do not assume the populations have these characteristics.
- The null hypothesis for nonparametric tests is that the population distributions are the same. For parametric tests, the null hypothesis is usually that the population means are the same (H_0: $\mu_1 = \mu_2$). Because distributions can differ in form, variability, central tendency, or all three, the interpretation of a rejection of the null hypothesis may not be quite so clear-cut after a nonparametric test.

Recommendations on how to choose between a parametric and a nonparametric test have varied over the years. Two of the issues involved in the debate are the (1) scale of measurement and (2) power of the tests.

In the 1950s and after, some texts recommended that nonparametric tests be used if the scale of measurement was nominal or ordinal. After a period of controversy, this consideration was dropped. Later, it resurfaced. Currently, most researchers do not use a "scale of measurement" criterion when deciding between nonparametric and parametric tests.

The other issue, power, is also complicated. Power, you may recall, comes up when the null hypothesis *should* be rejected (see pages 222–223). A test's power is the probability that the test will reject a false null hypothesis. It is clear to mathematical statisticians that if the populations being sampled from are normally distributed and have equal variances, then parametric tests are more powerful than nonparametric ones.

If the populations are not normal or do not have equal variances, then it is less clear what to recommend. Early work on this question showed that parametric tests were robust, meaning that they gave approximately correct probabilities even though populations were not normal or did not have equal variances. However, this robustness has been questioned. For example, Blair, Higgins, and Smitley (1980) showed that a nonparametric test (Mann–Whitney *U*) is generally more powerful than its parametric counterpart (*t* test) for the nonnormal distribution they tested. Blair and Higgins (1985) arrived at a similar conclusion when they compared the Wilcoxon matched-pairs signed-ranks *T* test to the *t* test. I am sorry to leave these issues without giving you specific advice, but no simple rule of thumb about superiority is possible except for one: If the data are ranks, use a nonparametric test.

For each of the four tests in this chapter, you will have to *understand* the steps. Formulas aren't available that permit a mechanical "plug in the raw numbers and grind out an answer" approach. Also, the interpretation of rank statistics requires that you know whether the rank of #1 is a good thing or a bad thing. To work and interpret nonparametric problems correctly, think your way through them.

PROBLEMS

15.1. Name the test that is the nonparametric relative of an independent-samples *t* test. Name the relative of a paired-samples *t* test.

15.2. How are nonparametric tests different from parametric tests?

15.3. Sketch out in your own words the rationale of nonparametric tests.

15.4. Two issues have dominated the discussion of how to choose between nonparametric and parametric tests. What are they?

The Mann–Whitney *U* Test

Mann–Whitney *U* test Nonparametric test that compares two independent samples.

The **Mann–Whitney *U* test** is a nonparametric test for data from an independent-samples design.[5] Thus, it is the appropriate test for the child psychologists to use to test the difference in ranks of tennis players.

The Mann–Whitney *U* test produces a statistic, *U*, that is evaluated by consulting the sampling distribution of *U*. When *U* is calculated from small samples (both samples have 20 or fewer scores), the sampling distribution is **Table H** in Appendix C. Table H has critical values for sample sizes of 1 to 20. If the number of scores in one of the samples is greater than 20, the normal curve is used to evaluate *U*. With larger samples, a *z* score is calculated, and the values of ±1.96 and ±2.58 are used as critical values for $\alpha = .05$ and $\alpha = .01$.

Mann–Whitney *U* Test for Small Samples

To provide data to illustrate the Mann–Whitney *U* test, I made up the numbers in Table 15.2 about the intensive early training of the top ten male singles tennis players.

In **Table 15.2** the players are listed by initials in the left column; the right column indicates that they had intensive early training ($N_{yes} = N_1 = 4$) or that they did not ($N_{no} = N_2 = 6$). Each player's rank is shown in the middle column. At the bottom of the table, the ranks of the four *yes* players are summed (27), as are the ranks of the six *no* players (28).

The sums of the ranks are used to calculate two *U* values. The smaller of the two *U* values is used to enter Table H, which yields a probability figure. For the *yes* group, the *U* value is

$$U = (N_1)(N_2) + \frac{N_1(N_1 + 1)}{2} - \Sigma R_1$$

$$= (4)(6) + \frac{(4)(5)}{2} - 27 = 7$$

[5] The Mann–Whitney *U* test is identical to the Wilcoxon rank-sum test. Frank Wilcoxon published his test first (1945), but when Henry Mann and Donald Whitney independently published a test based on the same logic (1947), they provided tables and a *name* for their statistic. Today, researchers usually call it the Mann–Whitney *U* test. The lesson: name your invention. [A hint of justice recently surfaced. The *Encyclopedia of Statistics in Behavioral Science* (2005) refers to the test as the Wilcoxon–Mann–Whitney test.]

TABLE 15.2 Early training of top ten male singles tennis players

Players	Rank	Intensive early training?
Y.O.	1	No
U.E.	2	No
X.P.	3	No
E.C.	4	Yes
T.E.	5	No
D.W.	6	Yes
O.R.	7	No
D.S.	8	Yes
H.E.	9	Yes
R.E.	10	No

$\Sigma R_{yes} = 4 + 6 + 8 + 9 = 27$

$\Sigma R_{no} = 1 + 2 + 3 + 5 + 7 + 10 = 28$

For the *no* group, the U value is

$$U = (N_1)(N_2) + \frac{N_2(N_2 + 1)}{2} - \Sigma R_2$$

$$= (4)(6) + \frac{(6)(7)}{2} - 28 = 17$$

error detection

The sum of the two U values in a Mann–Whitney U test is equal to the product of $(N_1)(N_2)$.

Applying the error detection hint to the tennis ranking calculations, you get $7 + 17 = 24 = (4)(6)$. The calculations check.

Now, please examine **Table H**. It appears on two pages. On the first page of the table, the *lightface* type gives the critical values for α levels of .01 for a one-tailed test and .02 for a two-tailed test. The numbers in *boldface* type are critical values for $\alpha = .005$ for a one-tailed test and $\alpha = .01$ for a two-tailed test. In a similar way, the second page shows larger α values for both one- and two-tailed tests. Having chosen the page you want, use N_1 and N_2 to determine the column and row for your problem. The commonly used two-tailed test with $\alpha = .05$ is on this second page (boldface type).

Now, you can test the value $U = 7$ from the tennis data. From the conversation of the two child psychologists, it is clear that a two-tailed test is appropriate; they are interested in knowing whether intensive early training *helps* or *hinders* players. Because an α level was not discussed, you should do what they would do—see if the difference is significant at the .05 level and, if it is, see if it is also significant at some smaller α level. Thus, in Table H begin by looking for the critical value of U for a two-tailed test with $\alpha = .05$. This number is on the second page in boldface type at the intersection of $N_1 = 4$, $N_2 = 6$. The critical value is 2.

With the U test, a small U value obtained from the data indicates a large difference between the samples. This is just the opposite from t tests, ANOVA, and chi square, in which large sample differences produce large t, F, and χ^2 values. Thus, for statistical significance with the Mann–Whitney U test, the obtained U value must be equal to or *smaller than* the critical value in Table H.

For the tennis training example, because the U value of 7 is larger than the critical value of 2, you must *retain* the null hypothesis and conclude that there is no evidence from the sample that the distribution of players trained early and intensively is significantly different from the distribution of those without such training.

Although you can easily find a U value using the preceding method and quickly go to Table H and reject or retain the null hypothesis, it would help your understanding of this test to think about small values of U. Under what conditions would you get a small U value? What kinds of samples would give you a U value of zero?

By examining the formula for U, you can see that $U = 0$ when the members of one sample all rank lower than every member of the other sample. Under such conditions, rejecting the null hypothesis seems reasonable. By playing with formulas in this manner, you can move from the rote memory level to the understanding level.

Assigning Ranks and Tied Scores

Sometimes you may choose a nonparametric test for data that are not already in ranks.[6] In such cases, you have to rank the scores. Two questions often arise. Is the largest or the smallest score ranked 1, and what should I do about the ranks for scores that are tied?

You will find the answer to the first question to be very satisfactory. For the Mann–Whitney test, it doesn't make any difference whether you call the largest or the smallest score 1. (This is not true for the test described next, the Wilcoxon T test.)

Ties are handled by giving all tied scores the same rank. This rank is the mean of the ranks the tied scores would have if no ties had occurred. For example, if a distribution of scores is 12, 13, 13, 15, and 18, the corresponding ranks are 1, 2.5, 2.5, 4, 5. The two scores of 13 would have been 2 and 3 if they had not been tied, and 2.5 is the mean of 2 and 3. As a slightly more complex example, the scores 23, 25, 26, 26, 26, 29 have ranks of 1, 2, 4, 4, 4, 6. Ranks of 3, 4, and 5 average out to be 4.

Ties do not affect the value of U if they are in the same group. If several ties involve both groups, a correction factor may be advisable.[7]

error detection

Assigning ranks is tedious, and it is easy to make errors. Pay careful attention to the examples, and practice by assigning ranks yourself to all the problems.

[6] Severe skewness or populations with very unequal variances are often reasons for such a choice.

[7] See Kirk (2008, p. 504) for the correction factor.

z test
Statistical test that uses the normal curve as the sampling distribution.

Mann–Whitney *U* Test for Larger Samples

If one sample (or both samples) has 21 scores or more, the normal curve is used to assess U. A z **test** value is obtained by the formula

$$z = \frac{(U + c) - \mu_U}{\sigma_U}$$

where U = the smaller of the two U values
$c = 0.5$, a correction factor explained below

$$\mu_U = \frac{(N_1)(N_2)}{2}$$

$$\sigma_U = \sqrt{\frac{(N_1)(N_2)(N_1 + N_2 + 1)}{12}}$$

The correction factor, c, is a correction for continuity. It is used because the normal curve is a continuous function but the values of z that are possible in this formula are discrete.

Once again, an NHST formula has a familiar form: the difference between a statistic based on data $(U + c)$ and the expected value of a parameter (μ_U) divided by a measure of variability, σ_U. After you have obtained z, critical values can be obtained from Table C, the normal curve table. For a two-tailed test, reject H_0 if $|z| \geq 1.96$ $(\alpha = .05)$. For a one-tailed test, reject H_0 if $z \geq 1.65$ $(\alpha = .05)$. The corresponding values for $\alpha = .01$ are $z \geq 2.58$ and $z \geq 2.33$.

Here is a problem for which the normal curve is necessary. An undergraduate psychology major was devoting a year to the study of memory. The principal independent variable was gender. Among her several experiments was one in which she asked the students in a general psychology class to write down everything they remembered that was unique to the previous day's class, during which a guest had lectured. Students were encouraged to write down every detail they remembered. This class was routinely videotaped so it was easy to check each recollection for accuracy and uniqueness. Because the samples indicated that the population data were severely skewed, the psychology major chose a nonparametric test. (If you plot the scores in Table 15.3, you can see the skew.)

The scores, their ranks, and the statistical analysis are presented in **Table 15.3**. The z score of -2.61 led to rejection of the null hypothesis, so the psychology major returned to the original data and their means to interpret the results. Because the mean rank of the women, 15 (258 ÷ 17), is greater than that of the men, 25 (603 ÷ 24), and because higher ranks (those closer to 1) mean more recollections, she concluded that women recalled significantly more items than men did.

Her conclusion singles out central tendency for emphasis. On the average, women did better than men. The Mann–Whitney test, however, compares distributions. What our undergraduate has done is what most researchers who use the Mann–Whitney do: Assume that the two populations have the same form but differ in central tendency. Thus, when a significant U value is found, it is common to attribute it to a difference in central tendency.

Table 15.4 shows SPSS output for a Mann–Whitney U test of the gender–memory experiment. The upper panel gives N and mean rank for the two groups; the lower panel gives the z score for the difference in ranks and the probability of such a z score,

TABLE 15.3 Numbers of items recalled by men and women, ranks, and a Mann–Whitney *U* test

Men ($N = 24$)		Women ($N = 17$)	
Items recalled	Rank	Items recalled	Rank
70	3	85	1
51	6	72	2
40	9	65	4
29	13	52	5
24	15	50	7
21	16.5	43	8
20	18.5	37	10
20	18.5	31	11
17	21	30	12
16	22	27	14
15	23	21	16.5
14	24.5	19	20
13	26.5	14	24.5
13	26.5	12	28.5
11	30.5	12	28.5
11	30.5	10	33
10	33	10	33
9	35.5		$\Sigma R_2 = 258$
9	35.5		
8	37.5		
8	37.5		
7	39		
6	40		
3	41		
	$\Sigma R_1 = 603$		

$$U = (N_1)(N_2) + \frac{N_1(N_1 + 1)}{2} - \Sigma R_1 = (24)(17) + \frac{(24)(25)}{2} - 603 = 105$$

$$\mu_U = \frac{(N_1)(N_2)}{2} = \frac{(24)(17)}{2} = 204$$

$$\sigma_U = \sqrt{\frac{(N_1)(N_2)(N_1 + N_2 + 1)}{12}} = \sqrt{\frac{(24)(17)(42)}{12}} = 37.79$$

$$z = \frac{(U + c) - \mu_U}{\sigma_U} = \frac{105 + 0.5 - 204}{37.79} = -2.61$$

if the null hypothesis is true. I don't know why the SPSS z score (-2.62) is different from the formula-based z score (-2.61).

error detection

Here are two checks you can make on your assignment of ranks. First, the last rank is the sum of the two N's. In Table 15.3, $N_1 + N_2 = 41$, which is the lowest rank.

Second, when ΣR_1 and ΣR_2 are added, the sum is equal to $N(N + 1)/2$, where N is the total number of scores. In Table 15.3, $603 + 258 = 861 = (41)(42)/2$.

TABLE 15.4 Results of an SPSS Mann–Whitney *U* test of the gender–memory experiment data

Ranks

	Gender	N	Mean Rank	Sum of Ranks
ItemsRecalled	Men	24	16.88	405.00
	Women	17	26.82	456.00
	Total	41		

Test Statistics[a]

	ItemsRecalled
Mann-Whitney U	105.000
Wilcoxon W	405.000
Z	−2.621
Asymp. Sig. (2-tailed)	.009

[a]Grouping Variable: Gender

From the information in the error detection box, you can see how I found the expected sum of 105 in the section on the rationale of nonparametric tests. That example had 20 scores, so the overall sum of the ranks is

$$\frac{(20)(21)}{2} = 210$$

Half of this total should be found in each group, so the *expected* sum of ranks of each group, both of which come from the same population, is 105.

PROBLEMS

15.5. Many studies show that noise affects cognitive functioning. The data that follow mirror the results of Hygge, Evans, and Bullinger (2002). One elementary school was near a noisy airport; the other was in a quiet area of the city. Fifth-grade students were given a difficult reading test at both schools under no-noise conditions. The errors of 17 students follow. Analyze the errors with a Mann–Whitney *U* test and write a conclusion about the effects of noise on reading.

Near airport	32	25	22	20	18	15	15	13	
Quiet area	23	19	16	14	12	11	10	8	7

***15.6.** A friend of yours is trying to convince a mutual friend that the ride is quieter in an automobile built by [Toyota or Honda (you choose)] than one built by [Honda or Toyota (no choice this time; you had just 1 degree of freedom—once the first choice was made, the second was determined)]. This friend has arranged to borrow six fairly new cars—three made by each company—and to drive your blindfolded mutual friend around in the six cars (labeled A–F) until a stable ranking for quietness is achieved. So convinced is your friend that he insists on adopting $\alpha = .01$, "so that only the recalcitrant will not be convinced."

As the statistician in the group, you decide that a Mann–Whitney U test is appropriate. However, being the type who thinks through the statistics *before* gathering experimental data, you look at the appropriate subtable in Table H and find that the experiment, as designed, is doomed to retain the null hypothesis. Write an explanation of why the experiment must be redesigned.

15.7. Suppose your friend in problem 15.6 arranged for three more cars, labeled the 9 cars A–I, changed α to .05, and conducted the "quiet" test. The results are shown in the accompanying table. Perform a Mann–Whitney U test and write a conclusion.

Car	Company *Y* ranks	Car	Company *Z* ranks
B	1	H	3
I	2	A	5
F	4	G	7
C	6	E	8
		D	9

15.8. Grackles, commonly referred to as blackbirds, are hosts for a parasitic worm that lives in the tissue around the brain. To see if the incidence of this parasite was changing, 24 blackbirds were captured and the number of brain parasites in each bird was recorded. These data were compared with the infestation of 16 birds captured from the same place 10 years earlier. Analyze the data with a Mann–Whitney test and write a conclusion. Be careful and systematic in assigning ranks. Errors here are quite frustrating.

Present day

20	16	12	11	51	8	23	68	23	44	0	78
0	28	53	20	44	20	36	32	64	16	101	0

Ten years earlier

16	19	43	90	16	72	29	62
103	39	70	29	110	32	87	57

The Wilcoxon Matched-Pairs Signed-Ranks *T* Test

The **Wilcoxon matched-pairs signed-ranks *T* test** (1945) is appropriate for testing the difference between two paired samples. In Chapter 10 you learned of three kinds of paired-samples designs: natural pairs, matched pairs, and repeated measures (before and after). In each of these designs, a score in one group is logically paired with a score in the other group. If you are not sure of the difference between a paired-samples and an independent-samples design, you should review the explanations and problems in Chapter 10. Knowing the difference is necessary if you are to decide correctly between a Mann–Whitney U test and a Wilcoxon matched-pairs signed-ranks T test.

Wilcoxon matched-pairs signed-ranks *T* test
Nonparametric test that compares two paired samples.

TABLE 15.5 **Illustration of how to calculate T**

Participant	Variable 1	Variable 2	D	Rank	Signed rank
Caitlin	16	21	−5	2	−2
Kevin	14	17	−3	1	−1
Selene	26	18	8	3	3
Ian	23	9	14	4	4
			Σ(positive ranks) = 7		
			Σ(negative ranks) = −3		
			$T = 3$		

The result of a Wilcoxon matched-pairs signed-ranks test is a T value,[8] which is interpreted using critical values from **Table J** in Appendix C. Finding T involves some steps that are different from finding U.

Table 15.5 provides you with a few numbers from a paired-samples design. Using **Table 15.5**, work through the following steps, which lead to a T value for a Wilcoxon matched-pairs signed-ranks T:

1. Find a difference, D, for every pair of scores. The order of subtraction doesn't matter, but it *must* be the same for all pairs.
2. Using the *absolute value* for each difference, rank the differences. The rank of 1 is given to the *smallest* difference, 2 goes to the next smallest, and so on.
3. Attach to each rank the sign of its difference. Thus, if a difference produces a negative value, the rank for that pair is negative.
4. Sum the positive ranks and sum the negative ranks.
5. T is the *smaller of the absolute values* of the two sums.[9]

Here are two cautions about the Wilcoxon matched-pairs signed-ranks T test: (1) It is the *differences* that are ranked, and *not the scores* themselves, and (2) a rank of 1 always goes to the smallest difference.

The rationale of the Wilcoxon matched-pairs signed-ranks T test is that *if* there is no difference between the two populations, the absolute value of the negative sum should be equal to the positive sum, with all deviations being due to sampling fluctuations.

Table J shows the critical values for both one- and two-tailed tests for several α levels. To enter the table, use N, the number of *pairs* of subjects. Reject H_0 when the obtained T is *equal to* or *less than* the critical value in the table. Like the Mann–Whitney test, the Wilcoxon T must be equal to or less than the tabled critical value if you are to reject H_0.

To illustrate the calculation and interpretation of a Wilcoxon matched-pairs signed-ranks T test, I'll describe an experiment based on some early work of Muzafer Sherif (1935). Sherif was interested in whether a person's basic perception could be influenced by others. The basic perception he used was a judgment of the size of the

[8] Be alert when you see a capital T in your outside readings; it has uses other than to symbolize the Wilcoxon matched-pairs signed-ranks test. Also note that this T is capitalized, whereas the t in the t test and t distribution is not capitalized except by some computer programs. Some writers avoid these problems by designating the Wilcoxon matched-pairs signed-ranks test with a W or a W_s.

[9] The Wilcoxon test is like the Mann–Whitney test in that you have a choice of two values for your test statistic. For both tests, choose the smaller value.

TABLE 15.6 Wilcoxon matched-pairs signed-ranks analysis of the effect of peers on perception

	Mean movement (in.)			
Participant	Before	After	D	Signed ranks
1	3.7	2.1	1.6	4.5
2	12.0	7.3	4.7	10
3	6.9	5.0	1.9	6
4	2.0	2.6	−0.6	−3
5	17.6	16.0	1.6	4.5
6	9.4	6.3	3.1	8
7	1.1	1.1	0.0	Eliminated
8	15.5	11.4	4.1	9
9	9.7	9.3	0.4	2
10	20.3	11.2	9.1	11
11	7.1	5.0	2.1	7
12	2.2	2.3	−0.1	−1

$\Sigma(\text{positive}) = 62$
$\Sigma(\text{negative}) = -4$
$T = 4$
$N = 11$

Check: $62 + 4 = 66$ and $\frac{11(12)}{2} = 66$

autokinetic effect. The autokinetic effect is obtained when a person views a stationary point of light in an otherwise dark room. After a few moments, the light appears to move erratically. Sherif asked his participants to judge how many inches the light moved. Under such conditions, judgments differ widely among individuals but they are fairly consistent for each individual.

In the first phase of Sherif's experiment, participants worked alone. They estimated the distance the light moved and, after a few judgments, their estimates stabilized. These are the before measurements. Next, additional observers were brought into the dark room and everyone announced aloud their perceived movement. These additional observers were confederates of Sherif, and they always judged the movement to be somewhat less than that of the participant. Finally, the confederates left and the participant made another series of judgments (the after measurements). The perceived movements (in inches) of the light are shown in Table 15.6 for 12 participants.

Table 15.6 also shows the calculation of a Wilcoxon matched-pairs signed-ranks T test. The before measurements minus the after measurements produce the D column. These D scores are then ranked by absolute size and the sign of the difference attached in the Signed ranks column. Notice that when $D = 0$, that pair of scores is dropped from further analysis and N is reduced by 1. The negative ranks have the smaller sum, so $T = 4$.

The obtained T is less than the T value of 5 shown in Table J under $\alpha = .01$ (two-tailed test) for $N = 11$. Thus, the null hypothesis is rejected. The after scores represent a distribution that is different from the before scores. Now let's interpret the result using the terms of the experiment.

TABLE 15.7 SPSS output of a Wilcoxon matched-pairs signed-ranks test of data in Table 15.6

Ranks

		N	Mean Rank	Sum of Ranks
Before-After	Negative Ranks	2[a]	2.00	4.00
	Positive Ranks	9[b]	6.89	62.00
	Ties	1[c]		
	Total	12		

[a]Before < After
[b]Before > After
[c]Before = After

Test Statistics[b]

	Before-After
Z	−2.580[a]
Asymp. Sig. (2-tailed)	.010

[a]Based on negative ranks
[b]Wilcoxon Signed Ranks Test

By examining the *D* column, you can see that all scores except two are positive. This means that after hearing others give judgments smaller than their own, the participants saw the amount of movement as less. Thus, you may conclude (as did Sherif) that even basic perceptions tend to conform to opinions expressed by others.

Table 15.7 shows the results of an SPSS Wilcoxon test of the autokinetic movement data in Table 15.7. Descriptive statistics are in the upper panel; a *z*-score test and its probability are in the lower panel. The probability figure of .010 is consistent with the statistical significance found for $T = 4$.

Tied Scores and *D* = 0

Ties among the *D* scores are handled in the usual way—that is, each tied score is assigned the mean of the ranks that would have been assigned if there had been no ties. Ties do not affect the probability of the rank sum unless they are numerous (10 percent or more of the ranks are tied). When there are numerous ties, the probabilities in Table J associated with a given critical *T* value may be too large. In this case, the test is described as too conservative because it may fail to ascribe significance to differences that are in fact significant (Wilcoxon and Wilcox, 1964).

As you already know from Table 15.6, when *one* of the *D* scores is zero, it is not assigned a rank and *N* is reduced by 1. When *two* of the *D* scores are tied at zero, each is given the average rank of 1.5. Each is kept in the computation; one is assigned a plus sign and the other a minus sign. If *three D* scores are zero, one is dropped, *N* is reduced by 1, and the remaining two are given signed ranks of +1.5 and −1.5. You can generalize from these three cases to situations with four, five, or more zeros.

Wilcoxon Matched-Pairs Signed-Ranks *T* Test for Large Samples

When the number of pairs exceeds 50, the T statistic may be evaluated using the z test and the normal curve. The test statistic is

$$z = \frac{(T + c) - \mu_T}{\sigma_T}$$

where T = smaller sum of the signed ranks

$c = 0.5$

$$\mu_T = \frac{N(N + 1)}{4}$$

$$\sigma_T = \sqrt{\frac{N(N + 1)(2N + 1)}{24}}$$

N = number of pairs

PROBLEMS

15.9. A private consultant was asked to evaluate a job-retraining program. As part of the evaluation, she determined the income of 112 individuals before and after retraining. She found a T value of 4077. Complete the analysis and draw a conclusion. Be especially careful in wording your conclusion.

15.10. Six industrial workers were chosen for a study of the effects of rest periods on production. Output was measured for one week before the new rest periods were instituted and again during the first week of the new schedule. Perform an appropriate nonparametric statistical test on the results shown in the following table.

	Output	
Worker	Without rests	With rests
1	2240	2421
2	2069	2260
3	2132	2333
4	2095	2314
5	2162	2297
6	2203	2389

15.11. Undergraduates from Canada and the United States responded to a questionnaire on attitudes toward government regulation of business. High scores indicate a favorable attitude. Analyze the data in the accompanying table with the appropriate nonparametric test and write a conclusion.

Canada	United States	Canada	United States
12	16	27	26
39	19	33	15
34	6	34	14
29	14	18	25
7	20	31	21
10	13	17	30
17	28	8	3
5	9		

15.12. A professor gave his general psychology class a 50-item true/false test on the first day of class to measure the students' beliefs about punishment, reward, mental breakdowns, and so forth—topics that would be covered during the course. Many items were phrased to represent a common misconception ("mental breakdowns are usually caused by defective genes"). At the end of the course, the same test was given again. High scores mean lots of misconceptions. Analyze the data and write a conclusion about the effect of the course on misconceptions.

Student	Before	After
1	18	4
2	14	14
3	20	10
4	6	9
5	15	10
6	17	5
7	29	16
8	5	4
9	8	8
10	10	4
11	26	15
12	17	9
13	14	10
14	12	12

15.13. A health specialist conducted an 8-week workshop on weight control during which all 17 of the people who completed the course lost weight. To assess the long-term effects of the workshop, she weighed the participants again 10 months later. The weight lost or gained is listed with a positive sign for those who continued to lose weight and a negative sign for those who gained some back. What can you conclude about the long-term effects of the workshop?

−5	24	0	13	9	6	−7	−3	2
−10	−16	7	12	−19	−4	8	15	

The Wilcoxon–Wilcox Multiple-Comparisons Test

So far in this chapter on the analysis of ranked data, you have studied techniques for independent-samples designs with two groups (Mann–Whitney U) and related-samples designs with two groups (Wilcoxon matched-pairs signed-ranks T). The next technique is for data from three or more independent groups. This method allows you to compare all possible *pairs* of groups, regardless of the number of groups in the experiment. This is the nonparametric equivalent of a one-way ANOVA followed by Tukey HSD tests.[10]

The **Wilcoxon–Wilcox multiple-comparisons test** (1964) allows you to compare all possible pairs of treatments, which is like having a Mann–Whitney test on each pair of treatments. However, the Wilcoxon–Wilcox multiple-comparisons test keeps the α level at .05 or .01, regardless of how many pairs you have. Like the Mann–Whitney U test, this test requires independent samples.[11] (Remember that Wilcoxon devised a test identical to the Mann–Whitney U test.)

Wilcoxon–Wilcox multiple-comparisons test
Nonparametric test of all possible pairs from an independent-samples design.

To create a Wilcoxon–Wilcox multiple-comparisons test, begin by ordering the scores from the K treatments into one overall ranking. Then, within each sample, add the ranks, which gives a ΣR for each sample. Finally, for each pair of treatments, subtract one ΣR from the other, which gives a difference. Finally, compare the absolute size of the difference to a critical value in Table K.

The rationale of the Wilcoxon–Wilcox multiple-comparisons test is that when the null hypothesis is true, the various ΣR values should be about the same. A *large* difference indicates that the two samples came from different populations. Of course, the larger K is, the greater the likelihood of large differences by chance alone, and this is taken into account in the sampling distribution that Table K is based on.

The Wilcoxon–Wilcox test can be used only when the N's for all groups are equal. A common solution to the problem of unequal N's is to reduce the too-large group(s) by dropping one or more randomly chosen scores. A better solution is to conduct the experiment so that you have equal N's.

The data in **Table 15.8** represent the results of a solar collector experiment by two designer-entrepreneurs. These two had designed and built a 4-foot by 8-foot solar collector they planned to market, and they wanted to know the optimal rate at which to pump water through the collector. The rule of thumb for pumping is 1/2 gallon per hour per square foot of collector, so they chose values of 14, 15, 16, and 17 gallons per hour for their experiment. Starting with the reservoir full of water at 0ºC, the water was pumped for 1 hour through the collector and back to the reservoir. At the end of the hour, the temperature of the water in the reservoir was measured. Then the water was replaced with 0ºC water, the flow rate was changed, and the process repeated. The temperature measurements (to the nearest tenth of a degree) are shown in Table 15.8.

In Table 15.8 ranks are given to each temperature, ignoring the group the temperature is in. The ranks of all those in a group are summed, producing ΣR values that range from 20 to 84 for flow rates of 14 to 17 gallons per hour. Note that you can

[10] The Kruskal–Wallis one-way ANOVA on ranks is a direct analogue of a one-way ANOVA. (See Howell, 2008, pp. 507–509, or Sprinthall, 2007, pp. 479–482.) Unfortunately, tests that compare treatments (as the Tukey HSD does) are not readily available.

[11] A nonparametric test for more than two related samples, the Friedman rank test for correlated samples, is explained in Howell (2008, pp. 509–511).

TABLE 15.8 Temperature increases (°C) for four flow rates in a solar collector

Experimental conditions								
14 gal/hr			15 gal/hr		16 gal/hr		17 gal/hr	
	Temp	Rank	Temp	Rank	Temp	Rank	Temp	Rank
	28.1	7	28.9	3	25.1	14	24.7	15
	28.8	4	27.7	8.5	25.3	13	23.5	18
	29.4	1	27.7	8.5	23.7	17	22.6	19
	29.0	2	28.6	5	25.9	11	21.7	20
	28.3	6	26.0	10	24.2	16	25.8	12
Σ (ranks)		20		35		71		84

Check: $20 + 35 + 71 + 84 = 210$

$$\frac{N(N+1)}{2} = \frac{20(21)}{2} = 210$$

confirm your arithmetic for the Wilcoxon–Wilcox test using the same checks you used for the Mann–Whitney U test. That is, the sum of the four group sums, 210, is equal to $N(N + 1)/2$, where N is the total number of observation (20). Also, the largest rank, 20, is equal to the total number of observations.

The next step is to make pairwise comparisons. With four groups, six pairwise comparisons are possible. The rate of 14 gallons per hour can be paired with 15, 16, and 17; the rate of 15 can be paired with 16 and 17; and the rate of 16 can be paired with 17. For each pair, a difference in the sum of ranks is found and the absolute value of that difference is compared with the critical value in Table K.

The upper half of **Table K** has critical values for $\alpha = .05$; the lower half has values for $\alpha = .01$. For the experiment in Table 15.8, where $K = 4$ and $N = 5$, the critical values are 48.1 ($\alpha = .05$) and 58.2 ($\alpha = .01$). To reject H_0, a difference in rank sums must be equal to or *greater than* the critical values.

It is convenient (and conventional) to arrange the differences in rank sums into a matrix summary table. **Table 15.9** is an example using the flow-rate data. The table

TABLE 15.9 Summary table for differences in the sums of ranks in the flow-rate experiment

		Flow rates		
		14	15	16
Flow rates	15	15		
	16	51*	36	
	17	64**	49*	13

* $p < .05$; for $\alpha = .05$, the critical value is 48.1.
** $p < .01$; for $\alpha = .01$, the critical value is 58.2.

displays, for each pair of treatments, the difference in the sum of the ranks. Asterisks are used to indicate differences that exceed various α values. Thus, in Table 15.9, with $\alpha = .05$, rates of 14 and 16 are significantly different from each other, as are rates of 15 and 17. In addition, a rate of 14 is significantly different from a rate of 17 at the .01 level. What does all this mean for our two designer-entrepreneurs? Let's listen to their explanation to their old statistics professor.

"How did the flow-rate experiment come out, fellows?" inquired the kindly old gentleman.

"OK, but we are going to have to do a follow-up experiment using different flow rates. We know that 16 and 17 gallons per hour are not as good as 14, but we don't know if 14 is optimal for our design. Fourteen was the best of the rates we tested, though. On our next experiment, we are going to test rates of 12, 13, 14, and 15."

The professor stroked his beard and nodded thoughtfully. "Typical experiment. You know more after it than you did before, . . . but not yet enough."

PROBLEMS

15.14. Farmer Marc, a Shakespearean devotee, delivered a plea at the County Faire, asking three different groups to lend him unshelled corn. (Perhaps you can figure out how he phrased his plea.) The numbers of bushels offered by the farmers are listed. What can you conclude about the three groups?

Friends	Romans	Countrymen
25	14	8
31	18	15
17	10	11
22	16	7
27	9	13
29	10	12

15.15. Many studies have investigated methods of reducing anxiety and depression. I've embedded some of their conclusions in this problem. College students who reported anxiety or depression were randomly assigned to one of four groups: no treatment (NO), relaxation (RE), solitary exercise (SE), and group exercise (GE). The numbers are the students' improvement scores after 10 weeks. Analyze the data and write a conclusion.

NO	RE	SE	GE
10	14	12	22
8	10	27	32
3	21	23	20
0	10	20	28
7	15	17	22

15.16. The effect of three types of leadership on group productivity and satisfaction was investigated.[12] Groups of five children were randomly constituted and assigned an authoritarian, a democratic, or a laissez-faire leader. Nine groups were formed—three with each type of leader. The groups worked on various projects for a week. Measures of each child's personal satisfaction were taken and pooled to give a score for each child. The data are presented in the accompanying table. Test all possible comparisons with a Wilcoxon–Wilcox test.

Authoritarian	Democratic	Laissez-faire
120	108	100
102	156	69
141	92	103
90	132	76
130	161	99
153	90	126
77	105	79
97	125	114
135	146	141
121	131	82
100	107	84
147	118	120
137	110	101
128	132	62
86	100	50

15.17. Given the summary data in the following table, test all possible comparisons. Each group had a sample of eight subjects.

Sum of ranks for six groups						
Group	1	2	3	4	5	6
ΣR	196	281	227	214	93	165

15.18. List the tests presented so far in this chapter and the design for which each is appropriate. Be sure you can do this from memory.

Correlation of Ranked Data

Here is a short review of what you learned about correlation in Chapter 6.

1. Correlation requires a bivariate distribution (a logical pairing of scores).
2. Correlation is a method of describing the degree of relationship between two variables—that is, the degree to which high scores on one variable are associated with low or high scores on the other variable.

[12] For a summary of a similar classic investigation, see Lewin (1958).

3. Correlation coefficients range in value from +1.00 (perfect positive) to −1.00 (perfect negative). A value of .00 indicates that there is no relationship between the two variables.
4. Statements about cause and effect may not be made on the basis of a correlation coefficient alone.

In 1901, Charles Spearman (1863–1945) was "inspired by a book by Galton" and began experimenting at a little village school in England. He thought there was a relationship between intellect (school grades) and sensory ability (detecting musical discord). Spearman worked to find a mathematical way to express the *degree* of this relationship. He came up with a coefficient that did this, although he later found that others were ahead of him in developing such a coefficient (Spearman, 1930).

Spearman's name is attached to the coefficient that is used to show the degree of correlation between two sets of *ranked* data. He used the Greek letter ρ (rho) as the symbol for his coefficient. Later statisticians began to reserve Greek letters to indicate parameters, so the modern symbol for Spearman's statistic has become r_s, the *s* honoring Spearman. **Spearman r_s** is a special case of the Pearson product-moment correlation coefficient and is most often used when the number of pairs of scores is small (less than 20).

Spearman r_s
Correlation coefficient for degree of relationship between two variables measured by ranks.

Actually, r_s is a *descriptive statistic* that could have been introduced in the first part of this book. I waited until now to introduce it because r_s is a rank-order statistic, and this is a chapter about ranks. The next section shows you how to calculate this descriptive statistic; determining the statistical significance of r_s follows.

Calculation of r_s

The formula for r_s is

$$r_s = 1 - \frac{6\Sigma D^2}{N(N^2 - 1)}$$

where D = difference in ranks of a pair of scores
N = number of pairs of scores

I started this chapter with speculation about male tennis players and a made-up data set. I will end it with speculation about female tennis players, but now I have actual data.

What is the progression in women's professional tennis? Do they work their way up through the ranks, advancing their ranking year by year? Or do the younger players flash to the top and then gradually lose their ranking year by year? An r_s might lend support to one of these hypotheses.

If a rank of 1 is assigned to the oldest player, a positive r_s means that the older the player, the higher her rank (supporting the first hypothesis). A negative r_s means that the older the player, the lower her rank (supporting the second hypothesis). A zero r_s means no relationship between age and rank.

Table 15.10 shows the 12 top-ranked women tennis players just before the U.S. Open in 2009 and their age as a rank score among the 12 (*www.sonyericssonwtatour.com*). The Spearman r_s is .57. Thus, the descriptive statistic lends support for the "work your way up"

TABLE 15.10 Top 12 women tennis players, their rank in age, and the calculation of Spearman r_s

Player	Rank in tennis	Rank in age	D	D^2
Safina (Russia)	1	8	−7	49
Williams, S. (USA)	2	2	0	0
Williams, V. (USA)	3	1	2	4
Dementieva (Russia)	4	3	1	1
Jankovic (Serbia)	5	6	−1	1
Kuznetsova (Russia)	6	7	−1	1
Zvonareva (Russia)	7	5	2	4
Wozniacki (Denmark)	8	12	−3	9
Azarenka (Belarus)	9	11	−3	9
Pennetta (Italy)	10	4	6	36
Ivanovic (Serbia)	11	9	2	4
Radwanska (Poland)	12	10	2	4
				$\Sigma = 122$

$$r_s = 1 - \frac{6\Sigma D^2}{N(N^2 - 1)} = 1 - \frac{6(122)}{12(143)} = 1 - \frac{732}{1716} = 1 - 0.43 = .57$$

Source: www.sonyericssonwtatour.com.

hypothesis. Perhaps this conclusion could be made stronger by ruling out chance as an explanation. A NHST test is called for.

error detection

For Spearman r_s, first assign ranks to the scores and then find the differences. The Wilcoxon matched-pairs signed-ranks T test is different. For it, first find the differences and then assign ranks to the differences.

Testing the Significance of r_s

At this point in your studies, it is probably easy for you to imagine a sampling distribution of r_s. It would consist of many r_s values all calculated from samples taken from a population in which the correlation coefficient is zero. With such a sampling distribution, you can determine the probability of any r_s. If the probability of your sample r_s is small, reject the null hypothesis and conclude that the relationship in the population that r_s came from is *not zero*. You probably recall that you conducted this same test for a Pearson product-moment r in Chapter 9.

Table L in Appendix C gives the critical values for r_s for the .05 and .01 levels of significance when the number of pairs is 16 or fewer. If the obtained r_s is *equal to* or *greater than* the value in Table L, reject the null hypothesis. The null hypothesis is that the population correlation coefficient is zero.

For the tennis data in Table 15.10, $r_s = .57$. Table L shows that a correlation of .587 (either positive or negative) is required for statistical significance at the .05 level for 12 pairs. Thus, a correlation of .57 is not statistically significant.

"But... but...," you might be thinking, "A correlation coefficient of $r_s = .57$ is almost .587. Surely, closeness counts for something." I see your point and can tell you that such a result does affect researchers. They are frustrated when their outcome is so close to statistical significance, but they are resigned to the rules of the game—NS is NS. Usually, in such cases, a researcher gathers more data so that that a larger sample will confirm (or erode) the notion that there is a relationship. Of course, some data are easier to gather than others, but some, such as the tennis data, are very easy (www.sonyericssonwtatour.com).

Notice in Table L that rather large correlations are required for significance. As with *r*, not much confidence can be placed in low or moderate correlation coefficients that are based on only a few pairs of scores.

For samples larger than 16, test the significance of r_s by using Table A. Note, however, that **Table A** requires *df* and not the number of pairs. The degrees of freedom for r_s is $N - 2$ (the number of pairs minus 2), which is the same formula used for *r*.

When $D = 0$ and Tied Ranks

In calculating r_s, you may get a *D* value of zero (as I did for Serena Williams in Table 15.10). Should a zero be dropped from further analysis as it is in a Wilcoxon matched-pairs signed-ranks *T* test? You can answer this question by taking a moment to decide what $D = 0$ means in an r_s correlation problem. (Decide now.)

A zero means there is a perfect correspondence between the two ranks. If all differences were zero, the r_s would be 1.00. Thus, differences of zero should *not* be dropped when calculating an r_s.

The formula for r_s that you are working with is not designed to handle ranks that are tied. With r_s, ties are troublesome. A tedious procedure has been devised to overcome ties, but your best solution is to arrange your data collection so there are no ties. Sometimes, however, you are stuck with tied ranks, perhaps as a result of working with someone else's data. Kirk (2008) recommends assigning average ranks to ties, as you did for the three other procedures in this chapter, and then computing a Pearson *r* on the data.

My Final Word

In the first chapter I said that the essence of applied statistics is to measure a phenomenon of interest, apply statistical techniques to the numbers that result, and write the story of the new understanding of the phenomenon. The question, What should we measure? was not raised in this book, but it is important nonetheless. Here is an anecdote by E. F. Schumacher (1979), an English economist, that highlights the issue of what to measure.

> I will tell you a moment in my life when I almost missed learning something. It was during the war and I was a farm laborer and my task was before breakfast to go to yonder hill and to a field there and count the cattle. I went and I counted the cattle—there were always thirty-two—and then I went back to the bailiff, touched my cap, and said "Thirty-two, sir," and went and had my breakfast. One day when I arrived at the field an old farmer was standing at the gate, and he said, "Young man, what do you do here every morning?" I said, "Nothing much. I just count the cattle." He shook his head and said, "If you count them every day they won't flourish." I went back, I reported thirty-two, and on the way back I thought, Well, after all, I am a professional statistician, this is only a country yokel, how stupid can he get. One day I went back, I counted

and counted again, there were only thirty-one. Well, I didn't want to spend all day there so I went back and reported thirty-one. The bailiff was very angry. He said, "Have your breakfast and then we'll go up there together." And we went together and we searched the place and indeed, under a bush, was a dead beast. I thought to myself, Why have I been counting them all the time? I haven't prevented this beast dying. Perhaps that's what the farmer meant. They won't flourish if you don't look and watch the quality of each individual beast. Look him in the eye. Study the sheen on his coat. Then I might have gone back and said, "Well, I don't know how many I saw but one looks mimsey."

PROBLEMS

15.19. For each situation described, tell whether you would use a Mann–Whitney test, a Wilcoxon matched-pairs signed-ranks test, a Wilcoxon–Wilcox multiple-comparisons test, or an r_s.

a. A limnologist (a scientist who studies freshwater streams and lakes) measured algae growth in a lake before and after the construction of a copper smelting plant to see what effect the plant had.

b. An educational psychologist compared the sociability scores of firstborn children with scores of their next-born brother or sister to see if the two groups differed in sociability.

c. A child psychologist wanted to determine the degree of relationship between eye–hand coordination scores obtained at age 6 and scores obtained from the same individuals at age 12.

d. A nutritionist randomly and evenly divided boxes of cornflakes cereal into three groups. One group was stored at 40°F, one group at 80°F, and one group alternated from day to day between the two temperatures. After 30 days, the vitamin context of each box was assayed.

e. The effect of STP gas treatment on gasoline mileage was assessed by driving six cars over a 10-mile course, adding STP, and then again driving over the same 10-mile course.

15.20. Fill in the cells of the table.

	Symbol of statistic (if any)	Appropriate for what design?	Reject H_0 when statistic is (greater, less) than critical value?
Mann–Whitney test			
Wilcoxon matched-pairs signed-ranks test			
Wilcoxon–Wilcox multiple-comparisons test			

15.21. Two members of the department of philosophy (Locke and Kant) were responsible for hiring a new professor. Each privately ranked from 1 to 10 the ten candidates who met the objective requirements (degree, specialty, and so forth). The rankings are shown in the table. Begin by writing an interpretation of a low correlation and a high correlation. Then calculate an r_s and choose the correct interpretation.

Candidates	Locke	Kant
A	7	8
B	10	10
C	3	5
D	9	9
E	1	1
F	8	7
G	5	3
H	2	4
I	6	6
J	4	2

15.22. Self-esteem is a widely discussed concept. (See Baumeister et al., 2003, for a review.) The two problems that follow are based on Diener, Wolsic, and Fujita (1995).

a. The researchers measured participants' self-esteem and then had them judge their own physical attractiveness. Calculate r_s and test for statistical significance.

Self-esteem scores	40	39	37	36	35	32	28	24	20	16	12	7
Self-judged attractiveness	97	99	100	93	59	88	83	96	90	94	70	78

b. As part of the same study, the researchers had others judge photographs of the participants for attractiveness. Calculate r_s and write a conclusion that is based on both data sets in the problem.

Self-esteem scores	28	39	24	40	37	16	32	36	20	7	35	12
Other-judged attractiveness	28	43	26	37	29	49	54	25	40	34	60	31

15.23. With $N = 16$ and $\Sigma D^2 = 308$, calculate r_s and test its significance at the .05 level.

15.24. You are once again asked to give advice to a friend who comes to you for criticism of an experimental design. This friend has four pairs of scores obtained randomly from a population. Her intention is to calculate a correlation coefficient r_s and decide whether there is any significant correlation in the population. Give advice.

15.25. One last time: Review the objectives at the beginning of the chapter.

ADDITIONAL HELP FOR CHAPTER 15

Visit *cengage.com/psychology/spatz*. At the Student Companion Site, you'll find multiple-choice tutorial quizzes and flashcards with definitions.

KEY TERMS

Assigning ranks (pp. 329, 334)
$D = 0$ (pp. 336, 345)
Distribution-free tests (p. 324)
Mann–Whitney U test (p. 327)
Nonparametric tests (p. 324)
Operational definition (p. 324)
Parametric tests (p. 325)
Power (p. 326)
Scales of measurement (p. 326)
Spearman r_s (p. 343)
T value (p. 334)
Tied scores (pp. 329, 336, 345)
U value (p. 327)
Wilcoxon matched-pairs signed-ranks T test (p. 333)
Wilcoxon–Wilcox multiple-comparisons test (p. 339)
z test (pp. 330, 337)

What Would You Recommend? Chapters 11–15

Here is the last set of *What would you recommend?* problems. Your task is to choose an appropriate statistical technique from among the several you have learned in the previous five chapters. No calculations are necessary. For each problem that follows, (1) recommend a statistic that will either answer the question or make the comparison, and (2) explain why you recommend that statistic.

a. Systematic desensitization and flooding are the names of two techniques that help people overcome anxiety about specific behaviors. One anxiety-arousing behavior is strapping on a parachute and jumping out of an airplane. To compare the two techniques, researchers had military trainees experience one technique or the other. Later, at the time of the first five actual jumps, each trainee's "delay time" was measured. (The assumption was that longer delay means more anxiety.) On the basis of the severely skewed distribution of delay times, each trainee was given a rank. What statistical test should be used to compare the two techniques?

b. Stanley Milgram researched the question, When do people submit to demands for compliance? In one experiment, participants were told to administer a shock to another participant, who was in another room. (No one was actually shocked in this experiment.) The experimenter gave directions to half the participants over a telephone; the other half received directions in person. The dependent variable was whether or not a participant complied with the directions.

c. Forensic psychologists study factors that influence people as they judge defendants in criminal cases. They often supply participants in their studies with fictitious "pretrial reports" and ask them to estimate the likelihood, on a scale of .00 to 1.00, that the defendant will be convicted. In one study, the pretrial reports contained a photograph showing a person who was either baby-faced or mature-faced. In addition, half the reports indicated that the defendant was simply negligent in the alleged incident and half indicated that the defendant had been intentionally deceptive in the incident.

d. As part of a class project, two students administered a creativity test to undergraduate seniors. They also obtained each person's rank in class. In what way could the two students determine the degree of relationship between the two variables?

e. Suppose a theory predicts that the proportion of people who participate in a rather difficult exercise will drop by one-half each time the exercise is repeated. When the data were gathered, 100 people participated on the first day. On each of the next three occasions, fewer and fewer people participated. What statistical test could be used to arrive at a judgment about the theory?

f. Participants in an experiment were told to "remember what you hear." A series of sentences followed. (An example was "The plumber slips $50 into his wife's purse.") Later, when the participants were asked to recall the sentences, one group was given disposition cues (such as "generous"), one was given semantic (associative) cues (such as "pipes"), and one was given no cues at all. Recall measures ranged between 0 and 100. The recall scores for the "no cues" condition were much more variable than those for the other two conditions. The researcher's question was: Which of the three conditions produces the best recall?

g. To answer a question about the effect of an empty stomach on eating, investigators in the 1930s studied dogs that ate a normal-sized meal (measured in ounces). Because of a surgical operation, however, no food reached the stomach. Thirty minutes later, the dogs were allowed to eat again. The data showed that the dogs, on the average, ate less the second time (even though their stomachs were just as empty). At both time periods, the data were positively skewed.

h. College cafeteria personnel are often dismayed by the amount of food that is left on plates. One way to reduce waste is to prepare the food using a tastier recipe. Suppose a particular vegetable is served on three different occasions, and each time a different recipe is used. As the plates are cleaned in the kitchen, the amount of the vegetable left on the plate is recorded. How might each pair of recipes be compared to determine if one recipe has resulted in significantly less waste?

i. Many programs seek to change behavior. One way to assess the effect of a program is to survey the participants about their behavior, administer the program, and give the survey again. To demonstrate the lasting value of the program, however, follow-up data some months later are required. If the data consist of each participant's three scores on the same 100-point survey (pretest, post-test, and follow-up), what statistical test is needed to determine the effect of the program?

CHAPTER 16

Choosing Tests and Writing Interpretations

OBJECTIVES FOR CHAPTER 16

After studying the text and working the problems in this chapter, you should be able to:

1. List the descriptive and inferential statistical techniques that you studied in this book
2. Describe, using a table of characteristics or decision trees or both, the descriptive and inferential statistics that you studied in this book
3. Read the description of a study and identify a statistical technique that is appropriate for such data
4. Read the description of a study and recognize that you have not studied a statistical technique that is appropriate for such data
5. Given the statistical analysis of a study, write an interpretation of what the study reveals

THIS CHAPTER IS designed to help you consolidate the many bits and pieces of statistics that you have learned. It will help you, regardless of the number of chapters your course covered (as long as you completed through Chapter 10, "Hypothesis Testing, Effect Size, and Confidence Intervals: Two-Sample Designs").

The first section of this chapter is a review of the major concepts of each chapter. The second section describes statistical techniques that you *did not* read about in this book. These two sections should help prepare you to apply what you have learned to the exercises in the last section of this chapter, which are designed to help you see the big picture of what you've learned.

A Review

If you have been reading for integration, the kind of reading in which you are continually asking the question, How does this stuff relate to what I learned before? you may have clearly in your head a major thread that ties this book together from Chapter 2 onward.

The name of that thread is *distributions*. It is the importance of distributions that justifies the subtitle of this book, *Tales of Distributions*. I want to summarize explicitly some of what you have learned about distributions. I hope this will help you understand not only the material in this book but also other statistical techniques that you might encounter in future courses or in future encounters with statistics.

In Chapter 2 you were confronted with a disorganized array of scores on one variable. You learned to organize those scores into a frequency distribution and graph the distribution. In Chapter 3 you learned about central tendency values and in Chapter 4, you studied the variability of distributions—especially the standard deviation. Chapter 5 introduced the z score, effect size index, and boxplot, which are descriptive statistics that are useful for a Descriptive Statistics Report.

In Chapter 6, where you were introduced to bivariate distributions, you learned to express the relationship between variables with a Pearson correlation coefficient and, using a linear regression equation, predict scores on one variable from scores on another. In Chapter 7 you learned about the normal distribution and how to use means and standard deviations to find probabilities associated with distributions. In Chapter 8 you studied the important concept of the sampling distribution. You used two sampling distributions, the normal distribution and the t distribution, to determine probabilities about sample means and to establish confidence intervals about a sample mean.

Chapter 9 presented you with the concept of null hypothesis statistical testing (NHST). You used the t distribution to test the hypothesis that a sample mean came from a population with a hypothesized mean, and you calculated an effect size index. In Chapter 10 you studied the design of a simple two-group experiment. You applied the NHST procedure to both independent-samples designs and paired-samples designs, using the t distribution as a source of probabilities. You calculated effect sizes and for a second time, found confidence intervals. In Chapter 11 you learned about still another sampling distribution (F distribution) that is used when there are two or more independent samples in an experiment. Chapter 12 covered the analysis of two or more repeated-measures samples. Chapter 13 discussed an even more complex experiment, one with *two* independent variables for which probability figures could be obtained with the F distribution.

Chapter 14 taught you to answer questions of independence and goodness of fit by using frequency counts and the chi square distribution. In Chapter 15 you analyzed data using nonparametric tests—techniques based on distributions that do not satisfy the assumptions that the dependent-variable scores be normally distributed and have equal variances.

Look back at the preceding paragraphs. You have learned a lot, even if your course did not cover *every* chapter. Allow me to offer my congratulations!

Future Steps

One short step you can take is to acknowledge that you are prepared to analyze data from a variety of sources and to write a Statistics Report. Similar to the Descriptive Statistics Report that you wrote for Chapter 5, a Statistics Report gives details about how the data were gathered, descriptive statistics (perhaps with graphs) that reveal the relationships between the variables, NHST test results, and an overall interpretation of the study.

As for additional NHST techniques, several paths might be taken from here. One logical next step is to analyze an experiment with three independent variables. The three main effects and the several interactions are tested with an *F* distribution.

A second possibility is to study techniques for the analysis of experiments that have more than one *dependent* variable. Such techniques are called multivariate statistics. Many of these techniques are analogous to those you have studied in this book, except there are two or more dependent variables instead of just one. For example, Hotelling's T^2 is analogous to the *t* test; one-way multivariate analysis of variance (MANOVA) is analogous to one-way ANOVA; and higher-order MANOVA is analogous to factorial ANOVA.

It turns out that the inferential statistics techniques described in this book and those mentioned in the previous paragraph are all special cases of a more general approach to finding relationships among variables. This approach, the General Linear Model, is a topic in advanced statistics courses (and sometimes it is *the* topic).

meta-analysis
A technique for reaching conclusions when the data consist of many separate studies done by different investigators.

Finally, there is the task of combining the results of many *separate* studies that all deal with the same topic. It is not unusual to have conflicting results among 10 or 15 or 50 studies, and researchers have been at the task of synthesizing since the beginning. This task is facilitated by a set of statistical procedures called **meta-analysis**.

What I've just described are advanced statistical techniques that are well established. As I mentioned in Chapter 1, however, statistics is a dynamic discipline; your future statistics studies will no doubt be influenced by the recent challenges to the central role of NHST in data analysis. For example, Killeen (2005) derived a new statistic to take the place of *p*, the probability of the data observed, if the null hypothesis is true. He proposed that data be analyzed using p_{rep}, the probability of replicating the effect if the experiment is conducted again. The statistic p_{rep} does not depend on a null hypothesis. In 2009, researchers were regularly using p_{rep} in articles in *Psychological Science,* a leading psychology journal. For now, however, let's finish this elementary statistics textbook with problems that help you consolidate what you have learned so far (especially the logic).

Choosing Tests and Writing Interpretations

In Chapter 1, I described the fourfold task of a statistics student:

1. Read a problem.
2. Decide what statistical procedure to use.
3. Perform that procedure.
4. Write an interpretation of the results.

Students tell me that the parts of this task they still need to work on at the end of the course are (2) deciding what statistical procedure to use and (4) writing an interpretation of the results.

Even though you may have some uncertainties, you are well on your way to being able to do all four tasks. What you may need at this point is exercises that help you "put it all together." (Of course, an ice cream cone couldn't hurt, either.) Pick and choose from the four exercises that follow; the emphasis is heavily on the side of deciding what statistical test to use.

Exercise A: Reread all the chapters you have studied and all the tests and homework you have. If you choose to do this exercise (or part of it), spend most of your time on concepts that you recognize as especially important (such as NHST). It will take a considerable amount of time for you to do this—probably 8 to 16 hours. However, the benefits are great, according to students who have invested the time. A typical quote is "Well, during the course I would read a chapter and work the problems without too much trouble, but I didn't relate it to the earlier stuff. When I went back over the whole book, though, I got the big picture. All the pieces do fit together."

Exercise B: Construct a table with all the statistical procedures and their characteristics. List the procedures down the left side. I suggest column heads such as *Description* (brief description of procedure), *Purpose* (descriptive or inferential), *Null hypothesis* (parametric or nonparametric), *Number of independent variables* (if applicable), and so forth.

I've included my version of this table (**Table 16.1**), but please don't look at what I've done until you have a copy of your own. However, after you have worked for a while on your own, it wouldn't hurt to peek at the column heads I used.

Exercise C: Construct two decision trees. For the first one, show all the descriptive statistics you have learned. To begin, you might divide these statistics into categories of univariate and bivariate distributions. Under univariate, you might list the different categories of descriptive statistics you have learned and then, under those categories, identify specific descriptive statistics. Under bivariate, choose categories under which specific descriptive statistics belong.

The second decision tree is for inferential statistics. You might begin with the categories of NHST and confidence intervals, or you might decide to divide inferential statistics into the type of statistics analyzed (means, frequency counts, and ranks). Work in issues such as the number of independent variables, whether the design is related samples or independent samples, and other issues that will occur to you.

There are several good ways you might organize your decision trees. My versions are Figure 16.1 and Figure 16.2, but please don't look at them now. Looking at the figures now would prevent you from going back over your text and digging out the information you learned earlier, which you now need for your decision trees. My experience has been that students who create their own decision trees learn more than those who are handed someone else's. After you have *your* trees in hand, *then* look at mine.

Exercise D: Work the two sets of problems that follow. The first set requires you to decide what statistical test or descriptive statistic answers the question the investigator is asking. The second set requires you to interpret the results of an analyzed experiment. You will probably be glad that no number crunching is required for these problems. You won't need your calculator or computer.

However, here is a word of warning. For a few of the problems, the statistical techniques necessary are not covered in this book. (In addition, perhaps your course did not cover all the chapters.) Thus, a part of your task is to recognize what problems you are prepared to solve and what problems would require digging into a statistics textbook.

TABLE 16.1 Summary table for elementary statistics

Procedure (text page numbers)	Purpose	Description	Null hypothesis	No. of IVs	No. of IV levels	Design
Frequency distributions (Chapter 2)	D	Picture of raw data	NA	NA	NA	NA
Boxplot (76)	D	Data picture that includes statistics	NA	NA	1, 2+	NA
Central tendency—Mean (41, 44, 50)	D	One number to represent all	NA	NA	NA	NA
Median (43, 45)	D	One number to represent all	NA	NA	NA	NA
Mode (44, 46)	D	One number to represent all	NA	NA	NA	NA
Variability—Range (56)	D	Dispersion of scores	NA	NA	NA	NA
Standard deviation (60, 65)	D	Dispersion of scores	NA	NA	NA	NA
Variance (68)	D	Dispersion of scores	NA	NA	NA	NA
Interquartile range (57)	D	Dispersion of scores	NA	NA	NA	NA
Outlier (75)	D	An extreme score	NA	NA	NA	NA
z Score (72)	D	Relative position of one score	NA	NA	NA	NA
Regression (114)	D	Linear equation; 2 variables	NA	NA	NA	Corr
Correlation—Pearson *r* (96)	D/I	Degree of relationship; 2 variables	P	NA	NA	Corr
Spearman r_s (343)	D/I	Degree of relationship; 2 variables	NP	NA	NA	Corr
Confidence interval (164, 218, 219)	D/I	Limits around mean or mean difference	P	1	NA	Indep
t test—One sample (181)	I	Test sample $\bar{X}$ from pop. μ	P	1	NA	Indep
Independent samples (206)	I	2 sample $\bar{X}$'s; same μ?	P	1	2	Indep
Paired samples (210)	I	2 sample $\bar{X}$'s; same μ?	P	1	2	Rel
Pearson *r* (191)	I	Test *r* from $\rho = .00$	P	NA	NA	Corr
d—One sample (189)	D	Assess size of IV effect	P	1	NA	Indep
Two independent samples (216)	D	Assess size of IV effect	P	1	2	Indep
Two paired samples (217)	D	Assess size of IV effect	P	1	2	Rel
ANOVA (251, 296)	D	Assess size of IV effect	P	1	2	Indep
F test—Two variances (245)	I	2 sample variances; same σ^2?	P	1	2	Indep
One-way ANOVA (243)	I	$\geq$2 sample $\bar{X}$'s; same μ?	P	1	2+	Indep
Repeated measures (262)	I	$\geq$2 sample $\bar{X}$'s; same μ?	P	1	2+	Rel
Factorial ANOVA (287)	I	2 or more IVs plus interactions	P	2+	2+	Indep
f—One-way ANOVA (252)	D	Assess size of IV effect	P	1	2+	Indep
Factorial ANOVA (296)	D	Assess size of IV effect	P	1	2+	Indep
Tukey HSD (248, 262, 295)	I	Tests all pairwise differences	P	1	2	Indep
Chi square—Goodness of fit (310)	I	Do data fit theory?	NP	NA	NA	Indep
Independence (304)	I	Are variables independent?	NP	NA	NA	Indep
φ: Chi square effect size (308)	D	Assess size of relationship	NP	NA	NA	Indep
Mann–Whitney *U* (327)	I	2 samples from same population?	NP	1	2	Indep
Wilcoxon matched-pairs signed-ranks *T* (333)	I	2 samples from same population?	NP	1	2	Rel
Wilcoxon–Wilcox multiple comparisons (339)	I	Tests all pairwise differences	NP	1	2+	Indep

Key: Corr = Correlated; D = Descriptive; I = Inferential; Indep = Independent; IV = Independent variable; NA = Not applicable; NP = Nonparametric; P = Parametric; Rel = Related

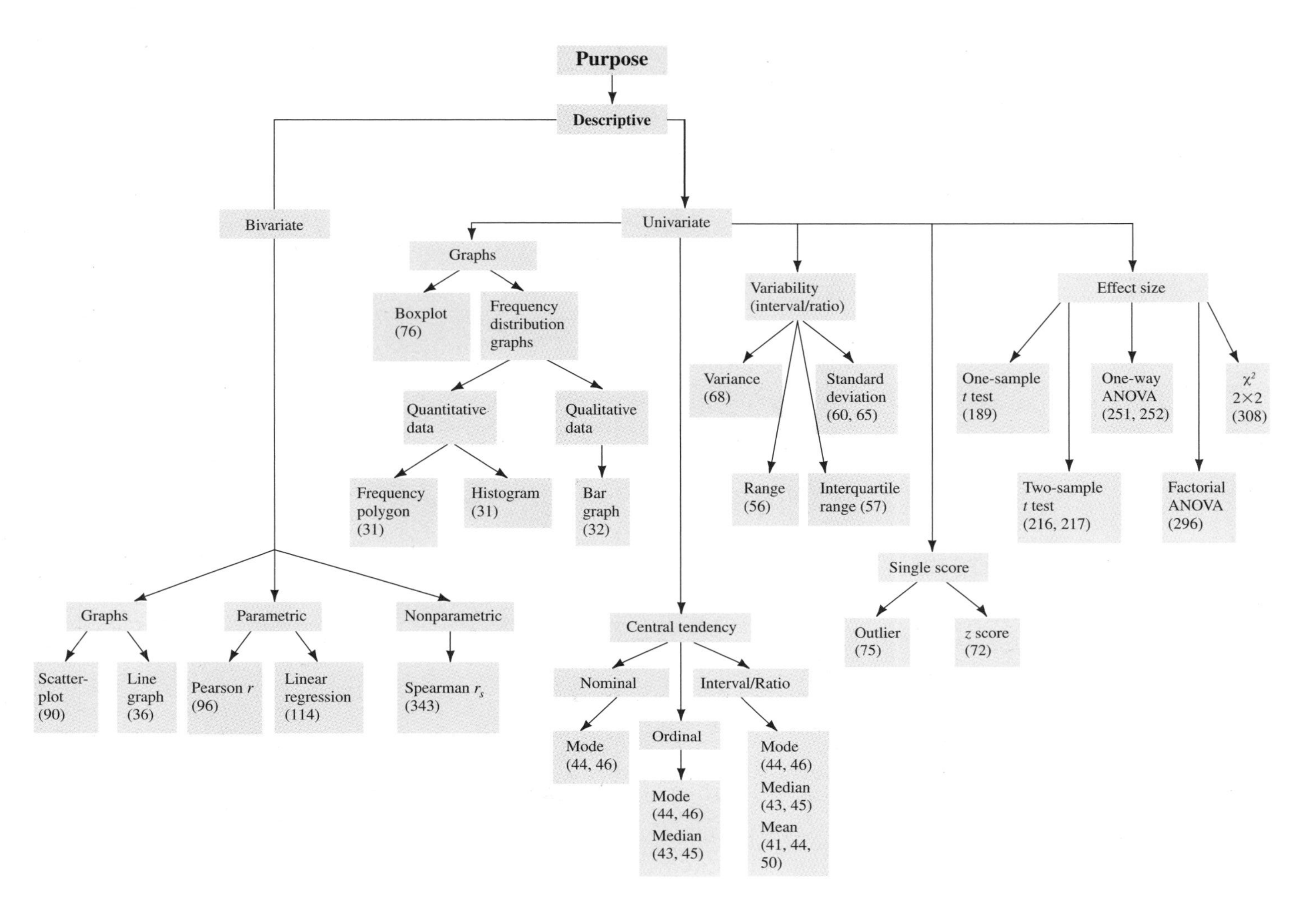

FIGURE 16.1 Decision tree for descriptive statistics. (Text page numbers are in parentheses.)

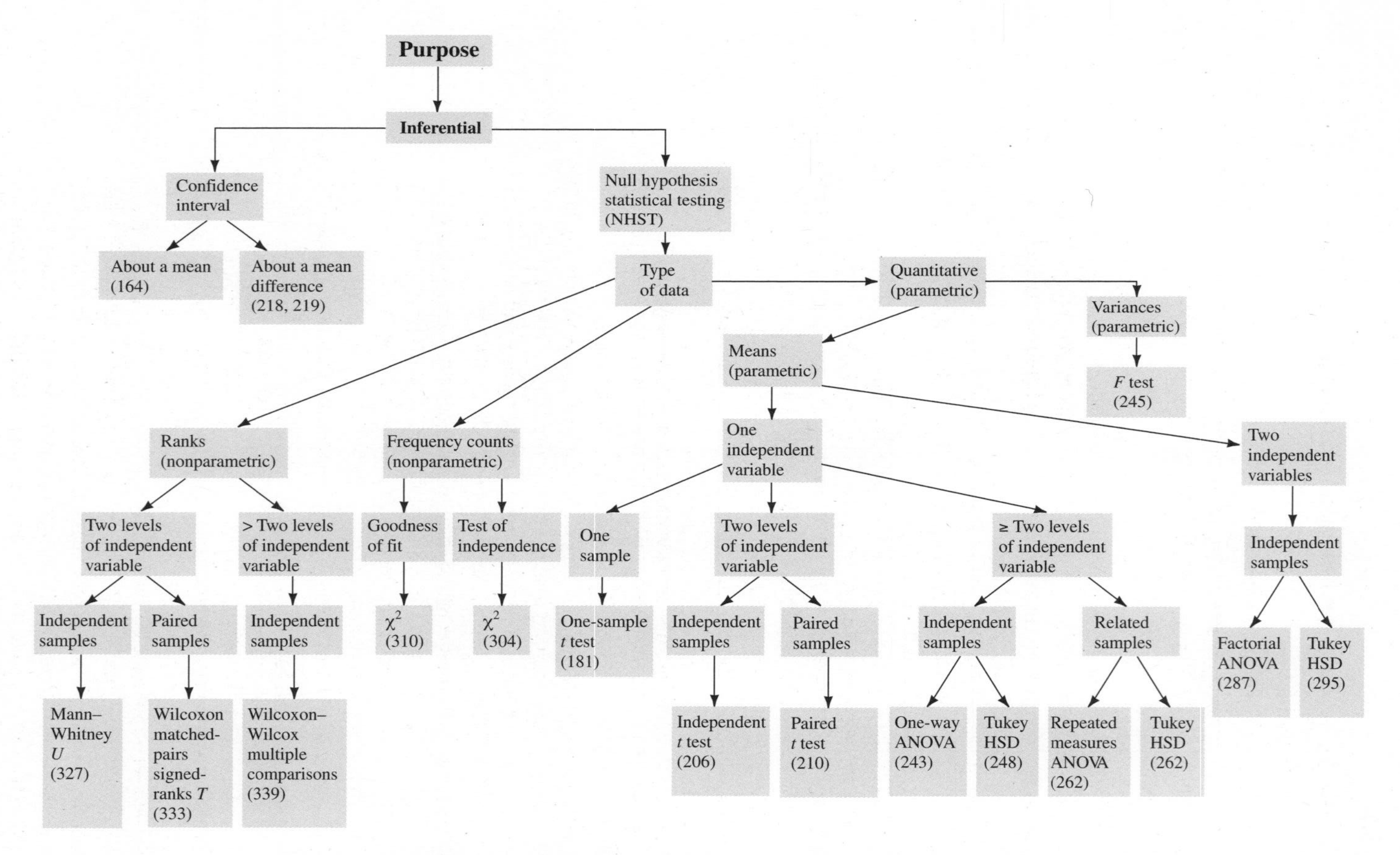

FIGURE 16.2 Decision tree for inferential statistics. (Text page numbers are in parentheses.)

PROBLEMS

Set A. Determine what descriptive statistic or inferential test is appropriate for each problem in this set.

16.1. A company has three separate divisions. Based on the capital invested, Division A made 10 cents per dollar, Division B made 20 cents per dollar, and Division C made 30 cents per dollar. How can the overall profit for the company be found?

16.2. Reaction time scores tend to be severely skewed. The majority of the responses are quick and there are diminishing numbers of scores in the slower categories. A student wanted to find out the effects of alcohol on reaction time, so he found the reaction time of each subject under both conditions—alcohol and no alcohol.

16.3. As part of a 2-week treatment for phobias, a therapist measured the general anxiety level of each client four times: before, after 1 week of treatment, at the end of treatment, and 4 months later.

16.4. "I want a number that will describe the typical score on this test. Most did very well, some scored in the middle, and a very few did quite poorly."

16.5. The city park benches are 16.7 inches high. The mean distance from the bottom of the shoe to the back of the knee is 18.1 inches for women. The standard deviation is 0.70 inch. What proportion of the women who use the bench will have their feet dangle (unless they sit forward on the bench)?

16.6. A measure of dopamine (a neurotransmitter) activity is available for 14 acute schizophrenic patients and 14 chronic schizophrenic patients. Do acute and chronic schizophrenics differ in dopamine activity?

16.7. A large school district needs to know the range of scores within which they can expect the mean reading achievement score of their sixth-grade students. A random sample of 50 reading achievement scores is available.

16.8. Based on several large studies, a researcher knew that 40 percent of the public in the United States favored capital punishment, 35 percent were against it, and 25 percent expressed no strong feelings. The researcher polled 150 Canadians on this question to compare the two countries in attitudes toward capital punishment.

16.9. What descriptive statistic should be used to identify the most typical choice of major at your college?

16.10. A music-minded researcher wanted to express with a correlation coefficient the strength of the relationship between stimulus intensity and pleasantness. She knew that both very low and very high intensities are not pleasant and that the middle range of intensity produces the highest pleasantness ratings. Would you recommend a Pearson r? An r_s?

16.11. A consumer testing group compared Boraxo and Tide to determine which got laundry whiter. White towels that had been subjected to a variety of filthy treatments were identified on each end and cut in half. Each was washed in either Boraxo or Tide. After washing, each half was tested with a photometer for the amount of light reflected.

16.12. An investigator wants to predict a male adult's height from his length at birth. He obtains records of both measures from a sample of male military personnel.

16.13. An experimenter is interested in the effect of expectations and drugs on alertness. Each participant is given an amphetamine (stimulant) or a placebo (an inert substance). In addition, half in each group are told that the drug taken was a stimulant and half that the drug was a depressant. An alertness score for each participant is obtained from a composite of measures that included a questionnaire and observation.

16.14. A professor told a class that she found a correlation coefficient of .20 between the alphabetical order of a person's last name and that person's cumulative point total for 50 students in a general psychology course. One student in the class wondered whether the correlation coefficient was reliable; that is, would a nonzero correlation be found in other classes? How can reliability be determined without calculating an r on another class?

16.15. As part of a test for some advertising copy, a company assembled 120 people who rated four toothpastes. They read several ads (including the ad being tested). Then they rated all four toothpastes again. The data to be analyzed consisted of the number of people who rated Crest highest before reading the ad and the number who rated it highest after reading the ad.

16.16. Most lightbulb manufacturing companies claim that their 60-watt bulbs have an average life of 1000 hours. A skeptic with some skill as an electrician wired up sockets and timers and lighted 40 bulbs. The life of each bulb recorded automatically. Is the companies' claim justified?

16.17. An investigator wondered whether a person's education is related to his or her satisfaction in life. Both of these variables can be measured using a quantitative scale.

16.18. To find out the effect of a drug on reaction time, an investigator administered 0, 25, 50, or 75 milligrams to four groups of volunteer rats. The quickness of their response to a tipping platform was measured in seconds.

16.19. The experimenters (a man and a woman) staged 140 "shoplifting" incidents in a grocery store. In each case, the "shoplifter" (one experimenter) picked up a carton of cigarettes in full view of a customer and then walked out. Half the time the experimenter was well dressed and half the time sloppily dressed. Half the incidents involved a male shoplifter and half involved a female. For each incident, the experimenters (with the cooperation of the checkout person) simply noted whether the shoplifter was reported or not.

16.20. A psychologist wanted to illustrate for a group of seventh-graders the fact that items in the middle of a series are more difficult to learn than those at either end. The students learned an eight-item series. The psychologist found the mean number of repetitions necessary to learn each of the items.

16.21. A nutritionist and an anthropologist gathered data on the incidence of cancer among 21 cultures. The incidence data were quite skewed, with many cultures having low numbers and the rest having various higher numbers. In eight of the cultures, red meat was a significant portion of the diet. For the other 13, diet consisted primarily of grains.

16.22. A military science teacher wanted to know if there was a relationship between the class rank of West Point graduates and their military rank in the service at age 45.

16.23. The boat dock operators at a lake sponsored a fishing contest. Prizes were to be awarded for the largest bass, crappie, and bream and to the overall winner. The problem in deciding the overall winner is that the bass are by nature larger than the other two species of fish. Describe an objective way to find the overall winner that will be fair to all three kinds of contestants.

16.24. *Yes* is the answer to the question, Is psychotherapy effective? Studies comparing a quantitative measure of groups who did or did not receive psychotherapy show that the therapy group improves significantly more than the nontherapy group. What statistical technique will answer the question, *How* effective is psychotherapy?

Set B. Read each problem, look at the statistical analysis, and write an appropriate conclusion using the terms in the problem; that is, rather than saying, "The null hypothesis was rejected at the .05 level," say, "Those with medium anxiety solved problems more quickly than those with high anxiety."

For some of the problems, the design of the study has flaws. There are uncontrolled extraneous variables that would make it impossible to draw precise conclusions about the experiment. It is good for you to be able to recognize such design flaws, but the task here is to interpret the statistics. Thus, don't let design flaws, keep you from drawing a conclusion based on the statistic.

For all problems, use a two-tailed test with $\alpha = .05$. If the results are significant at the .01 or .001 level, report that, but treat any difference that has a probability greater than .05 as not significant.

16.25. The interference theory of the serial position effect predicts that participants who are given extended practice will perform more poorly on the initial items of a series than will those not given extended practice. The mean numbers of errors on initial items are shown in the accompanying table. A t test with 54 df produced a value of 3.88.

	Extended practice	No extended practice
Mean number of errors	13.7	18.4

16.26. A company that offered instruction in taking college entrance examinations correctly claimed that its methods "produced improvement that was statistically significant." A parent with some skill as a statistician obtained the information the claim was based on (a one-sample t test) and calculated an effect size index. The effect size index was 0.10.

16.27. In a particular recycling process, the break-even point for each batch occurs when 54 kilograms of raw material remain as unusable foreign matter. An engineer developed a screening process that leaves a mean of 38 kilograms of foreign material. The upper and lower limits of a 95 percent confidence interval are 30 and 46 kilograms.

16.28. Four kinds of herbicides were compared for their weed-killing characteristics. Eighty plots were randomly but evenly divided, and one herbicide was applied to each. Because the effects of two of the herbicides were quite variable, the dependent variable was converted to ranks and a Wilcoxon–Wilcox multiple-comparisons test was used. The plot with the fewest weeds remaining was given a rank of 1. The summary table follows.

	A (936.5)	*B* (316.5)	*C* (1186)
B (316.5)	620		
C (1186)	249.5	869.5	
D (801)	135.5	482.5	385

16.29. A developmental psychologist advanced a theory that predicted the proportion of children who would, during a period of stress, cling to their mother, attack the mother, or attack a younger sibling. The stress situation was set up and the responses of 50 children recorded. The χ^2 value was 5.30.

16.30. An experimental psychologist at a Veterans Administration hospital obtained approval from the Institutional Review Board to conduct a study of the efficacy of Cymbalta for treating depression. The experiment lasted 60 days, during which one group of depressed patients was given placebos, one was given a low dose of Cymbalta, and one was given a high dose. At the end of the experiment, the patients were observed by two outside psychologists who rated several behaviors such as eye contact, posture, verbal output, and activity for degree of depression. The ratings went into a composite score for each patient, with high scores indicating depression. The means were: placebo, 18.86; low dose, 10.34; and high dose, 16.21. An ANOVA summary table and two Tukey HSD tests are shown in the accompanying table.

Source	*df*	*F*
Between treatments	2	21.60
Error	18	

HSD (placebo vs. low) = 13.14
HSD (placebo vs. high) = 0.65

16.31. One sample of 16 third-graders had been taught to read by the "look–say" method. A second sample of 18 had been taught by the phonics method. Both were given a reading achievement test, and a Mann–Whitney *U* test was performed on the results. The *U* value was 51.

16.32. Sociologists make a profile of the attitudes of a group by asking each individual to rate his or her tolerance of people in particular categories. For example, members might be asked to assess, on a scale of 1–7, their tolerance of prostitutes, atheists, former mental patients, intellectuals, college graduates, and so on. The mean score for each category is then calculated and the categories are ranked from low to high. When 18 categories of people were ranked by a group of people who owned businesses and also by a group of college students, an r_s of .753 between the rankings was found.

16.33. In a large high school, one group of juniors took an English course that included a 9-week unit on poetry. Another group of juniors studied plays during that 9-week period. Afterward, both groups completed a questionnaire on their attitudes toward poetry. High scores mean favorable attitudes. The means and variances are presented in the accompanying table. For the t test: t (78 df) = 0.48. To compare the variances, an F test was performed: F (37, 41) = 3.62. (See footnote 8 in Chapter 11.)

	Attitude toward poetry	
	Studied poetry	Studied plays
Mean	23.7	21.7
Variance	51.3	16.1

16.34. Here is an example of the results of experiments that asked the question, Can you get more attitude change from an audience by presenting just one side of an argument and claiming that it is correct, or should you present both sides and then claim that one side is correct? In this experiment, a second independent variable was also examined: amount of education. The first table contains the mean change in attitude, and the second is the ANOVA summary table.

	Presentation	
Amount of education	One side	Both sides
Less than 12 years	4.3	2.7
13+ years	2.1	4.5

Source	df	F
Between presentations	1	2.01
Between education	1	1.83
Presentations × education	1	7.93
Error	44	

16.35. A college student was interested in the relationship between handedness and verbal ability. She gave three classes of third-grade students a test of verbal ability that she had devised and then classified each child as right-handed, left-handed, or mixed. The obtained F value was 2.63.

	Handedness		
	Left	Right	Mixed
Mean verbal ability score	21.6	25.9	31.3
N	16	36	12

16.36. As part of a large-scale study on alcoholism, alcoholics and nonalcoholics were classified according to when they were toilet-trained as children. The three categories were early, normal, and late. A χ^2 of 7.90 was found.

ADDITIONAL HELP FOR CHAPTER 16

Visit *cengage.com/psychology/spatz*. At the Student Companion Site, you'll find multiple-choice tutorial quizzes and flashcards with definitions. For this chapter there is a Statistics Workshop on Choosing the Correct Statistical Test.

KEY TERMS

Meta-analysis (p. 352)

p_{rep} (p. 352)

appendixes

APPENDIX

Arithmetic and Algebra Review

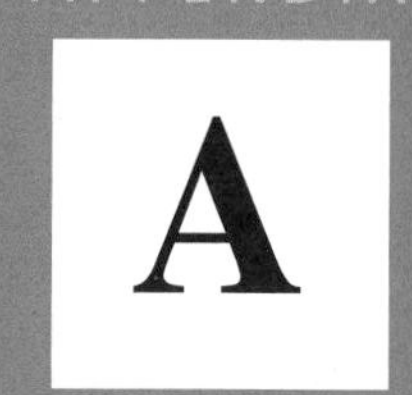

OBJECTIVES FOR APPENDIX A

After studying the text and working the problems, you should be able to:

1. Estimate answers
2. Round numbers
3. Find answers to problems with decimals, fractions, negative numbers, proportions, percents, absolute values, ± signs, exponents, square roots, complex expressions, and simple algebraic expressions

Statistical Symbols

As far as I know, there has never been a clinical case of neoiconophobia.[1] However, I know that some students show a mild form of this behavior. Symbols like $\bar{X}$, σ, μ, and Σ can cause a grimace, a frown, or a droopy eyelid. In severe cases, I suspect that the behavior involves avoiding a statistics course entirely. I'm sure that you don't have a severe case because you are still reading this textbook. Even so, if you are a typical beginning student in statistics, symbols like σ, μ, Σ, and maybe $\bar{X}$ are not very meaningful to you, and they may even elicit feelings of uneasiness. Soon, however, you will know what these symbols mean and be able to approach them with an unruffled psyche—and perhaps even with joy.

Some of the symbols stand for concepts you are already familiar with. For example, the capital letter X with a bar over it, $\bar{X}$, stands for the mean or average. ($\bar{X}$ is pronounced "mean" or sometimes "ex-bar.") You already know about means. You just add up the scores and divide the total by the number of scores. This verbal instruction can be put into symbols: $\bar{X} = \Sigma X/N$. The Greek uppercase sigma (Σ) is the instruction to add, and X is the symbol for scores. $\bar{X} = \Sigma X/N$ is something you already know about, even if you have not been using these symbols.

[1] An extreme and debilitating fear of new symbols.

Pay careful attention to symbols. They serve as shorthand notations for the ideas and concepts you are learning. Each time a new symbol is introduced, concentrate on it, learn it, memorize its definition and pronunciation. The more meaning a symbol has for you, the better you understand the concepts it represents and, of course, the easier the course will be.

Sometimes I need to distinguish between two different σ's or two X's. The convention is to use subscripts, and the results look like σ_1 and σ_2 or X_1 and X_2. In later chapters, subscripts other than numbers are used to identify a symbol. You will see both σ_X and $\sigma_{\bar{X}}$. The point here is that subscripts are for identification purposes only; they never indicate multiplication. Thus, $\sigma_{\bar{X}}$ does not mean $(\sigma)(\bar{X})$.

Two additional comments—to encourage and to caution you. I encourage you to do more in this course than just read the text, work the problems, and pass the tests, however exciting that may be. I hope you will occasionally get beyond this text and read journal articles or short portions of other statistics textbooks. I will provide recommendations in footnotes at appropriate places. The word of caution that goes with this encouragement is that reading statistics texts is like reading a Russian novel—the same characters have different names in different places. For example, the mean of a sample in some texts is symbolized M rather than $\bar{X}$. There is even more variety when it comes to symbolizing standard deviations. If you expect such differences, it will be less difficult for you to fit the new symbol into your established scheme of understanding.

Working Problems

This appendix covers the basic mathematical skills you need to work the problems in this course. This math is not complex. The highest level of mathematical sophistication is simple algebra.

Although the mathematical reasoning is not very complex, a good bit of arithmetic is required. To be good in statistics, you need to be good at arithmetic. A wrong answer is wrong whether the error is arithmetical or logical. For the beginning statistics student, the best way to prevent arithmetic errors is to use a calculator or a computer program.

Of course, even calculators can do more than simple arithmetic. Answers to many of the computational problems in this book can be found with calculators that produce means, standard deviations, correlation coefficients, and other statistics. By using a computer program, you can find numerical answers to all the computational problems. The question to ask yourself is: How can I use these aids so that I not only get the right answer but also *understand* what is going on and can tell the story that goes with my arithmetic? Your instructor will have some ideas about how to accomplish both goals.

Here is my advice. In the beginning, use a calculator to do the arithmetic for *each* part of a problem. Write down each step so you can see the progression from formula to final answer. When you know the steps and can explain them, use the special function keys (standard deviation, correlation, and so forth) to produce final answers directly from raw scores. Use computer programs when you understand a technique fairly well.

Most calculators produce values that are intermediate between raw scores and a final answer. For example, suppose a problem had three scores: 1, 2, and 3. For many techniques, you need to know the sum of the numbers (6) and the sum when each number is squared (14, which is 1 + 4 + 9). A calculator with a $\Sigma+$ key or a *stat* function gives you both the sum of the numbers and the sum of the squares—with just one entry of each of the three numbers. This feature saves time and reduces errors.

This appendix is designed to refresh your memory of arithmetic and algebra. It is divided into two parts, a pretest and a review of fundamentals. The pretest gives you problems that are similar to those you have to work later in the course. Take the pretest and then check your answers against the answers in Appendix G. If you find that you made any mistakes, work through those sections of the review that explain the problems you missed.

As stated earlier, to be good at statistics, you must be good at arithmetic. To be good at arithmetic, you must know the rules and be careful in your computations. The rules follow the pretest; it is up to you to be careful in your computations.

Pretest

Estimating answers (Estimate whole-number answers to the problems.)

A.1. $(4.02)^2$ **A.2.** 1.935×7.89 **A.3.** $31.219 \div 2.0593$

Rounding numbers (Round numbers to the nearest tenth.)

A.4. 6.06 **A.5.** 0.35 **A.6.** 10.348

Decimals

Add:

A.7. $3.12 + 6.3 + 2.004$ **A.8.** $12 + 8.625 + 2.316 + 4.2$

Subtract:

A.9. $28.76 - 8.91$ **A.10.** $3.2 - 1.135$

Multiply:

A.11. 6.2×8.06 **A.12.** 0.35×0.162

Divide:

A.13. $64.1 \div 21.25$ **A.14.** $0.065 \div 0.0038$

Fractions

Add:

A.15. $\frac{1}{8} + \frac{3}{4} + \frac{1}{2}$ **A.16.** $\frac{1}{3} + \frac{1}{2}$

Subtract:

A.17. $\frac{11}{16} - \frac{1}{2}$ **A.18.** $\frac{5}{9} - \frac{1}{2}$

Multiply:

A.19. $\frac{4}{5} \times \frac{3}{4}$ **A.20.** $\frac{2}{3} \times \frac{1}{4} \times \frac{3}{5}$

Divide:

A.21. $\frac{3}{8} \div \frac{1}{4}$ **A.22.** $\frac{7}{9} \div \frac{1}{4}$

Negative numbers

Add:

A.23. $(-5) + 16 + (-1) + (-4)$ **A.24.** $(-11) + (-2) + (-12) + 3$

Subtract:

A.25. $(-10) - (-3)$ **A.26.** $(-8) - (-2)$

Multiply:

A.27. $(-5) \times (-5)$ **A.28.** $(-8) \times (3)$

Divide:

A.29. $(-10) \div (-3)$ **A.30.** $(-21) \div 4$

Percents and proportions

A.31. 12 is what percent of 36?

A.32. Find 27 percent of 84.

A.33. What proportion of 112 is 21?

A.34. A proportion of .40 of the tagged birds were recovered. 150 were tagged in all. How many were recovered?

Absolute value

A.35. $|-5|$ **A.36.** $|8-12|$

± Problems

A.37. $8 \pm 2 =$ **A.38.** $13 \pm 9 =$

Exponents

A.39. 4^2 **A.40.** 2.5^2 **A.41.** 0.35^2

Square roots

A.42. $\sqrt{9.30}$ **A.43.** $\sqrt{0.93}$ **A.44.** $\sqrt{0.093}$

Complex problems (Round answers to two decimal places.)

A.45. $\dfrac{3 + 4 + 7 + 2 + 5}{5}$ **A.46.** $\dfrac{(5 - 3)^2 + (3 - 3)^2 + (1 - 3)^2}{3 - 1}$

A.47. $\dfrac{(8 - 6.5)^2 + (5 - 6.5)^2}{2 - 1}$ **A.48.** $\left(\dfrac{8 + 4}{8 + 4 - 2}\right)\left(\dfrac{1}{8} + \dfrac{1}{4}\right)$

A.49. $\left(\dfrac{3.6}{1.2}\right)^2 + \left(\dfrac{6.0}{2.4}\right)^2$ **A.50.** $\dfrac{(3)(4) + (6)(8) + (5)(6) + (2)(3)}{(4)(4 - 1)}$

A.51. $\dfrac{190 - (25^2/5)}{5 - 1}$ **A.52.** $\dfrac{12(50 - 20)^2}{(8)(10)(12)(11)}$

A.53. $\dfrac{6[(3 + 5)(4 - 1) + 2]}{5^2 - (6)(2)}$

A.54. For the numbers 1, 2, 3, and 4: Find the sum and the sum of the squared numbers.

A.55. For the numbers 2, 4, and 6: Find the sum and the sum of the squared numbers.

Simple algebra (Solve for x.)

A.56. $\dfrac{x - 3}{4} = 2.5$ **A.57.** $\dfrac{14 - 8.5}{x} = 0.5$

A.58. $\dfrac{20 - 6}{2} = 4x - 3$ **A.59.** $\dfrac{6 - 2}{3} = \dfrac{x + 9}{5}$

Review of Fundamentals

This section gives you a quick review of the pencil-and-paper rules of arithmetic and simple algebra. I assume that you once knew all these rules but that refresher exercises will be helpful. Thus, there isn't much explanation. To obtain basic explanations, ask a teacher in your school's mathematics department to recommend one of the many "refresher" books that are available.

Definitions

Sum. The answer to an addition problem is called a sum. In Chapter 11 you calculate a *sum of squares,* a quantity that is obtained by adding together squared numbers.

Difference. The answer to a subtraction problem is called a difference. Much of what you learn in statistics deals with differences and how to explain them.

Product. The answer to a multiplication problem is called a product. Chapter 6 is about the *product-moment correlation coefficient,* which requires multiplication. Multiplication problems are indicated either by a × or by parentheses. Thus, 6 × 4 and (6)(4) call for the same operation.

Quotient. The answer to a division problem is called a quotient. The three ways to indicate a division problem are ÷, / (slash mark), and — (division line). Thus, 9 ÷ 4, 9/4, and $\frac{9}{4}$ call for the same operation. It is a good idea to think of any common fraction as a division problem. The numerator is to be divided by the denominator.

Estimating Answers

It is a **very good idea** to just look at a problem and estimate the answer before you do any calculating. This is referred to as "eyeballing the data," and Edward Minium captured its importance with Minium's First Law of Statistics: "The eyeball is the statistician's most powerful instrument." (See Minium and King, 2002.)

very good idea
Examine a statistics problem until you understand it well enough to estimate the answer.

Estimating keeps you from making gross errors such as misplacing a decimal point. For example, $\frac{31.5}{5}$ can be estimated as a little more than 6. If you estimate before you divide, you are likely to recognize that an answer of 63 or 0.63 is wrong.

The estimated answer to the problem (21)(108) is 2000 because (20)(100) = 2000. The problem (0.47)(0.20) suggests an estimated answer of 0.10 because $\frac{1}{2}(0.20) = 0.10$. With 0.10 in mind, you are not likely to write 0.94 for the answer (which is 0.094). Estimating answers is also important if you are finding a square root. You can estimate that $\sqrt{95}$ is about 10 because $\sqrt{100} = 10$; $\sqrt{1.034}$ is about 1; $\sqrt{0.013}$ is about 0.1.

To calculate a mean, eyeball the numbers and estimate the mean. If you estimate a mean of 30 for a group of numbers that are primarily in the 20s, 30s, and 40s, a calculated mean of 60 will arouse your suspicion that you have made an error.

Rounding Numbers

Find the number that is to be rounded up or not rounded up. If the number to its right is 5 or greater, increase the number by 1. If the number to its right is less than 5, do not change the number. These rules are built into most calculators. Here are some illustrations of these rules.

Rounding to the nearest whole number:

6.2 = 6	4.5 = 5
12.7 = 13	163.5 = 164
6.49 = 6	9.5 = 10

Rounding to the nearest hundredth:

13.614 = 13.61	12.065 = 12.07
0.049 = 0.05	4.005 = 4.01
1.097 = 1.10	0.675 = 0.68
3.6248 = 3.62	1.995 = 2.00

A reasonable question is: How many decimal places should an answer in statistics have? A good rule of thumb in statistics is to carry all operations to three decimal places and then, for the final answer, round back to two decimal places.

Sometimes a rule of thumb can get you into trouble, though. For example, if halfway through a division problem of 0.0016 ÷ 0.0074 you dutifully round those four decimals to three (0.002 ÷ 0.007), you get an answer of 0.2857, which becomes 0.29. However, division without rounding gives you an answer of 0.2162 or 0.22. The difference between 0.22 and 0.29 may be substantial. I often give you cues if more than two decimal places are necessary, but you should always be alert to the problems of rounding.

Most calculators carry more decimal places in memory than they show in the display. If you have such a calculator, it will protect you from the problem of rounding too much or too soon.

PROBLEMS

A.60. Define (a) sum, (b) quotient, (c) product, and (d) difference.

A.61. Estimate answers to the expressions.

a. $\sqrt{103.48}$	**b.** 74.16 ÷ 9.87	**c.** $(11.4)^2$	**d.** $\sqrt{0.0459}$
e. 0.41^2	**f.** 11.92 × 4.60	**g.** $\sqrt{0.888}$	**h.** $\sqrt{0.0098}$

A.62. Round the numbers to the nearest whole number.

a. 13.9	**b.** 126.4	**c.** 9.0
d. 0.4	**e.** 127.5	**f.** 12.51
g. 12.49	**h.** 12.50	**i.** 9.46

A.63. Round the numbers to the nearest hundredth.

a. 6.3348	**b.** 12.997	**c.** 0.050
d. 0.965	**e.** 2.605	**f.** 0.3445
g. 0.003	**h.** 0.015	**i.** 0.9949

Decimals

1. *Addition and subtraction of decimals.* There is only one rule about the addition and subtraction of numbers that have decimals: *Keep the decimal points in a vertical*

line. The decimal point in the answer goes directly below those in the problem. This rule is illustrated in the five problems here.

Add			***Subtract***	
	0.004	6.0		
1.26	1.310	18.0	14.032	16.00
10.00	4.039	0.5	8.26	4.32
11.26	5.353	24.5	5.772	11.68

2. *Multiplication of decimals.* The basic rule for multiplying decimals is that the number of decimal places in the answer is found by adding the numbers of decimal places in the two numbers that are being multiplied. To place the decimal point in the product, count from the right.

$$\begin{array}{r} 1.3 \\ \times\ 4.2 \\ \hline 26 \\ 52 \\ \hline 5.46 \end{array} \qquad \begin{array}{r} 0.21 \\ \times 0.4 \\ \hline 0.084 \end{array} \qquad \begin{array}{r} 1.47 \\ \times\ 3.12 \\ \hline 294 \\ 147 \\ 441 \\ \hline 4.5864 \end{array}$$

3. *Division of decimals.* Two methods have been used to teach division of decimals. The older method required the student to move the decimal in the divisor (the number you are dividing by) enough places to the right to make the divisor a whole number. The decimal in the dividend is then moved to the right the same number of places, and division is carried out in the usual way. The new decimal places are identified with carets (∧), and the decimal place in the quotient is just above the caret in the dividend.

$$\begin{array}{r} 20. \\ 0.016_\wedge\overline{)0.320_\wedge} \end{array}$$

Decimal moved three places in both the divisor and the dividend

$$\begin{array}{r} 38.46 \\ 0.39_\wedge\overline{)15.00_\wedge} \end{array}$$

Decimal moved two places in both the divisor and the dividend

$$\begin{array}{r} 2.072 \\ 6\overline{)12.432} \end{array}$$

Divisor is already a whole number

$$\begin{array}{r} .004 \\ 9.1_\wedge\overline{)0.0_\wedge369} \end{array}$$

Decimal moved one place in both the divisor and the dividend

The newer method of teaching the division of decimals is to multiply both the divisor and the dividend by the number that will make both of them whole numbers. (Actually, this is the way the caret method works also.)

$$\frac{0.32}{0.016} \times \frac{1000}{1000} = \frac{320}{16} = 20.00$$

$$\frac{0.0369}{9.1} \times \frac{10{,}000}{10{,}000} = \frac{369}{91{,}000} = 0.004$$

$$\frac{15}{0.39} \times \frac{100}{100} = \frac{1500}{39} = 38.46$$

$$\frac{12.432}{6} \times \frac{1000}{1000} = \frac{12{,}432}{6000} = 2.072$$

PROBLEMS

Perform the operations indicated.

A.64. $0.001 + 10 + 3.652 + 2.5$

A.65. $14.2 - 7.31$

A.66. 0.04×1.26

A.67. $143.3 + 16.92 + 2.307 + 8.1$

A.68. $3.06 \div 0.04$

A.69. $\dfrac{24}{11.75}$

A.70. $152.12 - 127.4$

A.71. $(0.5)(0.07)$

Fractions

In general, there are two ways to deal with fractions.

1. Convert each fraction to a decimal.
2. Work directly with the fractions, using a set of rules for each operation. The rule for addition and subtraction is: Convert the fractions to ones with common denominators, add or subtract the numerators, and place the result over the common denominator. The rule for multiplication is: Multiply the numerators together to get the numerator of the answer, and multiply the denominators together for the denominator of the answer. The rule for division is: Invert the divisor and multiply the fractions.

For statistics problems, it is usually easier to convert the fractions to decimals and then work with the decimals, and this is the method I use. However, if you are a whiz at working directly with fractions, by all means continue with your method. To convert a fraction to a decimal, divide the lower number into the upper one. Thus, $\frac{3}{4} = 0.75$ and $\frac{13}{17} = 0.765$.

Addition of fractions

$$\frac{1}{2} + \frac{1}{4} = 0.50 + 0.25 = 0.75$$

$$\frac{13}{17} + \frac{21}{37} = 0.765 + 0.568 = 1.33$$

$$\frac{2}{3} + \frac{3}{4} = 0.667 + 0.75 = 1.42$$

Subtraction of fractions

$$\frac{1}{2} - \frac{1}{4} = 0.50 - 0.25 = 0.25$$

$$\frac{11}{12} - \frac{2}{3} = 0.917 - 0.667 = 0.25$$

$$\frac{41}{53} - \frac{17}{61} = 0.774 - 0.279 = 0.50$$

Multiplication of fractions

$$\frac{1}{2} \times \frac{3}{4} = (0.5)(0.75) = 0.38$$

$$\frac{10}{19} \times \frac{61}{90} = 0.526 \times 0.678 = 0.36$$

$$\frac{1}{11} \times \frac{2}{3} = (0.09)(0.67) = 0.06$$

Division of fractions

$$\frac{9}{21} \div \frac{13}{19} = 0.429 \div 0.684 = 0.63$$

$$14 \div \frac{1}{3} = 14 \div 0.33 = 42$$

$$\frac{7}{8} \div \frac{3}{4} = 0.875 \div 0.75 = 1.17$$

PROBLEMS

Perform the operations indicated.

A.72. $\frac{9}{10} + \frac{1}{2} + \frac{2}{5}$

A.73. $\frac{9}{20} \div \frac{19}{20}$

A.74. $\left(\frac{1}{3}\right)\left(\frac{5}{6}\right)$

A.75. $\frac{4}{5} - \frac{1}{6}$

A.76. $\frac{1}{3} \div \frac{5}{6}$

A.77. $\frac{3}{4} \times \frac{5}{6}$

A.78. $18 \div \frac{1}{3}$

Negative Numbers

1. *Addition of negative numbers.* (Any number without a sign is understood to be positive.)

   a. To add a series of negative numbers, add the numbers in the usual way and attach a negative sign to the total.

$$\begin{array}{r} -3 \\ -8 \\ -12 \\ -5 \\ \hline -28 \end{array} \qquad (-1) + (-6) + (-3) = -10$$

   b. To add two numbers, one positive and one negative, subtract the smaller number from the larger number and attach the sign of the larger number to the result.

$$\begin{array}{r} 140 \\ -55 \\ \hline 85 \end{array} \qquad \begin{array}{r} -14 \\ 8 \\ \hline -6 \end{array} \qquad \begin{array}{l} (28) + (-9) = 19 \\ \\ 74 + (-96) = -22 \end{array}$$

   c. To add a series of numbers, some positive and some negative, add all the positive numbers together, add all the negative numbers together (see 1a), and then combine the two sums (see 1b).

$$(-4) + (-6) + (12) + (-5) + (2) + (-9) = 14 + (-24) = -10$$
$$(-7) + (10) + (4) + (-5) = 14 + (-12) = 2$$

2. *Subtraction of negative numbers.* To subtract a negative number, change it to positive and add it.

$$\begin{array}{r} (-14) \\ -(-2) \\ \hline -12 \end{array} \qquad \begin{array}{r} 5 \\ -(-7) \\ \hline 12 \end{array} \qquad \begin{array}{r} 18 \\ -(-3) \\ \hline 21 \end{array} \qquad \begin{array}{r} (-7) \\ -(-5) \\ \hline -2 \end{array}$$

3. *Multiplication of negative numbers.* When the two numbers to be multiplied are both negative, the product is positive.

   $(-3)(-3) = 9 \qquad (-6)(-8) = 48$

   When one of the numbers is negative and the other is positive, the product is negative.

   $(-8)(3) = -24 \qquad 14 \times -2 = -28$

4. *Division of negative numbers.* The rule in division is the same as the rule in multiplication. If both the numbers are negative, the quotient is positive.

   $(-10) \div (-2) = 5 \qquad (-4) \div (-20) = 0.20$

   If one number is negative and the other positive, the quotient is negative.

   $(-10) \div 2 = -5 \qquad 6 \div (-18) = -0.33$

   $14 \div (-7) = -2 \qquad (-12) \div 3 = -4$

PROBLEMS

A.79. Add the numbers.
- **a.** −3, 19, −14, 5, −11
- **b.** −8, −12, −3
- **c.** −8, 11
- **d.** 3, −6, −2, 5, −7

A.80. $(-8)(5)$ **A.81.** $(-4)(-6)$ **A.82.** $(4)(12)$

A.83. $(11)(-3)$ **A.84.** $(-18) - (-9)$ **A.85.** $14 \div (-6)$

A.86. $12 - (-3)$ **A.87.** $(-6) - (-7)$ **A.88.** $(-9) \div (-3)$

A.89. $(-10) \div 5$ **A.90.** $4 \div (-12)$ **A.91.** $(-7) - 5$

Proportions and Percents

A **proportion** is a part of a whole and can be expressed as a fraction or as a decimal. Usually, proportions are expressed as decimals. If eight students in a class of 44 received A's, 8 is a proportion of the whole (44). Thus, $\frac{8}{44}$, or 0.18, is the proportion of the class that received A's.

proportion
A part of a whole.

To convert a proportion to a percent (per one hundred), multiply by 100. Thus, $0.18 \times 100 = 18$; 18 percent of the students received A's. As you can see, proportions and percents are two ways to express the same idea.

If you know a proportion (or percent) and the size of the original whole, you can find the number that the proportion represents. If 0.28 of the students were absent due to illness and there are 50 students in all, then 0.28 of the 50 were absent: $(0.28)(50) =$ 14 students who were absent. Here are some more problems. Cover the answers and work all four problems.

1. 26 out of 31 completed the course. What proportion completed the course?
2. What percent completed the course?

3. What percent of 19 is 5?
4. If 90 percent of the population agreed and the population consisted of 210 members, how many members agreed?

Answers:

1. $\frac{26}{31} = 0.83$
2. $0.83 \times 100 = 83$ percent
3. $\frac{5}{19} = 0.26 \quad 0.26 \times 100 = 26$ percent
4. $0.90 \times 210 = 189$ members

Absolute Value

The **absolute value** of a number ignores the sign of the number. Thus, the absolute value of -6 is 6. This is expressed with symbols as $|-6| = 6$. It is expressed verbally as "the absolute value of negative six is six." In a similar way, the absolute value of $4 - 7$ is 3; that is, $|4 - 7| = |-3| = 3$.

absolute value
A number without consideration of its algebraic sign.

± Signs

A ± sign ("plus or minus" sign) means to both add *and* subtract. A ± problem *always* has two answers.

$$10 \pm 4 = 6, 14$$

$$8 \pm (3)(2) = 8 \pm 6 = 2, 14$$

$$\pm (4)(3) + 21 = \pm 12 + 21 = 9, 33$$

$$\pm 4 - 6 = -10, -2$$

Exponents

In the expression 5^2 ("five squared"), 2 is the exponent. The 2 means that 5 is to be multiplied by itself. Thus, $5^2 = 5 \times 5 = 25$.

In elementary statistics, the only exponent used is 2, but it will be used frequently. When a number has an exponent of 2, the number is said to be squared. The expression 4^2 ("four squared") means 4×4, and the product is 16.

$$8^2 = 8 \times 8 = 64 \qquad (0.75)^2 = (0.75)(0.75) = 0.5625$$

$$1.2^2 = (1.2)(1.2) = 1.44 \qquad 12^2 = 12 \times 12 = 144$$

Square Roots

Statistics problems often require you to find the square root of a number. Use a calculator with a square root key.

PROBLEMS

A.92. 6 is what proportion of 13?

A.93. 6 is what percent of 13?

A.94. What proportion of 25 is 18?

A.95. What percent of 115 is 85?

A.96. Find 72 percent of 36.

A.97. 22 is what proportion of 72?

A.98. If .40 of the 320 members voted against the proposal, how many voted no?

A.99. 22 percent of the 850 coupons were redeemed. How many were redeemed?

A.100. $|-31|$

A.101. $|21 - 25|$

A.102. $12 \pm (2)(5)$

A.103. $\pm(5)(6) + 10$

A.104. $\pm(2)(2) - 6$

A.105. $(2.5)^2$

A.106. 9^2

A.107. $(0.3)^2$

A.108. $\left(\frac{1}{4}\right)^2$

A.109. Find the square root of the numbers to two decimal places (or more if appropriate).

a. 625 **b.** 6.25 **c.** 0.625
d. 0.0625 **e.** 16.85 **f.** 0.003
g. 181,476 **h.** 0.25 **i.** 22.51

Complex Expressions

Two rules are used for the complex expressions encountered in statistics.

1. Perform the operations within the parentheses. If there are brackets in the expression, perform the operations within the parentheses and then the operations within the brackets.
2. Perform the operations in the numerator separately from those in the denominator and, finally, carry out the division.

$$\frac{(8-6)^2 + (7-6)^2 + (3-6)^2}{3-1} = \frac{2^2 + 1^2 + (-3)^2}{2}$$

$$= \frac{4+1+9}{2} = \frac{14}{2} = 7.00$$

$$\left(\frac{10+12}{4+3-2}\right)\left(\frac{1}{4}+\frac{1}{3}\right) = \left(\frac{22}{5}\right)(0.25 + 0.333)$$

$$= (4.40)(0.583) = 2.57$$

$$\left(\frac{8.2}{4.1}\right)^2 + \left(\frac{4.2}{1.2}\right)^2 = (2)^2 + (3.5)^2 = 4 + 12.25 = 16.25$$

$$\frac{6[(13-10)^2 - 5]}{6(6-1)} = \frac{6(3^2 - 5)}{6(5)}$$

$$= \frac{6(9-5)}{30} = \frac{6(4)}{30} = \frac{24}{30} = 0.80$$

$$\frac{18 - \frac{6^2}{4}}{4 - 1} = \frac{18 - \frac{36}{4}}{3} = \frac{18 - 9}{3} = \frac{9}{3} = 3.00$$

Simple Algebra

To solve a simple algebra problem, isolate the unknown (x) on one side of the equal sign and combine the numbers on the other side. To do this, remember that the same number can be *added to* or *subtracted from* both sides of the equation without affecting the value of the unknown.

$$\begin{aligned} x - 5 &= 12 \\ x - 5 + 5 &= 12 + 5 \\ x &= 17 \end{aligned} \qquad \begin{aligned} x + 7 &= 9 \\ x + 7 - 7 &= 9 - 7 \\ x &= 2 \end{aligned}$$

In a similar way, you can *multiply* or *divide* both sides of the equation by the same number without affecting the value of the unknown.

$$\begin{aligned} \frac{x}{6} &= 9 \\ (6)\left(\frac{x}{6}\right) &= (6)(9) \\ x &= 54 \end{aligned} \qquad \begin{aligned} 11x &= 30 \\ \frac{11x}{11} &= \frac{30}{11} \\ x &= 2.73 \end{aligned} \qquad \begin{aligned} \frac{3}{x} &= 14 \\ (x)\left(\frac{3}{x}\right) &= (14)(x) \\ 3 &= 14x \\ \frac{3}{14} &= \frac{14x}{14} \\ 0.21 &= x \end{aligned}$$

I combine some of these steps in the problems that follow. Be sure you see what operation is being performed on both sides in each step.

$$\begin{aligned} \frac{x - 2.5}{1.3} &= 1.96 \\ x - 2.5 &= (1.3)(1.96) \\ x &= 2.548 + 2.5 \\ x &= 5.048 \end{aligned} \qquad \begin{aligned} \frac{21.6 - 15}{x} &= 0.04 \\ 6.6 &= 0.04x \\ \frac{6.6}{0.04} &= x \\ 165 &= x \end{aligned}$$

$$\begin{aligned} 4x - 9 &= 5^2 \\ 4x &= 25 + 9 \\ x &= \frac{34}{4} \\ x &= 8.50 \end{aligned} \qquad \begin{aligned} \frac{14 - x}{6} &= 1.9 \\ 14 - x &= (1.9)(6) \\ 14 &= 11.4 + x \\ 2.6 &= x \end{aligned}$$

PROBLEMS

Reduce these complex expressions to a single number rounded to two decimal places.

A.110. $\dfrac{(4-2)^2+(0-2)^2}{6}$

A.111. $\dfrac{(12-8)^2+(8-8)^2+(5-8)^2+(7-8)^2}{4-1}$

A.112. $\left(\dfrac{5+6}{3+2-2}\right)\left(\dfrac{1}{3}+\dfrac{1}{2}\right)$

A.113. $\left(\dfrac{13+18}{6+8-2}\right)\left(\dfrac{1}{6}+\dfrac{1}{8}\right)$

A.114. $\dfrac{8[(6-2)^2-5]}{(3)(2)(4)}$

A.115. $\dfrac{[(8-2)(5-1)]^2}{5(10-7)}$

A.116. $\dfrac{6}{1/2}+\dfrac{8}{1/3}$

A.117. $\left(\dfrac{9}{2/3}\right)^2+\left(\dfrac{8}{3/4}\right)^2$

A.118. $\dfrac{10-(6^2/9)}{8}$

A.119. $\dfrac{104-(12^2/6)}{5}$

Find the value of x.

A.120. $\dfrac{x-4}{2}=2.58$

A.121. $\dfrac{x-21}{6.1}=1.04$

A.122. $x=\dfrac{14-11}{2.5}$

A.123. $x=\dfrac{36-41}{8.2}$

APPENDIX

B

Grouped Frequency Distributions and Central Tendency

OBJECTIVES FOR APPENDIX B

After studying the text and working the problems, you should be able to:

1. Use four conventions for constructing grouped frequency distributions
2. Arrange raw data into a grouped frequency distribution
3. Find the mean, median, and mode of a grouped frequency distribution

(In writing this appendix, I assumed that you have studied Chapter 2, "Exploring Data: Frequency Distributions and Graphs," and Chapter 3, "Exploring Data: Central Tendency.")

As you know from Chapter 2, converting a batch of raw scores into a simple frequency distribution brings order out of apparent chaos. For some distributions, even more order can be obtained if the raw scores are arranged into a **grouped frequency distribution.** The order becomes even more apparent when grouped frequency distributions are graphed. In addition to grouping and graphing, this appendix covers the calculation of the mean, median, and mode of grouped frequency distributions.

Grouped frequency distributions are used when the range of scores is too large for a simple frequency distribution. How large is too large? A rule of thumb is that grouped frequency distributions are appropriate when the range of scores is greater than 20. At times, however, ignoring this rule of thumb produces an improved analysis.

Grouped Frequency Distributions

As you may recall from Chapter 2, the only difference between simple frequency distributions and grouped frequency distributions is that grouped frequency distributions have class intervals in the place of scores. Each class interval in a grouped frequency distribution covers the same number of scores. The number of scores in the interval is symbolized i (interval size).

Establishing Class Intervals

There are no hard-and-fast rules for establishing class intervals. The ones that follow are used by many researchers, but some computer programs do not follow them.

1. *The number of class intervals.* The number of class intervals should be 10 to 20. On the one hand, with fewer than 10 intervals, the extreme scores in the data are not as apparent because they are clustered with more frequently occurring scores. On the other hand, more than 20 class intervals often make it difficult to see the shape of the distribution.
2. *The size of i.* If i is odd, the midpoint of the class interval will be a whole number, and whole numbers look better on graphs than decimal numbers. Three and five often work well as interval sizes. You may find that $i = 2$ is needed if you are to have 10 to 20 class intervals. If an i of 5 produces more than 20 class intervals, data groupers usually jump to an i of 10 or some multiple of 10. An interval size of 25 is popular.
3. *The lower limit of a class interval.* Begin each class interval with a multiple of i. For example, if the lowest score is 5 and $i = 3$ (as happened with the Satisfaction With Life Scale (SWLS) scores in Table 2.4), the first class interval should be 3–5. An exception to this convention occurs when $i = 5$. When the interval size is 5, it is usually better to use a multiple of 5 as the midpoint because multiples of 5 are easier to read on graphs.
4. *The order of the intervals.* The largest scores go at the top of the table. (This is a convention not followed by some computer programs.)

Converting Unorganized Scores into a Grouped Frequency Distribution

With the conventions for establishing class intervals in mind, here are the steps for converting unorganized data into a grouped frequency distribution. As an example, I will use the raw data in Table 2.1 and describe converting it into Table 2.4.

1. Find the highest and lowest scores. In Table 2.1, the highest score is 35 and the lowest score is 5.
2. Find the range of the scores by subtracting the lowest score from the highest score ($35 - 5 = 30$).
3. Determine i by a trial-and-error procedure. Remember that there are to be 10 to 20 class intervals and that the interval size should be convenient (3, 5, 10, or a multiple of 10). Dividing the range by a potential i value gives the approximate number of class intervals. Dividing the range, 30, by 3 gives 10, which is a recommended number of class intervals.
4. Establish the lowest interval. Begin the interval with a multiple of i, *which may or may not be an actual raw score.* End the interval so that it contains

i scores (but not necessarily i frequencies). For Table 2.4, the lowest interval is 3–5. (Note that 3 is not an actual score but is a multiple of i.) Each interval above the lowest one begins with a multiple of i. Continue building the class intervals.

5. With the class intervals written, underline each score (Table 2.1) and put a tally mark beside its class interval (Table 2.4).
6. As a check on your work, add up the frequency column. The sum should be N, the number of scores in the unorganized data.

PROBLEMS

***B.1.** A sociology professor was deciding what statistics to present in her introduction to sociology classes. She developed a test that covered concepts such as the median, graphs, standard deviation, and correlation. She tested one class of 50 students, and on the basis of the results, planned a course syllabus for that class and the other six intro sections. Arrange the data into an appropriate rough-draft frequency distribution.

20	56	48	13	30	39	25	41	52	44
27	36	54	46	59	42	17	63	50	24
31	19	38	10	43	31	34	32	15	47
40	36	5	31	53	24	31	41	49	21
26	35	28	37	25	33	27	38	34	22

***B.2.** The measurements that follow are weights in pounds of a sample of college men in one study. Arrange them into a grouped frequency distribution. If these data are skewed, tell the direction of the skew.

164	158	156	148	180	176	171	150	152	155	
161	168	148	175	154	155	149	149	151	160	
157	158	161	167	152	168	151	157	150	154	189

Central Tendency of Grouped Frequency Distributions

Mean

Finding the mean of a grouped frequency distribution involves the same arithmetic as that for a simple frequency distribution. Setting up the problem, however, requires one additional step. Look at **Table B.1,** which has four columns (compared to the three in Table 3.3). For a grouped frequency distribution, the *midpoint* of the interval represents all the scores in the interval. Thus, multiplying the midpoint by its f value includes all the scores in that interval. As you can see at the bottom of Table B.1, summing the fX column gives ΣfX, which, when divided by N, yields the mean.

TABLE B.1 A grouped frequency distribution of Satisfaction With Life Scale scores with $i = 3$

SWLS scores (class interval)	Midpoint (X)	f	fX
33–35	34	5	170
30–32	31	11	341
27–29	28	23	644
24–26	25	24	600
21–23	22	14	308
18–20	19	8	152
15–17	16	5	80
12–14	13	3	39
9–11	10	5	50
6–8	7	0	0
3–5	4	2	8
		$N = 100$	2392

In terms of a formula,

$$\mu \text{ or } \overline{X} = \frac{\Sigma fX}{N}$$

For Table B.1,

$$\mu \text{ or } \overline{X} = \frac{\Sigma fX}{N} = \frac{2392}{100} = 23.92$$

Note that the mean of the grouped data is 23.92 but the mean of the simple frequency distribution is 24.00. The mean of grouped scores is often different, but seldom is this difference of any consequence.

Median

Finding the median of a grouped distribution is almost the same as finding the median of a simple frequency distribution. Of course, you are looking for a point that has as many frequencies above it as below it. To locate the median, use the formula

$$\text{Median location} = \frac{N + 1}{2}$$

For the data in Table B.1,

$$\text{Median location} = \frac{N + 1}{2} = \frac{100 + 1}{2} = 50.5$$

As before, look for a point with 50 frequencies above it and 50 frequencies below it. Adding frequencies from the bottom of the distribution, you find that there are 37 scores below the interval 24–26 and 24 scores in that interval. The 50.5th score is in the interval

24–26. The midpoint of the interval is the median. For the grouped SWLS scores in Table B.1, the median is 25.

Thus, to find the median of a grouped frequency distribution, locate the class interval that is the location of the middle score. The midpoint of that interval is the median.

Mode

The mode is the midpoint of the interval that has the highest frequency. In Table B.1 the highest frequency count is 24. The interval with 24 scores is 24–26. The midpoint of that interval, 25, is the mode.

PROBLEMS

B.3. Find the mean, median, and mode of the grouped frequency distribution you constructed from the statistics questionnaire data (problem B.1).

B.4. Find the mean, median, and mode of the weight data in problem B.2.

APPENDIX C

Tables

TABLE A Critical values for Pearson product-moment correlation coefficients, r*

df	α levels (two-tailed test) .10	.05	.02	.01	.001
	α levels (one-tailed test)				
(*df* = *N* − 2)	.05	.025	.01	.005	.0005
1	.98769	.99692	.999507	.999877	.9999988
2	.90000	.95001	.98000	.990000	.99900
3	.8053	.8783	.9343	.95874	.99113
4	.7293	.8114	.8822	.91720	.97407
5	.6694	.7545	.8329	.8745	.95089
6	.6215	.7067	.7888	.8343	.92491
7	.5823	.6664	.7498	.7976	.8983
8	.5495	.6319	.7154	.7646	.8721
9	.5214	.6020	.6850	.7348	.8471
10	.4972	.5760	.6581	.7079	.8233
11	.4762	.5529	.6339	.6836	.8010
12	.4574	.5324	.6120	.6614	.7800
13	.4409	.5139	.5922	.6411	.7604
14	.4258	.4973	.5742	.6226	.7419
15	.4124	.4821	.5577	.6055	.7247
16	.4000	.4683	.5425	.5897	.7084
17	.3888	.4556	.5285	.5750	.6932
18	.3783	.4438	.5154	.5614	.6788
19	.3687	.4329	.5033	.5487	.6652
20	.3599	.4227	.4921	.5368	.6524
21	.3516	.4133	.4816	.5256	.6402
22	.3438	.4044	.4715	.5151	.6287
23	.3365	.3961	.4623	.5051	.6177
24	.3297	.3883	.4534	.4958	.6073
25	.3233	.3809	.4451	.4869	.5974
26	.3173	.3740	.4372	.4785	.5880
27	.3114	.3673	.4297	.4706	.5790
28	.3060	.3609	.4226	.4629	.5703
29	.3009	.3550	.4158	.4556	.5620
30	.2959	.3493	.4093	.4487	.5541
40	.2573	.3044	.3578	.3931	.4896
50	.2306	.2733	.3218	.3542	.4432
60	.2109	.2500	.2948	.3248	.4078
80	.1829	.2172	.2565	.2830	.3568
100	.1638	.1946	.2301	.2540	.3211
120	.1496	.1779	.2104	.2324	.2943

* To be significant, the r obtained from the data must be *equal to or greater than* the value shown in the table.
Source: Entries computed by the author.

TABLE B Random digits

	00–04	05–09	10–14	15–19	20–24	25–29	30–34	35–39	40–45	45–49
00	54463	22662	65905	70639	79365	67382	29085	69831	47058	08186
01	15389	85205	18850	39226	42249	90669	96325	23248	60933	26927
02	85941	40756	82414	02015	13858	78030	16269	65978	01385	15345
03	61149	69440	11286	88218	58925	03638	52862	62733	33451	77455
04	05219	81619	10651	67079	92511	59888	84502	72095	83463	75577
05	41417	98326	87719	92294	46614	50948	64886	20002	97365	30976
06	28357	94070	20652	35774	16249	75019	21145	05217	47286	76305
07	17783	00015	10806	83091	91530	36466	39981	62481	49177	75779
08	40950	84820	29881	85966	62800	70326	84740	62660	77379	90279
09	82995	64157	66164	41180	10089	41757	78258	96488	88629	37231
10	96754	17676	55659	44105	47361	34833	86679	23930	53249	27083
11	34357	88040	53364	71726	45690	66334	60332	22554	90600	71113
12	06318	37403	49927	57715	50423	67372	63116	48888	21505	80182
13	62111	52820	07243	79931	89292	84767	85693	73947	22278	11551
14	47534	09243	67879	00544	23410	12740	02540	54440	32949	13491
15	98614	75993	84460	62846	59844	14922	48730	73443	48167	34770
16	24856	03648	44898	09351	98795	18644	39765	71058	90368	44104
17	96887	12479	80621	66223	86085	78285	02432	53342	42846	94771
18	90801	21472	42815	77408	37390	76766	52615	32141	30268	18106
19	55165	77312	83666	36028	28420	70219	81369	41493	47366	41067
20	75884	12952	84318	95108	72305	64620	91318	89872	45375	85436
21	16777	37116	58550	42958	21460	43910	01175	87894	81378	10620
22	46230	43877	80207	88877	89380	32992	91380	03164	98656	59337
23	42902	66892	46134	01432	94710	23474	20423	60137	60609	13119
24	81007	00333	39693	28039	10154	95425	39220	19774	31782	49037
25	68089	01122	51111	72373	06902	74373	96199	97017	41273	21546
26	20411	67081	89950	16944	93054	87687	96693	87236	77054	33848
27	58212	13160	06468	15718	82627	76999	05999	58680	96739	63700
28	70577	42866	24969	61210	76046	67699	42054	12696	93758	03283
29	94522	74358	71659	62038	79643	79169	44741	05437	39038	13163
30	42626	86819	85651	88678	17401	03252	99547	32404	17918	62880
31	16051	33763	57194	16752	54450	19031	58580	47629	54132	60631
32	08244	27647	33851	44705	94211	46716	11738	55784	95374	72655
33	59497	04392	09419	89964	51211	04894	72882	17805	21896	83864
34	97155	13428	40293	09985	58434	91412	69124	82171	59058	82859
35	98409	66162	95763	47420	20792	61527	29441	39435	11859	41567
36	45476	84882	65109	96597	25930	66790	65706	61203	53634	22557
37	89300	69700	50741	30329	11658	23166	05400	66669	48708	02306
38	50051	95137	91631	66315	91428	12275	24816	68091	71710	33258
39	31753	85178	31310	89642	98364	92396	24617	09609	83942	22716
40	79152	53829	77250	20190	56535	18760	69942	77448	33278	48805
41	44560	38750	83635	56540	64900	42912	13953	79149	18710	68618
42	68328	83378	63369	71381	39564	95615	42451	64559	97501	65747
43	46939	38689	58625	08342	30459	85863	20781	09284	26333	91777
44	83544	86141	15707	96256	23068	13782	08467	89469	93842	55349
45	91621	00881	04900	54224	46177	55309	17852	27491	89415	23466
46	91896	67126	04151	03795	59077	11848	12630	98375	52068	60142
47	55751	62515	21108	80830	02263	29303	37204	96926	30506	09808
48	85156	87689	95493	88842	00664	55017	55539	17771	69448	87530
49	07521	56898	12236	60277	39102	62315	12239	07105	11844	01117

(continued)

Source: From *Statistical Methods,* by G. W. Snedecor and W. G. Cochran, Seventh Edition. Copyright © 1980 Iowa State University Press. Reprinted by permission.

TABLE B *(continued)*

	50–54	55–59	60–64	65–69	70–74	75–79	80–84	85–89	90–94	95–99
00	59391	58030	52098	82718	87024	82848	04190	96574	90464	29065
01	99567	76364	77204	04615	27062	96621	43918	01896	83991	51141
02	10363	97518	51400	25670	98342	61891	27101	37855	06235	33316
03	86859	19558	64432	16706	99612	59798	32803	67708	15297	28612
04	11258	24591	36863	55368	31721	94335	34936	02566	80972	08188
05	95068	88628	35911	14530	33020	80428	39936	31855	34334	64865
06	54463	47237	73800	91017	36239	71824	83671	39892	60518	37092
07	16874	62677	57412	13215	31389	62233	80827	73917	82802	84420
08	92494	63157	76593	91316	03505	72389	96363	52887	01087	66091
09	15669	56689	35682	40844	53256	81872	35213	09840	34471	74441
10	99116	75486	84989	23476	52967	67104	39495	39100	17217	74073
11	15696	10703	65178	90637	63110	17622	53988	71087	84148	11670
12	97720	15369	51269	69620	03388	13699	33423	67453	43269	56720
13	11666	13841	71681	98000	35979	39719	81899	07449	47985	46967
14	71628	73130	78783	75691	41632	09847	61547	18707	65489	69944
15	40501	51089	99943	91843	41995	88931	73631	69361	05375	15417
16	22518	55576	98215	82068	10798	86211	36584	67466	69373	40054
17	75112	30485	62173	02132	14878	92879	22281	16783	86352	00077
18	80327	02671	98191	84342	90813	49268	95441	15496	20168	09271
19	60251	45548	02146	05597	48228	81366	34598	72856	66762	17002
20	57430	82270	10421	05540	43648	75888	66049	21511	47676	33444
21	73528	39559	34434	88596	54086	71693	43132	14414	79949	85193
22	25991	65959	70769	64721	86413	33475	42740	06175	82758	66248
23	78388	16638	09134	59880	63806	48472	39318	35434	24057	74739
24	12477	09965	96657	57994	59439	76330	24596	77515	09577	91871
25	83266	32883	42451	15579	38155	29793	40914	65990	16255	17777
26	76970	80876	10237	39515	79152	74798	39357	09054	73579	92359
27	37074	65198	44785	68624	98336	84481	97610	78735	46703	98265
28	83712	06514	30101	78295	54656	85417	43189	60048	72781	72606
29	20287	56862	69727	94443	64936	08366	27227	05158	50326	59566
30	74261	32592	86538	27041	65172	85532	07571	80609	39285	65340
31	64081	49863	08478	96001	18888	14810	70545	89755	59064	07210
32	05617	75818	47750	67814	29575	10526	66192	44464	27058	40467
33	26793	74951	95466	74307	13330	42664	85515	20632	05497	33625
34	65988	72850	48737	54719	52056	01596	03845	35067	03134	70322
35	27366	42271	44300	73399	21105	03280	73457	43093	05192	48657
36	56760	10909	98147	34736	33863	95256	12731	66598	50771	83665
37	72880	43338	93643	58904	59543	23943	11231	83268	65938	81581
38	77888	38100	03062	58103	47961	83841	25878	23746	55903	44115
39	28440	07819	21580	51459	47971	29882	13990	29226	23608	15873
40	63525	94441	77033	12147	51054	49955	58312	76923	96071	05813
41	47606	93410	16359	89033	89696	47231	64498	31776	05383	39902
42	52669	45030	96279	14709	52372	87832	02735	50803	72744	88208
43	16738	60159	07425	62369	07515	82721	37875	71153	21315	00132
44	59348	11695	45751	15865	74739	05572	32688	20271	65128	14551
45	12900	71775	29845	60774	94924	21810	38636	33717	67598	82521
46	75086	23527	49939	33595	13484	97588	28617	17979	70749	35234
47	99495	51434	29181	09993	38190	42553	68922	52125	91077	40197
48	26075	31671	45386	36583	93159	48599	52022	41330	60651	91321
49	13636	93596	23377	51133	95126	61496	42474	45141	46660	42338

(continued)

TABLE B *(continued)*

	00–04	05–09	10–14	15–19	20–24	25–29	30–34	35–39	40–45	45–49
50	64249	63664	39652	40646	97306	31741	07294	84149	46797	82487
51	26538	44249	04050	48174	65570	44072	40192	51153	11397	58212
52	05845	00512	78630	55328	18116	69296	91705	86224	29503	57071
53	74897	68373	67359	51014	33510	83048	17056	72506	82949	54600
54	20872	54570	35017	88132	25730	22626	86723	91691	13191	77212
55	31432	96156	89177	75541	81355	24480	77243	76690	42507	84362
56	66890	61505	01240	00660	05873	13568	76082	79172	57913	93448
57	48194	57790	79970	33106	86904	48119	52503	24130	72824	21627
58	11303	87118	81471	52936	08555	28420	49416	44448	04269	27029
59	54374	57325	16947	45356	78371	10563	97191	53798	12693	27928
60	64852	34421	61046	90849	13966	39810	42699	21753	76192	10508
61	16309	20384	09491	91588	97720	89846	30376	76970	23063	35894
62	42587	37065	24526	72602	57589	98131	37292	05967	26002	51945
63	40177	98590	97161	41682	84533	67588	62036	49967	01990	72308
64	82309	76128	93965	26743	24141	04838	40254	26065	07938	76236
65	79788	68243	59732	04257	27084	14743	17520	95401	55811	76099
66	40538	79000	89559	25026	42274	23489	34502	75508	06059	86682
67	64016	73598	18609	73150	62463	33102	45205	87440	96767	67042
68	49767	12691	17903	93871	99721	79109	09425	26904	07419	76013
69	76974	55108	29795	08404	82684	00497	51126	79935	57450	55671
70	23854	08480	85983	96025	50117	64610	99425	62291	86943	21541
71	68973	70551	25098	78033	98573	79848	31778	29555	61446	23037
72	36444	93600	65350	14971	25325	00427	52073	64280	18847	24768
73	03003	87800	07391	11594	21196	00781	32550	57158	58887	73041
74	17540	26188	36647	78386	04558	61463	57842	90382	77019	24210
75	38916	55809	47982	41968	69760	79422	80154	91486	19180	15100
76	64288	19843	69122	42502	48508	28820	59933	72998	99942	10515
77	86809	51564	38040	39418	49915	19000	58050	16899	79952	57849
78	99800	99566	14742	05028	30033	94889	53381	23656	75787	79223
79	92345	31890	95712	08279	91794	94068	49037	88674	35355	12267
80	90363	65162	32245	82279	79256	80834	06088	99462	56705	06118
81	64437	32242	48431	04835	39070	59702	31508	60935	22390	52246
82	91714	53662	28373	34333	55791	74758	51144	18827	10704	76803
83	20902	17646	31391	31459	33315	03444	55743	74701	58851	27427
84	12217	86007	70371	52281	14510	76094	96579	54863	78339	20839
85	45177	02863	42307	53571	22532	74921	17735	42201	80540	54721
86	28325	90814	08804	52746	47913	54577	47525	77705	95330	21866
87	29019	28776	56116	54791	64604	08815	46049	71186	34650	14994
88	84979	81353	56219	67062	26146	82567	33122	14124	46240	92973
89	50371	26347	48513	63915	11158	25563	91915	18431	92978	11591
90	53422	06825	69711	67950	64716	18003	49581	45378	99878	61130
91	67453	35651	89316	41620	32048	70225	47597	33137	31443	51445
92	07294	85353	74819	23445	68237	07202	99515	62282	53809	26685
93	79544	00302	45338	16015	66613	88968	14595	63836	77716	79596
94	64144	85442	82060	46471	24162	39500	87351	36637	42833	71875
95	90919	11883	58318	00042	52402	28210	34075	33272	00840	73268
96	06670	57353	86275	92276	77591	46924	60839	55437	03183	13191
97	36634	93976	52062	83678	41256	60948	18685	48992	19462	96062
98	75101	72891	85745	67106	26010	62107	60885	37503	55461	71213
99	05112	71222	72654	51583	05228	62056	57390	42746	39272	96659

(continued)

TABLE B *(continued)*

	50–54	55–59	60–64	65–69	70–74	75–79	80–84	85–89	90–94	95–99
50	32847	31282	03345	89593	69214	70381	78285	20054	91018	16742
51	16916	00041	30236	55023	14253	76582	12092	86533	92426	37655
52	66176	34047	21005	27137	03191	48970	64625	22394	39622	79085
53	46299	13335	12180	16861	38043	59292	62675	63631	37020	78195
54	22847	47839	45385	23289	47526	54098	45683	55849	51575	64689
55	41851	54160	92320	69936	34803	92479	33399	71160	64777	83378
56	28444	59497	91586	95917	68553	28639	06455	34174	11130	91994
57	47520	62378	98855	83174	13088	16561	68559	26679	06238	51254
58	34978	63271	13142	82681	05271	08822	06490	44984	49307	62717
59	37404	80416	69035	92980	49486	74378	75610	74976	70056	15478
60	32400	65482	52099	53676	74648	94148	65095	69597	52771	71551
61	89262	86332	51718	70663	11623	29834	79820	73002	84886	03591
62	86866	09127	98021	03871	27789	58444	44832	36505	40672	30180
63	90814	14833	08759	74645	05046	94056	99094	65901	32663	73040
64	19192	82756	20553	58446	55376	88914	75096	26119	83998	43816
65	77585	52593	56612	95766	10019	29531	73064	20953	53523	58136
66	23757	16364	05096	03192	62386	45389	85332	18877	55710	96459
67	45989	96257	23850	26216	23309	21526	07425	50254	19455	29315
68	92970	94243	07316	41467	64837	52406	25225	51553	31220	14032
69	74346	59596	40088	98176	17896	86900	20249	77753	19099	48885
70	87646	41309	27636	45153	29988	94770	07255	70908	05340	99751
71	50099	71038	45146	06146	55211	99429	43169	66259	97786	59180
72	10127	46900	64984	75348	04115	33624	68774	60013	35515	62556
73	67995	81977	18984	64091	02785	27762	42529	97144	80407	64525
74	26304	80217	84934	82657	69291	35397	98714	35104	08187	48109
75	81994	41070	56642	64091	31229	02595	13513	45148	78722	30144
76	59537	34662	79631	89403	65212	09975	06118	86197	58208	16162
77	51228	10937	62396	81460	47331	91403	95007	06047	16846	64809
78	31089	37995	29577	07828	42272	54016	21950	86192	99046	84864
79	38207	97938	93459	75174	79460	55436	57206	87644	21296	43395
80	88666	31142	09474	89712	63153	62333	42212	06140	42594	43671
81	53365	56134	67582	92557	89520	33452	05134	70628	27612	33738
82	89807	74530	38004	90102	11693	90257	05500	79920	62700	43325
83	18682	81038	85662	90915	91631	22223	91588	80774	07716	12548
84	63571	32579	63942	25371	90234	94592	98475	76884	37635	33608
85	68927	56492	67799	95392	77642	54613	91853	08424	81450	76229
86	56401	63186	39389	99798	31356	89235	97036	32341	33292	73757
87	24333	95603	02359	72942	46287	95382	08452	62862	97869	71775
88	17025	84202	95199	62272	06366	16175	97577	99304	41587	03686
89	02804	08253	52133	20224	68034	50865	57868	22343	55111	03607
90	08298	03879	20995	19850	73090	13191	18963	82244	78479	99121
91	59883	01785	82403	96062	03785	03488	12970	64896	38336	30030
92	46982	06682	62864	91837	74021	89094	39952	64158	79614	78235
93	31121	47266	07661	02051	67599	24471	69843	83696	71402	76287
94	97867	56641	63416	17577	30161	87320	37752	73276	48969	41915
95	57364	86746	08415	14621	49430	22311	15836	72492	49372	44103
96	09559	26263	69511	28064	75999	44540	13337	10918	79846	54809
97	53873	55571	00608	42661	91332	63956	74087	59008	47494	99581
98	35531	19162	86406	05299	77511	24311	57257	22826	77555	05941
99	28229	88629	25695	94932	30721	16197	78742	34974	97528	45447

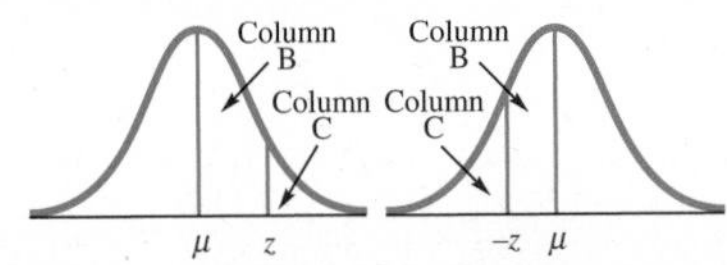

TABLE C
Normal curve areas

A	B	C	A	B	C	A	B	C
z	μ to z	z to ∞	z	μ to z	z to ∞	z	μ to z	z to ∞
0.00	.0000	.5000	0.55	.2088	.2912	1.10	.3643	.1357
0.01	.0040	.4960	0.56	.2123	.2877	1.11	.3665	.1335
0.02	.0080	.4920	0.57	.2157	.2843	1.12	.3686	.1314
0.03	.0120	.4880	0.58	.2190	.2810	1.13	.3708	.1292
0.04	.0160	.4840	0.59	.2224	.2776	1.14	.3729	.1271
0.05	.0199	.4801	0.60	.2257	.2743	1.15	.3749	.1251
0.06	.0239	.4761	0.61	.2291	.2709	1.16	.3770	.1230
0.07	.0279	.4721	0.62	.2324	.2676	1.17	.3790	.1210
0.08	.0319	.4681	0.63	.2357	.2643	1.18	.3810	.1190
0.09	.0359	.4641	0.64	.2389	.2611	1.19	.3830	.1170
0.10	.0398	.4602	0.65	.2422	.2578	1.20	.3849	.1151
0.11	.0438	.4562	0.66	.2454	.2546	1.21	.3869	.1131
0.12	.0478	.4522	0.67	.2486	.2514	1.22	.3888	.1112
0.13	.0517	.4483	0.68	.2517	.2483	1.23	.3907	.1093
0.14	.0557	.4443	0.69	.2549	.2451	1.24	.3925	.1075
0.15	.0596	.4404	0.70	.2580	.2420	1.25	.3944	.1056
0.16	.0636	.4364	0.71	.2611	.2389	1.26	.3962	.1038
0.17	.0675	.4325	0.72	.2642	.2358	1.27	.3980	.1020
0.18	.0714	.4286	0.73	.2673	.2327	1.28	.3997	.1003
0.19	.0753	.4247	0.74	.2704	.2296	1.29	.4015	.0985
0.20	.0793	.4207	0.75	.2734	.2266	1.30	.4032	.0968
0.21	.0832	.4168	0.76	.2764	.2236	1.31	.4049	.0951
0.22	.0871	.4129	0.77	.2794	.2206	1.32	.4066	.0934
0.23	.0910	.4090	0.78	.2823	.2177	1.33	.4082	.0918
0.24	.0948	.4052	0.79	.2852	.2148	1.34	.4099	.0901
0.25	.0987	.4013	0.80	.2881	.2119	1.35	.4115	.0885
0.26	.1026	.3974	0.81	.2910	.2090	1.36	.4131	.0869
0.27	.1064	.3936	0.82	.2939	.2061	1.37	.4147	.0853
0.28	.1103	.3897	0.83	.2967	.2033	1.38	.4162	.0838
0.29	.1141	.3859	0.84	.2995	.2005	1.39	.4177	.0823
0.30	.1179	.3821	0.85	.3023	.1977	1.40	.4192	.0808
0.31	.1217	.3783	0.86	.3051	.1949	1.41	.4207	.0793
0.32	.1255	.3745	0.87	.3078	.1922	1.42	.4222	.0778
0.33	.1293	.3707	0.88	.3106	.1894	1.43	.4236	.0764
0.34	.1331	.3669	0.89	.3133	.1867	1.44	.4251	.0749
0.35	.1368	.3632	0.90	.3159	.1841	1.45	.4265	.0735
0.36	.1406	.3594	0.91	.3186	.1814	1.46	.4279	.0721
0.37	.1443	.3557	0.92	.3212	.1788	1.47	.4292	.0708
0.38	.1480	.3520	0.93	.3238	.1762	1.48	.4306	.0694
0.39	.1517	.3483	0.94	.3264	.1736	1.49	.4319	.0681
0.40	.1554	.3446	0.95	.3289	.1711	1.50	.4332	.0668
0.41	.1591	.3409	0.96	.3315	.1685	1.51	.4345	.0655
0.42	.1628	.3372	0.97	.3340	.1660	1.52	.4357	.0643
0.43	.1664	.3336	0.98	.3365	.1635	1.53	.4370	.0630
0.44	.1700	.3300	0.99	.3389	.1611	1.54	.4382	.0618
0.45	.1736	.3264	1.00	.3413	.1587	1.55	.4394	.0606
0.46	.1772	.3228	1.01	.3438	.1562	1.56	.4406	.0594
0.47	.1808	.3192	1.02	.3461	.1539	1.57	.4418	.0582
0.48	.1844	.3156	1.03	.3485	.1515	1.58	.4429	.0571
0.49	.1879	.3121	1.04	.3508	.1492	1.59	.4441	.0559
0.50	.1915	.3085	1.05	.3531	.1469	1.60	.4452	.0548
0.51	.1950	.3050	1.06	.3554	.1446	1.61	.4463	.0537
0.52	.1985	.3015	1.07	.3577	.1423	1.62	.4474	.0526
0.53	.2019	.2981	1.08	.3599	.1401	1.63	.4484	.0516
0.54	.2054	.2946	1.09	.3621	.1379	1.64	.4495	.0505

(continued)

TABLE C *(continued)*

A	B	C	A	B	C	A	B	C
z	μ to z	z to ∞	z	μ to z	z to ∞	z	μ to z	z to ∞
1.65	.4505	.0495	2.22	.4868	.0132	2.79	.4974	.0026
1.66	.4515	.0485	2.23	.4871	.0129	2.80	.4974	.0026
1.67	.4525	.0475	2.24	.4875	.0125	2.81	.4975	.0025
1.68	.4535	.0465	2.25	.4878	.0122	2.82	.4976	.0024
1.69	.4545	.0455	2.26	.4881	.0119	2.83	.4977	.0023
1.70	.4554	.0446	2.27	.4884	.0116	2.84	.4977	.0023
1.71	.4564	.0436	2.28	.4887	.0113	2.85	.4978	.0022
1.72	.4573	.0427	2.29	.4890	.0110	2.86	.4979	.0021
1.73	.4582	.0418	2.30	.4893	.0107	2.87	.4979	.0021
1.74	.4591	.0409	2.31	.4896	.0104	2.88	.4980	.0020
1.75	.4599	.0401	2.32	.4898	.0102	2.89	.4981	.0019
1.76	.4608	.0392	2.33	.4901	.0099	2.90	.4981	.0019
1.77	.4616	.0384	2.34	.4904	.0096	2.91	.4982	.0018
1.78	.4625	.0375	2.35	.4906	.0094	2.92	.4982	.0018
1.79	.4633	.0367	2.36	.4909	.0091	2.93	.4983	.0017
1.80	.4641	.0359	2.37	.4911	.0089	2.94	.4984	.0016
1.81	.4649	.0351	2.38	.4913	.0087	2.95	.4984	.0016
1.82	.4656	.0344	2.39	.4916	.0084	2.96	.4985	.0015
1.83	.4664	.0336	2.40	.4918	.0082	2.97	.4985	.0015
1.84	.4671	.0329	2.41	.4920	.0080	2.98	.4986	.0014
1.85	.4678	.0322	2.42	.4922	.0078	2.99	.4986	.0014
1.86	.4686	.0314	2.43	.4925	.0075	3.00	.4987	.0013
1.87	.4693	.0307	2.44	.4927	.0073	3.01	.4987	.0013
1.88	.4699	.0301	2.45	.4929	.0071	3.02	.4987	.0013
1.89	.4706	.0294	2.46	.4931	.0069	3.03	.4988	.0012
1.90	.4713	.0287	2.47	.4932	.0068	3.04	.4988	.0012
1.91	.4719	.0281	2.48	.4934	.0066	3.05	.4989	.0011
1.92	.4726	.0274	2.49	.4936	.0064	3.06	.4989	.0011
1.93	.4732	.0268	2.50	.4938	.0062	3.07	.4989	.0011
1.94	.4738	.0262	2.51	.4940	.0060	3.08	.4990	.0010
1.95	.4744	.0256	2.52	.4941	.0059	3.09	.4990	.0010
1.96	.4750	.0250	2.53	.4943	.0057	3.10	.4990	.0010
1.97	.4756	.0244	2.54	.4945	.0055	3.11	.4991	.0009
1.98	.4761	.0239	2.55	.4946	.0054	3.12	.4991	.0009
1.99	.4767	.0233	2.56	.4948	.0052	3.13	.4991	.0009
2.00	.4772	.0228	2.57	.4949	.0051	3.14	.4992	.0008
2.01	.4778	.0222	2.58	.4951	.0049	3.15	.4992	.0008
2.02	.4783	.0217	2.59	.4952	.0048	3.16	.4992	.0008
2.03	.4788	.0212	2.60	.4953	.0047	3.17	.4992	.0008
2.04	.4793	.0207	2.61	.4955	.0045	3.18	.4993	.0007
2.05	.4798	.0202	2.62	.4956	.0044	3.19	.4993	.0007
2.06	.4803	.0197	2.63	.4957	.0043	3.20	.4993	.0007
2.07	.4808	.0192	2.64	.4959	.0041	3.21	.4993	.0007
2.08	.4812	.0188	2.65	.4960	.0040	3.22	.4994	.0006
2.09	.4817	.0183	2.66	.4961	.0039	3.23	.4994	.0006
2.10	.4821	.0179	2.67	.4962	.0038	3.24	.4994	.0006
2.11	.4826	.0174	2.68	.4963	.0037	3.25	.4994	.0006
2.12	.4830	.0170	2.69	.4964	.0036	3.30	.4995	.0005
2.13	.4834	.0166	2.70	.4965	.0035	3.35	.4996	.0004
2.14	.4838	.0162	2.71	.4966	.0034	3.40	.4997	.0003
2.15	.4842	.0158	2.72	.4967	.0033	3.45	.4997	.0003
2.16	.4846	.0154	2.73	.4968	.0032	3.50	.4998	.0002
2.17	.4850	.0150	2.74	.4969	.0031	3.60	.4998	.0002
2.18	.4854	.0146	2.75	.4970	.0030	3.70	.4999	.0001
2.19	.4857	.0143	2.76	.4971	.0029	3.80	.4999	.0001
2.20	.4861	.0139	2.77	.4972	.0028	3.90	.49995	.00005
2.21	.4864	.0136	2.78	.4973	.0027	4.00	.49997	.00003

TABLE D The *t* distribution*

	Confidence interval percents (two-tailed)					
	80%	90%	95%	98%	99%	99.9%
	α level for two-tailed test					
	.20	.10	.05	.02	.01	.001
	α level for one-tailed test					
df	.10	.05	.025	.01	.005	.0005
1	3.078	6.314	12.71	31.82	63.66	636.6
2	1.886	2.920	4.303	6.965	9.925	31.598
3	1.638	2.353	3.182	4.541	5.841	12.924
4	1.533	2.132	2.776	3.747	4.604	8.610
5	1.476	2.015	2.571	3.365	4.032	6.869
6	1.440	1.943	2.447	3.143	3.707	5.959
7	1.415	1.895	2.365	2.998	3.499	5.408
8	1.397	1.860	2.306	2.896	3.355	5.041
9	1.383	1.833	2.262	2.821	3.250	4.781
10	1.372	1.812	2.228	2.764	3.169	4.587
11	1.363	1.796	2.201	2.718	3.106	4.437
12	1.356	1.782	2.179	2.681	3.055	4.318
13	1.350	1.771	2.160	2.650	3.012	4.221
14	1.345	1.761	2.145	2.624	2.977	4.140
15	1.341	1.753	2.131	2.602	2.947	4.073
16	1.337	1.746	2.120	2.583	2.921	4.015
17	1.333	1.740	2.110	2.567	2.898	3.965
18	1.330	1.734	2.101	2.552	2.878	3.922
19	1.328	1.729	2.093	2.539	2.861	3.883
20	1.325	1.725	2.086	2.528	2.845	3.850
21	1.323	1.721	2.080	2.518	2.831	3.819
22	1.321	1.717	2.074	2.508	2.819	3.792
23	1.319	1.714	2.069	2.500	2.807	3.767
24	1.318	1.711	2.064	2.492	2.797	3.745
25	1.316	1.708	2.060	2.485	2.787	3.725
26	1.315	1.706	2.056	2.479	2.779	3.707
27	1.314	1.703	2.052	2.473	2.771	3.690
28	1.313	1.701	2.048	2.467	2.763	3.674
29	1.311	1.699	2.045	2.462	2.756	3.659
30	1.310	1.697	2.042	2.457	2.750	3.646
40	1.303	1.684	2.021	2.423	2.704	3.551
50	1.299	1.676	2.009	2.403	2.678	3.496
60	1.296	1.671	2.000	2.390	2.660	3.460
80	1.292	1.664	1.990	2.374	2.639	3.416
100	1.290	1.660	1.984	2.364	2.626	3.390
120	1.289	1.658	1.980	2.358	2.617	3.373
∞	1.282	1.645	1.960	2.326	2.576	3.291

* To be significant, the *t* obtained from the data must be *equal to or greater than* the value shown in the table.

Source: Reprinted by permission of Pearson Education Ltd.

TABLE E Chi square distribution*

	α levels				
df	.10	.05	.025	.01	.001
1	2.706	3.841	5.024	6.635	10.828
2	4.605	5.991	7.378	9.210	13.816
3	6.251	7.815	9.348	11.345	16.266
4	7.779	9.488	11.143	13.277	18.467
5	9.236	11.070	12.833	15.086	20.515
6	10.645	12.592	14.449	16.812	22.458
7	12.017	14.067	16.013	18.475	24.322
8	13.362	15.507	17.535	20.090	26.125
9	14.684	16.919	19.023	21.666	27.877
10	15.987	18.307	20.483	23.209	29.588
11	17.275	19.675	21.920	24.725	31.264
12	18.549	21.026	23.337	26.217	32.910
13	19.812	22.362	24.736	27.688	34.528
14	21.064	23.685	26.119	29.141	36.123
15	22.307	24.996	27.488	30.578	37.697
16	23.542	26.296	28.845	32.000	39.252
17	24.769	27.587	30.191	33.409	40.790
18	25.989	28.869	31.526	34.805	42.312
19	27.204	30.144	32.852	36.191	43.820
20	28.412	31.410	34.170	37.566	45.315
21	29.615	32.671	35.479	38.932	46.797
22	30.813	33.924	36.781	40.289	48.268
23	32.007	35.172	38.076	41.638	49.728
24	33.196	36.415	39.364	42.980	51.179
25	34.382	37.652	40.646	44.314	52.620
26	35.563	38.885	41.923	45.642	54.052
27	36.741	40.113	43.195	46.963	55.476
28	37.916	41.337	44.461	48.278	56.892
29	39.087	42.557	45.722	49.588	58.301
30	40.256	43.773	46.979	50.892	59.703

* To be significant, the χ^2 obtained from the data must be *equal to or greater than* the value shown in the table.
Source: Engineering Statistics Handbook, Retrieved from http://www.itl.nist.gov/div898/handbook/eda/section3/eda3674.htm.

TABLE F The *F* distribution*

	α levels of .05 (lightface) and .01 (**boldface**) for the distribution of *F*																								
	Degrees of freedom (for the numerator of *F* ratio)																								
Degrees of freedom (for the denominator of *F* ratio)	1	2	3	4	5	6	7	8	9	10	11	12	14	16	20	24	30	40	50	75	100	200	500	∞	
1	161	200	216	225	230	234	237	239	241	242	243	244	245	246	248	249	250	251	252	253	253	254	254	254	1
	4,052	**4,999**	**5,403**	**5,625**	**5,764**	**5,859**	**5,928**	**5,981**	**6,022**	**6,056**	**6,082**	**6,106**	**6,142**	**6,169**	**6,208**	**6,234**	**6,258**	**6,286**	**6,302**	**6,323**	**6,334**	**6,352**	**6,361**	**6,366**	
2	18.51	19.00	19.16	19.25	19.30	19.33	19.36	19.37	19.38	19.39	19.40	19.41	19.42	19.43	19.44	19.45	19.46	19.47	19.47	19.48	19.49	19.49	19.50	19.50	2
	98.49	**99.00**	**99.17**	**99.25**	**99.30**	**99.33**	**99.34**	**99.36**	**99.38**	**99.40**	**99.41**	**99.42**	**99.43**	**99.44**	**99.45**	**99.46**	**99.47**	**99.48**	**99.49**	**99.49**	**99.49**	**99.49**	**99.50**	**99.50**	
3	10.13	9.55	9.28	9.12	9.01	8.94	8.88	8.84	8.81	8.78	8.76	8.74	8.71	8.69	8.66	8.64	8.62	8.60	8.58	8.57	8.56	8.54	8.54	8.53	3
	34.12	**30.82**	**29.46**	**28.71**	**28.24**	**27.91**	**27.67**	**27.49**	**27.34**	**27.23**	**27.13**	**27.05**	**29.92**	**26.83**	**26.69**	**26.60**	**26.50**	**26.41**	**26.35**	**26.27**	**26.23**	**26.18**	**26.14**	**26.12**	
4	7.71	6.94	6.59	6.39	6.26	6.16	6.09	6.04	6.00	5.96	5.93	5.91	5.87	5.84	5.80	5.77	5.74	5.71	5.70	5.68	5.66	5.66	5.64	5.63	4
	21.20	**18.00**	**16.69**	**15.98**	**15.52**	**15.21**	**14.98**	**14.80**	**14.66**	**14.54**	**14.45**	**14.37**	**14.24**	**14.15**	**14.02**	**13.93**	**13.83**	**13.74**	**13.69**	**13.61**	**13.57**	**13.52**	**13.48**	**13.46**	
5	6.61	5.79	5.41	5.19	5.05	4.95	4.88	4.82	4.78	4.74	4.70	4.68	4.64	4.60	4.56	4.53	4.50	4.46	4.44	4.42	4.40	4.38	4.37	4.36	5
	16.26	**13.27**	**12.06**	**11.39**	**10.97**	**10.67**	**10.45**	**10.27**	**10.15**	**10.05**	**9.96**	**9.89**	**9.77**	**9.68**	**9.55**	**9.47**	**9.38**	**9.29**	**9.24**	**9.17**	**9.13**	**9.07**	**9.04**	**9.02**	
6	5.99	5.14	4.76	4.53	4.39	4.28	4.21	4.15	4.10	4.06	4.03	4.00	3.96	3.92	3.87	3.84	3.81	3.77	3.75	3.72	3.71	3.69	3.68	3.67	6
	13.74	**10.92**	**9.78**	**9.15**	**8.75**	**8.47**	**8.26**	**8.10**	**7.98**	**7.87**	**7.79**	**7.72**	**7.60**	**7.52**	**7.39**	**7.31**	**7.23**	**7.14**	**7.09**	**7.02**	**6.99**	**6.94**	**6.90**	**6.88**	
7	5.59	4.74	4.35	4.12	3.97	3.87	3.79	3.73	3.68	3.63	3.60	3.57	3.52	3.49	3.44	3.41	3.38	3.34	3.32	3.29	3.28	3.25	3.24	3.23	7
	12.25	**9.55**	**8.45**	**7.85**	**7.46**	**7.19**	**7.00**	**6.84**	**6.71**	**6.62**	**6.54**	**6.47**	**6.35**	**6.27**	**6.15**	**6.07**	**5.98**	**5.90**	**5.85**	**5.78**	**5.75**	**5.70**	**5.67**	**5.65**	
8	5.32	4.46	4.07	3.84	3.69	3.58	3.50	3.44	3.39	3.34	3.31	3.28	3.23	3.20	3.15	3.12	3.08	3.05	3.03	3.00	2.98	2.96	2.94	2.93	8
	11.26	**8.65**	**7.59**	**7.01**	**6.63**	**6.37**	**6.19**	**6.03**	**5.91**	**5.82**	**5.74**	**5.67**	**5.56**	**5.48**	**5.36**	**5.28**	**5.20**	**5.11**	**5.06**	**5.00**	**4.96**	**4.91**	**4.88**	**4.86**	
9	5.12	4.26	3.86	3.63	3.48	3.37	3.29	3.23	3.18	3.13	3.10	3.07	3.02	2.98	2.93	2.90	2.86	2.82	2.80	2.77	2.76	2.73	2.72	2.71	9
	10.56	**8.02**	**6.99**	**6.42**	**6.06**	**5.80**	**5.62**	**5.47**	**5.35**	**5.26**	**5.18**	**5.11**	**5.00**	**4.92**	**4.80**	**4.73**	**4.64**	**4.56**	**4.51**	**4.45**	**4.41**	**4.36**	**4.33**	**4.31**	
10	4.96	4.10	3.71	3.48	3.33	3.22	3.14	3.07	3.02	2.97	2.94	2.91	2.86	2.82	2.77	2.74	2.70	2.67	2.64	2.61	2.59	2.56	2.55	2.54	10
	10.04	**7.56**	**6.55**	**5.99**	**5.64**	**5.39**	**5.21**	**5.06**	**4.95**	**4.85**	**4.78**	**4.71**	**4.60**	**4.52**	**4.41**	**4.33**	**4.25**	**4.17**	**4.12**	**4.05**	**4.01**	**3.96**	**3.93**	**3.91**	
11	4.84	3.98	3.59	3.36	3.20	3.09	3.01	2.95	2.90	2.86	2.82	2.79	2.74	2.70	2.65	2.61	2.57	2.53	2.50	2.47	2.45	2.42	2.41	2.40	11
	9.65	**7.20**	**6.22**	**5.67**	**5.32**	**5.07**	**4.88**	**4.74**	**4.63**	**4.54**	**4.46**	**4.40**	**4.29**	**4.21**	**4.10**	**4.02**	**3.94**	**3.86**	**3.80**	**3.74**	**3.70**	**3.66**	**3.62**	**3.60**	
12	4.75	3.88	3.49	3.26	3.11	3.00	2.92	2.85	2.80	2.76	2.72	2.69	2.64	2.60	2.54	2.50	2.46	2.42	2.40	2.36	2.35	2.32	2.31	2.30	12
	9.33	**6.93**	**5.95**	**5.41**	**5.06**	**4.82**	**4.65**	**4.50**	**4.39**	**4.30**	**4.22**	**4.16**	**4.05**	**3.98**	**3.86**	**3.78**	**3.70**	**3.61**	**3.56**	**3.49**	**3.46**	**3.41**	**3.38**	**3.36**	
13	4.67	3.80	3.41	3.18	3.02	2.92	2.84	2.77	2.72	2.67	2.63	2.60	2.55	2.51	2.46	2.42	2.38	2.34	2.32	2.28	2.26	2.24	2.22	2.21	13
	9.07	**6.70**	**5.74**	**5.20**	**4.86**	**4.62**	**4.44**	**4.30**	**4.19**	**4.10**	**4.02**	**3.96**	**3.85**	**3.78**	**3.67**	**3.59**	**3.51**	**3.42**	**3.37**	**3.30**	**3.27**	**3.21**	**3.18**	**3.16**	

(continued)

* To be significant, the *F* obtained from the data must be *equal to or greater than* the value shown in the table.
Source: From *Statistical Methods,* by G. W. Snedecor and W. W. Cochran, Seventh Edition. Copyright © 1980 Iowa State University Press. Reprinted by permission. For critical values of *F* for α = .10, see *www.itl.nist.gov/div898/handbook/eda/section3/eda3673.htm.*

TABLE F *(continued)*

Degrees of freedom (for the denominator of F ratio) — rows; Degrees of freedom (for the numerator of F ratio) — columns

	1	2	3	4	5	6	7	8	9	10	11	12	14	16	20	24	30	40	50	75	100	200	500	∞	
14	4.60	3.74	3.34	3.11	2.96	2.85	2.77	2.70	2.65	2.60	2.56	2.53	2.48	2.44	2.39	2.35	2.31	2.27	2.24	2.21	2.19	2.16	2.14	2.13	14
	8.86	**6.51**	**5.56**	**5.03**	**4.69**	**4.46**	**4.28**	**4.14**	**4.03**	**3.94**	**3.86**	**3.80**	**3.70**	**3.62**	**3.51**	**3.43**	**3.34**	**3.26**	**3.21**	**3.14**	**3.11**	**3.06**	**3.02**	**3.00**	
15	4.54	3.68	3.29	3.06	2.90	2.79	2.70	2.64	2.59	2.55	2.51	2.48	2.43	2.39	2.33	2.29	2.25	2.21	2.18	2.15	2.12	2.10	2.08	2.07	15
	8.68	**6.36**	**5.52**	**4.89**	**4.56**	**4.32**	**4.14**	**4.00**	**3.89**	**3.80**	**3.73**	**3.67**	**3.56**	**3.48**	**3.36**	**3.29**	**3.20**	**3.12**	**3.07**	**3.00**	**2.97**	**2.92**	**2.89**	**2.87**	
16	4.49	3.63	3.24	3.01	2.85	2.74	2.66	2.59	2.54	2.49	2.45	2.42	2.37	2.33	2.28	2.24	2.20	2.16	2.13	2.09	2.07	2.04	2.02	2.01	16
	8.53	**6.23**	**5.29**	**4.77**	**4.44**	**4.20**	**4.03**	**3.89**	**3.78**	**3.69**	**3.61**	**3.55**	**3.45**	**3.37**	**3.25**	**3.18**	**3.10**	**3.01**	**2.96**	**2.89**	**2.86**	**2.80**	**2.77**	**2.75**	
17	4.45	3.59	3.20	2.96	2.81	2.70	2.62	2.55	2.50	2.45	2.41	2.38	2.33	2.29	2.23	2.19	2.15	2.11	2.08	2.04	2.02	1.99	1.97	1.96	17
	8.40	**6.11**	**5.18**	**4.67**	**4.34**	**4.10**	**3.93**	**3.79**	**3.68**	**3.59**	**3.52**	**3.45**	**3.35**	**3.27**	**3.16**	**3.08**	**3.00**	**2.92**	**2.86**	**2.79**	**2.76**	**2.70**	**2.67**	**2.65**	
18	4.41	3.55	3.16	2.93	2.77	2.66	2.58	2.51	2.46	2.41	2.37	2.34	2.29	2.25	2.19	2.15	2.11	2.07	2.04	2.00	1.89	1.95	1.93	1.92	18
	8.28	**6.01**	**5.09**	**4.58**	**4.25**	**4.01**	**3.85**	**3.71**	**3.60**	**3.51**	**3.44**	**3.37**	**3.27**	**3.19**	**3.07**	**3.00**	**2.91**	**2.83**	**2.78**	**2.71**	**2.68**	**2.62**	**2.59**	**2.57**	
19	4.38	3.52	3.13	2.90	2.74	2.63	2.55	2.48	2.43	2.38	2.34	2.31	2.26	2.21	2.15	2.11	2.07	2.02	2.00	1.96	1.94	1.91	1.90	1.88	19
	8.18	**5.93**	**5.01**	**4.50**	**4.17**	**3.94**	**3.77**	**3.63**	**3.52**	**3.43**	**3.36**	**3.30**	**3.19**	**3.12**	**3.00**	**2.92**	**2.84**	**2.76**	**2.70**	**2.63**	**2.60**	**2.54**	**2.51**	**2.49**	
20	4.35	3.49	3.10	2.87	2.71	2.60	2.52	2.45	2.40	2.35	2.31	2.28	2.23	2.18	2.12	2.08	2.04	1.99	2.96	1.92	1.90	1.87	1.85	1.84	20
	8.10	**5.85**	**4.94**	**4.43**	**4.10**	**3.87**	**3.71**	**3.56**	**3.45**	**3.37**	**3.30**	**3.23**	**3.13**	**3.05**	**2.94**	**2.86**	**2.77**	**2.69**	**2.63**	**2.56**	**2.53**	**2.47**	**2.44**	**2.42**	
21	4.32	3.47	3.07	2.84	2.68	2.57	2.49	2.42	2.37	2.32	2.28	2.25	2.20	2.15	2.09	2.05	2.00	1.96	1.93	1.89	1.87	1.84	1.82	1.81	21
	8.02	**5.78**	**4.87**	**4.37**	**4.04**	**3.81**	**3.65**	**3.51**	**3.40**	**3.31**	**3.24**	**3.17**	**3.07**	**2.99**	**2.88**	**2.80**	**2.72**	**2.63**	**2.58**	**2.51**	**2.47**	**2.42**	**2.38**	**2.36**	
22	4.30	3.44	3.05	2.82	2.66	2.55	2.47	2.40	2.35	2.30	2.26	2.23	2.18	2.13	2.07	2.03	1.98	1.93	1.91	1.87	1.84	1.81	1.80	1.78	22
	7.94	**5.72**	**4.82**	**4.31**	**3.99**	**3.76**	**3.59**	**3.45**	**3.35**	**3.26**	**3.18**	**3.12**	**3.02**	**2.94**	**2.83**	**2.75**	**2.67**	**2.58**	**2.53**	**2.46**	**2.42**	**2.37**	**2.33**	**2.31**	
23	4.28	3.42	3.03	2.80	2.64	2.53	2.45	2.38	2.32	2.28	2.24	2.20	2.14	2.10	2.04	2.00	1.96	1.91	1.88	1.84	1.82	1.79	1.77	1.76	23
	7.88	**5.66**	**4.76**	**4.26**	**3.94**	**3.71**	**3.54**	**3.41**	**3.30**	**3.21**	**3.14**	**3.07**	**2.97**	**2.89**	**2.78**	**2.70**	**2.62**	**2.53**	**2.48**	**2.41**	**2.37**	**2.32**	**2.28**	**2.26**	
24	4.26	3.40	3.01	2.78	2.62	2.51	2.43	2.36	2.30	2.26	2.22	2.18	2.13	2.09	2.02	1.98	1.94	1.89	1.86	1.82	1.80	1.76	1.74	1.73	24
	7.82	**5.61**	**4.72**	**4.22**	**3.90**	**3.67**	**3.50**	**3.36**	**3.25**	**3.17**	**3.09**	**3.03**	**2.93**	**2.85**	**2.74**	**2.66**	**2.58**	**2.49**	**2.44**	**2.36**	**2.23**	**2.27**	**2.33**	**2.21**	
25	4.24	3.38	2.99	2.76	2.60	2.49	2.41	2.34	2.28	2.24	2.20	2.16	2.11	2.06	2.00	1.96	1.92	1.87	1.84	1.80	1.77	1.74	1.72	1.71	25
	7.77	**5.57**	**4.68**	**4.18**	**3.86**	**3.63**	**3.46**	**3.32**	**3.21**	**3.13**	**3.05**	**2.99**	**2.89**	**2.81**	**2.70**	**2.62**	**2.54**	**2.45**	**2.40**	**2.32**	**2.29**	**2.23**	**2.19**	**2.17**	
26	4.22	3.37	2.98	2.74	2.59	2.47	2.39	2.32	2.27	2.22	2.18	2.15	2.10	2.05	1.99	1.95	1.90	1.85	1.82	1.78	1.76	1.72	1.70	1.69	26
	7.72	**5.53**	**4.64**	**4.14**	**3.82**	**3.59**	**3.42**	**3.29**	**3.17**	**3.09**	**3.02**	**2.96**	**2.86**	**2.77**	**2.66**	**2.58**	**2.50**	**2.41**	**2.36**	**2.28**	**2.25**	**2.19**	**2.15**	**2.13**	

(continued)

TABLE F *(continued)*

Degrees of freedom (for the denominator of F ratio)	Degrees of freedom (for the numerator of F ratio)																								
	1	2	3	4	5	6	7	8	9	10	11	12	14	16	20	24	30	40	50	75	100	200	500	∞	
27	4.21	3.35	2.96	2.73	2.57	2.46	2.37	2.30	2.25	2.20	2.16	2.13	2.08	2.03	1.97	1.93	1.88	1.84	1.80	1.76	1.74	1.71	1.68	1.67	27
	7.68	**5.49**	**4.60**	**4.11**	**3.79**	**3.56**	**3.39**	**3.26**	**3.14**	**3.06**	**2.98**	**2.93**	**2.83**	**2.74**	**2.63**	**2.55**	**2.47**	**2.38**	**2.33**	**2.25**	**2.21**	**2.16**	**2.12**	**2.10**	
28	4.20	3.34	2.95	2.71	2.56	2.44	2.36	2.29	2.24	2.19	2.15	2.12	2.06	2.02	1.96	1.91	1.87	1.81	1.78	1.75	1.72	1.69	1.67	1.65	28
	7.64	**5.45**	**4.57**	**4.07**	**3.76**	**3.53**	**3.36**	**3.23**	**3.11**	**3.03**	**2.95**	**2.90**	**2.80**	**2.71**	**2.60**	**2.52**	**2.44**	**2.35**	**2.30**	**2.22**	**2.18**	**2.13**	**2.09**	**2.06**	
29	4.18	3.33	2.93	2.70	2.54	2.43	2.35	2.28	2.22	2.18	2.14	2.10	2.05	2.00	1.94	1.90	1.85	1.80	1.77	1.73	1.71	1.68	1.65	1.64	29
	7.60	**5.42**	**4.54**	**4.04**	**3.73**	**3.50**	**3.33**	**3.20**	**3.08**	**3.00**	**2.92**	**2.87**	**2.77**	**2.68**	**2.57**	**2.49**	**2.41**	**2.32**	**2.27**	**2.19**	**2.15**	**2.10**	**2.06**	**2.03**	
30	4.17	3.32	2.92	2.69	2.53	2.42	2.34	2.27	2.21	2.16	2.12	2.09	2.04	1.99	1.93	1.89	1.84	1.79	1.76	1.72	1.69	1.66	1.64	1.62	30
	7.56	**5.39**	**4.51**	**4.02**	**3.70**	**3.47**	**3.30**	**3.17**	**3.06**	**2.98**	**2.90**	**2.84**	**2.74**	**2.66**	**2.55**	**2.47**	**2.38**	**2.29**	**2.24**	**2.16**	**2.13**	**2.07**	**2.03**	**2.01**	
32	4.15	3.30	2.90	2.67	2.51	2.40	2.32	2.25	2.19	2.14	2.10	2.07	2.02	1.97	1.91	1.86	1.82	1.76	1.74	1.69	1.67	1.64	1.61	1.59	32
	7.50	**5.34**	**4.46**	**3.97**	**3.66**	**3.42**	**3.25**	**3.12**	**3.01**	**2.94**	**2.86**	**2.80**	**2.70**	**2.62**	**2.51**	**2.42**	**2.34**	**2.25**	**2.20**	**2.12**	**2.08**	**2.02**	**1.98**	**1.96**	
34	4.13	3.28	2.88	2.65	2.49	2.38	2.30	2.23	2.17	2.12	2.08	2.05	2.00	1.95	1.89	1.84	1.80	1.74	1.71	1.67	1.64	1.61	1.59	1.57	34
	7.44	**5.29**	**4.42**	**3.93**	**3.61**	**3.38**	**3.21**	**3.08**	**2.97**	**2.89**	**2.82**	**2.76**	**2.66**	**2.58**	**2.47**	**2.38**	**2.30**	**2.21**	**2.15**	**2.08**	**2.04**	**1.98**	**1.94**	**1.91**	
36	4.11	3.26	2.86	2.63	2.48	2.36	2.28	2.21	2.15	2.10	2.06	2.03	1.98	1.93	1.87	1.82	1.78	1.72	1.69	1.65	1.62	1.59	1.56	1.55	36
	7.39	**5.24**	**4.38**	**3.89**	**3.58**	**3.35**	**3.18**	**3.04**	**2.94**	**2.86**	**2.78**	**2.72**	**2.62**	**2.54**	**2.43**	**2.35**	**2.26**	**2.17**	**2.12**	**2.04**	**2.00**	**1.94**	**1.90**	**1.87**	
38	4.10	3.25	2.85	2.62	2.46	2.35	2.26	2.19	2.14	2.09	2.05	2.02	1.96	1.92	1.85	1.80	1.76	1.71	1.67	1.63	1.60	1.57	1.54	1.53	38
	7.35	**5.21**	**4.34**	**3.86**	**3.54**	**3.32**	**3.15**	**3.02**	**2.91**	**2.82**	**2.75**	**2.69**	**2.59**	**2.51**	**2.40**	**2.32**	**2.22**	**2.14**	**2.08**	**2.00**	**1.97**	**1.90**	**1.86**	**1.84**	
40	4.08	3.23	2.84	2.61	2.45	2.34	2.25	2.18	2.12	2.07	2.04	2.00	1.95	1.90	1.84	1.79	1.74	1.69	1.66	1.61	1.59	1.55	1.53	1.51	40
	7.31	**5.18**	**4.31**	**3.83**	**3.51**	**3.29**	**3.12**	**2.99**	**2.88**	**2.80**	**2.73**	**2.66**	**2.56**	**2.49**	**2.37**	**2.29**	**2.20**	**2.11**	**2.05**	**1.97**	**1.94**	**1.88**	**1.84**	**1.81**	
42	4.07	3.22	2.83	2.59	2.44	2.32	2.24	2.17	2.11	2.06	2.02	1.99	1.94	1.89	1.82	1.78	1.73	1.68	1.64	1.60	1.57	1.54	1.51	1.49	42
	7.25	**5.15**	**4.29**	**3.80**	**3.49**	**3.26**	**3.10**	**2.96**	**2.86**	**2.77**	**2.70**	**2.64**	**2.54**	**2.46**	**2.35**	**2.26**	**2.17**	**2.08**	**2.02**	**1.94**	**1.91**	**1.85**	**1.80**	**1.78**	
44	4.06	3.21	2.82	2.58	2.43	2.31	2.23	2.16	2.10	2.05	2.01	1.98	1.92	1.88	1.81	1.76	1.72	1.66	1.63	1.58	1.56	1.52	1.50	1.48	44
	7.24	**5.12**	**4.26**	**3.78**	**3.46**	**3.24**	**3.07**	**2.94**	**2.84**	**2.75**	**2.68**	**2.62**	**2.52**	**2.44**	**2.32**	**2.24**	**2.15**	**2.06**	**2.00**	**1.92**	**1.88**	**1.82**	**1.78**	**1.75**	
46	4.05	3.20	2.81	2.57	2.42	2.30	2.22	2.14	2.09	2.04	2.00	1.97	1.91	1.87	1.80	1.75	1.71	1.65	1.62	1.57	1.54	1.51	1.48	1.46	46
	7.21	**5.10**	**4.24**	**3.76**	**3.44**	**3.22**	**3.05**	**2.92**	**2.82**	**2.73**	**2.66**	**2.60**	**2.50**	**2.42**	**2.30**	**2.22**	**2.13**	**2.04**	**1.98**	**1.90**	**1.86**	**1.80**	**1.76**	**1.72**	
48	4.04	3.19	2.80	2.56	2.41	2.30	2.21	2.14	2.08	2.03	1.99	1.96	1.90	1.86	1.79	1.74	1.70	1.64	1.61	1.56	1.53	1.50	1.47	1.45	48
	7.19	**5.08**	**4.22**	**3.74**	**3.42**	**3.20**	**3.04**	**2.90**	**2.80**	**2.71**	**2.64**	**2.58**	**2.48**	**2.40**	**2.28**	**2.20**	**2.11**	**2.02**	**1.96**	**1.88**	**1.84**	**1.78**	**1.73**	**1.70**	

(continued)

TABLE F *(continued)*

Degrees of freedom (for the denominator of F ratio)	Degrees of freedom (for the numerator of F ratio)																								
	1	2	3	4	5	6	7	8	9	10	11	12	14	16	20	24	30	40	50	75	100	200	500	∞	
50	4.03	3.18	2.79	2.56	2.40	2.29	2.20	2.13	2.07	2.02	1.98	1.95	1.90	1.85	1.78	1.74	1.69	1.63	1.60	1.55	1.52	1.48	1.46	1.44	50
	7.17	**5.06**	**4.20**	**3.72**	**3.41**	**3.18**	**3.02**	**2.88**	**2.78**	**2.70**	**2.62**	**2.56**	**2.46**	**2.39**	**2.26**	**2.18**	**2.10**	**2.00**	**1.94**	**1.86**	**1.82**	**1.76**	**1.71**	**1.68**	
55	4.02	3.17	2.78	2.54	2.38	2.27	2.18	2.11	2.05	2.00	1.97	1.93	1.88	1.83	1.76	1.72	1.67	1.61	1.58	1.52	1.50	1.46	1.43	1.41	55
	7.12	**5.01**	**4.16**	**3.68**	**3.37**	**3.15**	**2.98**	**2.85**	**2.75**	**2.66**	**2.59**	**2.53**	**2.43**	**2.35**	**2.23**	**2.15**	**2.06**	**1.96**	**1.90**	**1.82**	**1.78**	**1.71**	**1.66**	**1.64**	
60	4.00	3.15	2.76	2.52	2.37	2.25	2.17	2.10	2.04	1.99	1.95	1.92	1.86	1.81	1.75	1.70	1.65	1.59	1.56	1.50	1.48	1.44	1.41	1.39	60
	7.08	**4.98**	**4.13**	**3.65**	**3.34**	**3.12**	**2.95**	**2.82**	**2.72**	**2.63**	**2.56**	**2.50**	**2.40**	**2.32**	**2.20**	**2.12**	**2.03**	**1.93**	**1.87**	**1.79**	**1.74**	**1.68**	**1.63**	**1.60**	
65	3.99	3.14	2.75	2.51	2.36	2.24	2.15	2.08	2.02	1.98	1.94	1.90	1.85	1.80	1.73	1.68	1.63	1.57	1.54	1.49	1.46	1.42	1.39	1.37	65
	7.04	**4.95**	**4.10**	**3.62**	**3.31**	**3.09**	**2.93**	**2.79**	**2.70**	**2.61**	**2.54**	**2.47**	**2.37**	**2.30**	**2.18**	**2.09**	**2.00**	**1.90**	**1.84**	**1.76**	**1.71**	**1.64**	**1.60**	**1.56**	
70	3.98	3.13	2.74	2.50	2.35	2.23	2.14	2.07	2.01	1.97	1.93	1.89	1.84	1.79	1.72	1.67	1.62	1.56	1.53	1.47	1.45	1.40	1.37	1.35	70
	7.01	**4.92**	**4.08**	**3.60**	**3.29**	**3.07**	**2.91**	**2.77**	**2.67**	**2.59**	**2.51**	**2.45**	**2.35**	**2.28**	**2.15**	**2.07**	**1.98**	**1.88**	**1.82**	**1.74**	**1.69**	**1.62**	**1.56**	**1.53**	
80	3.96	3.11	2.72	2.48	2.33	2.21	2.12	2.05	1.99	1.95	1.91	1.88	1.82	1.77	1.70	1.65	1.60	1.54	1.51	1.45	1.42	1.38	1.35	1.32	80
	6.96	**4.88**	**4.04**	**3.56**	**3.25**	**3.04**	**2.87**	**2.74**	**2.64**	**2.55**	**2.48**	**2.41**	**2.32**	**2.24**	**2.11**	**2.03**	**1.94**	**1.84**	**1.78**	**1.70**	**1.65**	**1.57**	**1.52**	**1.49**	
100	3.94	3.09	2.70	2.46	2.30	2.19	2.10	2.03	1.97	1.92	1.88	1.85	1.79	1.75	1.68	1.63	1.57	1.51	1.48	1.42	1.39	1.34	1.30	1.28	100
	6.90	**4.82**	**3.98**	**3.51**	**3.20**	**2.99**	**2.82**	**2.69**	**2.59**	**2.51**	**2.43**	**2.36**	**2.26**	**2.19**	**2.06**	**1.98**	**1.89**	**1.79**	**1.73**	**1.64**	**1.59**	**1.51**	**1.46**	**1.43**	
125	3.92	3.07	2.68	2.44	2.29	2.17	2.08	2.01	1.95	1.90	1.86	1.83	1.77	1.72	1.65	1.60	1.55	1.49	1.45	1.39	1.36	1.31	1.27	1.25	125
	6.84	**4.78**	**3.94**	**3.47**	**3.17**	**2.95**	**2.79**	**2.65**	**2.56**	**2.47**	**2.40**	**2.33**	**2.23**	**2.15**	**2.03**	**1.94**	**1.85**	**1.75**	**1.68**	**1.59**	**1.54**	**1.46**	**1.40**	**1.37**	
150	3.91	3.06	2.67	2.43	2.27	2.16	2.07	2.00	1.94	1.89	1.85	1.82	1.76	1.71	1.64	1.59	1.54	1.47	1.44	1.37	1.34	1.29	1.25	1.22	150
	6.81	**4.75**	**3.91**	**3.44**	**3.14**	**2.92**	**2.76**	**2.62**	**2.53**	**2.44**	**2.37**	**2.30**	**2.20**	**2.12**	**2.00**	**1.91**	**1.83**	**1.72**	**1.66**	**1.56**	**1.51**	**1.43**	**1.37**	**1.33**	
200	3.89	3.04	2.65	2.41	2.26	2.14	2.05	1.98	1.92	1.87	1.83	1.80	1.74	1.69	1.62	1.57	1.52	1.45	1.42	1.35	1.32	1.26	1.22	1.19	200
	6.76	**4.71**	**3.88**	**3.41**	**3.11**	**2.90**	**2.73**	**2.60**	**2.50**	**2.41**	**2.34**	**2.28**	**2.17**	**2.09**	**1.97**	**1.88**	**1.79**	**1.69**	**1.62**	**1.53**	**1.48**	**1.39**	**1.33**	**1.28**	
400	3.86	3.02	2.62	2.39	2.23	2.12	2.03	1.96	1.90	1.85	1.81	1.78	1.72	1.67	1.60	1.54	1.49	1.42	1.38	1.32	1.28	1.22	1.16	1.13	400
	6.70	**4.66**	**3.83**	**3.36**	**3.06**	**2.85**	**2.69**	**2.55**	**2.46**	**2.37**	**2.29**	**2.23**	**2.12**	**2.04**	**1.92**	**1.84**	**1.74**	**1.64**	**1.57**	**1.47**	**1.42**	**1.32**	**1.24**	**1.19**	
1000	3.85	3.00	2.61	2.38	2.22	2.10	2.02	1.95	1.89	1.84	1.80	1.76	1.70	1.65	1.58	1.53	1.47	1.41	1.36	1.30	1.26	1.19	1.13	1.08	1000
	6.66	**4.62**	**3.80**	**3.34**	**3.04**	**2.82**	**2.66**	**2.53**	**2.43**	**2.34**	**2.26**	**2.20**	**2.09**	**2.01**	**1.89**	**1.81**	**1.71**	**1.61**	**1.54**	**1.44**	**1.38**	**1.28**	**1.19**	**1.11**	
∞	3.84	2.99	2.60	2.37	2.21	2.09	2.01	1.94	1.88	1.83	1.79	1.75	1.69	1.64	1.57	1.52	1.46	1.40	1.35	1.28	1.24	1.17	1.11	1.00	∞
	6.64	**4.60**	**3.78**	**3.32**	**3.02**	**2.80**	**2.64**	**2.51**	**2.41**	**2.32**	**2.24**	**2.18**	**2.07**	**1.99**	**1.87**	**1.79**	**1.69**	**1.59**	**1.52**	**1.41**	**1.36**	**1.25**	**1.15**	**1.00**	

TABLE G Critical values of the studentized range statistic (for Tukey HSD tests)

	$\alpha = .05$													
	Number of levels of the independent variable													
df_{error}	2	3	4	5	6	7	8	9	10	11	12	13	14	15
1	17.97	26.98	32.82	37.07	40.41	43.12	45.40	47.36	49.07	50.59	51.96	53.20	54.33	55.36
2	6.08	8.33	9.80	10.88	11.74	12.44	13.03	13.54	13.99	14.39	14.75	15.08	15.38	15.65
3	4.50	5.91	6.82	7.50	8.04	8.48	8.85	9.18	9.46	9.72	9.95	10.15	10.35	10.53
4	3.93	5.04	5.76	6.29	6.71	7.05	7.35	7.60	7.33	8.03	8.21	8.37	8.52	8.66
5	3.64	4.60	5.22	5.67	6.03	6.33	6.58	6.80	7.00	7.17	7.32	7.47	7.60	7.72
6	3.46	4.34	4.90	5.31	5.63	5.90	6.12	6.32	6.49	6.65	6.79	6.92	7.03	7.14
7	3.34	4.16	4.68	5.06	5.36	5.61	5.82	6.00	6.16	6.30	6.43	6.55	6.66	6.76
8	3.26	4.04	4.53	4.89	5.17	5.40	5.60	5.77	5.92	6.05	6.18	6.29	6.39	6.48
9	3.20	3.95	4.42	4.76	5.02	5.24	5.43	5.60	5.74	5.87	5.98	6.09	6.19	6.28
10	3.15	3.88	4.33	4.65	4.91	5.12	5.30	5.46	5.60	5.72	5.83	5.94	6.03	6.11
11	3.11	3.82	4.26	4.57	4.82	5.03	5.20	5.35	5.49	5.60	5.71	5.81	5.90	5.98
12	3.08	3.77	4.20	4.51	4.75	4.95	5.12	5.26	5.40	5.51	5.62	5.71	5.79	5.88
13	3.06	3.74	4.15	4.45	4.69	4.88	5.05	5.19	5.32	5.43	5.53	5.63	5.71	5.79
14	3.03	3.70	4.11	4.41	4.64	4.83	4.99	5.13	5.25	5.36	5.46	5.55	5.64	5.71
15	3.01	3.67	4.08	4.37	4.60	4.78	4.94	5.08	5.20	5.31	5.40	5.49	5.57	5.65
16	3.00	3.65	4.05	4.33	4.56	4.74	4.90	5.03	5.15	5.26	5.35	5.44	5.52	5.59
17	2.98	3.63	4.02	4.30	4.52	4.70	4.86	4.99	5.11	5.21	5.31	5.39	5.47	5.54
18	2.97	3.61	4.00	4.28	4.50	4.67	4.82	4.96	5.07	5.17	5.27	5.35	5.43	5.50
19	2.96	3.59	3.98	4.25	4.47	4.64	4.79	4.92	5.04	5.14	5.23	5.32	5.39	5.46
20	2.95	3.58	3.96	4.23	4.44	4.62	4.77	4.90	5.01	5.11	5.20	5.28	5.36	5.43
24	2.92	3.53	3.90	4.17	4.37	4.54	4.68	4.81	4.92	5.01	5.10	5.18	5.25	5.32
30	2.89	3.49	3.84	4.10	4.30	4.46	4.60	4.72	4.82	4.92	5.00	5.08	5.15	5.21
40	2.86	3.44	3.79	4.04	4.23	4.39	4.52	4.64	4.74	4.82	4.90	4.98	5.04	5.11
60	2.83	3.40	3.74	3.98	4.16	4.31	4.44	4.55	4.65	4.73	4.81	4.88	4.94	5.00
120	2.80	3.36	3.69	3.92	4.10	4.24	4.36	4.47	4.56	4.64	4.71	4.78	4.84	4.90
∞	2.77	3.31	3.63	3.86	4.03	4.17	4.29	4.39	4.47	4.55	4.62	4.68	4.74	4.80

(continued)

Source: From "Tables of Range and Studentized Range," by H. L. Harter, 1960, *Annals of Mathematical Statistics, 31,* 1122–1147.

TABLE G *(continued)*

	$\alpha = .01$													
	Number of levels of the independent variable													
df_{error}	2	3	4	5	6	7	8	9	10	11	12	13	14	15
1	90.03	135.00	164.30	185.60	202.20	215.80	227.20	237.00	245.60	253.20	260.00	266.20	271.80	277.00
2	14.04	19.02	22.29	24.72	26.63	28.20	29.53	30.68	31.69	32.59	33.40	34.13	34.81	35.43
3	8.26	10.62	12.17	13.33	14.24	15.00	15.64	12.60	16.69	17.13	17.53	17.89	18.22	18.52
4	6.51	8.12	9.17	9.96	10.58	11.10	11.55	11.93	12.27	12.57	12.84	13.09	13.32	13.53
5	5.70	6.98	7.80	8.42	8.91	9.32	9.67	9.97	10.24	10.48	10.70	10.89	11.08	11.24
6	5.24	6.33	7.03	7.56	7.97	8.32	8.62	8.87	9.10	9.30	9.48	9.65	9.81	9.95
7	4.95	5.92	6.54	7.00	7.37	7.68	7.94	8.17	8.37	8.55	8.71	8.86	9.00	9.12
8	4.75	5.64	6.20	6.62	6.96	7.24	7.47	7.68	7.86	8.03	8.18	8.31	8.44	8.55
9	4.60	5.43	5.96	6.35	6.66	6.92	7.13	7.32	7.50	7.65	7.78	7.91	8.02	8.13
10	4.48	5.27	5.77	6.14	6.43	6.67	6.88	7.06	7.21	7.36	7.48	7.60	7.71	7.81
11	4.39	5.15	5.62	5.97	6.25	6.48	6.67	6.84	6.99	7.13	7.25	7.36	7.46	7.56
12	4.32	5.05	5.50	5.84	6.10	6.32	6.51	6.67	6.81	6.94	7.06	7.17	7.26	7.36
13	4.26	4.96	5.40	5.73	5.98	6.19	6.37	6.53	6.67	6.79	6.90	7.01	7.10	7.19
14	4.21	4.90	5.32	5.63	5.88	6.08	6.26	6.41	6.54	6.66	6.77	6.87	6.96	7.05
15	4.17	4.84	5.25	5.56	5.80	5.99	6.16	6.31	6.44	6.56	6.66	6.76	6.84	6.93
16	4.13	4.79	5.19	5.49	5.72	5.92	6.08	6.22	6.35	6.46	6.56	6.66	6.74	6.82
17	4.10	4.74	5.14	5.43	5.66	5.85	6.01	6.15	6.27	6.38	6.48	6.57	6.66	6.73
18	4.07	4.70	5.09	5.38	5.60	5.79	5.94	6.08	6.20	6.31	6.41	6.50	6.58	6.66
19	4.05	4.67	5.05	5.33	5.55	5.74	5.89	6.02	6.14	6.25	6.34	6.43	6.51	6.58
20	4.02	4.64	5.02	5.29	5.51	5.69	5.84	5.97	6.09	6.19	6.28	6.37	6.45	6.52
24	3.96	4.55	4.91	5.17	5.37	5.54	5.69	5.81	5.92	6.02	6.11	6.19	6.26	6.33
30	3.89	4.46	4.80	5.05	5.24	5.40	5.54	5.65	5.76	5.85	5.93	6.01	6.08	6.14
40	3.82	4.37	4.70	4.93	5.11	5.26	5.39	5.50	5.60	5.69	5.76	5.84	5.90	5.96
60	3.76	4.28	4.60	4.82	4.99	5.13	5.25	5.36	5.45	5.53	5.60	5.67	5.73	5.78
120	3.70	4.20	3.50	4.71	4.87	5.01	5.12	5.21	5.30	5.38	5.44	5.51	5.56	5.61
∞	3.64	4.12	4.40	4.60	4.76	4.88	4.99	5.08	5.16	5.23	5.29	5.35	5.40	5.45

TABLE H Critical values for the Mann–Whitney U test*

One-tailed test
$\alpha = .01$ (lightface)
$\alpha = .005$ (**boldface**)

Two-tailed test
$\alpha = .02$ (lightface)
$\alpha = .01$ (**boldface**)

N_2 \ N_1	1	2	3	4	5	6	7	8	9	10	11	12	13	14	15	16	17	18	19	20
1	—	—	—	—	—	—	—	—	—	—	—	—	—	—	—	—	—	—	—	—
2	—	—	—	—	—	—	—	—	—	—	—	—	0	0	0	0	0	0	1	1
													—	—	—	—	—	—	**0**	**0**
3	—	—	—	—	—	—	0	0	1	1	1	2	2	2	3	3	4	4	4	5
							—	—	**0**	**0**	**0**	**1**	**1**	**1**	**2**	**2**	**2**	**2**	**3**	**3**
4	—	—	—	—	0	1	1	2	3	3	45	5	6	7	7	8	9	9	10	
					—	**0**	**0**	**1**	**1**	**2**	**2**	**3**	**3**	**4**	**5**	**5**	**6**	**6**	**7**	**8**
5	—	—	—	0	1	2	3	4	5	6	7	8	9	10	11	12	13	14	15	16
				—	**0**	**1**	**1**	**2**	**3**	**4**	**5**	**6**	**7**	**7**	**8**	**9**	**10**	**11**	**12**	**13**
6	—	—	—	1	2	3	4	6	7	8	9	11	12	13	15	16	18	19	20	22
				0	**1**	**2**	**3**	**4**	**5**	**6**	**7**	**9**	**10**	**11**	**12**	**13**	**15**	**16**	**17**	**18**
7	—	—	0	1	3	4	6	7	9	11	12	14	16	17	19	21	23	24	26	28
			—	**0**	**1**	**3**	**4**	**6**	**7**	**9**	**10**	**12**	**13**	**15**	**16**	**18**	**19**	**21**	**22**	**24**
8	—	—	0	2	4	6	7	9	11	13	15	17	20	22	24	26	28	30	32	34
			—	**1**	**2**	**4**	**6**	**7**	**9**	**11**	**13**	**15**	**17**	**18**	**20**	**22**	**24**	**26**	**29**	**30**
9	—	—	1	3	5	7	9	11	14	16	18	21	23	26	28	31	33	36	38	40
			0	**1**	**3**	**5**	**7**	**9**	**11**	**13**	**16**	**18**	**20**	**22**	**24**	**27**	**29**	**31**	**33**	**36**
10	—	—	1	3	6	8	11	13	16	19	22	24	27	30	33	36	38	41	44	47
			0	**2**	**4**	**6**	**9**	**11**	**13**	**16**	**18**	**21**	**24**	**26**	**29**	**31**	**34**	**37**	**39**	**42**
11	—	—	1	4	7	9	12	15	18	22	25	28	31	34	37	41	44	47	50	53
			0	**2**	**5**	**7**	**10**	**13**	**16**	**18**	**21**	**24**	**27**	**30**	**33**	**36**	**39**	**42**	**45**	**48**
12	—	—	2	5	8	11	14	17	21	24	28	31	35	38	42	46	49	53	56	60
			1	**3**	**6**	**9**	**12**	**15**	**18**	**21**	**24**	**27**	**31**	**34**	**37**	**41**	**44**	**47**	**51**	**54**
13	—	0	2	5	9	12	16	20	23	27	31	35	39	43	47	51	55	59	63	67
		—	**1**	**3**	**7**	**10**	**13**	**17**	**20**	**24**	**27**	**31**	**34**	**38**	**42**	**45**	**49**	**53**	**56**	**60**
14	—	0	2	6	10	13	17	22	26	30	34	38	43	47	51	56	60	65	69	73
		—	**1**	**4**	**7**	**11**	**15**	**18**	**22**	**26**	**30**	**34**	**38**	**42**	**46**	**50**	**54**	**58**	**63**	**67**
15	—	0	3	7	11	15	19	24	28	33	37	42	47	51	56	61	66	70	75	80
		—	**2**	**5**	**8**	**12**	**16**	**20**	**24**	**29**	**33**	**37**	**42**	**46**	**51**	**55**	**60**	**64**	**69**	**73**
16	—	0	3	7	12	16	21	26	31	36	41	46	51	56	61	66	71	76	82	87
		—	**2**	**5**	**9**	**13**	**18**	**22**	**27**	**31**	**36**	**41**	**45**	**50**	**55**	**60**	**65**	**70**	**74**	**79**
17	—	0	4	8	13	18	23	28	33	38	44	49	55	60	66	71	77	82	88	93
		—	**2**	**6**	**10**	**15**	**19**	**24**	**29**	**34**	**39**	**44**	**49**	**54**	**60**	**65**	**70**	**75**	**81**	**86**
18	—	0	4	9	14	19	24	30	36	41	47	53	59	65	70	76	82	88	94	100
		—	**2**	**6**	**11**	**16**	**21**	**26**	**31**	**37**	**42**	**47**	**53**	**58**	**64**	**70**	**75**	**81**	**87**	**92**
19	—	1	4	9	15	20	26	32	38	44	50	56	63	69	75	82	88	94	101	107
		0	**3**	**7**	**12**	**17**	**22**	**28**	**33**	**39**	**45**	**51**	**56**	**63**	**69**	**74**	**81**	**87**	**93**	**99**
20	—	1	5	10	16	22	28	34	40	47	53	60	67	73	80	87	93	100	107	114
		0	**3**	**8**	**13**	**18**	**24**	**30**	**36**	**42**	**48**	**54**	**60**	**67**	**73**	**79**	**86**	**92**	**99**	**105**

(continued)

* To be significant, the U obtained from data must be *equal to or less than* the value shown in the table. Dashes in the body of the table indicate that no decision is possible at the stated level of significance.

Source: From *Elementary Statistics,* Second Edition, by R. E. Kirk, Brooks/Cole Publishing, 1984.

TABLE H *(continued)*

One-tailed test
$\alpha = .05$ (lightface)
$\alpha = .025$ (**boldface**)

Two-tailed test
$\alpha = .10$ (lightface)
$\alpha = .05$ (**boldface**)

N_2 \ N_1	1	2	3	4	5	6	7	8	9	10	11	12	13	14	15	16	17	18	19	20
1	—	—	—	—	—	—	—	—	—	—	—	—	—	—	—	—	—	—	0	0
2	—	—	—	—	0	0	0	1	1	1	1	2	2	2	3	3	3	4	4	4
					—	—	—	**0**	**0**	**0**	**0**	**1**	**1**	**1**	**1**	**1**	**2**	**2**	**2**	**2**
3	—	—	0	0	1	2	2	3	3	4	5	5	6	7	7	8	9	9	10	11
	—	—	—	—	**0**	**1**	**1**	**2**	**2**	**3**	**3**	**4**	**4**	**5**	**5**	**6**	**6**	**7**	**7**	**8**
4	—	—	0	1	2	3	4	5	6	7	8	9	10	11	12	14	15	16	17	18
	—	—	—	**0**	**1**	**2**	**3**	**4**	**4**	**5**	**6**	**7**	**8**	**9**	**10**	**11**	**11**	**12**	**13**	**13**
5	—	0	1	2	4	5	6	8	9	11	12	13	15	16	18	19	20	22	23	25
	—	—	**0**	**1**	**2**	**3**	**5**	**6**	**7**	**8**	**9**	**11**	**12**	**13**	**14**	**15**	**17**	**18**	**19**	**20**
6	—	0	2	3	5	7	8	10	12	14	16	17	19	21	23	25	26	28	30	32
	—	—	**1**	**2**	**3**	**5**	**6**	**8**	**10**	**11**	**13**	**14**	**16**	**17**	**19**	**21**	**22**	**24**	**25**	**27**
7	—	0	2	4	6	8	11	13	15	17	19	21	24	26	28	30	33	35	37	39
	—	—	**1**	**3**	**5**	**6**	**8**	**10**	**12**	**14**	**16**	**18**	**20**	**22**	**24**	**26**	**28**	**30**	**32**	**34**
8	—	1	3	5	8	10	13	15	18	20	23	26	28	31	33	36	39	41	44	47
	—	**0**	**2**	**4**	**6**	**8**	**10**	**13**	**15**	**17**	**19**	**22**	**24**	**26**	**29**	**31**	**34**	**36**	**38**	**41**
9	—	1	3	6	9	12	15	18	21	24	27	30	33	36	39	42	45	48	51	54
	—	**0**	**2**	**4**	**7**	**10**	**12**	**15**	**17**	**20**	**23**	**26**	**28**	**31**	**34**	**37**	**39**	**42**	**45**	**48**
10	—	1	4	7	11	14	17	20	24	27	31	34	37	41	44	48	51	55	58	62
	—	**0**	**3**	**5**	**8**	**11**	**14**	**17**	**20**	**23**	**26**	**29**	**33**	**36**	**39**	**42**	**45**	**48**	**52**	**55**
11	—	1	5	8	12	16	19	23	27	31	34	38	42	46	50	54	57	61	65	69
	—	**0**	**3**	**6**	**9**	**13**	**16**	**19**	**23**	**26**	**30**	**33**	**37**	**40**	**44**	**47**	**51**	**55**	**58**	**62**
12	—	2	5	9	13	17	21	26	30	34	38	42	47	51	55	60	64	68	72	77
	—	**1**	**4**	**7**	**11**	**14**	**18**	**22**	**26**	**29**	**33**	**37**	**41**	**45**	**49**	**53**	**57**	**61**	**65**	**69**
13	—	2	6	10	15	19	24	28	33	37	42	47	51	56	61	65	70	75	80	84
	—	**1**	**4**	**8**	**12**	**16**	**20**	**24**	**28**	**33**	**37**	**41**	**45**	**50**	**54**	**59**	**63**	**67**	**72**	**76**
14	—	2	7	11	16	21	26	31	36	41	46	51	56	61	66	71	77	82	87	92
	—	**1**	**5**	**9**	**13**	**17**	**22**	**26**	**31**	**36**	**40**	**45**	**50**	**55**	**59**	**64**	**67**	**74**	**78**	**83**
15	—	3	7	12	18	23	28	33	39	44	50	55	61	66	72	77	83	88	94	100
	—	**1**	**5**	**10**	**14**	**19**	**24**	**29**	**34**	**39**	**44**	**49**	**54**	**59**	**64**	**70**	**75**	**80**	**85**	**90**
16	—	3	8	14	19	25	30	36	42	48	54	60	65	71	77	83	89	95	101	107
	—	**1**	**6**	**11**	**15**	**21**	**26**	**31**	**37**	**42**	**47**	**53**	**59**	**64**	**70**	**75**	**81**	**86**	**92**	**98**
17	—	3	9	15	20	26	33	39	45	51	57	64	70	77	83	89	96	102	109	115
	—	**2**	**6**	**11**	**17**	**22**	**28**	**34**	**39**	**45**	**51**	**57**	**63**	**67**	**75**	**81**	**87**	**93**	**99**	**105**
18	—	4	9	16	22	28	35	41	48	55	61	68	75	82	88	95	102	109	116	123
	—	**2**	**7**	**12**	**18**	**24**	**30**	**36**	**42**	**48**	**55**	**61**	**67**	**74**	**80**	**86**	**93**	**99**	**106**	**112**
19	0	4	10	17	23	30	37	44	51	58	65	72	80	87	94	101	109	116	123	130
	—	**2**	**7**	**13**	**19**	**25**	**32**	**38**	**45**	**52**	**58**	**65**	**72**	**78**	**85**	**92**	**99**	**106**	**113**	**119**
20	0	4	11	18	25	32	39	47	54	62	69	77	84	92	100	107	115	123	130	138
	—	**2**	**8**	**13**	**20**	**27**	**34**	**41**	**48**	**55**	**62**	**69**	**76**	**83**	**90**	**98**	**105**	**112**	**119**	**127**

TABLE J Critical values for the Wilcoxon matched-pairs signed-ranks T test*

No. of pairs	α levels for a one-tailed test					α levels for a one-tailed test			
	.05	.025	.01	.005		.05	.025	.01	.005
	α levels for a two-tailed test					α levels for a two-tailed test			
N	.10	.05	.02	.01	N	.10	.05	.02	.01
5	0	—	—	—	28	130	116	101	91
6	2	0	—	—	29	140	126	110	100
7	3	2	0	—	30	151	137	120	109
8	5	3	1	0	31	163	147	130	118
9	8	5	3	1	32	175	159	140	128
10	10	8	5	3	33	187	170	151	138
11	13	10	7	5	34	200	182	162	148
12	17	13	9	7	35	213	195	173	159
13	21	17	12	9	36	227	208	185	171
14	25	21	15	12	37	241	221	198	182
15	30	25	19	15	38	256	235	211	194
16	35	29	23	19	39	271	249	224	207
17	41	34	27	23	40	286	264	238	220
18	47	40	32	27	41	302	279	252	233
19	53	46	37	32	42	319	294	266	247
20	60	52	43	37	43	336	310	281	261
21	67	58	49	42	44	353	327	296	276
22	75	65	55	48	45	371	343	312	291
23	83	73	62	54	46	389	361	328	307
24	91	81	69	61	47	407	378	345	322
25	100	89	76	68	48	426	396	362	339
26	110	98	84	75	49	446	415	379	355
27	119	107	92	83	50	466	434	397	373

* To be significant, the T obtained from the data must be *equal to or less than* the value shown in the table.
Source: From *Elementary Statistics,* Second Edition, by R. E. Kirk, Brooks/Cole Publishing, 1984.

TABLE K Critical differences for the Wilcoxon–Wilcox multiple-comparisons test*
(Column N represents the number in one group.)

				$\alpha = .05$ (two-tailed)				
N	$K = 3$	$K = 4$	$K = 5$	$K = 6$	$K = 7$	$K = 8$	$K = 9$	$K = 10$
1	3.3	4.7	6.1	7.5	9.0	10.5	12.0	13.5
2	8.8	12.6	16.5	20.5	24.7	28.9	33.1	37.4
3	15.7	22.7	29.9	37.3	44.8	52.5	60.3	68.2
4	23.9	34.6	45.6	57.0	68.6	80.4	92.4	104.6
5	33.1	48.1	63.5	79.3	95.5	112.0	128.8	145.8
6	43.3	62.9	83.2	104.0	125.3	147.0	169.1	191.4
7	54.4	79.1	104.6	130.8	157.6	184.9	212.8	240.9
8	66.3	96.4	127.6	159.6	192.4	225.7	259.7	294.1
9	78.9	114.8	152.0	190.2	229.3	269.1	309.6	350.6
10	92.3	134.3	177.8	222.6	268.4	315.0	362.4	410.5
11	106.3	154.8	205.0	256.6	309.4	363.2	417.9	473.3
12	120.9	176.2	233.4	292.2	352.4	413.6	476.0	539.1
13	136.2	198.5	263.0	329.3	397.1	466.2	536.5	607.7
14	152.1	221.7	293.8	367.8	443.6	520.8	599.4	679.0
15	168.6	245.7	325.7	407.8	491.9	577.4	664.6	752.8
16	185.6	270.6	358.6	449.1	541.7	635.9	732.0	829.2
17	203.1	269.2	392.6	491.7	593.1	696.3	801.5	907.9
18	221.2	322.6	427.6	535.5	646.1	758.5	873.1	989.0
19	239.8	349.7	463.6	580.6	700.5	822.4	946.7	1072.4
20	258.8	377.6	500.5	626.9	756.4	888.1	1022.3	1158.1
21	278.4	406.1	538.4	674.4	813.7	955.4	1099.8	1245.9
22	298.4	435.3	577.2	723.0	872.3	1024.3	1179.1	1335.7
23	318.9	465.2	616.9	772.7	932.4	1094.8	1260.3	1427.7
24	339.8	495.8	657.4	823.5	993.7	1166.8	1343.2	1521.7
25	361.1	527.0	698.8	875.4	1056.3	1240.4	1427.9	1617.6

				$\alpha = .01$ (two-tailed)				
N	$K = 3$	$K = 4$	$K = 5$	$K = 6$	$K = 7$	$K = 8$	$K = 9$	$K = 10$
1	4.1	5.7	7.3	8.9	10.5	12.2	13.9	15.6
2	10.9	15.3	19.7	24.3	28.9	33.6	38.3	43.1
3	19.5	27.5	35.7	44.0	52.5	61.1	69.8	78.6
4	29.7	41.9	54.5	67.3	80.3	93.6	107.0	120.6
5	41.2	58.2	75.8	93.6	111.9	130.4	149.1	168.1
6	53.9	76.3	99.3	122.8	146.7	171.0	195.7	220.6
7	67.6	95.8	124.8	154.4	184.6	215.2	246.3	277.7
8	82.4	116.8	152.2	188.4	225.2	262.6	300.6	339.0
9	98.1	139.2	181.4	224.5	268.5	313.1	358.4	404.2
10	114.7	162.8	212.2	262.7	314.2	366.5	419.5	473.1
11	132.1	187.6	244.6	302.9	362.2	422.6	483.7	545.6
12	150.4	213.5	278.5	344.9	412.5	481.2	551.0	621.4
13	169.4	240.6	313.8	388.7	464.9	542.4	621.0	700.5
14	189.1	268.7	350.5	434.2	519.4	606.0	693.8	782.6
15	209.6	297.6	388.5	481.3	575.8	671.9	769.3	867.7
16	230.7	327.9	427.9	530.1	634.2	740.0	847.3	955.7
17	252.5	359.0	468.4	580.3	694.4	810.2	927.8	1046.5
18	275.0	391.0	510.2	632.1	756.4	882.6	1010.6	1140.0
19	298.1	423.8	553.1	685.4	820.1	957.0	1095.8	1236.2
20	321.8	457.6	597.2	740.0	885.5	1033.3	1183.3	1334.9
21	346.1	492.2	642.4	796.0	952.6	1111.6	1273.0	1436.0
22	371.0	527.6	688.7	853.4	1021.3	1191.8	1364.8	1539.7
23	396.4	563.8	736.0	912.1	1091.5	1273.8	1458.8	1645.7
24	422.4	600.9	784.4	972.1	1163.4	1357.6	1554.8	1754.0
25	449.0	638.7	833.8	1033.3	1236.7	1443.2	1652.8	1864.6

* To be significant, the difference obtained from the data must be *equal to or greater than* the tabled value.
Source: From *Some Rapid Approximate Statistical Procedures,* by F. Wilcoxon and R. Wilcox.

TABLE L Critical values for Spearman r_s*

No. of pairs N	α levels			
	One-tailed test		Two-tailed test	
	.01	.05	.01	.05
4	—	1.000	—	—
5	1.000	.900	—	1.000
6	.943	.829	1.000	.886
7	.893	.714	.929	.786
8	.833	.643	.881	.738
9	.783	.600	.833	.700
10	.745	.564	.794	.648
11	.709	.536	.755	.618
12	.677	.503	.727	.587
13	.648	.484	.703	.560
14	.626	.464	.679	.538
15	.604	.446	.654	.521
16	.582	.429	.635	.503

For samples larger than 16, use Table A, which requires *df*.

* To be significant, the r_s obtained from the data must be *equal to or greater than* the value shown in the table.
Source: From "Testing the Significance of Kendall's τ and Spearman's r_s" by M. Nijsse, 1988, *Psychological Bulletin, 103,* 235–237. Copyright © 1988 American Psychological Association. Reprinted by permission.

APPENDIX D

Glossary of Words

***a priori* test** Multiple-comparisons test that must be planned before examination of the data.

abscissa The horizontal, or X, axis of a graph.

absolute value A number without consideration of its algebraic sign.

alpha (α) The probability of a Type I error.

alternative hypothesis (H_1) A hypothesis about population parameters that is accepted if the null hypothesis is rejected.

analysis of variance (ANOVA) An inferential statistics technique for comparing means, comparing variances, and assessing interactions.

asymptotic A line that continually approaches but never reaches a specified limit.

bar graph A graph of the frequency distribution of nominal or qualitative data.

beta (β) The probability of a Type II error.

biased sample A sample selected in such a way that not all samples from the population have an equal chance of being chosen.

bimodal distribution A distribution with two modes.

binomial distribution A distribution of the frequency of events that can have only two possible outcomes.

bivariate distribution A joint distribution of two variables. The individual scores of the variables are paired in some logical way.

boxplot A graph that shows a distribution's range, interquartile range, skew, median, and sometimes other statistics.

cell In a table of ANOVA data, those scores that receive the same combination of treatments.

Central Limit Theorem The sampling distribution of the mean approaches a normal curve as N gets larger. This normal curve has a mean equal to μ and a standard deviation equal to $\sigma/\sqrt{N}$.

central tendency Descriptive statistics that indicate a typical or representative score; mean, median, mode.

chi square distribution A theoretical sampling distribution of chi square values. Chi square distributions vary with degrees of freedom.

chi square test An NHST technique that compares observed frequencies to expected frequencies.

class interval A range of scores in a grouped frequency distribution.

coefficient of determination A squared correlation coefficient, which is an estimate of common variance of the two variables.

confidence interval A range of scores that is expected, with specified confidence, to capture a parameter.

control group A no-treatment group to which other groups are compared.

correlation coefficient A descriptive statistic calculated on bivariate data that expresses the degree of relationship between two variables.

critical region Synonym for rejection region.

critical value Number from a sampling distribution that determines whether the null hypothesis is rejected.

degrees of freedom A concept in mathematical statistics that determines the distribution that is appropriate for a particular set of sample data.

dependent variable The observed variable that is expected to change as a result of changes in the independent variable in an experiment.

descriptive statistic A number that conveys a particular characteristic of a set of data. Graphs and tables are sometimes included in this category. (Congratulations to you if you have checked this entry after reading footnote 1 in Chapter 1. Very few students make the effort to check out their authors' claims, as you just did. You have one of the makings of a scholar.)

deviation score A raw score minus the mean of its distribution.

dichotomous variable A variable that has only two values.

effect size index The amount or degree of separation between two distributions.

empirical distribution A set of scores that comes from observations.

epistemology The study or theory of the nature of knowledge.

error term Variance due to factors not controlled in the experiment; within-treatment or within-cell variance.

expected frequency A theoretical value in a chi square analysis that is derived from the null hypothesis.

expected value The mean value of a random variable over an infinite number of samplings. The expected value of a statistic is the mean of the sampling distribution of the statistic.

experimental group A group that receives treatment in an experiment and whose dependent-variable scores are compared to those of a control group.

extraneous variable A variable other than the independent variable that may affect the dependent variable.

***F* distribution** A theoretical sampling distribution of *F* values. There is a different *F* distribution for each combination of degrees of freedom.

***F* test** A test of the statistical significance of differences among means, or variances, or of an interaction.

factor Independent variable.

factorial ANOVA An experimental design with two or more independent variables; allows *F* tests for main effects and interactions.

frequency polygon A frequency distribution graph of a quantitative variable with frequency points connected by lines.

goodness-of-fit test A chi square test that compares observed frequencies to frequencies predicted by a theory.

grand mean The mean of all scores, regardless of treatment.

grouped frequency distribution A compilation of scores into equal-sized ranges, called class intervals. Includes frequencies for each interval; may include midpoints of class intervals.

histogram A graph of frequencies of a quantitative variable constructed with contiguous vertical bars.

independent The occurrence of one event does not affect the outcome of a second event.

independent-samples design An experimental design with samples whose dependent-variable scores cannot logically be paired.

independent variable A variable controlled by the researcher; changes in this variable may produce changes in the dependent variable.

inferential statistics A method that uses sample evidence and probability to reach conclusions about unmeasurable populations.

inflection point A point on a curve that separates a concave upward arc from a concave downward arc, or vice versa.

interaction In a factorial ANOVA, the effect of one independent variable on the dependent variable depends on the level of another independent variable.

interpolation A method to determine an intermediate value.

interquartile range A range of scores that contains the middle 50 percent of a distribution.

interval scale A measurement scale in which equal differences between numbers represent equal differences in the thing measured. The zero point is arbitrarily defined.

least squares A method of fitting a regression line such that the sum of the squared deviations from the straight regression line is a minimum.

level One value of the independent variable.

line graph A graph that shows the relationship between two variables with lines.

linear regression Method that produces a straight line that best fits the data.

lower limit The bottom of the range of possible values that a measurement on a quantitative variable can have.

main effect In a factorial ANOVA, a significance test of the deviations of the mean levels of one independent variable from the grand mean.

Mann–Whitney *U* test A nonparametric test that compares two independent samples.

matched pairs A paired-samples design in which individuals are paired by the researcher before the experiment.

mean The arithmetic average; the sum of the scores divided by the number of scores.

mean square (*MS*) The variance; a sum of squares divided by its degrees of freedom. ANOVA terminology.

median The point that divides a distribution of scores into equal halves; half the scores are above the median and half are below it. The 50th percentile.

meta-analysis A technique that combines separate studies into an overall conclusion about effect size.

mode The score that occurs most frequently in a distribution.

multiple-comparisons tests Tests for statistical significance between treatment means or combinations of means.

multiple correlation A correlation coefficient that expresses the degree of relationship between one variable and a set of two or more other variables.

natural pairs A paired-samples design in which pairing occurs without intervention by the researcher.

negative skew A graph with a great preponderance of large scores.

nominal scale A measurement scale in which numbers serve only as labels and do not indicate a quantitative relationship.

nonparametric tests Statistical techniques that do not require assumptions about the sampled populations.

normal distribution (normal curve) A mathematically defined, theoretical distribution with a particular bell shape. An empirical distribution of similar shape.

NS Not statistically significant.

null hypothesis (H_0) A hypothesis about a population or the relationship among populations.

null hypothesis statistical testing (NHST) A statistical technique that produces probabilities that are accurate when the null hypothesis is true.

observed frequency The count of actual events in categories in a chi square test.

one-sample *t* test A statistical test of the hypothesis that a sample with mean $\bar{X}$ came from a population with mean μ.

one-tailed test of significance A directional statistical test that can detect a positive difference in population means, or a negative difference, but not both.

one-way ANOVA A statistical test of the hypothesis that two or more population means in an independent-samples design are equal.

operational definition A definition that tells how to measure a variable.

ordinal scale A measurement scale in which numbers are ranks, but equal differences between numbers do not represent equal differences between the things measured.

ordinate The vertical axis of a graph; the *Y* axis.

outlier A very high or very low score, separated from other scores. A score 1.5 × IQR or more beyond the 25th or 75th percentile.

paired-samples design An experimental design in which scores from each group are logically matched.

parameter A numerical or nominal characteristic of a population.

partial correlation Technique that allows the separation or partialing out of the effects of one variable from the correlation of two other variables.

percentile The point below which a specified percentage of the distribution falls.

phi (ϕ) An effect size index for a 2 × 2 chi square test of independence.

population All measurements of a specified group.

positive skew A graph with a great preponderance of low scores.

***post hoc* test** Multiple-comparisons test that is appropriate after examination of the data.

power Power = $1 - \beta$. Power is the probability of rejecting a false null hypothesis.

proportion A part of a whole.

qualitative variable A variable whose levels are different kinds rather than different amounts.

quantification Translating a phenomenon into numbers promotes a better understanding of the phenomenon.

quantitative variable A variable whose levels indicate different amounts.

random sample A subset of a population chosen in such a way that all samples of the specified size have an equal probability of being selected.

range The highest score minus the lowest score.

ratio scale A measurement scale that has all the characteristics of an interval scale; in addition, zero means that none of the thing measured is present.

raw score A score obtained by observation or from an experiment.

rectangular distribution A distribution in which all scores have the same frequency; also called a uniform distribution.

regression coefficients The constants *a* (point where the regression line intersects the *Y* axis) and *b* (slope of the regression line) in a regression equation.

regression equation An equation that predicts values of *Y* for specific values of *X*.

regression line A line of best fit for a scatterplot.

rejection region The area of a sampling distribution that corresponds to test statistic values that lead to rejection of the null hypothesis.

reliability The dependability or consistency of a measure.

repeated measures An experimental design in which each subject contributes to more than one treatment.

repeated-measures ANOVA A statistical technique for designs with multiple measures on subjects or with subjects who are matched.

residual variance ANOVA term for variability due to unknown or uncontrolled variables; the error term.

sample A subset of a population; it may or may not be representative.

sampling distribution A theoretical distribution of a statistic based on all possible random samples drawn from the same population; used to determine probabilities.

scatterplot A graph of the scores of a bivariate frequency distribution.

significance level A probability (α) chosen as the criterion for rejecting the null hypothesis.

simple frequency distribution Scores arranged from highest to lowest, with the frequency of each score in a column beside the score.

skewed distribution An asymmetrical distribution; may be positive or negative.

Spearman r_s A correlation coefficient for the degree of relationship between two variables measured by ranks.

standard deviation A descriptive measure of the dispersion of scores around the mean of the distribution.

standard error The standard deviation of a sampling distribution.

standard error of a difference The standard deviation of a sampling distribution of differences between means.

standard error of estimate The standard deviation of the differences between predicted outcomes and actual outcomes.

standard score A score expressed in standard deviation units; z score is an example.

statistic A numerical or nominal characteristic of a sample.

statistically significant A difference so large that chance is not a likely explanation for the difference.

***t* distribution** Theoretical distribution used to determine probabilities when σ is unknown.

***t* test** Test of a null hypothesis that uses t distribution probabilities.

theoretical distribution Hypothesized scores based on mathematical formulas and logic.

treatment One value (or level) of the independent variable.

truncated range The range of the sample is smaller than the range of its population.

Tukey Honestly Significant Difference (HSD) test Significance test for all possible pairs of treatments in a multitreatment experiment.

two-tailed test of significance A nondirectional statistical test that can detect either a positive or a negative difference in population means.

Type I error Rejection of a null hypothesis when it is true.

Type II error Failure to reject a null hypothesis that is false.

univariate distribution A frequency distribution of one variable.

upper limit The top of the range of possible values that a measurement on a quantitative variable can have.

variability Having more than one value.

variable Something that exists in more than one amount or in more than one form.

variance The square of the standard deviation; also, mean square in ANOVA.

very good idea Examine a statistics problem until you understand it well enough to estimate the answer.

weighted mean Overall mean calculated from two or more samples.

Wilcoxon matched-pairs signed-ranks *T* test A nonparametric test that compares two paired samples.

Wilcoxon rank-sum test A nonparametric test of the difference between two independent samples.

Wilcoxon–Wilcox multiple-comparisons test A nonparametric test of all possible pairs from an independent-samples design.

Yates's correction A correction for a 2×2 chi square test with one small expected frequency; considered obsolete.

***z* score** A score expressed in standard deviation units; describes the relative standing of a score in its distribution.

***z* test** A statistical test that uses the normal curve as the sampling distribution.

APPENDIX

Glossary of Symbols

Greek Letter Symbols

α	The probability of a Type I error
β	The probability of a Type II error
μ	The mean of a population
ρ	Population correlation coefficient
Σ	The sum; an instruction to add
σ	Standard deviation of a population
ϕ	Effect size index for 2×2 chi square design
σ^2	Variance of a population
$\sigma_{\overline{X}}$	Standard error of the mean (population σ known)
χ^2	The chi square statistic

Mathematical and Latin Letter Symbols

∞	Infinity
$>$	Greater than
$<$	Less than
a	Point where the regression line intersects the Y axis
b	The slope of the regression line
D	The difference between two paired scores
$\overline{D}$	The mean of a set of difference scores
d	Effect size index for one-sample and two-sample comparisons
df	Degrees of freedom
E	In chi square, the expected frequency
$E(\overline{X})$	The expected value of the mean; the mean of a sampling distribution
F	The F statistic in ANOVA
f	Frequency; the number of times a score occurs
f	Effect size index for ANOVA
H_0	The null hypothesis
H_1	A hypothesis that is an alternative to the null hypothesis
HSD	Tukey Honestly Significant Difference; makes pairwise comparisons
i	The interval size; the number of score points in a class interval
IQR	Interquartile range
K	The number of levels of the independent variable
LL	Lower limit of a confidence interval
MS	Mean square; ANOVA term for the variance

N	The number of scores or observations
O	In chi square, the observed frequency
r	Pearson product-moment correlation coefficient
r_s	A correlation coefficient for ranked data; named for Spearman
r^2	The coefficient of determination
S	The standard deviation of a sample; describes the sample
$\hat{s}$	The standard deviation of a sample; estimates σ
$\hat{s}^2$	Variance of a sample; estimates σ^2
$\hat{s}_D$	Standard deviation of a distribution of differences between paired scores
$s_{\bar{D}}$	Standard error of the difference between paired means
$s_{\bar{X}_1-\bar{X}_2}$	Standard error of a difference between means
$s_{\bar{X}}$	Standard error of the mean
SS	Sum of squares; the sum of the squared deviations from the mean
T	Wilcoxon matched-pairs signed-ranks T statistic for paired samples
t	t test statistic
t_α	Critical value of t; level of significance $= \alpha$
U	Mann-Whitney statistic for independent samples
UL	Upper limit of a confidence interval
X	A score
$\bar{X}$	The mean of a sample
$\bar{X}_{\mathrm{W}}$	The weighted overall mean of two or more means
X_{H}	The highest score in a distribution
X_{L}	The lowest score in a distribution
$\hat{Y}$	The Y value predicted for some X value
$\bar{Y}$	The mean of the Y variable
z	A score expressed in standard deviation units; a standard score
z	Test statistic when the sampling distribution is normal
z_X	A z value for a score on variable X
z_Y	A z value for a score on variable Y

APPENDIX

Glossary of Formulas

Analysis of Variance

degrees of freedom in one-way ANOVA, p. 243

$$df_{\text{treat}} = K - 1$$
$$df_{\text{error}} = N_{\text{tot}} - K$$
$$df_{\text{tot}} = N_{\text{tot}} - 1$$

degrees of freedom in factorial ANOVA, p. 287

$$df_A = A - 1$$
$$df_B = B - 1$$
$$df_{AB} = (A - 1)(B - 1)$$
$$df_{\text{error}} = N_{\text{tot}} - (A)(B)$$
$$df_{\text{tot}} = N_{\text{tot}} - 1$$

degrees of freedom in one-factor repeated-measures ANOVA, p. 261

$$df_{\text{subjects}} = N_t - 1$$
$$df_{\text{treat}} = N_k - 1$$
$$df_{\text{error}} = (N_t - 1)(N_k - 1)$$
$$df_{\text{tot}} = N_{\text{tot}} - 1$$

F value in one-way ANOVA and repeated-measures ANOVA, p. 244

$$F = \frac{MS_{\text{treat}}}{MS_{\text{error}}}$$

F values in factorial ANOVA, p. 287

$$F_A = \frac{MS_A}{MS_{\text{error}}}$$
$$F_B = \frac{MS_B}{MS_{\text{error}}}$$
$$F_{AB} = \frac{MS_{AB}}{MS_{\text{error}}}$$

mean square, p. 243

$$MS = \frac{SS}{df}$$

total sum of squares, p. 239

$$SS_{\text{tot}} = \Sigma X_{\text{tot}}^2 - \frac{(\Sigma X_{\text{tot}})^2}{N_{\text{tot}}}$$

between-treatments sum of squares, p. 239

$$SS_{\text{treat}} = \Sigma\left[\frac{(\Sigma X_t)^2}{N_t}\right] - \frac{(\Sigma X_{\text{tot}})^2}{N_{\text{tot}}}$$

between-cells sum of squares, p. 282

$$SS_{\text{cells}} = \Sigma\left[\frac{(\Sigma X_{\text{cell}})^2}{N_{\text{cell}}}\right] - \frac{(\Sigma X_{\text{tot}})^2}{N_{\text{tot}}}$$

between-subjects sum of squares, p. 261

$$SS_{\text{subjects}} = \Sigma\left[\frac{(\Sigma X_k)^2}{N_k}\right] - \frac{(\Sigma X_{\text{tot}})^2}{N_{\text{tot}}}$$

error sum of squares for one-way ANOVA, p. 240

$$SS_{\text{error}} = \Sigma\left[\Sigma X_t^2 - \frac{(\Sigma X_t)^2}{N_t}\right]$$

error sum of squares for factorial ANOVA, p. 284

$$SS_{\text{error}} = \Sigma\left[\Sigma X_{\text{cell}}^2 - \frac{(\Sigma X_{\text{cell}})^2}{N_{\text{cell}}}\right]$$

error sum of squares for repeated-measures ANOVA, p. 261

$$SS_{\text{error}} = SS_{\text{tot}} - SS_{\text{subjects}} - SS_{\text{treat}}$$

sum of squares for the interaction effect in factorial ANOVA, p. 283

$$SS_{AB} = N_{\text{cell}}\Sigma[(\overline{X}_{AB} - \overline{X}_A - \overline{X}_B + \overline{X}_{\text{gm}})^2]$$

Check:

$$SS_{AB} = SS_{\text{cells}} - SS_A - SS_B$$

sum of squares for a main effect in factorial ANOVA, p. 282

$$SS = \frac{(\Sigma X_1)^2}{N_1} + \frac{(\Sigma X_2)^2}{N_2} + \cdots + \frac{(\Sigma X_k)^2}{N_k} - \frac{(\Sigma X_{\text{tot}})^2}{N_{\text{tot}}}$$

where 1 and 2 denote levels of a factor and k denotes the last level of a factor

Chi Square

basic formula, p. 303

$$\chi^2 = \Sigma\left[\frac{(O - E)^2}{E}\right]$$

degrees of freedom for a chi square table, p. 315

$df = (R - 1)(C - 1)$
where R = number of rows
C = number of columns

shortcut formula for a 2 × 2 table, p. 308

$$\chi^2 = \frac{N(AD - BC)^2}{(A + B)(C + D)(A + C)(B + D)}$$

where A, B, C, and D designate the four cells of the table, moving left to right across the top row and then across the bottom row

Coefficient of Determination

p. 102 $\qquad r^2$

Confidence Intervals

about a population mean, p. 164

$$\text{LL} = \overline{X} - t_\alpha(s_{\overline{X}})$$
$$\text{UL} = \overline{X} + t_\alpha(s_{\overline{X}})$$

about a mean difference (independent samples), p. 218

$$\text{LL} = (\overline{X}_1 - \overline{X}_2) - t_\alpha(s_{\overline{X}_1-\overline{X}_2})$$
$$\text{UL} = (\overline{X}_1 - \overline{X}_2) + t_\alpha(s_{\overline{X}_1-\overline{X}_2})$$

about a mean difference (paired samples), p. 219

$$\text{LL} = (\bar{X} - \bar{Y}) - t_\alpha(s_{\bar{D}})$$
$$\text{UL} = (\bar{X} - \bar{Y}) + t_\alpha(s_{\bar{D}})$$

Correlation

Pearson product-moment definition formula, p. 96

$$r = \frac{\Sigma(z_X z_Y)}{N}$$

computation formulas, pp. 96, 97

$$r = \frac{\dfrac{\Sigma XY}{N} - (\bar{X})(\bar{Y})}{(S_X)(S_Y)}$$

$$r = \frac{N\Sigma XY - (\Sigma X)(\Sigma Y)}{\sqrt{[N\Sigma X^2 - (\Sigma X)^2][N\Sigma Y^2 - (\Sigma Y)^2]}}$$

testing significance from .00 (or use Table A), p. 191

$$t = (r)\sqrt{\frac{N-2}{1-r^2}}$$
$$df = N - 2$$

Spearman r_s computation formula, p. 344

$$r_s = 1 - \frac{6\Sigma D^2}{N(N^2 - 1)}$$

Degrees of Freedom

See specific statistical tests.

Deviation Score, p. 59

$$X - \bar{X} \quad \text{or} \quad X - \mu$$

Effect Size Index

definition, p. 79

$$d = \frac{\mu_1 - \mu_2}{\sigma}$$

one-sample design, p. 189

$$d = \frac{\bar{X} - \mu}{\hat{s}}$$

two independent samples, $N_1 = N_2$, p. 216

$$d = \frac{\bar{X}_1 - \bar{X}_2}{\sqrt{\dfrac{N_1}{2}}(s_{\bar{X}_1 - \bar{X}_2})}$$

two independent samples, $N_1 \neq N_2$, p. 216

$$d = \frac{\bar{X}_1 - \bar{X}_2}{\sqrt{\dfrac{\hat{s}_1^2(df_1) + \hat{s}_2^2(df_2)}{df_1 + df_2}}}$$

two paired samples, p. 217

$$d = \frac{\bar{X} - \bar{Y}}{\sqrt{N}(s_{\bar{D}})}$$

ANOVA for all means, pp. 253, 296

$$f = \frac{\hat{s}_{\text{treat}}}{\hat{s}_{\text{error}}} = \frac{\sqrt{\dfrac{K-1}{N_{\text{tot}}}(MS_{\text{treat}} - MS_{\text{error}})}}{\sqrt{MS_{\text{error}}}}$$

for two means, pp. 252, 290

$$d = \frac{\overline{X}_1 - \overline{X}_2}{\sqrt{MS_{\text{error}}}}$$

chi square (2 × 2), p. 308

$$\phi = \sqrt{\frac{\chi^2}{N}}$$

Mann–Whitney U Test

value for U, p. 328

$$U_1 = (N_1)(N_2) + \frac{N_1(N_1 + 1)}{2} - \Sigma R_1$$

where ΣR_1 = sum of the ranks of the N_1 group

testing significance for larger samples, $N \geq 21$, p. 331

$$z = \frac{(U_1 + c) - \mu_U}{\sigma_U}$$

where $c = 0.5$

$$\mu_U = \frac{N_1 N_2}{2}$$

$$\sigma_U = \sqrt{\frac{(N_1)(N_2)(N_1 + N_2 + 1)}{12}}$$

Mean

from raw data, p. 41

$$\mu \text{ or } \overline{X} = \frac{\Sigma X}{N}$$

from a frequency distribution, p. 44

$$\mu \text{ or } \overline{X} = \frac{\Sigma f X}{N}$$

weighted mean, p. 50

$$\overline{X}_W = \frac{\Sigma(N_1\overline{X}_1 + N_2\overline{X}_2 + \cdots + N_K\overline{X}_K)}{\Sigma N}$$

where N_1, N_2, and so on are the numbers of scores associated with their respective means, and K is the last of the means being averaged

Median Location

p. 44

$$\frac{N + 1}{2}$$

Outlier

p. 75

Lower outlier = 25th percentile − (1.5 × IQR)
Upper outlier = 75th percentile + (1.5 × IQR)

Range

of a distribution, p. 56

Range = $X_H - X_L$
where X_H = highest score
X_L = lowest score

interquartile range, p. 57

IQR = 75th percentile − 25th percentile

Regression

predicting Y from X, p. 116

$$\hat{Y} = r\frac{S_Y}{S_X}(X - \overline{X}) + \overline{Y}$$

straight line, p. 114

$$\hat{Y} = a + bX$$

where a = value at the Y intercept
b = slope of the regression line

Y intercept of a regression line, p. 114

$$a = \overline{Y} - b\overline{X}$$

slope of a regression line, p. 114

$$b = r\frac{S_Y}{S_X}$$

$$b = \frac{N\Sigma XY - (\Sigma X)(\Sigma Y)}{N\Sigma X^2 - (\Sigma X)^2}$$

Standard Deviation of a Population or Sample (for description)

deviation-score method from ungrouped data, p. 60

$$S = \sqrt{\frac{\Sigma(X - \overline{X})^2}{N}}$$

raw-score method from ungrouped data, p. 62

$$\sigma \quad \text{or} \quad S = \sqrt{\frac{\Sigma X^2 - \frac{(\Sigma X)^2}{N}}{N}}$$

Standard Deviation of a Sample (an estimator of σ)

deviation-score method from ungrouped data, p. 64

$$\hat{s} = \sqrt{\frac{\Sigma(X - \overline{X})^2}{N - 1}}$$

raw-score method from ungrouped data, p. 65

$$\hat{s} = \sqrt{\frac{\Sigma X^2 - \frac{(\Sigma X)^2}{N}}{N - 1}}$$

raw-score method from grouped data, p. 65

$$\hat{s} = \sqrt{\frac{\Sigma fX^2 - \frac{(\Sigma fX)^2}{N}}{N - 1}}$$

paired samples, p. 211

$$\hat{s}_D = \sqrt{\frac{\Sigma D^2 - \frac{(\Sigma D)^2}{N}}{N - 1}}$$

Standard Error

of the mean

where σ is known, p. 157 $\sigma_{\overline{X}} = \frac{\sigma}{\sqrt{N}}$

estimated from a single sample, p. 164 $s_{\overline{X}} = \frac{\hat{s}}{\sqrt{N}}$

of a difference between means
equal-N samples, p. 206

$$s_{\overline{X}_1-\overline{X}_2} = \sqrt{s_{\overline{X}_1}{}^2 + s_{\overline{X}_2}{}^2}$$

$$= \sqrt{\left(\frac{\hat{s}_1}{\sqrt{N_1}}\right)^2 + \left(\frac{\hat{s}_2}{\sqrt{N_2}}\right)^2}$$

$$= \sqrt{\frac{\Sigma X_1^2 - \frac{(\Sigma X_1)^2}{N_1} + \Sigma X_2^2 - \frac{(\Sigma X_2)^2}{N_2}}{N_1(N_2 - 1)}}$$

samples with
unequal N's, p. 206

$$s_{\overline{X}_1-\overline{X}_2} = \sqrt{\left(\frac{\Sigma X_1^2 - \frac{(\Sigma X_1)^2}{N_1} + \Sigma X_2^2 - \frac{(\Sigma X_2)^2}{N_2}}{N_1 + N_2 - 2}\right)\left(\frac{1}{N_1} + \frac{1}{N_2}\right)}$$

paired samples by the
direct-difference method, p. 211

$$s_{\overline{D}} = \frac{\hat{s}_D}{\sqrt{N}}$$

$$\text{where } \hat{s}_D = \sqrt{\frac{\Sigma D^2 - \frac{(\Sigma D)^2}{N}}{N - 1}}$$

paired samples
when r is known, p. 210

$$s_{\overline{D}} = \sqrt{s_{\overline{X}}^2 + s_{\overline{Y}}^2 - 2r_{XY}(s_{\overline{X}})(s_{\overline{Y}})}$$

t Test

one sample μ, p. 181

$$t = \frac{\overline{X} - \mu}{s_{\overline{X}}}$$

$$df = N - 1$$

independent samples, p. 206

$$t = \frac{\overline{X}_1 - \overline{X}_2}{s_{\overline{X}_1-\overline{X}_2}}$$

$$df = N_1 + N_2 - 2$$

paired samples
where r is known, p. 210

$$t = \frac{\overline{X} - \overline{Y}}{\sqrt{s_{\overline{X}}^2 + s_{\overline{Y}}^2 - 2r_{XY}(s_{\overline{X}})(s_{\overline{Y}})}}$$

paired samples
using the direct-difference
method, p. 210

$$t = \frac{\overline{X} - \overline{Y}}{s_{\overline{D}}} = \frac{\overline{D}}{s_{\overline{D}}}$$

df = number of pairs minus 1

testing whether a correlation
coefficient is significantly
different from .00, p. 191

$$t = (r)\sqrt{\frac{N - 2}{1 - r^2}}$$

df = number of pairs minus 2

Tukey HSD

one-way, factorial, and repeated-measures ANOVA, pp. 248, 262, 295

$$\text{HSD} = \frac{\overline{X}_1 - \overline{X}_2}{s_{\overline{X}}}$$

$$\text{where } s_{\overline{X}} \text{ (for } N_1 = N_2 = N_t) = \sqrt{\frac{MS_{\text{error}}}{N_t}}$$

$$s_{\overline{X}} \text{ (for } N_1 \neq N_2) = \sqrt{\frac{MS_{\text{error}}}{2}\left(\frac{1}{N_1} + \frac{1}{N_2}\right)}$$

Variance

Use the formulas for the standard deviation. For $\hat{s}^2$ square $\hat{s}$. For σ^2 square σ.

Wilcoxon Matched-Pairs Signed-Ranks *T* Test

value for *T*, p. 335

T = smaller sum of the signed ranks

testing significance for larger samples, $N \geq 50$, p. 338

$$z = \frac{(T + c) - \mu_T}{\sigma_T}$$

where $c = 0.5$

$$\mu_T = \frac{N(N + 1)}{4}$$

$$\sigma_T = \sqrt{\frac{N(N + 1)(2N + 1)}{24}}$$

N = number of pairs

z-Score Formulas

descriptive, pp. 72, 131

$$z = \frac{X - \overline{X}}{S}; \qquad \text{also } z = \frac{X - \mu}{\sigma}$$

probability of a sample mean, p. 157

$$z = \frac{\overline{X} - \mu}{\sigma_{\overline{X}}}$$

(See also Mann–Whitney *U* test and Wilcoxon Matched-Pairs Signed-Ranks *T* test.)

APPENDIX

Answers to Problems

G

CHAPTER 1

1.1. **a.** 64.5–65.5 **b.** Qualitative **c.** Qualitative
d. 3.5–4.5 **e.** 80.5–81.5

1.2. Many paragraphs can qualify as good answers to this question. Your paragraph should include variations of these definitions:
Population: a defined set of scores that is of interest to an investigator
Sample: a subset of the population
Parameter: a numerical or nominal characteristic of a population
Statistic: a numerical or nominal characteristic of a sample

1.3. **a.** Inferential
b. Inferential
c. Descriptive

1.4. Nominal, ordinal, interval, and ratio

1.5. *Nominal:* Different numbers are assigned to different classes of things.
Ordinal: Nominal properties, plus the numbers carry information about greater than and less than.
Interval: Ordinal properties, plus the distance between units is equal.
Ratio: Interval properties, plus a zero point that means a zero amount of the thing being measured.

1.6. **a.** Ordinal **b.** Ratio **c.** Nominal **d.** Nominal
e. Ordinal **f.** Ratio **g.** Ordinal

1.7. **a.** *Dependent variable:* rating of qualification; *independent variable:* gender; *number of levels:* two; *names of levels:* female and male
b. *Dependent variable:* narcissism; *independent variable:* birth order; *number of levels:* three; *names of levels:* first, second, and later born
c. *Dependent variable:* jail sentence; *independent variable:* occupation; *number of levels:* three; *names of levels:* vice president, janitor, and unspecified

1.8. **a.** *Dependent variable:* number of suggestions complied with
Independent variable, number of levels, and their names: hypnosis; two; yes and no
i. *Nominal variables:* hypnosis, suggestions; *statistic:* mean number of suggestions complied with
ii. Interpretation: Barber's study shows that hypnotized participants do not acquire powers greater than normal (see Barber, 1976).
b. *Dependent variable:* answer to the question Did you see a barn?
Independent variable, number of levels, and their names: mention of the barn in the question about how fast the car was going; two; mentioned and not mentioned
i. Several answers can be correct here. The most general population answer is "people" or "people's memory." Less general is "the memory of people given misinformation about an event." One parameter is the percent of all people who say they see a barn even though no barn was in the film.
ii. Interpretation: If people are given misinformation after they have witnessed an event, they sometimes incorporate the misinformation into their memory. [See Loftus's book, *Eyewitness Testimony* (1979), which relates this phenomenon to courts of law.]
c. *Dependent variable:* weight (grams) of crackers consumed
Independent variable, number of levels, and their names: the time shown on the clock on the wall; three; slow, correct, and fast
i. Weight (grams)
ii. Interpretation: Eating by obese men is affected by what time they think it is; that is, consumption by obese men does not depend entirely on internal physiological cues of hunger (see Schachter and Gross, 1968).

1.9. Besides differing in the textbook used, the two classes differed in (1) the professor, one of whom might be a better teacher; (2) the time the class met; and (3) the length of the class period. A 10:00 A.M. class probably has younger students who have fewer outside commitments. Students probably concentrate less during a 3-hour class than they do in three 1-hour classes. Any of these extraneous variables might be responsible for the difference in scores. Also, the investigator needs assurance that the comprehensive test was appropriate for both textbooks.

1.10. Epistemology

1.11. **a.** Reason (or rationalism) and experience (or empiricism)

b. Reason (or rationalism)

1.12. Step 1 is an exploratory step that involves reading the information, descriptive statistics, and attention to differences. Step 2 involves inferential statistics and answers the question of whether the differences could be due to chance. Step 3 is a story (a paragraph) written in journal style that explains the data set.

CHAPTER 2

2.1.

Women		
Height (in.)	Tally marks	f
72	/	1
71		0
70	/	1
69	/	1
68	/	1
67	///	3
66	ⅢⅠ /	6
65	ⅢⅠ ⅢⅠ	10
64	ⅢⅠ ////	9
63	ⅢⅠ //	7
62	////	4
61	/	1
60	////	4
59	//	2

Men		
Height (in.)	Tally marks	f
77	/	1
76	/	1
75	//	2
74	/	1
73	////	4
72	ⅢⅠ //	7
71	ⅢⅠ /	6
70	ⅢⅠ //	7
69	ⅢⅠ ///	8
68	////	4
67	//	2
66	//	2
65	///	3
64	/	1
63		0
62	/	1

2.2. The order of candidates is arbitrary; no order is wrong.

Candidate	f
Attila	13
Bolivar	5
Gandhi	19
Lenin	8
Mao	11

2.3.

Temperature intervals	Tally marks	f
99.3–99.5	///	3
99.0–99.2	//	2
98.7–98.9	ⅢⅠ /	6
98.4–98.6	ⅢⅠ	5
98.1–98.3	ⅢⅠ /	6
97.8–98.0	ⅢⅠ ⅢⅠ	10
97.5–97.7	//	2
97.2–97.4	///	3
96.9–97.1	//	2
96.6–96.8		0
96.3–96.5	/	1

2.4.

Errors	Tally marks	f
16	//	2
15		0
14	//	2
13	///	3
12	///	3
11	ⅢⅠ ////	9
10	ⅢⅠ /	6
9	ⅢⅠ //	7
8	////	4
7	//	2
6	/	1

Interpretation: Here are some points that would be appropriate: (1) Surprisingly, no one recognized that all 20 statements were new. (2) All the students thought they had heard more than a fourth of the statements before. (3) Most students thought they had heard about half the statements before, but, as the number of statements increased or decreased from half, fewer and fewer thought that.

2.5. **a.** 25 is the midpoint of the interval 24–26

b. 5 is a frequency count; for example, there were 5 scores in the interval with 10 as the midpoint

c. 0

2.6.

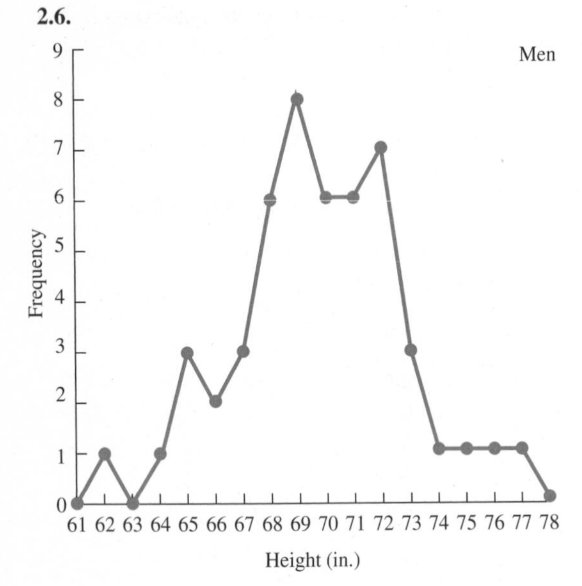

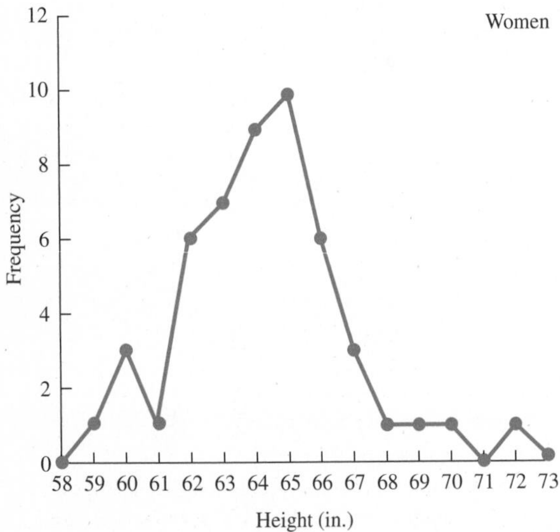

2.7. A bar graph is appropriate. I arranged the countries in the order of their scores so that *more than* and *less than* comparisons are easier.

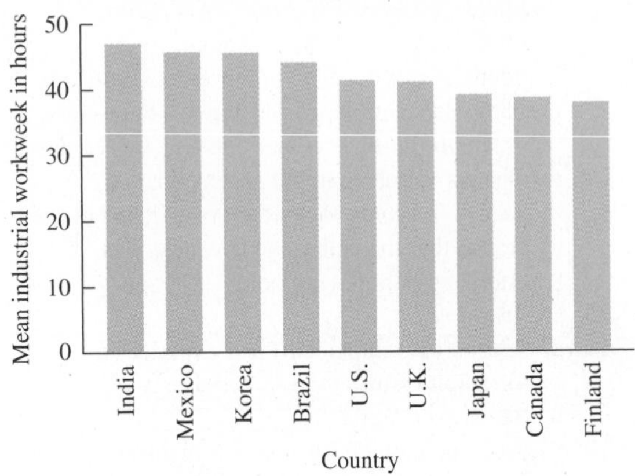

2.8. They should be graphed as frequency polygons. Note that the midpoint of the class interval is used as the *X*-axis score.

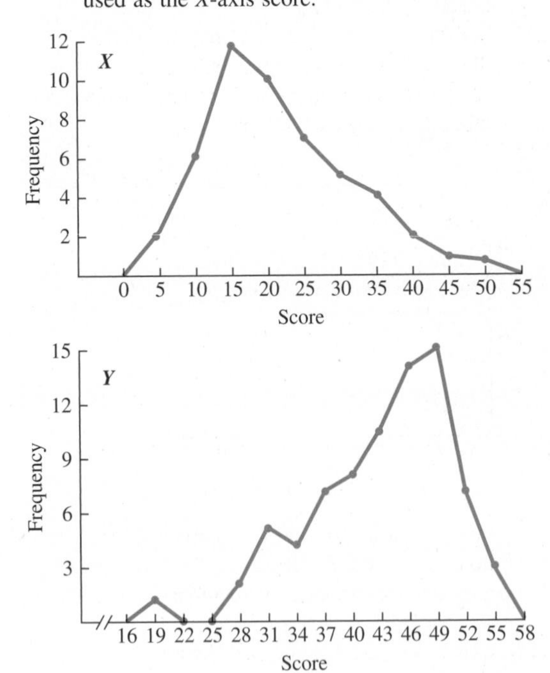

2.9. A bar graph is the proper graph for the qualitative data in problem 2.2. I ordered the candidates according to votes received, but other orders could be justified.

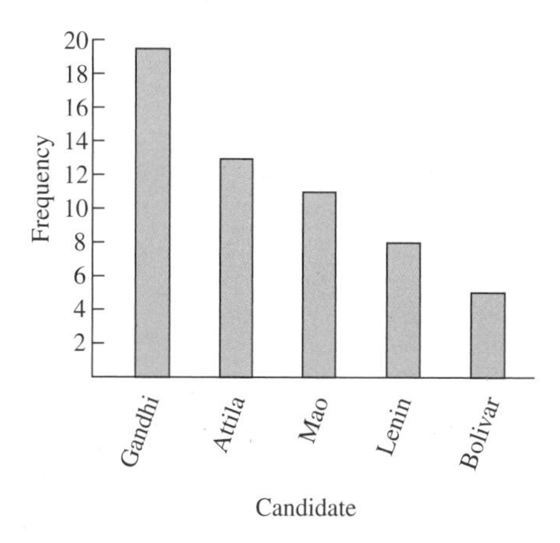

2.10. Distribution x is positively skewed; distribution y is negatively skewed.

2.11. A frequency polygon has a frequency count on the Y axis; the variable that is being counted is on the X axis. A line graph has some variable other than frequency on the Y axis.

2.12.

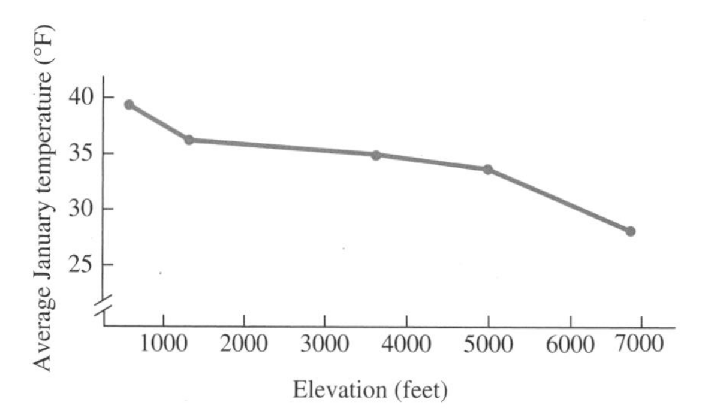

Interpretation: Temperature is affected by elevation. The higher the elevation, the lower the temperature, although the relationship is not a straight line.

2.13. Check your sketches against Figures 2.6 and 2.8.

2.14. Right

2.15. **a.** Positively skewed **b.** Symmetrical
c. Positively skewed **d.** Positively skewed
e. Negatively skewed **f.** Symmetrical

2.16. Of course many paragraphs can qualify for A's. Here are some points you might have included:

- Graphs are persuasive.
- Graphs can guide research, not just convey the results.
- Graphs are a means of communication among scientists.
- Graphs can help you understand.
- Designing graphs takes time.
- Attitudes toward graphs have changed in recent years.
- New kinds of graphs are being created.

2.17. **a.**

Body temperature	f
99.5	2
99.4	1
99.2	1
99.1	1
98.9	2
98.8	2
98.7	2
98.6	2
98.5	1
98.4	2
98.3	3
98.2	2
98.1	1
98.0	3
97.9	5
97.8	2
97.7	1
97.5	1
97.4	2
97.2	1
97.0	1
96.9	1
96.4	1
	40

b. Either a frequency polygon or a histogram is appropriate. The form is negatively skewed.

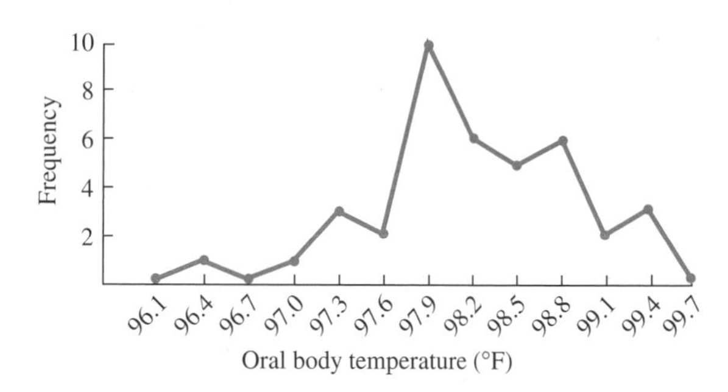

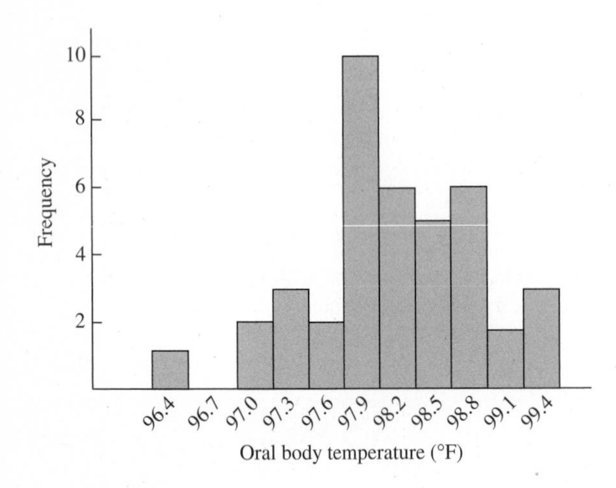

CHAPTER 3

3.1. To find the median, arrange the scores in descending order.
a. 5 **b.** 12.5, halfway between 12 and 13
c. 9.5, halfway between 8 and 11

3.2. It may help to arrange the scores in descending order. Only distribution c has two modes. They are 14 and 18.

3.3. **a.** For these nominal data, only the mode is appropriate. The mode is Gandhi, who accounted for 34 percent of the signs.
b. Because only one precinct was covered and the student's interest was citywide, this is a sample mode, a statistic.
c. Interpretation: Because there are more Gandhi yard signs than signs for any other candidate, Gandhi can be expected to get the most votes.

3.4.

Women

Height (in.)	Tally marks	f	fX
72	/	1	72
71		0	
70	/	1	70
69	/	1	69
68	/	1	68
67	///	3	201
66	~~////~~ /	6	396
65	~~////~~ ~~////~~	10	650
64	~~////~~ ////	9	576
63	~~////~~ //	7	441
62	////	4	248
61	/	1	61
60	////	4	240
59	//	2	118
		50	3210

$$Mean = \overline{X}_{\text{women}} = \frac{\Sigma fX}{N} = \frac{3210}{50} = 64.2 \text{ inches}$$

$$\text{Median location} = \frac{N+1}{2} = \frac{50+1}{2} = 25.5$$

Counting from the bottom of the distribution, you find 18 scores below 64 inches, so the median is located among the 9 scores of 64. *Median* = 64 inches. The most frequently occurring score (10 times) is 65 inches. *Mode* = 65 inches, which is found for 20 percent of the women.

Men

Height (in.)	Tally marks	f	fX
77	/	1	77
76	/	1	76
75	//	2	150
74	/	1	74
73	////	4	292
72	~~////~~ //	7	504
71	~~////~~ /	6	426
70	~~////~~ //	7	490
69	~~////~~ ///	8	552
68	////	4	272
67	//	2	134
66	//	2	132
65	///	3	195
64	/	1	64
63		0	
62	/	1	62
		50	3500

$$Mean = \overline{X}_{\text{men}} = \frac{\Sigma fX}{N} = \frac{3500}{50} = 70.0 \text{ inches}$$

$$\text{Median location} = \frac{N+1}{2} = \frac{50+1}{2} = 25.5$$

Counting from the top of the distribution, you find 22 scores above 70 inches. The median is among the 7 scores of 70. *Median* = 70 inches. A height of 69 inches occurs most frequently. *Mode* = 69 inches, which is the height of 16 percent of the men.

3.5. **a.** $\text{Median location} = \frac{N+1}{2} = \frac{19+1}{2} = 10$

Counting from the bottom, you reach 7 when you include the score of 12. The 10th score is located among the 5 scores of 13. *Median* = 13.

b. $\text{Median location} = \frac{N+1}{2} = \frac{20+1}{2} = 10.5$

Arrange the scores in descending order. Counting from the top, you reach 6 when you include the score of 2. The location of the 10.5th score is among the 5 scores of 1. *Median* = 1.

c. Median location $= \frac{N+1}{2} = \frac{6+1}{2} = 3.5$

The median is the average of 26 and 21.
Median = 23.5

3.6. $\Sigma(X - \bar{X}) = 0$ and $\Sigma(X - \bar{X})^2$ is a minimum.

3.7. **a.** The mode is appropriate because nominal variable events are being observed.

b. The median or mode is appropriate because helpfulness is measured with an ordinal variable.

c. The median and mode are appropriate for data with an open-ended category.

d. The median and mode are appropriate measures. It is conventional to use the median for income data because the distribution is often severely skewed. (About half of the frequencies are in the \$0–\$20,000 range.)

e. The mode is appropriate because these are nominal data.

f. The mean is appropriate because the data are not severely skewed.

3.8. This problem calls for a weighted mean.

$74 \times 12 = 888$

$69 \times 31 = 2139$

$75 \times \underline{17} = \underline{1275}$

$\Sigma \quad 60 \quad 4302 \quad \bar{X}_W = \frac{4302}{60} = 71.7$ percent correct

3.9. **a.** By inspection, the expenditures in Table 3.2 are positively skewed; the elevations in Table 3.4 are also positively skewed. For Table 3.2, the mean (\$2.90) is larger than the median (\$2.75), which is characteristic of positive skew. In Table 3.4, the mean (78.5 feet) is larger than the median (12.5 feet), again an indication of positive skew.

b. For distribution *x*, mean = 21.80 and median = 20. The mean is larger, so the distribution is positively skewed.

For distribution *y*, mean = 43.19 and median = 46. The mean is smaller, so the distribution is negatively skewed.

3.10. Interpretation: This is not correct. To find his lifetime batting average he should add his hits for the three years and divide by the total number of at-bats, a weighted mean. It appears that his lifetime average is less than .317.

3.11. Mean: $\bar{X} = \frac{\Sigma X}{N} = \frac{398}{15} = \$26.53$

Median location $= \frac{N+1}{2} = \frac{15+1}{2} = 8$. When the expenditures are arranged in order, the 8th score is \$24; median = \$24.

There are more scores of \$20 than any other; mode = \$20.

Skew: 26.53 − 24 = 2.53, which indicates a positive skew.

CHAPTER 4

4.1. **a.** Range = 17 − 1 = 16

b. Range = 0.45 − 0.30 = 0.15

4.2. With $N = 100$, the 25th percentile score is the 25th score from the bottom. There are 23 scores less than 21 and 3 scores of 21. The 25th percentile score is 21. The 75th percentile score is 25 frequencies from the top. There are 23 scores greater than 28 and 6 scores of 28. The 75th percentile score is 28.
IQR = 28 − 21 = 7.
Interpretation: Fifty percent of the SWLS scores are in the 7-point range of 21 to 28.

4.3. *Women:* Because 0.25 × 50 = 12.5, the 25th percentile score is 12.5 frequencies up from the bottom and the 75th percentile score is 12.5 frequencies down from the top. For the 25th percentile, there are 11 scores of 62 or less, so the 12.5th score is among the 7 scores of 63. 25th percentile = 63 inches. There are 7 scores of 67 or more so the 12.5th score from the top is among the 6 scores of 66. 75th percentile = 66 inches. IQR = 66 − 63 = 3 inches; 50 percent of women are 63 to 66 inches tall.

Men: With $N = 50$ for men also, the same 12.5 frequencies from the bottom and top identify the 25th and 75th percentile scores. The 25th percentile score is among the 6 scores of 68 inches (there are 9 shorter than 68). The 75th percentile score is among the 7 scores of 72 inches (there are 9 taller than 72). IQR = 72 − 68 = 4 inches for men; 50 percent are from 68 to 72 inches tall.

4.4.
- σ is used to describe the variability of a population.
- $\hat{s}$ is used to estimate σ from a sample of the population.
- S is used to describe the variability of a sample when you do not want to estimate σ.

4.5. a.

	X	$X - \bar{X}$	$(X - \bar{X})^2$
	7	2	4
	6	1	1
	5	0	0
	2	−3	9
Σ	20	0	14

$\bar{X} = 5$

$$S = \sqrt{\frac{\Sigma(X - \bar{X})^2}{N}} = \sqrt{\frac{14}{4}} = \sqrt{3.5} = 1.87$$

b.

	X	$X - \bar{X}$	$(X - \bar{X})^2$
	14	3.8	14.44
	11	0.8	0.64
	10	−0.2	0.04
	8	−2.2	4.84
	8	−2.2	4.84
Σ	51	0	24.80

$\bar{X} = 10.2$

$$S = \sqrt{\frac{\Sigma(X - \bar{X})^2}{N}} = \sqrt{\frac{24.80}{5}} = \sqrt{4.96} = 2.23$$

c.

	X	$X - \bar{X}$	$(X - \bar{X})^2$
	107	2	4
	106	1	1
	105	0	0
	102	−3	9
Σ	420	0	14

$\bar{X} = 105$

$$S = \sqrt{\frac{\Sigma(X - \bar{X})^2}{N}} = \sqrt{\frac{14}{4}} = \sqrt{3.5} = 1.87$$

4.6. **a.** The size of the numbers has no effect on the standard deviation. It is affected by the size of the *differences* among the numbers.
b. The mean is affected by the size of the numbers.

4.7.

City	Mean	Standard deviation
San Francisco	56.75°F	3.96°F
Albuquerque	56.75°F	16.24°F

Interpretation: Although the mean temperature is the same for the two cities, Albuquerque has a wider variety of temperatures. (If your mean scores are 56.69°F for San Francisco and 56.49°F for Albuquerque, you are *very* alert. These are weighted means based on 30 days in June and September and 31 days in March and December.)

4.8. In eyeballing data for variability, use the range as a quick index.
a. Set II is more variable.
b. Equal variability
c. Set I is more variable.
d. Set II is more variable; however, most of the variability is due to one extreme score, 15.
e. Equal variability

4.9. The second distribution (b) is more variable than the first (a).

a.

	X	X^2	$X - \bar{X}$	$(X - \bar{X})^2$
	5	25	2.5	6.25
	4	16	1.5	2.25
	3	9	0.5	0.25
	2	4	−0.5	0.25
	1	1	−1.5	2.25
	0	0	−2.5	6.25
Σ	15	55	0	17.50

$$S = \sqrt{\frac{\Sigma X^2 - \frac{(\Sigma X)^2}{N}}{N}} = \sqrt{\frac{55 - \frac{(15)^2}{6}}{6}}$$

$$= \sqrt{2.917} = 1.71$$

$$S = \sqrt{\frac{\Sigma(X - \bar{X})^2}{N}} = \sqrt{\frac{17.50}{6}} = \sqrt{2.917} = 1.71$$

b.

	X	X^2	$X - \bar{X}$	$(X - \bar{X})^2$
	5	25	2.5	6.25
	5	25	2.5	6.25
	5	25	2.5	6.25
	0	0	−2.5	6.25
	0	0	−2.5	6.25
	0	0	−2.5	6.25
Σ	15	75	0	37.50

$$S = \sqrt{\frac{\Sigma X^2 - \frac{(\Sigma X)^2}{N}}{N}} = \sqrt{\frac{75 - \frac{(15)^2}{6}}{6}}$$

$$= \sqrt{6.25} = 2.50$$

(Continued)

$$S = \sqrt{\frac{\Sigma(X - \bar{X})^2}{N}} = \sqrt{\frac{37.50}{6}} = \sqrt{6.25} = 2.50$$

How did your estimate compare to your computation?

4.10. For problem 4.9a, $\frac{5}{1.71} = 2.92$. For problem 4.9b, $\frac{5}{2.50} = 2.00$. Yes, these results are between 2 and 5.

4.11. **a.**

$$\sigma = \sqrt{\frac{\Sigma X^2 - \frac{(\Sigma X)^2}{N}}{N}} = \sqrt{\frac{262 - \frac{(34)^2}{5}}{5}} = 2.48$$

b. $$\sigma = \sqrt{\frac{294 - \frac{(38)^2}{5}}{5}} = 1.02$$

4.12.

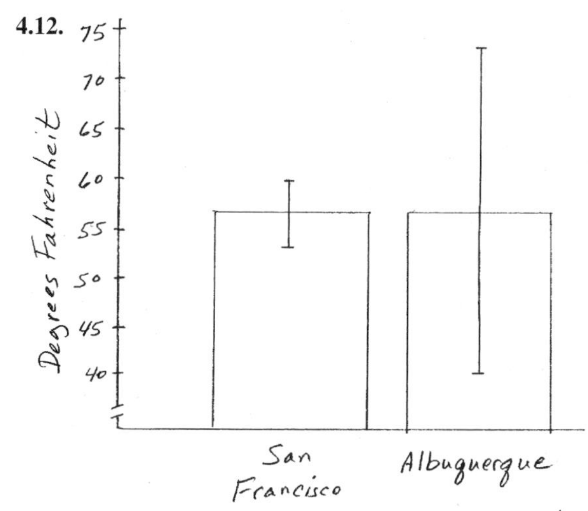

4.13. Interpretation: For $\hat{s}$ = \$0.02: Because you spend almost exactly \$2.90 each day in the Student Center, any day you don't go to the center will save you about \$2.90.
Interpretation: For $\hat{s}$ = \$2.50: Because you spend widely varying amounts in the Student Center, you can best reduce the total by reducing the number of big-ticket items you buy (Goodbye ice cream.). (The actual value of $\hat{s}$ for the data in Table 4.1 is \$2.87.)

4.14. The variance is the square of the standard deviation.

4.15. Because the researcher wanted to estimate the standard deviation of the freshman class scores from a sample, $\hat{s}$ is the correct standard deviation.

$$\hat{s} = \sqrt{\frac{\Sigma X^2 - \frac{(\Sigma X)^2}{N}}{N - 1}} = \sqrt{\frac{5064 - \frac{(304)^2}{21}}{20}}$$

$$= \sqrt{33.16} = 5.76$$

$\hat{s}^2 = 33.162$

4.16.

Women

Height (in.)	f	fX	fX^2
72	1	72	5,184
70	1	70	4,900
69	1	69	4,761
68	1	68	4,624
67	3	201	13,467
66	6	396	26,136
65	10	650	42,250
64	9	576	36,864
63	7	441	27,783
62	4	248	15,376
61	1	61	3,721
60	4	240	14,400
59	2	118	6,962
	Σ	3210	206,428

$$\hat{s} = \sqrt{\frac{206{,}428 - \frac{(3210)^2}{50}}{49}} = 2.66 \text{ inches}$$

Men

Height (in.)	f	fX	fX^2
77	1	77	5,929
76	1	76	5,776
75	2	150	11,250
74	1	74	5,476
73	4	292	21,316
72	7	504	36,288
71	6	426	30,246
70	7	490	34,300
69	8	552	38,088
68	4	272	18,496
67	2	134	8,978
66	2	132	8,712
65	3	195	12,675
64	1	64	4,096
62	1	62	3,844
	Σ	3500	245,470

$$\hat{s} = \sqrt{\frac{245{,}470 - \frac{(3500)^2}{50}}{49}} = 3.10 \text{ inches}$$

Interpretation: The heights of men are more variable than the heights of women.

4.17. Before: $\hat{s} = \sqrt{1.429} = 1.20, \bar{X} = 5.0$

After: $\hat{s} = \sqrt{14.286} = 3.78, \bar{X} = 5.0$

Interpretation: It appears that students were homogeneous and neutral before studying poetry. After studying poetry for 9 weeks, some students were turned on and some were turned off; they were no longer neutral.

4.18. $\hat{s}$ is the more appropriate standard deviation; generalization to the manufacturing process is expected.

$$\text{Process A:} \quad \hat{s} = \sqrt{\frac{18 - \frac{0^2}{6}}{5}} = 1.90 \text{ millimeters}$$

$$\text{Process B:} \quad \hat{s} = \sqrt{\frac{12 - \frac{0^2}{6}}{5}} = 1.55 \text{ millimeters}$$

Interpretation: Process B produces more consistent gizmos. Note that both processes have an average error of zero.

4.19. With 40 scores, the 25th percentile is located at the 10th score from the bottom of the distribution ($0.25 \times 40 = 10$) and the 75th percentile is at the 10th score from the top. 25th percentile = 97.8°F and 75th percentile = 98.7°F.

Interpretation: 50 percent of the population has a normal oral body temperature between 97.8°F and 98.7°F.

4.20.

$$\hat{s} = \sqrt{\frac{\Sigma fX^2 - \frac{(\Sigma fX)^2}{N}}{N-1}}$$

$$= \sqrt{\frac{385{,}749.24 - \frac{(3928)^2}{40}}{39}} = \sqrt{0.504} = 0.71°\text{F}$$

4.21.

Number of errors			
Heard before	f	fX	fX^2
16	2	32	512
15	0	0	0
14	2	28	392
13	3	39	507
12	3	36	432
11	9	99	1089
10	6	60	600
9	7	63	567
8	4	32	256
7	2	14	98
6	1	6	36
Σ	39	409	4489

$$Mean = \frac{\Sigma X}{N} = \frac{409}{39} = 10.49 \text{ errors}$$

$$\text{Median location} = \frac{N+1}{2} = \frac{39+1}{2} = 20$$

From the bottom of the distribution, there are 20 scores of 10 or less. From the top there are 19 scores of 11 or more. The 20th score is at 10. *Median* = 10 errors. *Mode* = 11 errors. *Range* = 16 − 6 = 10 errors. The 25th percentile is 9 errors; the 75th percentile is 12. The *interquartile range* is 3 errors (12 − 9 = 3).

Interpretation: Bransford and Franks are interested in the way human beings process information and remember it. Thus, these data are a sample and the purpose is to generalize to all humans. $\hat{s}$ is the appropriate standard deviation, and $\hat{s}^2$ is the appropriate variance.

$$\hat{s} = \sqrt{\frac{\Sigma fX^2 - \frac{(\Sigma fX)^2}{N}}{N-1}} = \sqrt{\frac{4489 - \frac{(409)^2}{39}}{38}}$$

$$= \sqrt{5.256} = 2.29 \text{ errors}$$

$\hat{s}^2 = 5.26$ errors

Interpretation: People are poor at recognizing what they have heard if the test includes concepts similar to what they heard originally. On the average, about half of the 20 sentences that they had never heard were identified as having been heard earlier.

(Continued)

(Mean = 10.49 errors; median = 10 errors; mode = 11 errors.) People are fairly consistent in their mistakes; the IQR was only three errors, although the range of errors was from 6 to 16. The standard deviation ($\hat{s}$) was 2.29 errors; the variance was 5.26 errors.

4.22. What is your response? Do you know this stuff? Can you work the problems?

CHAPTER 5

5.1. Zero

5.2. $\Sigma z = 0$. Because $\Sigma(X - \bar{X}) = 0$, it follows that

$$\frac{\Sigma(X - \bar{X})}{S} = 0.$$

5.3. Negative z scores are preferable when lower scores are preferable to higher scores. Examples include measures such as errors, pollution levels, and finish times.

5.4. Harriett: $z = \dfrac{37 - 39.5}{1.803} = -1.39$;

Heslope: $z = \dfrac{24 - 26.25}{1.479} = -1.52$

Interpretation: Of course, in timed events, the more negative the z score, the better the score. Heslope's -1.52 is superior to Harriett's -1.39.

5.5. Tobe's apple: $z = \dfrac{9 - 5}{1} = 4.00$;

Zeke's orange: $z = \dfrac{10 - 6}{1.2} = 3.33$

Interpretation: Tobe's z score is larger, so the answer to Hamlet must be a resounding "To be." Notice that each fruit varies from its group mean by the same amount. It is the smaller variability of the apple weights that makes Tobe's fruit a winner.

5.6. First test: $z = \dfrac{79 - 67}{4} = 3.00$;

Second test: $z = \dfrac{125 - 105}{15} = 1.33$;

Third test: $z = \dfrac{51 - 45}{3} = 2.00$

Interpretation: Drop Ableson's second test.

5.7. An efficient way to solve this problem is to first find the value of (1.5 × IQR). Then subtract from and add to the 25th percentile and the 75th percentile, respectively. Thus, 1.5 × IQR = 1.5 × 0.9 = 1.35. The lowest outlier temperature is 97.8 − 1.35 = 96.45°F.

The highest outlier temperature is 98.7 + 1.35 = 100.05°F

a. 98.6°F no
b. 99.9°F no
c. 96.0°F yes
d. 96.6°F no
e. 100.5°F yes

5.8.
- Distribution X, which is positively skewed, has a mean of about 3 and a median of about 1 1/2. The range of scores is 7 and the interquartile range is 2.
- Distribution Y. The mean of this distribution is near 6 1/2 with a median near 7 1/2. The range is 6, the interquartile range is 2, and the distribution is negatively skewed.
- Distribution Z is symmetrical with a mean and median of about 6 1/2. The range is 3 and the interquartile range is 1.

5.9.

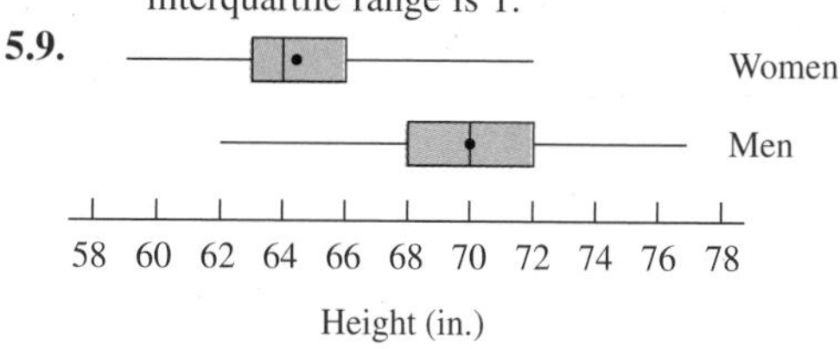

5.10. For normal oral body temperatures: highest = 99.5°F; lowest = 96.4°F; 25th percentile = 97.8°F; 75th percentile = 98.7°F; median = 98.2°F

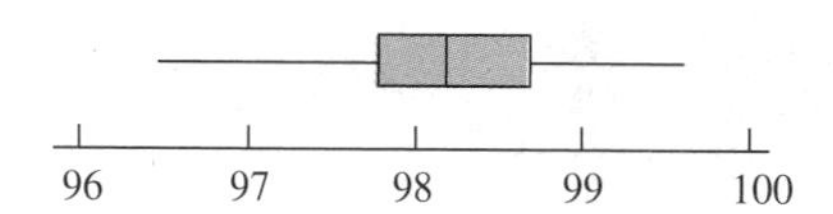

5.11. $d = \dfrac{\mu_1 - \mu_2}{\sigma} = \dfrac{\bar{X}_1 - \bar{X}_2}{\sigma}$

a. $d = 0.50$. Interpretation: A d of 0.50 is a medium effect size index.

b. $d = 0.90$. Interpretation: A d of 0.90 indicates a quite large effect size, greater even than the 0.80 that is considered large.

c. $d = 0.30$. Interpretation: A d of 0.30 is intermediate in size between values that are considered small and medium; I'd call it "somewhat larger than a small effect size." (Note that the means are the same in problems **b** and **c.** Data variability is always important.)

5.12. $d = \frac{\mu_1 - \mu_2}{\sigma} = \frac{\overline{X}_1 - \overline{X}_2}{\sigma} = \frac{469 - 456}{110} = 0.12$

Interpretation: Although first-born children score higher on cognitive ability than second-born children, the size of the effect is quite small, being about half the size usually designated as "small."

5.13. For cabbages:

$$\overline{X} = \frac{\Sigma X}{N} = \frac{300}{3} = 100 \text{ pounds}$$

$$S = \sqrt{\frac{\Sigma X^2 - \frac{(\Sigma X)^2}{N}}{N}}$$

$$= \sqrt{\frac{30{,}200 - \frac{(300)^2}{3}}{3}} = 8.165 \text{ pounds}$$

For pumpkins:

$$\overline{X} = \frac{\Sigma X}{N} = \frac{3600}{3} = 1200 \text{ pounds}$$

$$S = \sqrt{\frac{\Sigma X^2 - \frac{(\Sigma X)^2}{N}}{N}}$$

$$= \sqrt{\frac{4{,}400{,}000 - \frac{(3600)^2}{3}}{3}} = 163.3 \text{ pounds}$$

Lavery's cabbage:

$$z = \frac{X - \overline{X}}{S} = \frac{124 - 100}{8.165} = 2.94$$

Jutras's pumpkin:

$$z = \frac{X - \overline{X}}{S} = \frac{1689 - 1200}{163.3} = 2.99$$

Interpretation: Jutras's pumpkin is the *BIG* winner.

5.14. The lowest outlier score is *25th percentile* $-(1.5 \times \text{IQR}) = 21 - (1.5 \times 7) = 10.5$.
The highest outlier score is *75th percentile* $+(1.5 \times \text{IQR}) = 28 + (1.5 \times 7) = 38.5$.

The outliers in Table 2.3 are all low scores: 10, 9, and 5.

5.15. Compile the raw data into simple frequency distributions:

Psychotherapy	f	Control	f
34	1	22	1
28	1	21	1
25	1	18	1
23	1	13	1
18	1	10	1
15	1	9	1
13	2	5	1
11	1	4	2
10	2	3	2
9	1	2	1
7	2	1	1
5	1	0	1
0	1	−2	1
−2	1	−3	1
−5	1	−7	1
−10	1	−9	1
−15	1	−12	1
		−22	1
	Σ 20		Σ 20

$$\overline{X}_{\text{therapy}} = \frac{\Sigma X}{N} = \frac{196}{20} = 9.8$$

$$\overline{X}_{\text{control}} = \frac{\Sigma X}{N} = \frac{60}{20} = 3.0$$

$$\text{Median location} = \frac{N + 1}{2} = \frac{20 + 1}{2} = 10.5$$

$\text{Median}_{\text{therapy}}$: The 10.5th score is between the two scores of 10. $\text{Median}_{\text{therapy}} = 10$.

$\text{Median}_{\text{control}}$: The 10.5th score is between the two scores of 3. $\text{Median}_{\text{control}} = 3$.

Therapy: 25th percentile: $0.25 \times N = 5$. The 5th score is 0, so 0 is the 25th percentile.
75th percentile: The 5th score from the top is 18, which is the 75th percentile.

Control: 25th percentile: $0.25 \times N = 5$. -3 is the 25th percentile.
75th percentile: 10 is the 75th percentile.

$$d = \frac{\mu_1 - \mu_2}{\sigma} = \frac{\overline{X}_1 - \overline{X}_2}{\sigma} = \frac{9.8 - 3.0}{10} = 0.68$$

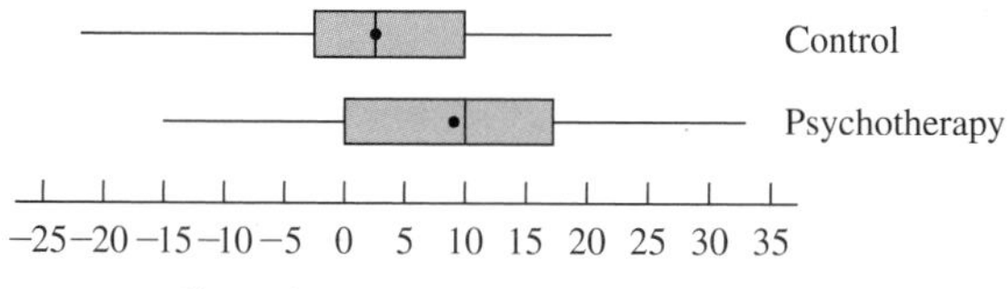

Interpretation: Psychotherapy helps. On the average, the psychological health of those who received psychotherapy was 9.8 points higher after treatment. Those not receiving treatment improved only 3.0 points. The difference in the two means produced an effect size index of 0.68, which is intermediate between "medium" and "large." Both distributions of scores were approximately symmetrical.

CHAPTER 6

6.1. A bivariate distribution has two variables. The scores of the variables are paired in some logical way.

6.2. Variation from high to low in one variable is accompanied by predictable variation in the other variable. Saying that two variables are correlated does not tell you if the correlation is positive or negative.

6.3. In a positive correlation, as X increases, Y increases. (It is equivalent to say that as X decreases, Y decreases.) In a negative correlation, as X increases, Y decreases. (Equivalent statement: As X decreases, Y increases in negative correlation.)

6.4. **a.** Yes; positive; taller people usually weigh more than shorter people.

b. No; these scores cannot be correlated because there is no basis for pairing a particular first-grader with a particular fifth-grader.

c. Yes, negative; as temperatures go up, less heat is needed and fuel bills go down.

d. Yes, positive; people with higher IQs score higher on reading comprehension tests.

e. Yes, positive; those who score well on quiz 1 are likely to score well on quiz 2. The fact that some students are in section 1 and some are in section 2 can be ignored.

f. No; there is no basis for pairing the score of a student in section 1 with that of a student in section 2.

6.5.

	Fathers	Daughters
Mean	67.50	62.50
S	3.15	2.22
Σ (score)	405	375
Σ (score)^2	27,397	23,467

$\Sigma XY = 25,334$
$r = .513$

r by the blanched formula:

$$r = \frac{\frac{25,334}{6} - (67.5)(62.5)}{(3.15)(2.22)}$$

$$= \frac{4222.33 - 4218.75}{6.99} = .512$$

By the raw-score formula:

$$r = \frac{(6)(25,334) - (405)(375)}{\sqrt{[(6)(27,397) - (405)^2][(6)(23,467) - (375)^2]}} = .513$$

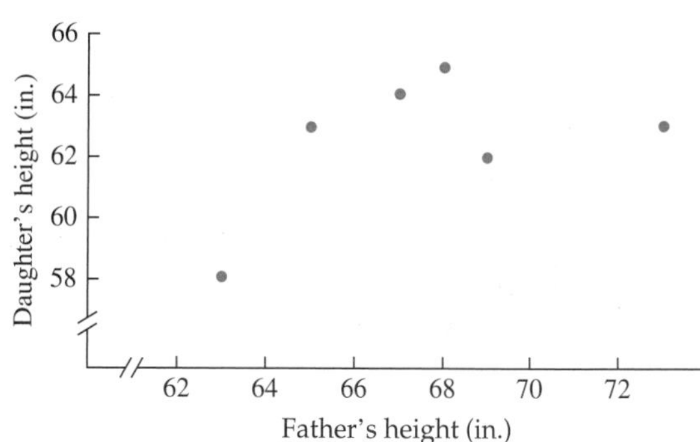

6.6. $\bar{X} = 105.00 \quad \bar{Y} = 103.00$

$$S_X = \sqrt{\frac{235,800 - \frac{(2205)^2}{21}}{21}} = \sqrt{203.571} = 14.268$$

$$S_Y = \sqrt{\frac{227,200 - \frac{(2163)^2}{21}}{21}} = \sqrt{210.048} = 14.493$$

$$r = \frac{\frac{231,100}{21} - (105)(103)}{(14.268)(14.493)}$$

$$= \frac{11,004.76 - 10,815.00}{206.786} = \frac{189.760}{206.786} = .92$$

(Continued)

Interpretation: A correlation of .92 indicates that the WAIS and the Wonderlic are measuring almost the same thing.

6.7.

$$r = \frac{(10)(8524) - (190)(444)}{\sqrt{[(10)(3940) - (190)^2][(10)(20096) - (444)^2]}} = .25$$

$$r = \frac{\frac{8524}{10} - (19.0)(44.4)}{(5.744)(6.184)} = .25$$

6.8. The actual correlation coefficients are:

a. −.60 **b.** −.95 **c.** .50 **d.** .25

Would you be interested in knowing how statistics teachers did on this question? I asked 31 of them to estimate r for these four scatterplots. These are the means and standard deviations of their estimates:

a. $\bar{X} = -.50; \hat{s} = .19$ **b.** $\bar{X} = -.87; \hat{s} = .06$
c. $\bar{X} = .42; \hat{s} = .17$ **d.** $\bar{X} = .13; \hat{s} = .11$

These data suggest to me that estimating correlation coefficients from scatterplots is not easy.

6.9. Coefficient of determination $= .92^2 = .85$.
Interpretation: The two measures of intelligence have about 85 percent of their variance in common, which shows that there is a great deal of overlap in what they are measuring.

6.10. $r^2 = (.25)^2 = .0625$
Interpretation: This means that 6.25 percent of the variation that is seen in infectious diseases is related to stress; 93.75 percent is due to other factors. Thus, to "explain" illness (prevalence, severity, and such), factors other than stress are required.

6.11. Your shortest answer here is *no*. Although you have two distributions, you do not have a bivariate distribution. There is no pairing of a height score for a woman with a height score for a man.

6.12. $(.10)^2 = .01$, or 1 percent; $(.40)^2 = .16$, or 16 percent. Note that a 4-fold increase in the correlation coefficient produces a 16-fold increase in the common variance.

6.13.

	Cigarette consumption	Death rate
Mean	604.27	205.00
S	367.865	113.719
Σ (score)	6,647	2,255
Σ (score)^2	5,505,173	604,527

$\Sigma XY = 1{,}705{,}077$
$N = 11$
$r = .74$

Interpretation: With $r = .74$, these data show that there is a strong relationship between per capita cigarette consumption and male death rate 20 years later. Data such as these do not permit cause-and-effect conclusions.

6.14. **a.** Interpretation: Vocational interests tend to remain stable from age 20 to age 40.
b. Identical twins raised together have very similar IQs.
c. There is a slight tendency for IQ to be lower as family size increases.
d. There is a slight tendency for taller men to have higher IQs than shorter men.
e. The lower a person's income level is, the greater the probability that he or she will be diagnosed as schizophrenic.

6.15. Interpretation: As humor scores increase, insight scores increase. The effect size index, 0.83, is very large, which indicates that there is a very strong relationship between humor and insight. Humor and insight have 69 percent of their variance in common, leaving only 31 percent not accounted for. The caveats (warnings) that go with r are that the scatterplot must show a linear relationship and that r by itself does not justify cause-and-effect statements.

6.16. Interpretation: A correlation of −.37 means that there is a medium-sized tendency for children with more older siblings to accept less credit or blame for their own successes or failures than children with fewer older siblings. The coefficient of determination, .1369, means that only about 14 percent of the variance in acceptance of responsibility is predictable from knowledge of the number of older siblings; 86 percent is not. No cause-and-effect statements are justified by a correlation coefficient alone.

6.17. A Pearson r is not a useful statistic for data such as these because the relationship is a curved one. You might note that the scatterplot is similar to the curved relationship you saw in Figure 2.9.

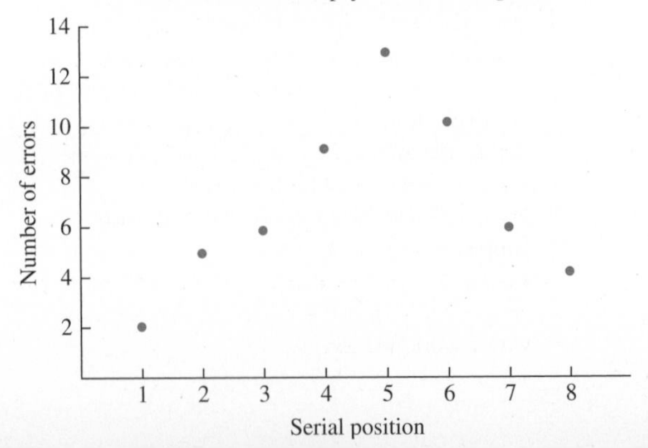

6.18.

$$b = r\frac{S_Y}{S_X} = (.724)\frac{90.42}{93.54} = (.724)(0.967) = 0.700$$

$$a = \overline{Y} - b\overline{X} = 515 - (0.700)(500) = 515 - 350 = 165$$

$$\hat{Y} = a + bX = 165 + 0.700X$$

6.19. a.

$$b = r\frac{S_Y}{S_X} = (.513)\frac{2.217}{3.149} = (.513)(.704) = 0.3612$$

$$a = \overline{Y} - b\overline{X} = 62.5 - (0.3612)(67.5) = 38.12$$

b. To draw the regression line, I used two points: $(\overline{X}, \overline{Y})$ and (62, 60.5). I got the point (62, 60.5) by using the formula $\hat{Y} = a + bX$. Other X values would work just as well.

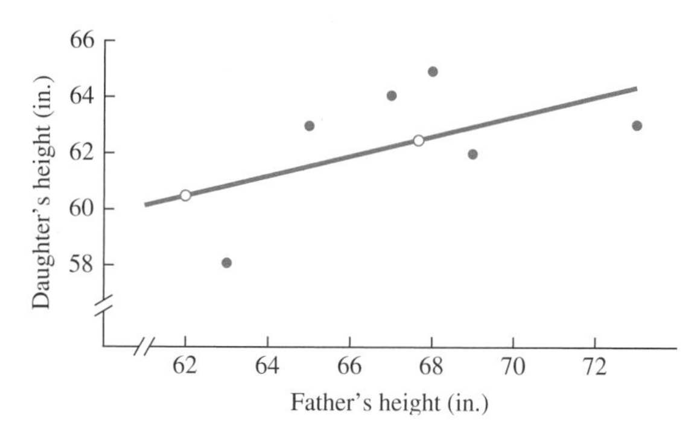

6.20. a.

$$b = r\frac{S_Y}{S_X} = (.92)\frac{14.493}{14.268} = (.92)(1.016) = 0.935$$

$$a = \overline{Y} - b\overline{X} = 103 - (0.935)(105)$$

$$= 103 - 98.123 = 4.877$$

b. $\hat{Y} = a + bX = 4.877 + (0.935)(130) = 126.4$

6.21. a.

	Advertising X	Sales Y
Mean	4.286	108.571
S	1.030	20.304
Σ (score)	30	760
Σ (score)^2	136	85,400
	$\Sigma XY = 3360$	
	$r = .703$	

b. $\hat{Y} = a + bX = 49.176 + 13.858X$

c. One point that I used for the regression line on the scatterplot was for $\overline{X} = 4.29$ and $\overline{Y} = 108.57$, the two means. The other point was $X = 6.00$ and $Y = 132.33$. Other X values would work just as well.

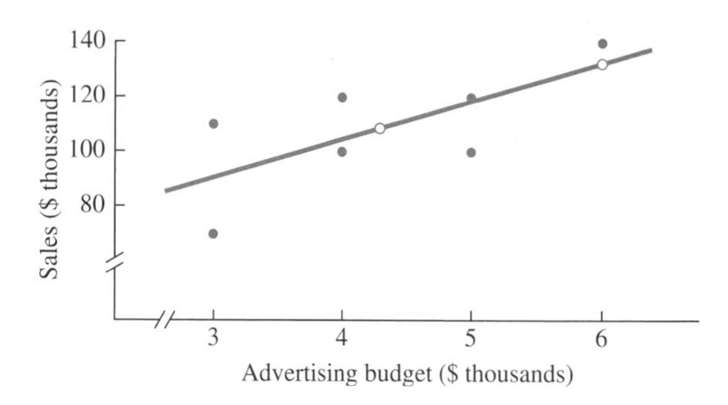

d. $\hat{Y} = a + bX$. For $X = 10$, $\hat{Y} = 187.756$ or, in terms of sales, $187,756.

e. Interpretation: You were warned that the interpretation of r can be a tricky business. Your answer might contain one of those common errors I mentioned. With $r = .70$, you have a fair degree of confidence that the prediction will be close. However, if your thinking (and your answer) was that increasing advertising would increase sales, you made the error of inferring a causal relationship from a correlation coefficient.

As a matter of fact, some analysts say that sales *cause* advertising; they say that advertising budgets are determined by *last year's* sales. In any case, don't infer more than the data permit. These data allow you to estimate with some confidence what sales were by knowing the amount spent for advertising. In this problem, the percent of sales spent on advertising (just less than 4 percent) is about the same as the percent spent nationally in the United States.

Another caution: The $10,000 figure is much more than any of the values used to find the regression line. It is not safe to predict outcomes for points far from the data.

6.22.

$$\hat{Y} = (.80)\left(\frac{15}{16}\right)(65 - 100) + 100 = 73.75 = 74$$

Interpretation: An IQ score depends to some degree on the test administered; a person with an IQ of 65 on the Stanford–Binet is predicted to have an IQ of 74 on the WAIS.

6.23. Designate the time number X and the graduates Y.

$\Sigma X = 15.0000 \quad \Sigma Y = 6.9700 \quad N = 5$

$\Sigma X^2 = 55.0000 \quad \Sigma Y^2 = 9.7403 \quad \Sigma XY = 21.4000$

$\overline{X} = 3.0000$ $\overline{Y} = 1.3940$ $S_X = 1.4142$

$S_Y = 0.0695$

$r = .998$ $a = 1.2469$ $b = 0.0490$

$\hat{Y} = a + bX = 1.2469 + 0.0490X$

For the year 2011, $X = 10$. Thus,

$\hat{Y} = 1.2469 + 0.0490(10) = 1.737$

Adjusting the decimals, the number of predicted graduates with bachelor's degrees in the year 2011 is approximately 1,737,000 or 1.7 million. Depending on when you are working this problem, you can check the accuracy of this prediction in the *Statistical Abstract of the United States.* (Also, because you already have the regression equation, you can easily predict the number of graduates for the particular year that *you* will graduate.)

6.24. On the big important questions like this one, you are on your own!

What Would You Recommend? Chapters 1–6

a. The mode is appropriate because this is a distribution of nominal data.

b. z scores can be used.

c. A regression analysis produces a prediction.

d. The median is appropriate because the distribution is severely skewed (the class interval with zero has the highest frequency). The range, interquartile range, and standard deviation would all be informative. The appropriate standard deviation is σ because the 50 states represent the entire population of states.

e. The applicant with the more consistent judgments is the better candidate. The range, interquartile range, and standard deviation would all be helpful.

f. A line graph is appropriate for the relationship that is described. A Pearson product-moment correlation coefficient is not appropriate because the relationship is curvilinear (much forgetting at first, and much less after 4 days).

g. The median is the appropriate central tendency statistic when one or more of the categories is open-ended.

h. A correlation coefficient will describe this relationship. (What direction and size would you guess for this coefficient? The answer, based on several studies, is $-.10$.)

CHAPTER 7

7.1. There are seven cards between the 3 and jack, each with a probability of $\frac{4}{52} = .077$. So $(7)(.077) = .539$.

7.2. The probability of drawing a 7 is $\frac{4}{52}$, and there are 52 opportunities to get a 7. Thus, $(52)(\frac{4}{52}) = 4$.

7.3. There are four cards that are higher than a jack or lower than a 3. Each has a probability of $\frac{4}{52}$. Thus, $(4)(\frac{4}{52}) = \frac{16}{52} = .308$.

7.4. The probability of a 5 or 6 is $\frac{4}{52} + \frac{4}{52} = \frac{8}{52} = .154$. In 78 draws, $(78)(.154) = 12$ cards that are 5s or 6s.

7.5. **a.** .7500; adding the probability of one head (.3750) to that of two heads (.3750), you get .7500.

b. $.1250 + .1250 = .2500$

7.6. $16(.1250) = 2$ times

7.7. **a.** .0668

b. Interpretation: The test scores are normally distributed.

7.8. **a.** .0832

b. .2912

c. .4778

7.9. Empirical; the scores are based on observations.

7.10. $z = (X - \mu)/\sigma$.

a. z score $= -3.00$

b. z score $= .67$

c. z score $= .20$

d. z score $= .00$

7.11. To make a quick check of your answer, compare it with the proportion of students who have IQs of 120 or higher, .0918. The proportion with IQs of 70 or lower will be less than the proportion with IQs of 120 or higher. Is your calculated proportion less? The figure that follows shows a normal distribution in which the proportion of the population with IQs of 70 or lower is shaded. For IQ $= 70$, $z = (70 - 100)/15 = -30/15 = -2.00$. The proportion beyond $z = 2.00$ is .0228 (column C in Table C). Thus, .0228, or 2.28 percent, of the population is expected to have IQs of 70 or lower.

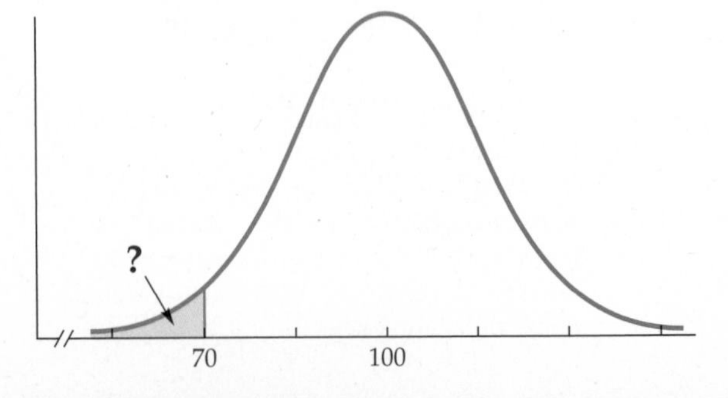

7.12. (.0228)(4000) = 91.2, or 91 students.

7.13. $z = (110 - 100)/15 = 10/15 = .67$. The proportion beyond $z = .67$ is .2514, which is the expected proportion of people with IQs of 110 or higher.

7.14. **a.** .2514 × 250 = 62.85, or 63 students
b. 250 − 62.85 = 187.15, or 187 students
c. $\frac{1}{2} \times 250 = 125$. I hope you were able to get this one immediately by thinking about the symmetrical nature of the normal distribution.

7.15. The z score that includes a proportion of .02 is 2.06. If $z = 2.06$, then $X = 100 + (2.06)(15) = 100 + 30.9 = 130.9 = 131$. In fact, Mensa requires an IQ of 130 on tests that have a standard deviation of 15, which permits somewhat more than 2 percent of the population to qualify.

7.16. **a.** The z score you need is 1.65. (1.64 will include more than the tallest 5 percent.) $1.65 = (X - 64.2)/2.5$; $X = 64.2 + 4.1 = 68.3$ inches.
b. $z = (58 - 64.2)/2.5 = -6.2/2.5 = -2.48$. The proportion excluded is .0066.

7.17. **a.** $z = (60 - 70.0)/3.0 = -10.0/3.0 = -3.33$. The proportion excluded is .0005.
b. $z = (62 - 70.0)/3.0 = -8.0/3.0 = -2.67$. The proportion taller than Napoleon is .9962 (.5000 + .4962).

7.18. **a.** $z = \dfrac{3.20 - 3.11}{.05} = 1.80$, proportion = .0359
b. The z score corresponding to a proportion of .1000 is 1.28. Thus,

$$X = 3.11 + (1.28)(.05) = 3.11 + .064 = 3.17 \text{ grams}$$

$$X = 3.11 - .064 = 3.05 \text{ grams}$$

Thus, the middle 80 percent of the pennies weigh between 3.05 and 3.17 grams.

7.19.

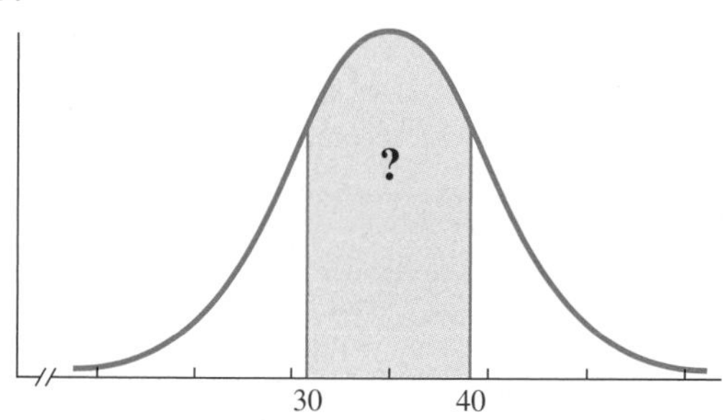

a. $z = \dfrac{30 - 35}{6} = -.83$, proportion = .2967;

$z = \dfrac{40 - 35}{6} = .83$, proportion = .2967;

(2)(.2967) = .5934 = the proportion of students with scores between 30 and 40.
b. The probability is also .5934.

7.20. Interpretation: No, because the proportion between 30 and 40 straddles the mean, where scores that occur frequently are found. The proportion between 20 and 30 is all in one tail of the curve, where scores occur less frequently.

7.21.

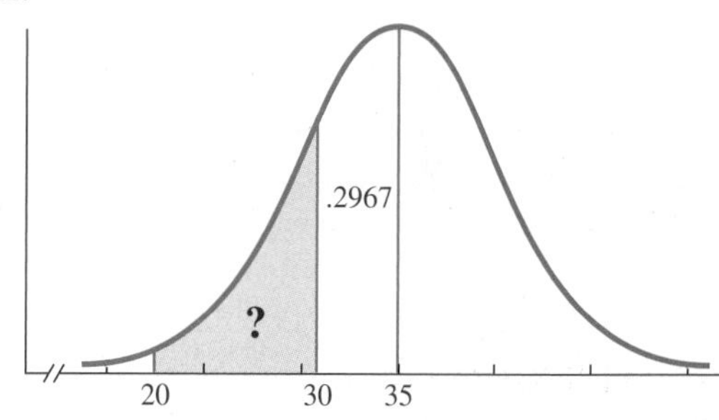

$z = \dfrac{20 - 35}{6} = -2.50$, proportion = .4938

$z = \dfrac{30 - 35}{6} = -.83$, proportion = .2967

.4938 − .2967 = .1971. The proportion of students with scores between 35 and 30 is subtracted from the proportion with scores between 35 and 20. There are other ways to set up this problem that yield the same proportion.

7.22. 800 × .1971 = 157.7 = 158 students

7.23.

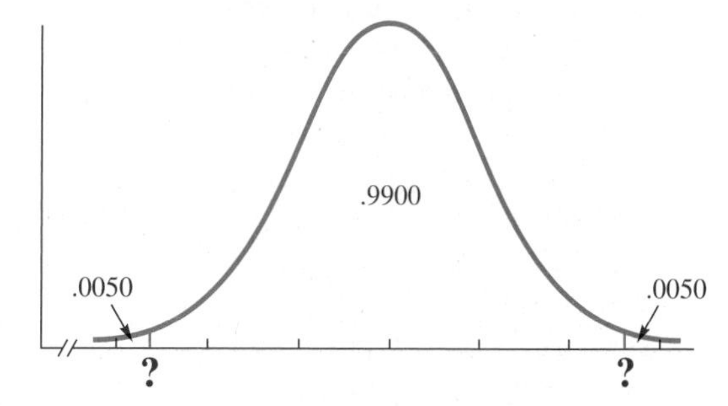

The z score that is conventionally used for a proportion of .0050 is 2.58 (column C). Thus,

$$X = \mu + (z)(\sigma) = 100 + (2.58)(15)$$
$$= 138.7, \text{ or } 139$$

and

$$X = 100 + (-2.58)(15) = 61.3, \text{ or } 61$$

IQ scores of 139 or higher or 61 or lower are achieved by 1 percent of the population.

7.24. Because .01 of the curve lies outside the limits of the scores 61 and 139, the probability is .01. The probability is .005 that the randomly selected person has an IQ of 139 or higher and .005 that it is 61 or lower.

7.25. $z = \frac{.85 - .99}{.17} = \frac{-.14}{.17} = -.82$

The proportion associated with $z = -.82$ is .2061 (column C), and (.2061)(185,822) = 38,298 truck drivers. Because the normal curve corresponds only approximately to the empirical curve, the usual procedure is to report this result as "About 38,000 truck drivers would get raises."

7.26. This problem has two parts: determining the number of trees over 8 inches DBH on 1 acre and multiplying by 100.

$$z = \frac{8 - 13.68}{4.83} = -1.18$$

proportion = .5000 + .3810 = .8810
.8810(199) = 175.319 trees on 1 acre
175.319(100) = 17,532 trees on 100 acres
(about 17,500 trees)

7.27. $z = \frac{300 - 268}{14} = \frac{32}{14} = 2.29$

The proportion of gestation periods expected to last 10 months or longer is .0110. In modern obstetrical care, pregnancies are normally brought to term (by drugs or surgery) earlier than 300 days.

7.28. Converting this problem to inches, you find the z score is

$$z = \frac{80 - 70.0}{3.0} = 3.33$$

The proportion of the curve above 3.33 is .0004, so only 4 in 10,000 have to duck to enter a residential room.

7.29. $\mu = \frac{3600}{100} = 36.0$ inches

$$\sigma = \sqrt{\frac{130{,}000 - \frac{(3600)^2}{100}}{100}} = 2.00 \text{ inches}$$

a. $z = \frac{32 - 36}{2} = -2.00$, proportion = .5000 + .4772 = .9772

b. $z = \frac{39 - 36}{2} = 1.50$, proportion = .5000 + .4332 = .9332
.9332(300) = 279.96, or 280 hobbits

c. $z = \frac{46 - 36}{2} = 5.00$,
proportion < .00003, or practically zero

CHAPTER 8

8.1. $\overline{X} = \frac{\Sigma X}{N} = \frac{68}{8} = 8.50$

8.2. For this problem, it is not the numbers in your sample that are "right" or "wrong" but the method you used to get them. Create an identifying number (01–20) for every population member. (Table 8.2 works fine.) Haphazardly find a starting place in Table B and record in sequence 10 two-digit identifying numbers (ignoring numbers over 20 and duplications). Write the 10 population members that correspond to the 10 selected identifying numbers. How does your mean compare to the mean of 8.50?

8.3. As in problem 8.2, the method, not the numbers, determines a correct answer. You should give each number an identifying number of 01 to 38, begin at some chance starting place in Table B, and select 12 numbers. The starting place for every random sample taken from Table B should be haphazardly chosen. If you start at the same place in the table every time, you will have the same identifying numbers every time.

8.4. Interpretation: Those with greater educational accomplishments are more likely to return the questionnaire. Such a biased sample will overestimate the accomplishments of the population.

8.5. Interpretation: The sample will be biased. Many of the general population will not even read the article and thus cannot be in the sample. Of those who read it, some will have no opinion and some will not bother to mark the ballot and return it. In short, the results of such a questionnaire cannot be generalized to the larger population.

8.6. **a.** Biased, although such a bias is not considered serious. Every sample does not have an equal chance of being selected. For example, two brothers named Kotki, next to each other on the list, could not both be in one sample. Thus, samples with both Kotki brothers are not possible.

b. Biased, because none of the samples would have freshmen, sophomores, or seniors, although they are members of the population.

c. Random, assuming the names in the box were thoroughly mixed.

d. Random; though small, this sample fits the definition of a random sample.

e. Biased; some members of the population have completed the English course and no samples could include these people.

8.7. For many important problems, random samples are impractical or impossible to obtain. Researchers often use convenient samples to conduct experiments. If the experiments produce a consistent pattern of results, researchers conclude that this pattern is true for the population, and they are usually correct.

8.8. Standard error; expected value

8.9. **a.** Select a set of numbers as the population. Draw many samples of the same size, find the range for each sample, and arrange those ranges into a frequency distribution.

b. The standard error of the range

c. The expected value of the range, which is the mean of the sampling distribution of the range, will be *less* than the population range. Only a very few sample ranges (maybe only one) can have a value equal to the population range. Thus, almost all the sample ranges will be less than the population range. Statistics (such as the range) with expected values that are not equal to the corresponding population parameters are called *biased estimators* and are not popular among mathematical statisticians.

8.10. Please check your answer with the definition and explanation in the chapter.

8.11. If this standard error (which is called the *standard error of estimate*) is very small, it means that the actual Y will be close to the predicted $\hat{Y}$. A very large standard error indicates that you cannot put much confidence in a predicted $\hat{Y}$. That is, the actual Y is subject to other influences besides X.

8.12.

	σ			
N	1	2	4	8
1	1	2	4	8
4	0.50	1	2	4
16	0.25	0.50	1	2
64	0.125	0.25	0.50	1

8.13. Interpretation: As N increases by a factor of 4, $\sigma_{\bar{X}}$ is reduced by one-half. More simply but less accurately, as N increases, $\sigma_{\bar{X}}$ decreases.

8.14. 16 times

8.15. $z = \frac{8.5 - 9}{2/\sqrt{8}} = \frac{-0.5}{0.707} = -.71$; proportion = .2389

8.16. For sample means of 8.5 or less,

$$z = \frac{8.5 - 9}{2/\sqrt{16}} = \frac{-0.5}{0.5} = -1.00;$$

proportion = .1587

For sample means of 10 or greater,

$$z = \frac{10 - 9}{2/\sqrt{16}} = \frac{1}{0.5} = 2.00; \text{ probability} = .0228$$

8.17. For mean IQs of 105 or greater,

$$z = \frac{105 - 100}{15/\sqrt{25}} = \frac{5}{3} = 1.67; \text{ probability} = .0475$$

For mean IQs of 90 or less,

$$z = \frac{90 - 100}{15/\sqrt{25}} = \frac{-10}{3} = -3.33;$$

probability = .0004

Interpretation: Classrooms with 25 students will have a mean IQ fairly close to 100; departures of as much as 5 points will happen less than 5 percent of the time; departures of as much as 10 points will be very rare, occurring less than one time in a thousand. This analysis is based on the premise that no systematic factors are influencing a child's assignment to a classroom.

8.18. $\sigma_{\bar{X}} = \frac{20{,}000}{\sqrt{40}} = 3162$. Therefore,

$$z = \frac{73{,}900 - 77{,}000}{3162} = -.98; p = .1635$$

(Continued)

Interpretation: The junior's suspicion that the \$73,900 sample mean might be just a chance fluctuation from a true mean of \$77,000 has some foundation. Such a sample mean (or one even smaller) would be expected in about 16 percent (one-sixth) of a large number of random samples if the true campus mean is \$77,000.

8.19. **a.** Normal; *t* **b.** *t;* normal **c.** *t;* normal

8.20. Degrees of freedom

8.21. W. S. Gosset, who wrote under the pseudonym "Student," invented the *t* distribution so that he could assess probabilities for small samples from populations for which σ was not known.

8.22. $\bar{X} = \frac{\Sigma X}{N} = \frac{156{,}000}{16} = 9750$ hours

$$\hat{s} = \sqrt{\frac{1{,}524{,}600{,}000 - \frac{(156{,}000)^2}{16}}{15}}$$

$= 489.90$ hours

$$s_{\bar{X}} = \frac{\hat{s}}{\sqrt{N}} = \frac{489.90}{\sqrt{16}} = 122.47 \text{ hours}$$

8.23. $\bar{X} = 27.00$ points; $\hat{s} = 3.803$ points; $s_{\bar{X}} = 1.016$ points; $df = 13$

The *t* value for 95 percent confidence for 13 *df* is 2.160.

$$LL = \bar{X} - t_{\alpha}(s_{\bar{X}})$$
$$= 27.00 - 2.160(1.016) = 24.80 \text{ points}$$
$$UL = \bar{X} + t_{\alpha}(s_{\bar{X}})$$
$$= 27.00 + 2.160(1.016) = 29.20 \text{ points}$$

Interpretation: The workshop is effective. The national norm of the DoorMAT test is 24.0 points. Based on a sample of 14, we are 95 percent confident that the average score of all who take the workshop is between 24.80 and 29.20 points. Because the national mean is less than the lower limit of the confidence interval, we can conclude that the workshop increases assertiveness scores. (This conclusion assumes that the clients were not more assertive than average before the workshop.)

8.24. $\bar{X} = 1.20$ inches, $\hat{s} = 0.10$ inch, $s_{\bar{X}} = 0.0577$ inch, $t = 31.598$

$$LL = 1.20 - 31.598(0.0577) = -0.62 \text{ inch}$$
$$UL = 1.20 + 31.598(0.0577) = 3.02 \text{ inches}$$

Interpretation: The 120 Krups coffee pots meet the specifications. The company can have 99.9 percent confidence that the true mean burn rate for this box of pots is less than 4 inches per minute. (The lower limit is not of interest to Hawker-Beechcraft, and of course, there is no such thing as −0.62 inches.)

8.25. **a.** $\pm 3.182(2) = \pm 6.364$ (12.728 units wide)

b. $\pm 2.131(1) = \pm 2.131$ (4.262 units wide). The fourfold increase in *N* produces a confidence interval about one-third as wide. The effect of changing the sample size depends on the relative change in *N*, not the absolute change.

c. $\pm 3.182(1) = \pm 3.182$ (6.364 units wide). Reducing $\hat{s}$ by half reduces the size of the confidence interval by half. This relationship is true regardless of the size of $\hat{s}$. Data gatherers reduce $\hat{s}$ by making careful measurements, using precise instruments, and increasing *N*.

8.26. The size of the standard deviation of each sample. If you happen to get a sample that produces a large standard deviation, a large standard error will result and the confidence interval will be wide.

8.27. **a.** 19

b. Narrower; because *N* was increased by four, $s_{\bar{X}}$ is reduced to half of its size. In addition, increasing *N* decreases the *t* value from Table D. Therefore, the lines are less than half as long.

8.28. **a.** 18

b. Narrower; the *t* value is smaller.

8.29. The actual values for the two 95 percent confidence intervals are:

females: 10.9–14.6 years

males: 14.0–15.5 years

8.30. The standard error of the mean for each of the four trial means is 1.0; the *t* value (24 *df*) for the 95 percent confidence interval is 2.064. The lower and upper limits for each mean are ± 2.064.

Look at the graph that accompanies this problem. The graph in problem 7.30 has been modified by adding a 95 percent confidence interval to each trial mean. The overlap of the trial 2 confidence interval with that of the trial 1 mean indicates that the dip at trial 2 may not be reliable. That is, you are 95 percent confident that the population mean for trial 2 is in the range of 3.94 to 8.06. Because the upper half of this range is also included in the trial 1 confidence interval, you aren't confident that a second sample of

(Continued)

data will show the dip at trial 2. [Note that the figure that shows only the means invites you to conclude that performance (always) goes down on trial 2. The figure with the confidence intervals leads to a more cautious interpretation.]

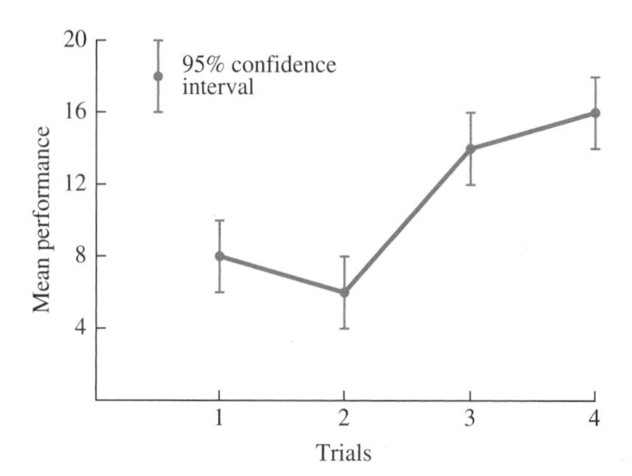

Look at the *lack* of overlap between the confidence intervals of trial 2 and trial 3. You can conclude that trial 3 performance is *reliably* better than that of trial 2; that is, you are more than 95 percent confident that a second sample of data will also show that the trial 3 mean is greater than the trial 2 mean.

CHAPTER 9

9.1. Null hypothesis statistical test (or testing)

9.2. Interpretation: .20 is the probability of data such as those actually observed, if the equality hypothesis is true.

9.3. The null hypothesis is a statement about a parameter.

9.4. The null hypothesis (symbolized H_0) is a tentative hypothesis about the value of some parameter (often the mean) of the population that the sample is taken from. The alternative hypothesis (symbolized H_1) specifies other values for the parameter.

9.5. H_1: $\mu_1 \neq 269.3$ grams; H_1: $\mu_1 < 269.3$ grams; H_1: $\mu_1 > 269.3$ grams

9.6. Your outline should include the following points:

a. Gather sample data from the population you are interested in and calculate a statistic.

b. Recognize two logical possibilities for the population:

- H_0: a statement specifying an exact value for the parameter of the population
- H_1: a statement specifying all other values for the parameter

c. Using a sampling distribution that assumes H_0 is correct, find the probability of the statistic you calculated.

d. If the probability is very small, reject H_0 and accept H_1. If the probability is large, retain both H_0 and H_1, acknowledging that the data do not allow you to reject H_0.

9.7. Disagree

9.8. False. Although the researcher probably thinks that H_0 is false, the sampling distribution is based on the assumption that H_0 is true.

9.9. A significance level (α level) is chosen by the researcher as the point that separates "This is an unlikely event if H_0 is true and therefore reject H_0" from "This could reasonably happen if H_0 is true and therefore retain H_0." The largest acceptable significance level without justification is .05.

9.10. Phrases that should be circled are "*p* is small" and "Accept H_1."

9.11. **a.** $t_{.01}$ (17 *df*) = 2.898
b. $t_{.001}$ (40 *df*) = 3.551
c. $t_{.05}$ (∞ *df*) = 1.96

9.12. A good interpretation explains the results with the terms and concepts used in the description of the problem. It also tells the direction of any difference that was observed.

9.13. α is the probability of a Type I error, which occurs when a researcher mistakenly rejects H_0. A statistical test of the data produces a *p* value, which is the probability of obtaining the sample statistic actually obtained, if H_0 is true.

9.14. You retain a false null hypothesis.

9.15. Interpretation: The probability of making a Type I error is zero. If at first you don't agree with this answer, keep thinking until you understand.

9.16. The probability of a Type I error decreases from .05 to .01, and the probability of a Type II error increases.

9.17. If an answer here surprises you, make a note of what surprised you. Students, like researchers, have expectations and surprises. A surprise often means that you are about to learn something. Figure out why you were surprised.

(Continued)

	Critical value	t-test value	Reject or retain H_0?	p value
a.	2.262	2.25	Retain	$p > .05$
b.	2.539	2.57	Reject	$p < .01$
c.	2.042	2.03	Retain	$p > .05$
d.	6.869	6.72	Retain	$p > .001$
e.	1.714	1.72	Reject	$p < .05$
f.	2.423	2.41	Retain	$p > .02$

9.18. $t_{.05}$ (23 df) = 2.069

a. Interpretation: The mean paranoia score on the MMPI-2 for the 24 police officers was 54.3, which is significantly greater than the population mean of 50 ($p < .05$). Police officers are significantly more paranoid than the general population.

b. Interpretation: The mean paranoia score on the MMPI-2 for the 24 police officers was 54.3, which is not significantly different from the population mean of 50 ($p > .05$). Police officers are not significantly more paranoid than the general population.

9.19. $t_{.05}$ (9 df) = 2.262

a. Interpretation: The reduction in return rate from 5.3 percent to 4.9 percent was not statistically significant ($p > .05$). Including testimonials with orders does not reduce returns by a statistically significant amount.

b. Interpretation: Including testimonials with orders reduced the return rate from 5.3 percent to 4.9 percent, a statistically significant reduction ($p < .05$).

9.20. The null hypothesis is H_0: $\mu_0 = 100$.

$$\overline{X} = \frac{5250}{50} = 105.00$$

$$\hat{s} = \sqrt{\frac{560{,}854 - \frac{(5250)^2}{50}}{49}} = 14.00$$

$$s_{\overline{X}} = \frac{14.00}{\sqrt{50}} = 1.98$$

$$t = \frac{\overline{X} - \mu_0}{s_{\overline{X}}} = \frac{105 - 100}{1.98} = 2.53; \quad 49\ df$$

$t_{.02}$ (40 df) = 2.423

Interpretation: The null hypothesis can be rejected; $p < .02$. The conclusion is that the Head Start participants had IQs that were significantly above average.

The actual studies that evaluated Head Start were more sophisticated than the one described here, but the conclusions were the same. The IQ differences, however, weren't lasting, but other benefits were.

9.21. The null hypothesis is H_0: $\mu_0 = 10$. The best alternative hypothesis is H_1: $\mu_1 \neq 10$. A two-tailed alternative hypothesis will allow you to conclude that odometers indicate too much or too little distance.

$$\overline{X} = \frac{121.2}{12} = 10.1$$

$$\hat{s} = \sqrt{\frac{1224.5 - \frac{(121.2)^2}{12}}{11}} = 0.186$$

$$s_{\overline{X}} = \frac{0.186}{\sqrt{12}} = 0.0537$$

$$t = \frac{10.1 - 10}{0.0537} = 1.86; \quad 11\ df$$

$t_{.05}$ (11 df) = 2.201

Interpretation: Retain the null hypothesis. Although the sample mean (10.1 miles) is greater than the actual distance traveled (10 miles), the t test does not provide evidence that odometers record a significantly greater distance.

9.22.

$$t = \frac{\overline{X} - \mu_0}{s_{\overline{X}}} = \frac{56.10 - 50.00}{2.774} = 2.199; \quad df = 12$$

$t_{.05}$ (12 df) = 2.179

Interpretation: Reject the null hypothesis and conclude that people who were successful in used car sales have significantly higher extraversion scores than the general population.

9.23. $$d = \frac{\overline{X} - \mu}{\sigma} = \frac{270.675 - 269.3}{0.523} = 2.63$$

Interpretation: The effect size index, 2.63, is much larger than the conventional designation of "large." Even though the sample mean was only 1.375 grams more than the Frito-Lay company claimed, the effect was large because the bag-filling machinery was so precise (the standard deviation was small).

9.24. $d = \frac{\bar{X} - \mu}{\sigma} = \frac{56.1 - 50.0}{10} = 0.61$

Interpretation: The difference in extraversion between successful used car salespeople and the population mean has a medium effect size index.

9.25. $d = \frac{\bar{X} - \mu}{\sigma} = \frac{98.2 - 98.6}{0.7} = -0.57$

Interpretation: An effect size index of −0.57 shows that a medium-size effect existed undetected for a long time and that the true value is less than the hypothesized value.

9.26. $t = (r)\sqrt{\frac{N-2}{1-r^2}}$

a. $t = 2.24$; $df = 8$. Because $t_{.05}$ (8 df) = 2.31, $r = .62$ is not significantly different from .00.

b. $t = -2.12$; $df = 120$. Because $t_{.05}$ (120 df) = 1.98, $r = -.19$ is significantly different from .00.

c. $t = 2.08$; $df = 13$; $t_{.05}$ (13 df) = 2.16. Retain the null hypothesis.

d. $t = -2.85$; $df = 62$; $t_{.05}$ (60 df) = 2.00. Reject the null hypothesis.

9.27. Interpretation: Here's my response: Chance is not a likely explanation for a correlation coefficient of .74 based on an N of 11 pairs. If there was really no relationship between cigarette smoking and lung cancer, chance would produce an $r = \pm.7348$ only 1 time in 100 (Table A, $df = 9$). An $r = .74$ is even less likely than an $r = .7348$, so I conclude that there is a statistically significant correlation between cigarette smoking and lung cancer ($p < .01$).

9.28. $r = \frac{N\Sigma XY - (\Sigma X)(\Sigma Y)}{\sqrt{[N\Sigma X^2 - (\Sigma X)^2][N\Sigma Y^2 - (\Sigma Y)^2]}}$

$$= \frac{42(46{,}885) - (903)(2079)}{\sqrt{[42(25{,}585) - (903)^2][42(107{,}707) - (2079)^2]}}$$

$= .40$

From Table A, $r_{.01}$ (40 df) = .3932.

Interpretation: Because .40 > .3932, reject the null hypothesis and conclude that there is a positive correlation between the time a student works on a test and the grade, $p < .01$.

If, as you finished up this problem, you thought about waiting until the end of the period to turn in your next test (so that you would get a better grade), you drew a cause-and-effect conclusion from a correlation coefficient. Better go back and reread pages 105–106.

9.29. H_0: $\mu_0 = 58.7$ $\quad \Sigma X = 360.8$

$\Sigma X^2 = 21{,}703.86 \quad N = 6 \quad \bar{X} = 60.1333$

$\hat{s} = 1.2453 \quad s_{\bar{X}} = 0.5084$

$$t = \frac{\bar{X} - \mu_0}{s_{\bar{X}}} = \frac{60.1333 - 58.7}{0.5084} = 2.82; \quad 5\ df$$

$t_{.05}$ (5 df) = 2.571

$$d = \frac{\bar{X} - \mu}{\hat{s}} = \frac{60.1333 - 58.7}{1.2453} = 1.15$$

Interpretation: Snickers candy bars weigh more than the company claims ($p < .05$). The effect size index for the difference is huge ($d = 1.15$).

9.30. H_0: $\mu_0 = 6 \quad \Sigma X = 216 \quad \Sigma X^2 = 1368$

$N = 40 \quad \bar{X} = 5.40 \quad \hat{s} = 2.274$

$s_{\bar{X}} = 0.3595$

$$t = \frac{\bar{X} - \mu_0}{s_{\bar{X}}} = \frac{5.40 - 6}{0.3595} = -1.67; \quad df = 39$$

$$d = \frac{5.4 - 6}{2.274} = -0.26$$

$t_{.05}$ (30 df) = 2.042

Interpretation: For number choices between 1 and 10, there is no significant difference between the numbers chosen by people subjected to subliminal stimuli and those who were not. The effect size index, 0.26, is small.

CHAPTER 10

10.1. Your outline should include at least the following points:

a. Two samples are selected using random assignment (or random selection).

b. The samples are treated the same except for one thing.

c. The samples are measured.

d. The difference in measurements either is attributed to the difference in treatments or is about what would be expected if the treatment does not have an effect.

e. Write an explanation of the results.

10.2. Random assignment of participants to groups

10.3. **a.** *Independent variable and levels:* satisfaction with life; high SWLS scorers and low SWLS scorers

Dependent variable: annual income

Null hypothesis: Mean annual income is the same for the population of people with high SWLS scores as for the population with low SWLS scores.

b. *Independent variable and levels:* type of crime; robbers and embezzlers
Dependent variable: years served in prison
Null hypothesis: Convicted robbers and convicted embezzlers spend, on the average, the same number of years in jail.

c. *Independent variable and levels:* amount (or weeks) of treatment; 1 week and 6 weeks
Dependent variable: flexibility scores
Null hypothesis: One week of physical therapy produces a mean flexibility score that is equal to the mean score after 6 weeks of treatment.

10.4. Your outline should include the following points:

a. Recognize two logical possibilities:
1. The treatment has no effect (null hypothesis, H_0).
2. The treatment has an effect (alternative hypothesis, H_1).

b. Tentatively assume H_0 to be correct and establish an α level.

c. Calculate an appropriate test statistic from the sample data. Compare the test statistic to the critical value for α (which comes from a sampling distribution based on the assumption that H_0 is true).

d. If the test statistic has a probability less than α, reject H_0, adopt H_1, and write a conclusion about the effects of the treatment. If the test statistic has a probability greater than α, retain H_0 and conclude that the data did not allow you to choose between H_0 and H_1.

10.5. Experiments are conducted so that researchers can reach a conclusion about the effect of an independent variable on a dependent variable.

10.6. Degrees of freedom are always equal to the number of observations minus either

- The number of necessary relationships that exist among the observations
- The number of parameters that are estimated from the observations

10.7. **a.** Paired samples. With twins divided between the two groups, there is a logical pairing (natural pairs); *independent variable:* environment; *levels:* good and bad; *dependent variable:* attitude toward education.

b. Paired samples. This is a before-and-after experiment (repeated measures). For each group, the amount of aggression before watching the other children is paired with the amount afterward; *independent variable:* exposure to better toys; *levels:* yes and no; *dependent variable:* number of aggressive encounters.

c. Paired samples. This is a repeated-measures design; *independent variable:* bright artificial light; *levels:* yes and no; *dependent variable:* depression score.

d. Paired samples (yoked control design). Participants are paired so that the amount of interrupted sleep is equal for the two. One, however, is deprived of REM sleep, the other is not; *independent variable:* REM deprivation; *levels:* yes and no; *dependent variable:* mood questionnaire score.

e. Independent samples. The random assignment of individuals and the subsequent procedures give you no reason to pair two particular participants' mood questionnaire scores; *independent variable:* REM deprivation; *levels:* yes and no; *dependent variable:* mood questionnaire score.

f. Independent samples. The instructor randomly assigned individuals to one of the two groups; *independent variable:* sophomore honors course; *levels:* yes and no; *dependent variable:* grade point average.

This paragraph is really not about statistics, and you may skip it if you wish. Were you somewhat more anxious about your decisions on experiments c and d than on experiments a and b? Was this anxiety based on the expectation that surely it was time for an answer to be "independent samples"? Did you base your answer on your expectation rather than on the problem? If your answers are yes, you were exhibiting a *response bias.* A response bias occurs when you make a response on the basis of previous responses rather than on the basis of the current stimulus. Response biases often lead to a correct answer in textbooks, and you may learn to make decisions you are not sure about based on irrelevant cues (such as what your response was on the last question). To the extent it rewards your response biases, a textbook is doing you a disservice. So, be forewarned; recognize response bias and resist it. (I'll try to do my part.)

10.8. A two-tailed test is appropriate; if one of the two kinds of animals is superior, you want to know which one.

H_0: $\mu_1 = \mu_2$
H_1: $\mu_1 \neq \mu_2$

	Fellow humans	Pigeons
ΣX	147	75
ΣX^2	7515	1937
N	3	3
$\bar{X}$	49	25
$\hat{s}$	12.4900	5.5678

$$t = \frac{\bar{X}_1 - \bar{X}_2}{\sqrt{\left(\frac{\hat{s}_1}{\sqrt{N_1}}\right)^2 + \left(\frac{\hat{s}_2}{\sqrt{N_2}}\right)^2}} = \frac{49 - 25}{\sqrt{\left(\frac{12.49}{\sqrt{3}}\right)^2 + \left(\frac{5.5678}{\sqrt{3}}\right)^2}} = \frac{24}{7.895} = 3.04$$

$df = 4$

Interpretation: Reject the null hypothesis and conclude that pigeons can spot a person in the ocean more quickly than humans can, $p < .05$.

10.9. *Independent variable:* percent of cortex removed; *levels:* 0 percent and 20 percent; *dependent variable:* number of errors

$$\hat{s}_{0\text{ percent}} = \sqrt{\frac{1706 - \frac{(208)^2}{40}}{40 - 1}} = 4.00$$

$$\hat{s}_{20\text{ percent}} = \sqrt{\frac{2212 - \frac{(252)^2}{40}}{40 - 1}} = 4.00$$

$$s_{\bar{X}_1 - \bar{X}_2} = \sqrt{\left(\frac{4.00}{\sqrt{40}}\right)^2 + \left(\frac{4.00}{\sqrt{40}}\right)^2} = 0.894$$

$$t = \frac{\bar{X}_0 - \bar{X}_{20}}{s_{\bar{X}_1 - \bar{X}_2}} = \frac{5.2 - 6.3}{0.894} = -1.23; \; df = 78$$

$t_{.05}$ (60 df) = 2.00.

Interpretation: Retain the null hypothesis. Conclude that a 20 percent loss of cortex did not produce a statistically significant reduction in the memory of rats that were learning a simple maze. *Note:* Some brain operations *improve* performance on some tasks, so a two-tailed test is appropriate.

10.10.

	New package	Old package
ΣX	45.9	52.8
ΣX^2	238.73	354.08
$\bar{X}$	5.1	6.6
N	9	8
$\hat{s}$	0.76	0.89

$$t = \frac{6.6 - 5.1}{\sqrt{\left(\frac{10.24}{15}\right)(0.24)}} = \frac{1.5}{0.40} = 3.75; \quad df = 15$$

$t_{.005}$ (15 df) = 2.947 (one-tailed test). Thus, $p < .005$.

Interpretation: Reject the null hypothesis and conclude that the search report can be compiled more quickly with the new software than with the old. A one-tailed test is appropriate here because the only interest is whether the new software is better than the one on hand. (See the section "One-Tailed and Two-Tailed Tests" in Chapter 9.)

10.11. First of all, you might point out to the sisters that if the populations are the two freshman classes, no statistics are necessary; you have the population data and there is no sampling error. State U. is one-tenth of a point higher than The U. If the question is not about those freshman classes but about the two schools, and the two freshman classes can be treated as representative samples, a two-tailed test is called for because superiority of either school would be of interest.

$$s_{\bar{X}_1 - \bar{X}_2} = \sqrt{\left(\frac{\hat{s}_1}{\sqrt{N_1}}\right)^2 + \left(\frac{\hat{s}_2}{\sqrt{N_2}}\right)^2} = 0.0474$$

$$t = \frac{\bar{X}_1 - \bar{X}_2}{s_{\bar{X}_1 - \bar{X}_2}} = \frac{23.4 - 23.5}{0.0474} = -2.11; \quad df > 120$$

$t_{.05}$ (120 df) = 1.980. Thus, $p < .05$.

Interpretation: Students at State U. have statistically significantly higher ACT admission scores than do students at The U. You may have noted how small the difference actually is—only one-tenth of a point, an issue that is discussed in a later section in the text.

10.12. **a.** $$\hat{s}_D = \sqrt{\frac{\Sigma D^2 - \frac{(\Sigma D)^2}{N}}{N - 1}}$$

$\hat{s}_D$ is the standard deviation of the distribution of differences between paired scores.

b. $D = X - Y$. D is the difference between two paired scores.

c. $s_{\bar{D}} = \hat{s}_D/\sqrt{N}$. $s_{\bar{D}}$ is the standard error of the difference between means for a paired set of scores.

d. $\bar{Y} = \Sigma Y/N$. $\bar{Y}$ is the mean of a set of scores that is paired with another set.

10.13.

	X	Y	D	D^2
	16	18	−2	4
	10	11	−1	1
	17	19	−2	4
	4	6	−2	4
	9	10	−1	1
	12	14	−2	4
Σ	68	78	−10	18
Mean	11.3	13.0		

$$\hat{s}_D = \sqrt{\frac{\Sigma D^2 - \frac{(\Sigma D)^2}{N}}{N-1}} = \sqrt{\frac{18 - \frac{(-10)^2}{6}}{5}}$$

$$= \sqrt{0.2667} = 0.5164$$

$$s_{\bar{D}} = \frac{\hat{s}_D}{\sqrt{N}} = \frac{0.5164}{\sqrt{6}} = 0.2108$$

$$t = \frac{11.3333 - 13.0000}{0.2108} = \frac{-1.6667}{0.2108} = -7.91; \quad df = 5$$

$t_{.001}$ (5 df) = 6.869. Therefore, $p < .001$.
Interpretation: The conclusion reached by the researchers was that frustration (produced by seeing others treated better) leads to aggression.

Notice that the small difference between means (1.67) is highly significant even though the data consist of only six pairs of scores. This illustrates the power that a large correlation can have in reducing the standard error.

10.14. On this problem you cannot use the direct-difference method because you do not have the raw data. This leaves you with the definition formula, $t = (\bar{X} - \bar{Y})/\sqrt{s_{\bar{X}}^2 + s_{\bar{Y}}^2 - 2r_{XY}(s_{\bar{X}})(s_{\bar{Y}})}$, which requires a correlation coefficient. Fortunately, you have the data necessary to calculate r. (The initial clue for many students is the term ΣXY.)

Women	Men
$\bar{X} = \$12{,}782.25$	$\bar{Y} = \$15{,}733.25$
$\hat{s}_X = 2{,}358.99$	$\hat{s}_Y = 3{,}951.27$
$s_{\bar{X}} = 589.75$	$s_{\bar{Y}} = 987.82$
$r = .693$	

$$t = \frac{12{,}782.25 - 15{,}733.25}{\sqrt{(589.75)^2 + (987.82)^2 - (2)(.693)(589.75)(987.82)}}$$

$$= -4.11$$

$t_{.001}$ (15 df) = 4.073. Thus, $p < .001$.
Interpretation: Here is the conclusion written for the court.

> Chance is not a likely explanation for the $2951 annual difference in favor of men. Chance would be expected to produce such a difference less than one time in a thousand if there is really no difference. In addition, the difference cannot be attributed to education because both groups were equally educated. Likewise, the two groups were equal in their work experience at the rehabilitation center. One explanation that has not been eliminated is discrimination, based on sex.

(These data are 1979 salary data submitted in *Hartman and Hobgood* v. *Hot Springs Rehabilitation Center,* HS-76-23-C.)

10.15. This is a paired-samples study. It helps to carry four or five decimal places.

$\bar{X} = 0.1629$ second $\quad \bar{Y} = 0.1814$ second

$\Sigma D = -0.1300 \quad \Sigma D^2 = 0.0053$

$$\bar{X} - \bar{Y} = \frac{\Sigma D}{N} = -0.0186 \qquad N = 7$$

$$\hat{s}_D = \sqrt{\frac{\Sigma D^2 - \frac{(\Sigma D)^2}{N}}{N-1}} = \sqrt{0.00048} = 0.02193$$

$$s_{\bar{D}} = \frac{0.02193}{\sqrt{7}} = 0.00829$$

$$t = \frac{-0.0186}{0.00829} = -2.24; \quad df = 6$$

$t_{.05}$ (6 df) = 2.447. Thus, $p > .05$.
Interpretation: The obtained t does not indicate statistical significance. (This is a case in which a small sample led to a Type II error. Auditory RT is faster than visual RT. For practiced subjects, auditory RT = 0.16 second and visual RT = 0.18 second.)

10.16. I hope that you did not treat these data as paired samples. If you did, you exhibited a response bias that led you astray. If you recognized the design as independent samples, you are on your (mental) toes.

Primacy	Recency
$\Sigma X_2 = 29$	$\Sigma X_1 = 23$
$\Sigma X_2^2 = 225$	$\Sigma X_1^2 = 133$
$N_2 = 5$	$N_1 = 5$

$$s_{\bar{X}_1-\bar{X}_2} = \sqrt{\frac{\Sigma X_1^2 - \frac{(\Sigma X_1)^2}{N} + \Sigma X_2^2 - \frac{(\Sigma X_2)^2}{N}}{N(N-1)}}$$

$$= 2.049$$

$$t = \frac{\bar{X}_1 - \bar{X}_2}{s_{\bar{X}_1-\bar{X}_2}} = \frac{5.80 - 4.60}{2.049} = 0.59; df = 8$$

$t_{.05}$ (8 df) = 2.306. Thus, $p > .05$.
Interpretation: Retain the null hypothesis. This experiment does not provide evidence that either primacy or recency is more powerful.

10.17. $$t = \frac{\bar{X}_1 - \bar{X}_2}{s_{\bar{X}_1-\bar{X}_2}} = \frac{60 - 36}{\sqrt{\left(\frac{12}{\sqrt{45}}\right)^2 + \left(\frac{6}{\sqrt{45}}\right)^2}}$$

$$= \frac{24}{2} = 12.00; \quad df = 88$$

$t_{.001}$ (60 df) = 3.460. Thus, $p < .001$.
Interpretation: The meaning of very large t-test values is quite clear; they indicate a very, very small probability. Thus, conclude that children whose illness begins before age 3 show more problems than do those whose illness begins later.

10.18. The p is the probability of obtaining such a difference in samples (or a larger difference) if the two populations they come from (total physical response method scores and lecture–discussion scores) are identical. That is, if the populations are the same, then chance would produce such a difference (or one larger) less than 1 time in 100.

10.19. Here is one version: Significant differences are due to the independent variable and not to chance; important differences are ones that change our understanding about something. Important differences are significant, but significant differences may or may not be important.

10.20. $$d = \frac{\bar{X}_1 - \bar{X}_2}{\hat{s}} = \frac{5.2 - 6.3}{4.00} = -0.28$$

Interpretation: An effect size index of 0.28 is small. Thus, in these data d is small and a t test using large samples did not find a significant difference. This combination leads to the conclusion that the effect (if any) of removal of 20 percent of the cortex is small. If there is an effect, to detect it will require very large samples.

10.21. $$d = \frac{\bar{X} - \bar{Y}}{\hat{s}} = \frac{28.36 - 34.14}{5.74} = -1.01$$

Interpretation: An effect size index of 1.01 is large; thus, the effect of the multiracial camp on racial attitudes was both reliable (statistically significant) and sizable. Note that the effect size index gives you information that the means do not. That is, $d = 1.01$ tells you much more than does the fact that the girls scored an average of 5.78 points higher after the camp.

10.22. This is a paired-samples design.

$\bar{X}_{\text{asleep}} = 3.75 \quad \bar{Y}_{\text{awake}} = 2.50 \quad \Sigma D = 10$
$\Sigma D^2 = 28 \quad \bar{D} = 1.25$

$$\hat{s}_D = \sqrt{\frac{\Sigma D^2 - \frac{(\Sigma D)^2}{N}}{N-1}} = \sqrt{\frac{28 - \frac{(10)^2}{8}}{7}} = 1.488$$

$$s_{\bar{D}} = \frac{\hat{s}_D}{\sqrt{N}} = \frac{1.488}{\sqrt{8}} = 0.5261$$

$$\text{LL} = (\bar{X} - \bar{Y}) - t_\alpha(s_{\bar{D}})$$
$$= (3.75 - 2.50) - 2.365(0.5261)$$
$$= 1.25 - 1.24 = 0.01$$

$$\text{UL} = (\bar{X} - \bar{Y}) + t_\alpha(s_{\bar{D}})$$
$$= (3.75 - 2.50) + 2.365(0.5261)$$
$$= 1.25 + 1.24 = 2.49$$

Interpretation: We are 95 percent confident that sleep improved memory by 0.01 to 2.49 words in this experiment. With $\alpha = .05$, the null hypothesis is rejected. (Note how much more informative "a 95 percent confidence interval of 0.01 to 2.49 more words" is compared to "the null hypothesis was rejected.")

10.23. This is an independent-samples study. Note that problem 10.7c had the same independent and dependent variables. The procedures used, not the variables, determine the design.

$$\text{LL} = (\bar{X}_1 - \bar{X}_2) - t_\alpha(s_{\bar{X}_1-\bar{X}_2})$$
$$= (11 - 6) - 2.878(1.20) = 5 - 3.45 = 1.55$$

$$\text{UL} = (\bar{X}_1 - \bar{X}_2) + t_\alpha(s_{\bar{X}_1-\bar{X}_2})$$
$$= (11 - 6) + 2.878(1.20) = 5 + 3.45 = 8.45$$

Interpretation: These data support (with 99 percent confidence) the conclusion that 2 hours a day in bright light reduces depression by 1.55

to 8.45 units on the scale used to measure depression. Also note that these data allow the null hypothesis to be rejected, $p < .01$.

10.24. You can be confident that probabilities are accurate when the dependent-variable scores are (1) normally distributed and (2) have equal variances.

10.25. Power is the probability of rejecting the null hypothesis when it is false.

10.26. My version of the list is (1) effect size, (2) sample size, (3) α level, and (4) preciseness of measuring the dependent variable.

10.27. The null hypothesis is that experience has no effect on performance. $H_0: \mu_{exp} = \mu_{no\ exp}$.
A two-tailed test allows a conclusion about the advantage *or* the disadvantage of experience. The experimental design is independent samples.

$$t = \frac{7.40 - 5.05}{\sqrt{\left(\frac{2.13}{\sqrt{20}}\right)^2 + \left(\frac{2.31}{\sqrt{20}}\right)^2}} = \frac{2.35}{0.7026} = 3.34; \quad df = 38$$

$t_{.01}$ (30 df) = 2.750. Thus, $p < .01$. Reject the null hypothesis.

$$d = \frac{\bar{X}_1 - \bar{X}_2}{\hat{s}} = \frac{7.40 - 5.05}{2.22} = 1.06$$

Interpretation: Because the experienced group took longer, you should conclude that the previous experience with the switch *retarded* the subject's ability to recognize the solution. An effect size index of 1.06 is large. Thus, the effect of experience produced a large delay in recognizing a solution to the problem.

10.28. This is a paired-samples design.

$\bar{X}_{no\ cell} = 0.95$ second $\quad \bar{Y}_{cell} = 1.11$ second

$\Sigma D = -1.28 \quad \Sigma D^2 = 0.438$

$$\bar{X} - \bar{Y} = \frac{\Sigma D}{N} = -0.160 \quad N = 8$$

$$s_{\bar{D}} = \frac{\hat{s}_D}{\sqrt{N}} = \frac{\sqrt{\frac{\Sigma D^2 - \frac{(\Sigma D)^2}{N}}{N - 1}}}{\sqrt{N}}$$

$$= \frac{\sqrt{\frac{1.6384 - \frac{0.438}{8}}{8 - 1}}}{\sqrt{8}} = 0.0645$$

$$t = \frac{\bar{X} - \bar{Y}}{s_{\bar{D}}} = \frac{\bar{D}}{s_{\bar{D}}} = \frac{-0.160}{0.0645} = -2.48; \quad df = 7$$

$t_{.05}$ (7 df) = 2.365; $p < .05$

$$d = \frac{\bar{X} - \bar{Y}}{\hat{s}_D} = \frac{0.95 - 1.11}{0.182} = -0.88$$

Interpretation: Engaging in a hands-free cell phone conversation significantly increased the reaction time to brake lights of a lead car, $p < .05$. The effect size index (0.88) shows that conversing on the phone has a large effect on reaction time.

These data were constructed so that the means of the two groups (0.95 and 1.11 seconds) and the effect size index (0.88) would be identical to those found by Strayer, Drews, and Johnston (2003). Incidentally, simulated crashes occurred during phone conversations but not during the driving-only condition.

10.29. I hope you recognized that these data could be arranged as matched pairs. Each pair represents a level of achievement in French II.

	Listened to recorder					
	In car		In laboratory			
	Rank	Errors	Rank	Errors	D	D^2
	1	7	2	7	0	0
	4	7	3	15	−8	64
	5	12	6	22	−10	100
	8	12	7	24	−12	144
	9	21	10	32	−11	121
Sum		59		100	−41	429
Mean		11.8		20.0	−8.2	

$$\hat{s}_D = \sqrt{\frac{\Sigma D^2 - \frac{(\Sigma D)^2}{N}}{N - 1}} = \sqrt{\frac{429 - \frac{(-41)^2}{5}}{4}} = 4.817$$

$$s_{\bar{D}} = \frac{\hat{s}_D}{\sqrt{N}} = \frac{4.817}{\sqrt{5}} = 2.154$$

$$t = \frac{\bar{X} - \bar{Y}}{s_{\bar{D}}} = \frac{11.8 - 20.0}{2.154} = \frac{-8.2}{2.154} = -3.81; \quad df = 4$$

$$d = \frac{\bar{X} - \bar{Y}}{\hat{s}_D} = \frac{11.8 - 20.0}{4.817} = -1.70$$

$t_{.02}$ (4 df) = 3.747. Thus, $p < .02$. Reject the null hypothesis at the .02 level.

Interpretation: Because the car group made fewer errors, conclude that concrete, immediate experience facilitates learning vocabulary. A two-tailed test is appropriate because *disproving* a claim is always of interest to the scientifically minded. To use your data to support *or* refute a claim requires you to choose a two-tailed test (before the data are gathered). An effect size index of 1.70 is very large. (It is true that concrete, immediate experience has a great effect on learning foreign vocabulary words.)

Perhaps you analyzed these data with a confidence interval. Here is one for 95 percent confidence.

$$\text{LL} = (\overline{X} - \overline{Y}) - t_\alpha(s_{\overline{D}})$$
$$= (11.8 - 20.0) - 2.776(2.154)$$
$$= -8.2 - 5.98 = -14.18$$
$$\text{UL} = (\overline{X} - \overline{Y}) + t_\alpha(s_{\overline{D}})$$
$$= (11.8 - 20.0) + 2.776(2.154)$$
$$= -8.2 + 5.98 = -2.22$$

Interpretation: With 95 percent confidence, you can be sure that the effect of concrete, immediate experience is to reduce the number of errors on a 25-item vocabulary test by 2.22 to 14.18 items.

10.30. These data are independent samples. Answers and interpretations for a t test and a 95 percent confidence interval follow.

	Malpractice suits	
	0	2
ΣX	24	42
ΣX^2	88	218
N	8	9
$\overline{X}$	3.000	4.667
$\hat{s}$	1.512	1.658

$$t = \frac{3.000 - 4.667}{\sqrt{\left(\frac{1.512}{\sqrt{8}}\right)^2 + \left(\frac{1.658}{\sqrt{9}}\right)^2}} = \frac{-1.667}{0.769} = -2.16$$

$df = 15$

$$d = \frac{3.000 - 4.667}{\sqrt{\frac{(1.512)^2(7) + (1.658)^2(8)}{7 + 8}}} = \frac{-1.667}{1.592} = 1.05$$

$t_{.05}$ (15 df) = 2.131. Thus, $p < .05$.

Interpretation: Using audio recordings, judges gave significantly higher dominance scores to surgeons who had been sued twice as compared to surgeons who had never been sued, $p < .05$. The difference in the number of suits (0 or 2) produced a large effect on the judgments, $d = 1.05$.

$$\text{LL} = (\overline{X}_1 - \overline{X}_2) - t_\alpha(s_{\overline{X}_1 - \overline{X}_2})$$
$$= (3.000 - 4.667) - 2.131(0.769) = -3.31$$
$$\text{UL} = (\overline{X}_1 - \overline{X}_2) + t_\alpha(s_{\overline{X}_1 - \overline{X}_2})$$
$$= (3.000 - 4.667) + 2.131(0.769) = -0.03$$

Interpretation: With 95 percent confidence, conclude that the tone of dominance used by surgeons who had been sued twice was 0.03 to 3.31 units greater than that used by surgeons who had never been sued. Units represent judgments on a 1 to 7 scale. The difference in the number of suits (0 or 2) produced a large effect on the judgments, $d = 1.05$.

What Would You Recommend? Chapters 7–10

a. Statistics appropriate for this problem are a paired-samples t test, a confidence interval about a mean difference, and d, an effect size index. The independent variable is audience (present and not present) and the dependent variable is smiles. The samples are paired because each infant contributes data to both the audience and the no-audience condition.

b. A z-score analysis using the normal curve gives the percent of the population that is less gregarious than the social worker who has a score of 73.

c. A one-sample t test gives the probability of obtaining a sample with a mean of 55 from a population with a mean of 50. If this probability is very low, conclude that the workshop increased gregariousness scores.

d. The lower and upper limits of a confidence interval will capture a population mean with a certain level of confidence (usually 95 or 99 percent).

e. The state's population mean is not captured in problem *d* because the sample is not a random sample from the state (or even a representative one). A better interpretation is that the interval captures the mean of "those students planning to go to college."

f. To determine the probability that an $r = -.39$ could be a chance fluctuation, test the null hypothesis H_0: $\rho = 0$. If H_0 is rejected, conclude that there is a reliable relationship between the two variables. (See Elgin and Pritchard, 2006.)

g. The probability that a student, chosen at random, comes from one of the six categories is found by dividing the number of students in that category by the total number of students. (See the most recent issue of *Statistical Abstract of the United States* for current data.)

h. Statistics that would be helpful for this problem are a paired-samples *t* test, a confidence interval about a mean difference, and the effect size index, *d*. The independent variable is diet (high sugar and low sugar) and the dependent variables are measures of cognition and behavior, which are not specified. The design is a paired-samples one; each child serves in both conditions.

CHAPTER 11

11.1. *Example 2: Independent variable and levels:* schedule of reinforcement; four levels. *Dependent variable:* persistence. The null hypothesis is that the mean persistence scores of the four populations are equal; that is, the schedule of reinforcement has no effect on persistence.

Example 3: Independent variable and levels: degree of modernization; three levels. *Dependent variable:* suicide rate. The null hypothesis is that suicide rates are not dependent on degree of modernization, that the mean suicide rate for countries with low, medium, and high degrees of modernization is the same. In formal terms, the null hypothesis is H_0: $\mu_{\text{low}} = \mu_{\text{med}} = \mu_{\text{high}}$.

11.2. Larger

11.3. $F = \dfrac{\sigma^2}{\sigma^2}$

11.4. The between-treatments estimate of variance is a variance calculated from the sample means; it is the numerator of the *F* ratio. The within-treatments estimate of variance is calculated from each set of treatment scores; it is the denominator of the *F* ratio.

11.5. Interpretation: There are several ways to write an answer to this question, but any writing is better than just reading my answer. As for elements, you could use graphs, algebra, words, or all three. A good answer includes such elements as: (a) *F* is a ratio of two variances—a between-treatments variance over an error variance; (b) when the null hypothesis is true, both variances are good estimators of the population variance and the ratio is about 1.00; (c) when the null hypothesis is false, the between-treatments variance overestimates the population variance, resulting in a ratio that is greater than 1.00; (d) the *F* value calculated from the data is compared to a critical value from the *F* distribution, a distribution that assumes the null hypothesis is true.

11.6. **a.** Retain the null hypothesis and conclude that the data do not provide evidence that the methods of therapy differ in reducing fear responses.

b. Reject the null hypothesis and conclude that there is a relationship between degree of modernization and suicide rate.

11.7.

	X_1	X_2	X_3	Σ
ΣX	12	12	24	48
ΣX^2	50	56	224	330
$\bar{X}$	4	4	8	

$$SS_{\text{tot}} = 330 - \frac{48^2}{9} = 330 - 256 = 74.00$$

$$SS_{\text{treat}} = \frac{12^2}{3} + \frac{12^2}{3} + \frac{24^2}{3} - \frac{48^2}{9}$$

$$= 48 + 48 + 192 - 256$$

$$= 288 - 256 = 32.00$$

$$SS_{\text{error}} = \left(50 - \frac{12^2}{3}\right) + \left(56 - \frac{12^2}{3}\right) + \left(224 - \frac{24^2}{3}\right)$$

$$= 2 + 8 + 32 = 42.0$$

Check: $32.00 + 42.00 = 74.00$

11.8. *Independent variable:* degree of modernization, three levels; *dependent variable:* suicide rate.

	Degree of modernization			
	Low	Medium	High	Σ
ΣX	24	48	84	156
ΣX^2	154	614	1522	2290
$\bar{X}$	6.0	12.0	16.8	
N	4	4	5	13

$$SS_{\text{tot}} = 2290 - \frac{(156)^2}{13} = 418.00$$

$$SS_{\text{mod}} = \frac{(24)^2}{4} + \frac{(48)^2}{4} + \frac{(84)^2}{5} - \frac{(156)^2}{13} = 259.20$$

$$SS_{\text{error}} = \left(154 - \frac{(24)^2}{4}\right) + \left(614 - \frac{(48)^2}{4}\right) + \left(1522 - \frac{(84)^2}{5}\right) = 158.80$$

Check: 259.20 + 158.80 = 418.00
Incidentally, the rate per 100,000 is about 11 in the United States and about 12 in Canada.

11.9. *Independent variable:* kind of therapy; *levels:* model, film, desensitization, and control; *dependent variable:* snake-approach responses.

	Model	Film	Desensitization	Control	Σ
ΣX	104	72	68	40	284
ΣX^2	2740	1322	1182	430	5674
$\bar{X}$	26	18	17	10	
N	4	4	4	4	16

$$SS_{\text{tot}} = 5674 - \frac{284^2}{16} = 5674 - 5041 = 633$$

$$SS_{\text{treat}} = \frac{104^2}{4} + \frac{72^2}{4} + \frac{68^2}{4} + \frac{40^2}{4} - \frac{284^2}{16}$$

$$= 2704 + 1296 + 1156 + 400 - 5041$$

$$= 5556 - 5041 = 515$$

$$SS_{\text{error}} = \left(2740 - \frac{104^2}{4}\right) + \left(1322 - \frac{72^2}{4}\right) + \left(1182 - \frac{68^2}{4}\right) + \left(430 - \frac{40^2}{4}\right)$$

$$= 36 + 26 + 26 + 30 = 118$$

Check: 515 + 118 = 633

11.10. Six groups. The critical values of F are based on 5, 70 *df* and are 2.35 and 3.29 at the .05 and .01 levels, respectively. If $\alpha = .01$, retain the null hypothesis; if $\alpha = .05$, reject the null hypothesis.

11.11. Data from problem 11.7:

a.

Source	*SS*	*df*	*MS*	*F*	*p*
Treatments	32.00	2	16.00	2.29	>.05
Error	42.00	6	7.00		
Total	74.00	8			

$F_{.05}$ (2, 6 *df*) = 5.14. Interpretation: Because 2.29 does not reach the critical value, retain the null hypothesis and conclude that you do not have strong evidence against the hypothesis that these three groups came from the same population.

b.

$$SS_{\text{tot}} = 280 - \frac{36^2}{6} = 64.00$$

$$SS_{\text{treat}} = \frac{12^2}{3} + \frac{24^2}{3} + \frac{36^2}{6} = 24.00$$

$$SS_{\text{error}} = \left(56 - \frac{12^2}{3}\right) + \left(224 - \frac{24^2}{3}\right) = 40.00$$

Source	*SS*	*df*	*MS*	*F*	*p*
Treatments	24.00	1	24.00	2.40	>.05
Error	40.00	4	10.00		
Total	64.00	5			

$F_{.05}(1, 4\ df) = 7.71$. Retain the null hypothesis.

$$t = \frac{\bar{X}_1 - \bar{X}_2}{s_{\bar{X}_1 - \bar{X}_2}} = \frac{8 - 4}{2.582} = 1.55; \quad df = 4$$

Note that $(1.55)^2 = 2.40$; that is, $t^2 = F$.

11.12. Durkheim's modernization and suicide data from problem 11.8: For *df*, use

$$df_{\text{tot}} = N_{\text{tot}} - 1 = 13 - 1 = 12$$

$$df_{\text{mod}} = K - 1 = 3 - 1 = 2$$

$$df_{\text{error}} = N_{\text{tot}} - K = 13 - 3 = 10$$

Source	*SS*	*df*	*MS*	*F*	*p*
Modernization	259.20	2	129.60	8.16	<.01
Error	158.80	10	15.88		
Total	418.00	12			

$F_{.01}$ (2, 10 *df*) = 7.56. Interpretation: Reject the null hypothesis and conclude that there is a relationship between suicide rate and a country's degree of modernization.

11.13.

Source	*SS*	*df*	*MS*	*F*	*p*
Treatments	515	3	171.667	17.46	<.01
Error	118	12	9.833		
Total	633	15			

$F_{.01}$ (3, 12 *df*) = 5.95. Interpretation: Reject the null hypothesis at the .01 level and conclude that the treatment method had an effect on the number of snake-approach responses. By the end of this chapter, you will be able to compare the means to determine which method is best (or worst).

11.14. *A priori* tests are those planned in advance. They are tests based on the logic of the design of an experiment. *Post hoc* tests are chosen after examining the data.

11.15. Recall that the mean suicide rates for the different degrees of modernization were: low—6.0; medium—12.0; high—16.8. Comparing countries with low and medium degrees of modernization:

$$\text{HSD} = \frac{12.0 - 6.0}{\sqrt{15.88/4}} = \frac{6.0}{1.99} = 3.02; \quad \text{NS};$$

$\text{HSD}_{.05} = 3.88$

Comparing countries with medium and high degrees of modernization:

$$\text{HSD} = \frac{16.8 - 12.0}{\sqrt{\frac{15.88}{2}\left(\frac{1}{4} + \frac{1}{5}\right)}} = \frac{4.8}{1.890} = 2.54; \quad \text{NS};$$

$\text{HSD}_{.05} = 3.88$

Comparing countries with low and high degrees of modernization:

$$\text{HSD} = \frac{16.8 - 6.0}{1.890} = \frac{10.8}{1.890} = 5.71; \quad p < .01;$$

$\text{HSD}_{.01} = 5.27$

Interpretation: Suicide rates are significantly higher in countries with a high degree of modernization than they are in countries with a low degree of modernization. Although the other differences were not significant, the trend is in the direction predicted by Durkheim's hypothesis.

11.16. I calculated HSDs for each of the six pairwise differences. For some of these HSDs the equal-*N* formula was appropriate, and for some the unequal-*N* formula was necessary. The two values of $s_{\bar{X}}$ are

$$s_{\bar{X}} = \sqrt{\frac{MS_{\text{error}}}{N_t}} = \sqrt{\frac{3.93}{5}} = 0.887$$

$$s_{\bar{X}} = \sqrt{\frac{MS_{\text{error}}}{2}\left(\frac{1}{N_1} + \frac{1}{N_2}\right)}$$

$$= \sqrt{\frac{3.93}{2}\left(\frac{1}{5} + \frac{1}{7}\right)} = 0.821$$

The table shows all six HSD values. The significance level of each comparison is indicated with one or two asterisks. $\text{HSD}_{.05} = 4.00$, $\text{HSD}_{.01} = 5.09$.

	crf	FR2	FR4
FR2	4.63*		
FR4	7.44**	3.41	
FR8	9.47**	5.60**	2.03

* $p < .05$
** $p < .01$

Interpretation: The trend of the means is consistent; the higher the ratio of response to reinforcement during learning, the greater the persistence during extinction. The crf schedule produced the least persistence, significantly less than any other schedule. The FR2 schedule produced less persistence than the FR4 schedule (NS) and significantly less than the FR8 schedule ($p < .01$).

11.17. Comparing the best treatment (model) with the second best (film):

$$\text{HSD} = \frac{\bar{X}_1 - \bar{X}_2}{\sqrt{MS_{\text{error}}/N_t}} = \frac{26 - 18}{1.568} = 5.10; \quad p < .05;$$

$\text{HSD}_{.05} = 4.20$

Comparing the poorest treatment (desensitization) with the control group:

$$\text{HSD} = \frac{\bar{X}_3 - \bar{X}_4}{s_{\bar{X}}} = \frac{17 - 10}{1.568} = 4.46; \; p < .05;$$

$\text{HSD}_{.05} = 4.20$

Interpretation: For the treatment of phobias, using a live model and encouraging participation are significantly better than using a film or desensitization. All three of the treatments produced significant improvement compared to an untreated control group.

11.18. Normally distributed dependent variable and homogeneity of variance.

11.19. Random assignment permits strong statements of cause and effect between the independent and dependent varable.

11.20. $$f = \frac{\sqrt{\dfrac{K-1}{N_{\text{tot}}}(MS_{\text{treat}} - MS_{\text{error}})}}{\sqrt{MS_{\text{error}}}}$$

$$= \frac{\sqrt{\dfrac{4-1}{16}(171.667 - 9.833)}}{\sqrt{9.833}} = 1.76$$

Interpretation: An effect size index of 1.76 is much greater than the value $f = 0.40$ that signifies a large effect; thus, the type of therapy given to those suffering from a phobia has a very large effect on improvement.

11.21.

	Disorder			
	None	Schizophrenia	Depression	Σ
ΣX	178	138	72	388
ΣX^2	4162	2796	890	7848
$\bar{X}$	22.25	19.71	12.00	
N	8	7	6	21

Source	SS	df	MS	F	p
Disorder categories	376.31	2	188.16	11.18	<.01
Error	302.93	18	16.83		
Total	679.24	20			

$F_{.01}$ (2, 18 df) = 6.01

Depression and schizophrenia:

$$\text{HSD} = \frac{12.00 - 19.71}{\sqrt{\dfrac{16.83}{2}\left(\dfrac{1}{7} + \dfrac{1}{6}\right)}} = \frac{-7.71}{1.614}$$

$$= -4.78; \quad p < .01; \quad \text{HSD}_{.01} = 4.70$$

Depression and no disorder:

$$\text{HSD} = \frac{12.00 - 22.25}{\sqrt{\dfrac{16.83}{2}\left(\dfrac{1}{6} + \dfrac{1}{8}\right)}} = \frac{-10.25}{1.567}$$

$$= -6.54; \quad p < .01; \quad \text{HSD}_{.01} = 4.70$$

No disorder and schizophrenia:

$$\text{HSD} = \frac{22.25 - 19.71}{\sqrt{\dfrac{16.83}{2}\left(\dfrac{1}{8} + \dfrac{1}{7}\right)}} = \frac{2.54}{1.501}$$

$$= 1.69; \quad \text{NS}; \quad \text{HSD}_{.05} = 3.61$$

$$d = \frac{\bar{X}_{\text{sch}} - \bar{X}_{\text{dep}}}{\hat{s}_{\text{error}}} = \frac{19.71 - 12.00}{\sqrt{16.83}} = 1.88$$

Interpretation: Those diagnosed with depression have significantly lower R scores on the Rorschach test than those with schizophrenia ($p < .01$). The effect of these two diagnoses on R scores was huge ($d = 1.88$).

Additional note: Depressed patients' R scores were also significantly lower than those with no diagnosis ($p < .01$). Schizophrenics and those with no diagnosis were not significantly different. For a review of the Rorschach test and an explanation of the problem with R scores, see Lilienfeld, Wood, and Garb (2000).

11.22.

	ACS	4MAT	Control
$\bar{X}$	8.9	18.1	3.1

Source	SS	df	MS	F	p
Instruction methods	2288.533	2	1144.267	20.81	<.01
Error	3134.400	57	54.989		
Total	5422.933	59			

$F_{.01}$ (2, 55 df) = 5.01

HSD values calculated for the three pairwise comparisons are:

ACS and 4MAT:	HSD = 5.55**
ACS and control:	HSD = 3.50*
4MAT and control:	HSD = 9.05**

*$p < .05$ **$p < .01$
$\text{HSD}_{.05} = 3.44$ $\text{HSD}_{.01} = 4.37$

(Continued)

$$f = \frac{\sqrt{\dfrac{K-1}{N_{tot}}(MS_{treat} - MS_{error})}}{\sqrt{MS_{error}}}$$

$$= \frac{\sqrt{\dfrac{3-1}{60}(1144.267 - 54.989)}}{\sqrt{54.989}} = 0.81$$

Interpretation: Three months after BSE training, women retained significantly more information if they had learned with the 4MAT method than if they were taught with the ACS method. Either method is better than no training. The effect size index of $f = 0.81$ shows that BSE training has a very large effect on information retained 3 months later.

11.23.

Source	*SS*	*df*	*MS*	*F*	*p*
Between embarrassments	141.733	2	70.867	5.76	<.05
Error	147.600	12	12.300		
Total	289.333	14			

$F_{.05}(2, 12\ df) = 3.88$

$$f = \frac{\sqrt{\dfrac{K-1}{N_{tot}}(MS_{treat} - MS_{error})}}{\sqrt{MS_{error}}}$$

$$= \frac{\sqrt{\dfrac{3-1}{15}(70.867 - 12.300)}}{\sqrt{12.300}} = 0.80$$

HSD values are calculated from pairwise comparisons with

$$s_{\bar{X}} = \sqrt{\frac{12.300}{5}} = 1.568$$

Severe and mild embarrassment: HSD = 3.95*

Severe and no embarrassment: HSD = 4.34*

Mild and no embarrassment: HSD = 0.38

$*p < .05$, $HSD_{.05} = 3.77$

Interpretation: The participants who were subjected to severe embarrassment in order to be in the experiment judged the uninteresting discussion more favorably than did those who experienced mild embarrassment ($p < .05$) or no embarrassment ($p < .05$). The effect size index for the embarrassment procedures in this experiment is quite large ($f = 0.80$).

Although common sense says that a valuable goal is worth suffering for, these data (and many more) say that if we have suffered, then whatever we get is valuable. This balance that we create in our minds is called *effort justification* by social psychologists.

CHAPTER 12

12.1. Obviously, there are several solutions to this problem. Before reading further, you might ask yourself if your solution is a practical one. If not, try to create a more practical solution.

One solution is to select participants in groups of three. Within each group the three are matched in some way related to the material that is on the multiple-choice test. If the material is unfamiliar and the participants are college students, I would match on grade point average. Assignment of treatments within a group would be random. This is a three-group version of the matched-pairs design (page 203).

12.2.

SS_{tot} → $SS_{subjects}$, SS_{treat}, SS_{error}

12.3.

Subject	Rest interval (weeks) 2	4	8	Σ
1	40	50	58	148
2	58	56	65	179
3	44	70	69	183
4	57	61	74	192
Σ	199	237	266	702
$\bar{X}$	49.75	59.25	66.50	

Source	*SS*	*df*	*MS*	*F*
Subjects	365.667	3		
Intervals	564.500	2	282.250	7.21
Error	234.833	6	39.139	
Total	1165.000	11		

$F_{.05}(2, 6\ df) = 5.14$

Tukey HSD tests; $HSD_{.05} = 4.34$;

$$s_{\bar{X}} = \sqrt{\frac{39.139}{4}} = 3.128$$

$HSD_{8v4} = 2.32$; NS

$HSD_{8v2} = 5.35$; $p < .05$

$HSD_{4v2} = 3.04$; NS

Interpretation: These data show that as the rest interval increases from 2 weeks to 4 weeks to 8 weeks, there is an increase in percent of recall. That is, the longer the rest interval, the better the percent of recall. Only the difference between 2 weeks and 8 weeks is statistically significant, $p < .05$.

12.4.

$$HSD_{1st\ v.\ cont} = \frac{74 - 75}{\sqrt{1.333/4}} = \frac{-1}{0.577} = -1.73;\quad p > .05$$

$HSD_{.05} = 4.34$. Although always staying with your first choice resulted in an average loss of 1 point, the difference is not statistically significant.

12.5. The comparison of the therapy techniques does not lend itself to a repeated-measures design using the same participants. Clearly, it doesn't make sense to provide therapy for someone for a period sufficient to help and then use another therapy on that same person.

12.6. (1) Carryover effects, (2) levels of the independent variable are chosen by researcher, (3) assumptions of the test are true for the populations the samples are from.

12.7. A carryover effect occurs when the effect of one level of the independent variable continues to influence dependent-variable scores during the administration of a second level of the independent variable.

12.8.

	Oxygen consumption (cc/min)		
	Before mediation	During mediation	After mediation
$\bar{X}$	247	206	249

Source	*SS*	*df*	*MS*	*F*	*p*
Subjects	34,543.33	4			
Periods	5,890.00	2	2945.00	44.73	<.01
Error	526.67	8	65.83		
Total	40,960.00	14			

$F_{.01}(2, 8\ df) = 8.65$

Tukey HSD tests; $HSD_{.01} = 5.64$;

$$s_{\bar{X}} = \sqrt{\frac{65.833}{5}} = 3.629$$

$HSD_{before\ v\ during} = 11.30$; $p < .01$

$HSD_{before\ v\ after} = 0.55$; NS

$HSD_{after\ v\ during} = 11.85$; $p < .01$

Interpretation: During meditation, oxygen consumption drops to about 80 percent of the level observed before and after meditation. The drop is statistically significant, $p < .01$. The oxygen consumption rates before and after meditation are not significantly different.

12.9. If the conclusion that meditation lowers oxygen consumption is wrong, it is a Type I error. If the conclusion that oxygen consumption before and after meditation is not significantly different is wrong, it is a Type II error.

12.10.

Matching characteristic	Placebo	Interpersonal therapy	Cognitive-behavioral therapy	Drug therapy	Σ
Women at age 20	18	11	8	10	47
Men at age 30	21	9	15	10	55
Women at age 45	24	16	10	16	66
Σ	63	36	33	36	168
$\bar{X}$	21	12	11	12	

Source	SS	df	MS	F
Subjects	45.500	2		
Therapies	198.000	3	66.000	8.16*
Error	48.500	6	8.083	
Total	292.000	11		

$F_{.05}(3, 6\ df) = 4.76 \qquad *p < .05$

Tukey HSD tests: $\text{HSD}_{.05} = 4.90$;

$$s_{\bar{X}} = \sqrt{\frac{8.083}{3}} = 1.641$$

$\text{HSD}_{\text{no v interpersonal}} = 5.48; p < .05$

$\text{HSD}_{\text{no v drug}} = 5.48; p < .05$

$\text{HSD}_{\text{no v cognitive}} = 6.09; p < .01$

$\text{HSD}_{\text{drug v cognitive}} = 0.61$; NS

Interpretation: Each of the three therapies is significantly better than the placebo control group. There are no significant differences among the three therapy groups.

CHAPTER 13

13.1. **a.** 2×2 **b.** 3×2 **c.** 4×3
d. $4 \times 3 \times 3$; this is a factorial design with three independent variables.

13.2. a, b, c, and e all have one factor.

13.3. a, b, and d all have an independent variable that has two levels.

13.4. My version: An interaction occurs when the effect of changing levels of one factor depends on which level of the other factor you are looking at.

13.5. Interpretation: The effect on problem solving of time of day depends on gender. There are several ways to write a specific answer; here are two:
a. Women solve problems better at 7:00 A.M.; men solve them better at 7:00 P.M.
b. A 7:00 A.M. session in problem solving produces more correct answers from men; a 7:00 P.M. session produces more correct answers from women.

13.6. A main effect is a comparison of the different levels of one factor using dependent variable scores.

13.7. **i. a.**

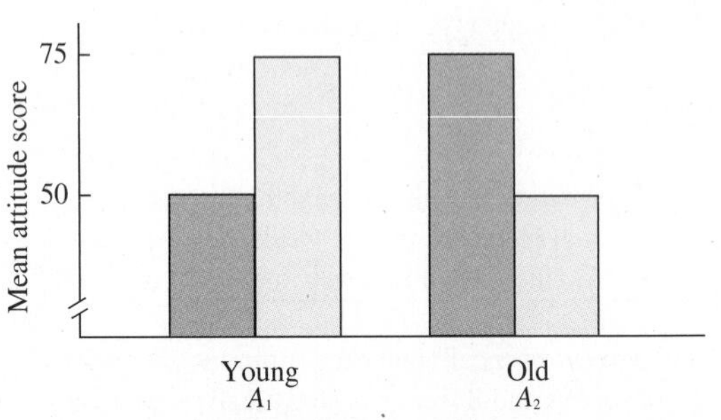

b. There appears to be an interaction.
c. Interpretation: Attitudes toward games depend on age. The young prefer computer games, and the old prefer cards.
d. There appear to be no main effects. Margin means for games are cards = 62.5 and computers = 62.5. Margin means for age are young = 62.5 and old = 62.5.
e. Later in this chapter you will learn that when there is a significant interaction, the interpretation of main effects isn't simple. Although all four means are exactly the same, there is a difference in attitudes toward the two games, but this difference depends on the age of the participant.

ii. a.

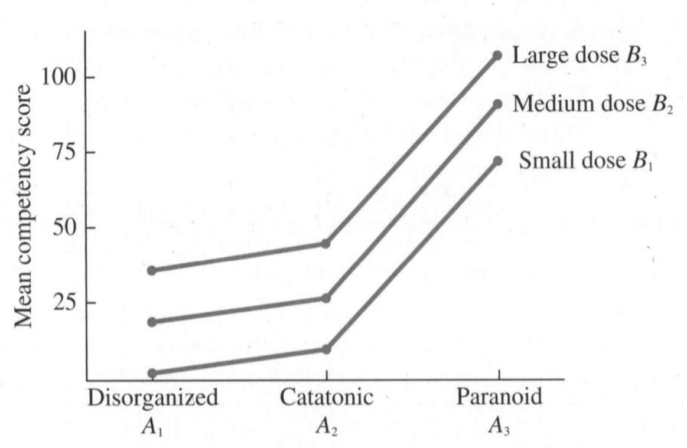

b. There appears to be no interaction.
c. Interpretation: The effect of the drug dose does not depend on the diagnosis of the patient.

d. There appear to be main effects for both factors.

e. Interpretation: For people with all types of schizophrenia, the larger the dose, the higher the competency score (Factor *B*, main effect). People with paranoia have higher competency scores than the others at all dose levels (Factor *A*, main effect).

iii. a.

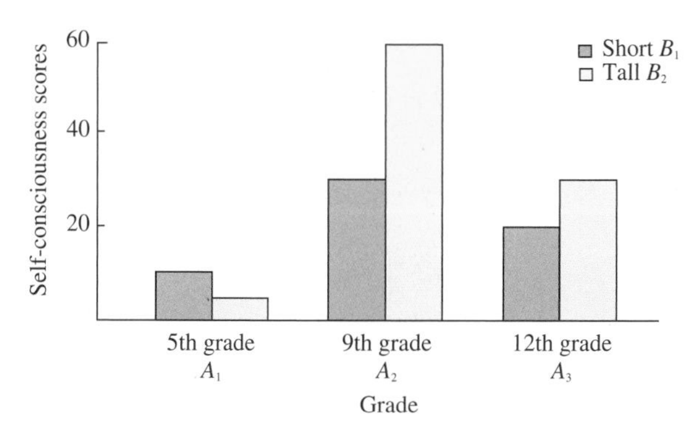

b. There appears to be an interaction.

c. Interpretation: The effect of a person's height on self-consciousness depends on the person's grade in school. Height seems to make little difference in the 5th grade. In the 9th grade, tall students are much more self-conscious than short students, but this difference is much smaller in the 12th grade.

d. The main effects appear to be significant. The margin means are 20 and 30 for height; 7.5, 45, and 25 for grade level.

e. Interpretation: Here's an interpretation that is appropriate for the information you've been given so far. (More on interpretation of main effects when the interaction is significant comes later.) Tall students are more self-conscious than short students (mean scores of 35 and 30, respectively). Ninth-graders ($\bar{X} = 45$) are more self-conscious than 12th-graders ($\bar{X} = 25$) or 5th-graders ($\bar{X} = 7.5$).

iv. a.

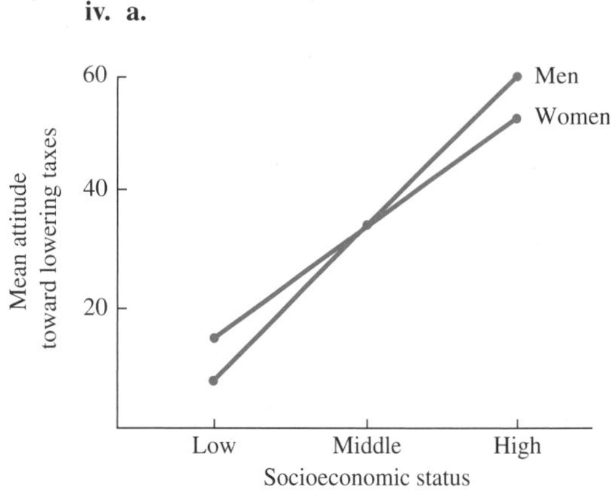

b. There appears to be no interaction. Although the lines in the graph are not parallel—they even cross—the departure from parallel is slight and may be due to sampling variation.

c. Interpretation: Gender differences in attitudes toward lowering taxes on investments do not depend on socioeconomic status.

d. There appears to be a main effect for socioeconomic status but not for gender.

e. Interpretation: Men and women do not differ in their attitudes toward lowering taxes on investments (Factor *B*). The higher the socioeconomic status, the more positive are attitudes toward lowering taxes on investments.

13.8.

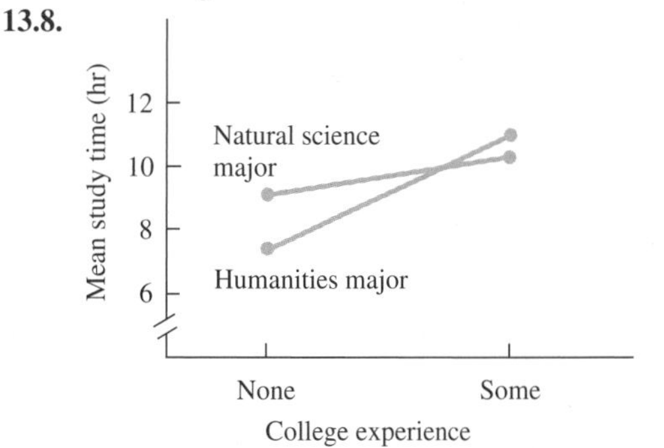

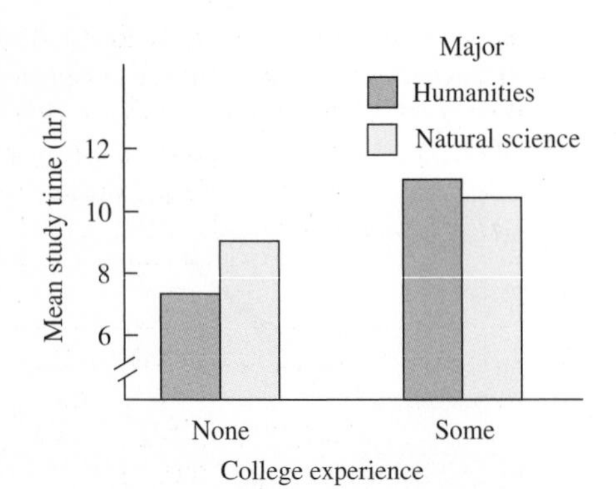

Interpretation: At this stage in the chapter, most any opinion will do. There appears to be an interaction between students' experience in college and their major. Both main effects might be significant because there is a difference in each set of margin means. Right or wrong, a preliminary analysis of a study improves your understanding (and may prepare you to recognize gross computational errors if they occur later).

13.9. **a.** 2×2

b. *Independent variables and levels:* gender: male and female; handedness: left and right
Dependent variable: age at death

c. $SS_{\text{tot}} = 61{,}455 - \dfrac{855^2}{12}$

$$= 61{,}455 - 60{,}918.75 = 536.25$$

$$SS_{\text{cells}} = \frac{219^2}{3} + \frac{186^2}{3} + \frac{234^2}{3} + \frac{216^2}{3} - \frac{855^2}{12}$$

$$= 61{,}323 - 60{,}918.75 = 404.25$$

$$SS_{\text{gender}} = \frac{453^2}{6} + \frac{402^2}{6} - 60{,}918.75$$

$$= 61{,}135.50 - 60{,}918.75 = 216.75$$

$$SS_{\text{handedness}} = \frac{405^2}{6} + \frac{450^2}{6} - 60{,}918.75$$

$$= 61{,}087.50 - 60{,}918.75 = 168.75$$

$$SS_{AB} = 3[(73 - 75.5 - 67.5 + 71.25)^2 + (62 - 67.5 - 67 + 71.25)^2 + (78 - 75 - 75.5 + 71.25)^2 + (72 - 75 - 67 + 71.25)^2]$$

$$= 3(6.25) = 18.75$$

Check: $SS_{AB} = 404.25 - 216.75 - 168.75 = 18.75$

$$SS_{\text{error}} = \left[16{,}013 - \frac{219^2}{3}\right] + \left[11{,}574 - \frac{186^2}{3}\right] + \left[18{,}284 - \frac{234^2}{3}\right] + \left[15{,}584 - \frac{216^2}{3}\right]$$

$$= 26.00 + 42.00 + 32.00 + 32.00 = 132.00$$

Check: $404.25 + 132.00 = 536.25$

13.10. **a.** 2×3

b. *Independent variables and levels:* therapy: psychodynamic, interpersonal, cognitive-behavioral; gender: women, men
Dependent variable: improvement scores

c.

$$SS_{\text{tot}} = 87{,}250 - \frac{1692^2}{36} = 87{,}250 - 79{,}524 = 7726$$

$$SS_{\text{cells}} = \frac{192^2}{6} + \frac{348^2}{6} + \frac{300^2}{6} + \frac{228^2}{6} + \frac{300^2}{6} + \frac{324^2}{6} - \frac{1692^2}{36} = 2964$$

$$SS_{\text{therapy}} = \frac{420^2}{12} + \frac{648^2}{12} + \frac{624^2}{12} - \frac{1692^2}{36} = 2616$$

$$SS_{\text{gender}} = \frac{840^2}{18} + \frac{852^2}{18} - \frac{1692^2}{36} = 4$$

$$SS_{AB} = 6[(32 - 35 - 46.667 + 47)^2 + (58 - 54 - 46.667 + 47)^2 + (50 - 52 - 46.667 + 47)^2 + (38 - 35 - 47.333 + 47)^2 + (50 - 54 - 47.333 + 47)^2 + (54 - 52 - 47.333 + 47)^2]$$

$$= 344$$

Check: $SS_{AB} = 2964 - 2616 - 4 = 344$

$$SS_{\text{error}} = \left(6930 - \frac{192^2}{6}\right) + \left(20{,}964 - \frac{348^2}{6}\right) + \left(15{,}742 - \frac{300^2}{6}\right) + \left(9514 - \frac{228^2}{6}\right) + \left(15{,}748 - \frac{300^2}{6}\right) + \left(18{,}352 - \frac{324^2}{6}\right)$$

$$= 4762$$

Check: $4762 + 2964 = 7726$

13.11.

$$SS_{\text{tot}} = 1405 - \frac{145^2}{20} = 353.75$$

$$SS_{\text{cells}} = \frac{25^2}{5} + \frac{57^2}{5} + \frac{43^2}{5} + \frac{20^2}{5} - \frac{145^2}{20} = 173.35$$

$$SS_{\text{learn}} = \frac{68^2}{10} + \frac{77^2}{10} - \frac{145^2}{20} = 4.05$$

$$SS_{\text{recall}} = \frac{63^2}{10} + \frac{82^2}{10} - \frac{145^2}{20} = 18.05$$

$$\begin{aligned} SS_{AB} &= 5[(8.60 - 6.80 - 6.30 + 7.25)^2 \\ &\quad + (4.00 - 7.70 - 6.30 + 7.25)^2 \\ &\quad + (5.00 - 6.80 - 8.20 + 7.25)^2 \\ &\quad + (11.40 - 7.70 - 8.20 + 7.25)^2] \\ &= 151.25 \end{aligned}$$

Check: $SS_{AB} = 173.35 - 4.05 - 18.05 = 151.25$

$$\begin{aligned} SS_{\text{error}} &= \left(400 - \frac{43^2}{5}\right) + \left(100 - \frac{20^2}{5}\right) \\ &\quad + \left(155 - \frac{25^2}{5}\right) + \left(750 - \frac{57^2}{5}\right) \\ &= 180.40 \end{aligned}$$

Check: $173.35 + 180.40 = 353.75$

13.12.

Source	SS	df	MS	F	p
Gender (A)	216.75	1	216.75	13.14	<.01
Handedness (B)	168.75	1	168.75	10.23	<.05
A × B	18.75	1	18.75	1.14	>.05
Error	132.00	8	16.50		
Total	536.25	11			

$F_{.05}(1, 8\ df) = 5.32$; $F_{.01}(1, 8\ df) = 11.26$

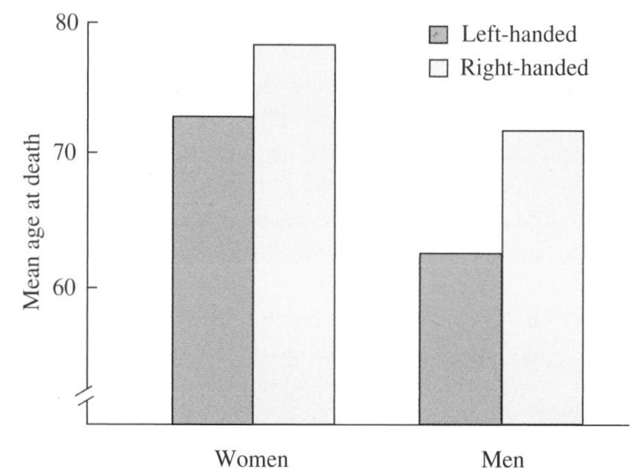

Interpretation: The interaction is not significant, so the main effects can be interpreted directly. Both gender and handedness produced significant differences. Right-handed people live longer than left-handed people ($p < .05$) and women live longer than men ($p < .01$).

13.13.

Source	SS	df	MS	F	p
Therapy (A)	2616	2	1308	8.24	<.01
Gender (B)	4	1	4	0.03	>.05
AB	344	2	172	1.08	>.05
Error	4762	30	158.73		
Total	7726	35			

$F_{.05}(2, 30\ df) = 3.32$; $F_{.01}(2, 30\ df) = 5.39$

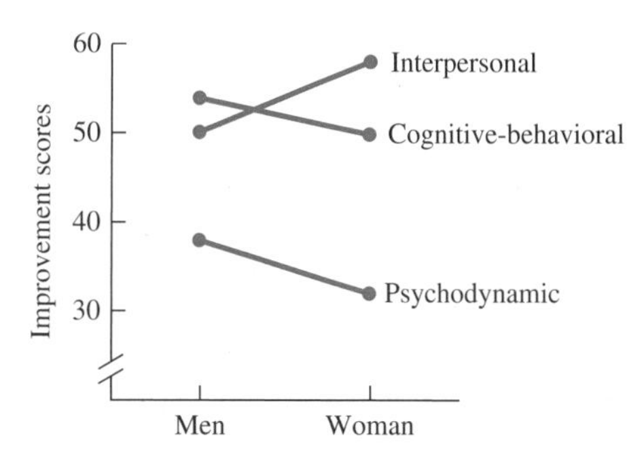

Interpretation: The interaction between therapy and gender was not statistically significant. The main effect of gender was also not significant. The differences among the three kinds of therapy did produce a significant F, $p < .01$. The amount of improvement among depressed patients depended on the kind of therapy received. [Hollon, Thrase, and Markowitz's article on treatment and prevention of depression (2002) explains these psychotherapies and also other psychotherapies and antidepressant medications.]

13.14.

Source	SS	df	MS	F	p
Learning (A)	4.05	1	4.05	0.36	>.05
Recall (B)	18.05	1	18.05	1.60	>.05
AB	151.25	1	151.25	13.41	<.01
Error	180.40	16	11.28		
Total	353.75	19			

$F_{.05}(1, 16\ df) = 4.49$; $F_{.01}(1, 16\ df) = 8.53$

(Continued)

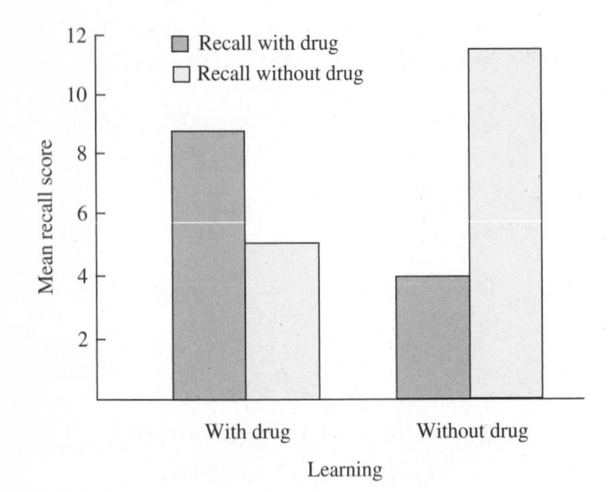

13.15. **i. a.** *Independent variables:* (1) treatment the participant received (insulted or treated neutrally) and (2) administrator of test (same or different). *Dependent variable:* number of captions produced.

b. $F_{\text{treatments}} = 9.08, p < .01$; $F_{\text{experimenters}} = 6.82, p < .05$; $F_{AB} = 4.88, p < .05$.

c. Interpretation: The significant interaction means that the effect of the experimenter *depended* on whether the participant was insulted or treated neutrally. If the participant was insulted, taking the humor test from the person who had done the insulting caused a larger number of captions to be written—larger, that is, than if a new experimenter had administered the test. If the participant was not insulted, there was little difference between the two kinds of experimenters.

Both main effects are statistically significant. For both main effects, the difference in margin means depends heavily on one cell, Cell A_1B_1. Interpreting the main effects for this problem isn't advisable.

ii. a. *Independent variables:* (1) name (high prestige or low prestige) and (2) occupation of father (research chemist or unemployed). *Dependent variable:* score assigned to the theme.

b. $F_{\text{name}} = 0.44, p > .05$; $F_{\text{occupation}} = 0.05, p > .05$; $F_{AB} = 8.20, p < .01$

c. Interpretation: The significant interaction means that the effect of the name depends on the occupation of the father. The effect of the name David is good if his father is a research chemist (Cell A_1B_1) and bad if his father is unemployed (Cell A_1B_2). The effect of the name Elmer is good if his father is unemployed (Cell A_2B_2) and bad if his father is a research chemist (Cell A_2B_1).

The main effects are not significant. The average "David" theme is about equal to the average "Elmer" theme. Similarly, the themes do not differ significantly according to whether the father was unemployed or a research chemist. Notice that this interpretation of the main effects is misleading; that is, *names do have an effect,* but the effect depends on the occupation of the father.

iii. a. *Independent variables:* (1) questions (used and not used) and (2) demonstrations (used and not used). *Dependent variable:* score on a comprehension test.

b. $F_{\text{questions}} = 4.59, p < .05$; $F_{\text{demos}} = 9.80, p < .01$; $F_{AB} = 0.24, p > .05$

c. Interpretation: Because the interaction is not significant, the main effects can be interpreted in a straightforward manner. Questions are important; when questions are incorporated, students learn significantly more than when questions are not used. Demonstrations make a difference; presentations that have demonstrations produce comprehension scores that are significantly better than presentations without demonstrations.

13.16. Interpretation: Only the interaction is significant. The conclusion is that when conditions of learning and recall are the same, memory is enhanced. When conditions of learning and recall are different, memory is retarded.

13.17. **a.** The number of scores in each group (cell) must be the same.

b. The scores in each cell must be independent.

c. The levels of both factors must be chosen by the experimenter and not selected randomly.

d. The samples are drawn from normally distributed populations.
e. The variances of the populations are equal.
f. The participants are assigned randomly to the group, or the samples are drawn randomly from the populations.

13.18.

Source	*SS*	*df*	*MS*	*F*	*p*
Rewards (*A*)	199.31	3	66.44	3.90	<.05
Classes (*B*)	2.25	1	2.25	0.13	>.05
AB	0.13	3	0.04	0.00	>.05
Error	953.25	56	17.02		
Total	1154.94	63			

$F_{.05}(3, 55\ df) = 2.78$

Interpretation: There is no significant interaction between rewards and class standing, and freshmen and seniors do not differ significantly. There is a significant difference among the rewards; that is, rewards have an effect on attitudes toward the police. Pairwise comparisons are called for.

The table that follows shows the four treatment means, *N*'s, and a matrix of HSD values calculated from the formula,

$$\text{HSD} = \frac{\bar{X}_1 - \bar{X}_2}{\sqrt{17.02/16}}$$

	\$20	\$10	\$5	\$1
Means	8.938	10.250	12.125	13.563
N	16	16	16	16
\$20		1.27	3.09	4.48*
10			1.82	3.21
5				1.39

* $\text{HSD}_{.05} = 3.79$

Interpretation: The less students were paid to defend the police, the more positive their attitude toward the police. Those who paid 1 dollar had significantly more positive attitudes toward the police than those who paid 20 dollars, $p < .05$.

13.19. $SS_{\text{tot}} = 480 - \frac{(84)^2}{18} = 88.00$

$$SS_{\text{cells}} = \frac{(12)^2}{3} + \frac{(15)^2}{3} + \frac{(24)^2}{3} + \frac{(9)^2}{3} + \frac{(9)^2}{3} + \frac{(15)^2}{3} - \frac{(84)^2}{18} = 52.00$$

$$SS_{\text{marital}} = \frac{(51)^2}{9} + \frac{(33)^2}{9} - \frac{(84)^2}{18} = 18.00$$

$$SS_{\text{religious}} = \frac{(21)^2}{6} + \frac{(24)^2}{6} + \frac{(39)^2}{6} - \frac{(84)^2}{18} = 31.00$$

$$\begin{aligned} SS_{AB} = 3[&(4.0 - 3.5 - 5.667 + 4.667)^2 \\ &+ (5.0 - 4.0 - 5.667 + 4.667)^2 \\ &+ (8.0 - 6.5 - 5.667 + 4.667)^2 \\ &+ (3.0 - 3.5 - 3.667 + 4.667)^2 \\ &+ (3.0 - 4.0 - 3.667 + 4.667)^2 \\ &+ (5.0 - 6.5 - 3.667 + 4.667)^2] \\ = 3&(1.00) = 3.00 \end{aligned}$$

Check: $SS_{\text{cells}} = 18.00 + 31.00 + 3.00 = 52.00$

$$\begin{aligned} SS_{\text{error}} = &\left(56 - \frac{(12)^2}{3}\right) + \left(83 - \frac{(15)^2}{3}\right) \\ &+ \left(194 - \frac{(24)^2}{3}\right) + \left(29 - \frac{(9)^2}{3}\right) \\ &+ \left(35 - \frac{(9)^2}{3}\right) + \left(83 - \frac{(15)^2}{3}\right) \\ = &\ 36.00 \end{aligned}$$

Check: $52.00 + 36.00 = 88.00$

Source	*df*	*MS*	*F*	*p*
Marital status	1	18.00	6.00	<.05
Religious behavior	2	15.50	5.17	<.05
AB	2	1.50	0.50	>.05
Error	12	3.00		
Total	17			

$F_{.05}(1, 12\ df) = 4.75$; $F_{.05}(2, 12\ df) = 3.88$

(Continued)

The interaction is not significant; HSD tests are appropriate for comparing the three levels of frequency of religious behavior. For this problem,

$$\text{HSD} = \frac{\bar{X}_1 - \bar{X}_2}{\sqrt{3.00/6}}; \quad \text{HSD}_{.05} = 3.77$$

$$\text{HSD}_{\text{often \& occasionally}} = 3.54; \quad \text{NS}$$

$$\text{HSD}_{\text{never \& occasionally}} = 0.71; \quad \text{NS}$$

$$\text{HSD}_{\text{often \& never}} = 4.24; \quad p < .05$$

Interpretation: Both marital status and frequency of religious behavior have an effect on reports of happiness. Married people reported being more happy than unmarried people ($p < .05$). Those frequently engaged in religious behavior reported higher happiness scores than those who never engaged in religious behavior ($p < .05$). Happiness was intermediate (but not significantly different) for those who occasionally practiced religious behavior. There was not an interaction between marital status and frequency of religious behavior.

CHAPTER 14

14.1.

	Received program	Control	Total
Police record	114	101	215
No police record	211	224	435
Total	325	325	650

The expected frequencies are:

$$\frac{(215)(325)}{650} = 107.5 \qquad \frac{(215)(325)}{650} = 107.5$$

$$\frac{(435)(325)}{650} = 217.5 \qquad \frac{(435)(325)}{650} = 217.5$$

O	E	$O - E$	$(O - E)^2$	$\frac{(O-E)^2}{E}$
114	107.5	6.5	42.25	0.393
101	107.5	−6.5	42.25	0.393
211	217.5	−6.5	42.25	0.194
224	217.5	6.5	42.25	0.194
Σ 650	650.0			$\chi^2 = 1.174$
$\chi^2_{.05}$ (1 df) = 3.84				$df = 1$

By the shortcut method:

$$\chi^2 = \frac{650[(114)(224) - (101)(211)]^2}{(215)(435)(325)(325)} = 1.17$$

Interpretation: Retain the null hypothesis and conclude that the number who had police records in the two groups did not differ significantly. When a statistical test does not show a statistically significant effect, an effect size index usually isn't calculated. (Did you notice that the proportion with police records was *higher* for the boys who participated in the program?)

The Cambridge–Somerville Youth study, discussed in many sociology texts, was first reported by Powers and Witmer (1951).

14.2. $$\chi^2 = \frac{100[(9)(34) - (26)(31)]^2}{(35)(65)(40)(60)} = 4.579$$

Calculations of the expected frequencies are as follows:

$$\frac{(35)(40)}{100} = 14 \qquad \frac{(35)(60)}{100} = 21$$

$$\frac{(65)(40)}{100} = 26 \qquad \frac{(65)(60)}{100} = 39$$

O	E	$O - E$	$(O - E)^2$	$\frac{(O-E)^2}{E}$
9	14	−5	25	1.786
26	21	5	25	1.191
31	26	5	25	0.962
34	39	−5	25	0.641
Σ 100	100			$\chi^2 = 4.58$
$\chi^2_{.05}$ (1 df) = 3.84				$df = 1$

Interpretation: Reject the hypothesis that group size and joining are independent. Conclude that passersby are more likely to join a group of five than a group of two.

$$\phi = \sqrt{\frac{\chi^2}{N}} = \sqrt{\frac{4.58}{100}} = \sqrt{.0458} = 0.21$$

The effect that group size has on joining is midway between a small effect and a medium effect.

14.3. a.

Recapture site

Capture site		Issaquah	East Fork
	Issaquah	46 (34.027)	0 (11.973)
	East Fork	8 (19.973)	19 (7.027)

$$\chi^2 = \frac{73[(46)(19) - (0)(8)]^2}{(46)(27)(54)(19)} = 43.76$$

$$\phi = \sqrt{\frac{\chi^2}{N}} = \sqrt{\frac{43.76}{73}} = \sqrt{.5995} = 0.77$$

$\chi^2_{.001}$ (1 df) = 10.83. Interpretation: Reject the null hypothesis (which is that the second choice is independent of the first) and conclude that choices are very consistent; salmon tend to choose the same stream each time. The effect size index is 0.77, which shows that streams have a very large effect on choice.

b.

Recapture site

Capture site		Issaquah	East Fork
	Issaquah	39 (40.071)	12 (10.929)
	East Fork	16 (14.929)	3 (4.071)

$$\chi^2 = \frac{70[(39)(3) - (12)(16)]^2}{(51)(19)(55)(15)} = 0.49$$

$\chi^2_{.05}$ (1 df) = 3.84. Interpretation: Hasler's hypothesis is supported; fish with plugged nasal openings do not make a consistent choice of streams, but those that get olfactory cues consistently choose the same stream. For further confirmation and a short summary of Hasler's work, see Hasler, Scholz, and Horrall (1978).

14.4. The work that needs to be done on the data is to divide each set of applicants into two independent categories, hired and not hired. The following table results. Expected values are in parentheses.

	White	Black	Total
Hired	390 (339.327)	18 (68.673)	408
Not hired	3810 (3860.673)	832 (781.327)	4642
Total	4200	850	5050

$$\chi^2 = \frac{5050[(390)(832) - (3810)(18)]^2}{(408)(4642)(4200)(850)}$$

$$= \frac{3.3070 \times 10^{14}}{6.7514 \times 10^{12}} = 48.91$$

Interpretation: Given that $\chi^2_{.001}$ (1 df) = 10.83, the null hypothesis can be rejected. Because 9.29 percent of the white applicants were hired compared to 2.12 percent of the black applicants, the statistical consultant and the company in question concluded that discrimination had occurred.

$$\phi = \sqrt{\frac{\chi^2}{N}} = \sqrt{\frac{48.91}{5050}} = \sqrt{.00969} = 0.10$$

The effect size index is 0.10, which is a small effect. The lesson here is that a small effect practiced on a large sample produces a reliable difference between the two groups.

14.5.

$$\chi^2 = \Sigma\left[\frac{(O - E)^2}{E}\right] = \frac{(316 - 300)^2}{300} + \frac{(84 - 100)^2}{100}$$

$$= 0.8533 + 2.5600$$

$$= 3.4133; \quad df = 1$$

$\chi^2_{.05}$ (1 df) = 3.84. Interpretation: Retain the null hypothesis and conclude that these data are consistent with a 3:1 hypothesis.

14.6. The expected frequencies are 20 correct and 40 incorrect: $(\frac{1}{3})(60) = 20$ and $(\frac{2}{3})(60) = 40$.

$$\chi^2 = \frac{(32 - 20)^2}{20} + \frac{(28 - 40)^2}{40} = 10.80;$$

$df = 1; \quad p < .01$

Interpretation: The three emotions can be distinguished from each other. Observers did not respond in a chance fashion.

There is an interesting sequel to this experiment. Subsequent researchers did not find the simple, clear-cut results that Watson reported. One experimenter (Sherman, 1927) found that if only the infants' reactions were observed, there was a great deal of disagreement. However, if the observers also knew the stimuli (patting, dropping, and so forth), they agreed with each other. This seems to be a case in which Watson's design did

(Continued)

not permit him to separate the effect of the infant's reaction (the independent variable) from the effect of knowing what caused the reaction (an extraneous variable).

14.7. To get the expected frequencies, multiply the percentages given by Professor Stickler by the 340 students. Then enter these expected frequencies in the usual table.

O	E	$O - E$	$(O - E)^2$	$\frac{(O-E)^2}{E}$
20	23.8	−3.8	14.44	0.607
74	81.6	−7.6	57.76	0.708
120	129.2	−9.2	84.64	0.655
88	81.6	6.4	40.96	0.502
38	23.8	14.2	201.64	8.472
Σ 340	340.0			$\chi^2 = 10.94$

$\chi^2_{.05}$ (4 df) = 9.49 $\qquad df = 5 - 1 = 4$

(The only restriction on these theoretical frequencies is that $\Sigma E = \Sigma O$. Thus, $df = 4$.)

Interpretation: Reject the contention of the professor that his grades conform to "the curve." By examining the data, you can also reject the contention of the colleague that the professor is too soft. The primary reason the data do not fit the curve is that there were too many "flunks."

14.8. Goodness of fit

14.9.

O	E	$O - E$	$(O - E)^2$	$\frac{(O-E)^2}{E}$
195	200	−5	25	0.125
200	200	0	0	0.000
220	200	20	400	2.000
215	200	15	225	1.125
190	200	−10	100	0.500
180	200	−20	400	2.000
Σ 1200	1200			$\chi^2 = 5.75$

$\chi^2_{.05}$ (5 df) = 11.07 $\qquad df = 6 - 1 = 5$

Interpretation: Each of the six sides of a die is equally likely. With 1200 as the total number of throws, the expected value for any one of them is $\frac{1}{6}$ times 1200, or 200. With $\chi^2 = 5.75$, retain the null hypothesis. The results of the evening do not differ significantly from the "unbiased dice" model.

14.10. Goodness of fit

14.11.

O	E	$O - E$	$(O - E)^2$	$\frac{(O-E)^2}{E}$
21	12.735	8.265	68.310	5.365
5	13.265	−8.265	68.310	5.150
27	28.408	−1.408	1.982	0.070
31	29.592	1.408	1.982	0.067
0	6.857	−6.857	47.018	6.857
14	7.143	6.857	47.018	6.583
Σ 98	98.000			$\chi^2 = 24.09$

$\chi^2_{.001}$ (2 df) = 13.82 $\qquad df = (3 - 1)(2 - 1) = 2$

Interpretation: Reject the null hypothesis and conclude that Madison used the word *by* significantly more frequently than did Hamilton. The stage is now set to examine the 12 disputed papers for their use of *by*, an examination that Mosteller and Wallace carried out. They describe the rates they found as "Madisonian." In addition, other words that distinguished between Madison and Hamilton (*upon*, *also*, and others) were used at Madisonian rates in the 12 papers.

14.12. Independence

14.13. To convert the information you have into frequencies that can be analyzed with χ^2, multiply the percentages (as proportions) by the number of observations. Thus, for *observed frequencies:*

.193(140) = 27 schizophrenic offspring
.807(140) = 113 nonschizophrenic offspring

For the *expected frequencies:*

.25(140) = 35 schizophrenic offspring
.75(140) = 105 nonschizophrenic offspring

O	E	$O - E$	$(O - E)^2$	$\frac{(O-E)^2}{E}$
27	35	−8	64	1.829
113	105	−8	64	0.610
Σ 140	140			$\chi^2 = 2.439$

$\chi^2_{.05}$ (1 df) = 3.84 $\qquad df = 2 - 1 = 1$

Interpretation: Retain the null hypothesis and conclude that these data are consistent with the 1:3 ratio predicted by the theory.

This problem illustrates that data that are consistent with the theory do not prove that the theory is true. Current explanations of schizophrenia include a genetic component, but not the simple one envisioned by the early hypothesis.

14.14. In the goodness-of-fit test, expected frequencies are predicted by a theory, whereas in the independence test, they are obtained from the data.

14.15. The first task in this problem is to determine the expected frequencies. To do this, multiply each color's population percentage by the total number of M&Ms, 57. Next, construct a summary table of O, E, and χ^2 values.

Color	O	E	$O - E$	$(O - E)^2$	$\frac{(O-E)^2}{E}$
Blue	9	13.7	−4.7	22.09	1.612
Orange	11	11.4	−0.4	0.16	0.014
Green	17	9.1	7.9	62.41	6.858
Yellow	8	7.6	0.4	0.16	0.021
Brown	5	7.6	−2.6	6.76	0.900
Red	7	7.6	−0.6	0.36	0.047
Σ	57	57.0			$\chi^2 = 9.45$

$\chi^2_{.05}$ (5 df) = 11.07 $\qquad df = 6 - 1 = 5$

Interpretation: Retain the null hypothesis. The observed frequencies are not significantly different from the expected frequencies, which are based on the company's claims.

Fortunately, this is the kind of problem you can gather your own data on. Of course, you *must* use a larger sample (and having a research partner would be nice, too). Should you and your partner find yourselves somehow forced to test peanut M&Ms, the percentages are 23 percent blue, 23 percent orange, 15 percent green, 15 percent yellow, 12 percent brown, and 12 percent red. Also, I want to acknowledge my debt to Randolph A. Smith, who explained to me the value of M&Ms as a statistical tool.

14.16. You should be gentle with your friend but explain that the observations in his study are not independent and cannot be analyzed with χ^2. Explain that because one person is making five observations, the observations are not independent; the choice of one female candidate may cause the subject to pick a male candidate next (or vice versa).

14.17.

		Candidates: Hill	Candidates: Dale	Σ
Signaled turn	Yes	57 (59.248)	31 (28.752)	88
	No	11 (8.752)	2 (4.248)	13
	Σ	68	33	101

$$\chi^2 = \frac{101[(57)(2) - (31)(11)]^2}{(88)(13)(68)(33)} = 2.03$$

$\chi^2_{.05}$ (1 df) = 3.84. Interpretation: Cars with bumper stickers for candidates Hill and Dale were observed as they turned left at a busy intersection. Sixteen percent of the 68 Hill cars failed to signal their left turn; 6 percent of the 33 Dale cars turned left without signaling. This difference is not statistically significant, which means that a difference this big is expected more than 5 percent of the time if, in actuality, there is no difference in the percentage of traffic signal violators among supporters of the two candidates.

14.18. I hope you would say something like, "Gee, I've been able to analyze data like these since I learned the t test. I will need to know the standard deviations for those means, though." If you attempted to analyze these data using χ^2, you erred (which, according to Alexander Pope, is human). If you erred, forgive yourself and reread pages 301 and 319.

14.19.

		Houses: Brick	Houses: Frame	Σ
Candidates	Hill	17 (37.116)	59 (38.884)	76
	Dale	88 (67.884)	51 (71.116)	139
	Σ	105	110	215

$$\chi^2 = \frac{N(AD - BC)^2}{(A + B)(C + D)(A + C)(B + D)}$$

$$= \frac{215(867 - 5192)^2}{(76)(139)(105)(110)} = 32.96; \quad df = 1$$

$$\phi = \sqrt{\frac{\chi^2}{N}} = \sqrt{\frac{32.96}{215}} = \sqrt{.1533} = 0.39$$

$\chi^2_{.001}$ (1 df) = 10.83; $p < .001$

Interpretation: Of the 105 affluent (brick) houses, 16 percent had Hill signs and 84 percent had Dale signs. At the 110 less affluent (frame) houses, the relationship was reversed; 54 percent had Hill signs and 46 percent had Dale signs. These differences are statistically significant, which means that if there is really no difference between the two candidates, percentages like this would be rare, occurring less than one time in a

(Continued)

thousand. These data support the conclusion that Dale's supporters are more affluent than Hill's supporters. The effect size index ($\phi = 0.39$) indicates that the relationship between candidate support and affluence is between medium and large.

14.20. $df = (R - 1)(C - 1)$, except for a table with only one row.
a. 3 **b.** 12 **c.** 3 **d.** 10

CHAPTER 15

15.1. The relative of the independent-samples t test is the Mann–Whitney U test. Corresponding to a paired-samples t test is the Wilcoxon matched-pairs signed-ranks T test.

15.2. There are two differences: (a) Parametric tests give accurate probabilities of a Type I error if the populations the samples are drawn from have the characteristics that parametric tests assume (e.g., ANOVA assumes the populations are normally distributed and have equal variances). Nonparametric tests do not have assumptions about the populations built in.
(b) The null hypothesis for parametric tests is that the population means are equal; for nonparametric tests, the null hypothesis is that the population distributions are the same.

15.3. Compare your answer with the material in the section "The Rationale of Nonparametric Tests."

15.4. Scales of measurement and power

15.5. I ranked the fewest errors (7) as #1 so that high ranks would indicate many errors. If you gave 32 errors a rank of #1, the sum of ranks is shown in parentheses, but the U values and conclusion are the same.

Near airport $N_1 = 8$ Sum of ranks = 94 (50)

Quiet area $N_2 = 9$ Sum of ranks = 59 (103)

$$U_1 = (8)(9) + \frac{(8)(9)}{2} - R_1 = 108 - 94 = 14$$

$$U_2 = (8)(9) + \frac{(9)(10)}{2} - R_2 = 117 - 59 = 58$$

Thus, $U = 14$

From Table H for a two-tailed test with $\alpha = .05$ (the second page), $U = 15$.

Interpretation: The obtained U is smaller than the tabled U, so reject the null hypothesis. The mean rank in errors for those near the airport was 11.75 (94/8), which is higher than the mean rank for the other group, 6.56 (59/9). Thus, conclude that children whose school is near a busy airport make significantly more errors on a difficult reading test than children whose school is in a quiet area of the same city. The study by Hygge et al. (2002) was conducted in Munich, Germany, when a new airport opened and the old airport closed.

15.6. With $\alpha = .01$, there are *no* possible results that would allow H_0 to be rejected. The friend must find more cars.

15.7. $\Sigma R_Y = 13$, $\Sigma R_Z = 32$

$$\text{Smaller } U = (4)(5) + \frac{(5)(6)}{2} - 32 = 3$$

Even with $\alpha = .05$ and a two-tailed test, a U of 1 or less is required to reject H_0.
Interpretation: With an obtained $U = 3$, conclude that the quietness test did not produce evidence that Y cars are quieter than Z cars.

15.8. Giving a score of 0 a rank of 2 and 110 a rank of 40, the sum of the ranks for present day is 409.5; the sum of the ranks for 10 years earlier is 410.5. If you ranked in reverse order, the sum of the ranks for present day is 574.5 and the sum for 10 years earlier is 245.5. Either way, the smaller U is 109.5 and the larger U is 274.5.

$$U_1 = (24)(16) + \frac{(24)(25)}{2} - 409.5 = 274.5$$

$$U_2 = (24)(16) + \frac{(16)(17)}{2} - 410.5 = 109.5$$

$$\mu_U = \frac{(24)(16)}{2} = 192$$

$$\sigma_U = \sqrt{\frac{(24)(16)(41)}{12}} = 36.22$$

$$z = \frac{(109.5 + 0.5) - 192}{36.22} = -2.26$$

The critical value of z is $|1.96|$ for a two-tailed test.

(Continued)

Interpretation: Reject the null hypothesis that the distributions are the same. By examining the average ranks (23.9 for present-day birds and 15.3 for 10-years-ago birds), you can conclude that present-day birds have significantly fewer brain parasites. (If the 0 score birds rank 2, the average ranks are 17.1 for present-day birds and 25.7 for 10-years-ago birds. The same conclusion results.)

15.9. $$\mu_T = \frac{N(N+1)}{4} = \frac{(112)(113)}{4} = 3164$$

$$\sigma_T = \sqrt{\frac{N(N+1)(2N+1)}{24}}$$

$$= \sqrt{\frac{(112)(113)(225)}{24}} = 344.46$$

$$z = \frac{(4077 + 0.5) - 3164}{344.46} = 2.65$$

Interpretation: Because $2.65 > 1.96$, reject H_0. The incomes were significantly different after the program. Because you do not have the actual data, you cannot tell whether the incomes were higher or lower than before.

15.10.

Worker	Without rests	With rests	D	Signed ranks
1	2240	2421	181	2
2	2069	2260	191	4
3	2132	2333	201	5
4	2095	2314	219	6
5	2162	2297	135	1
6	2203	2389	186	3

Check: $21 + 0 = 21$ $\qquad \Sigma$ (positive) $= 21$

$\frac{(6)(7)}{2} = 21$ $\qquad \Sigma$ (negative) $= 0$

$T = 0$

$N = 6$

Interpretation: The critical value of T at the .05 level for a two-tailed test is 0. Because 0 (obtained) is equal to or less than 0 (table), reject the null hypothesis and conclude that the output with rests is greater than the output without rests. Here is the story behind this study.

From 1927 to 1932 the Western Electric Company conducted a study on a group of workers at the Hawthorne Works plant near Chicago. In the study, workers who assembled telephone relays were separated from the rest of the workers. A variety of changes in their daily routine followed, one at a time, with the following results: 5-minute rest periods increased output; 10-minute rest periods increased output; company-provided snacks during rest periods increased output; and quitting 30 minutes early increased output. Going back to the original no-rest schedule increased output again, as did the reintroduction of the rest periods. Finally, management concluded that it was the special attention paid to the workers, rather than the changes, that increased output. Today, we call it the Hawthorne Effect when there is an improvement in performance that is due to being in an experiment (getting special attention) rather than to the specific manipulation in the experiment. For a summary of this study, see Mayo (1946).

15.11. The experimental design has two independent samples. A Mann–Whitney U test is needed. If high scores are given high rank (that is, 39 ranks 1):

Σ (Canadians) = 208; Σ (people from U.S.) = 257 (If low scores are given high ranks, the sums are reversed.)

$$U_1 = (15)(15) + \frac{(15)(16)}{2} - 257 = 88; \; U_2 = 137$$

Consulting the second page of Table H (boldface type), a U value of 64 is required to reject the null hypothesis ($\alpha = .05$, two-tailed test). The lower of the two calculated U values for this problem is 88. Because $88 > 64$, the null hypothesis must be retained.

Interpretation: Conclude that there is no strong evidence that Canadians and people from the United States have different attitudes toward the regulation of business.

Once again you must be cautious in writing the conclusion when the null hypothesis is retained. You have not demonstrated that the two groups are the same; you have shown only that the groups are not significantly different.

15.12.

Student	Before	After	D	Signed rank
1	18	4	14	13
2	14	14	0	1.5
3	20	10	10	9
4	6	9	−3	−4
5	15	10	5	6
6	17	5	12	11
7	29	16	13	12
8	5	4	1	3
9	8	8	0	−1.5
10	10	4	6	7
11	26	15	11	10
12	17	9	8	8
13	14	10	4	5
14	12	120		Eliminated

Check: $85.5 + 5.5 = 91$ $\quad \Sigma\text{ (positive)} = 85.5$

$\frac{(13)(14)}{2} = 91$ $\quad \Sigma\text{ (negative)} = -5.5$

$T = 5.5$

$N = 13$

The tabled value for T for a two-tailed test with $\alpha = .01$ is 9, so reject H_0.

Interpretation: Conclude that the after-distribution is from a different population than the before-distribution. Except for one person, the number of misconceptions stayed the same or decreased, so the course *reduced* the number of misconceptions.

I would like to remind you here of the distinction made in Chapter 10 between statistically significant and important. There is a statistically significant decrease in the number of misconceptions, but a professor might be quite dismayed at the number of misconceptions that remain.

15.13. $\Sigma\text{ (positive)} = 81.5$ *Check:* $81.5 + 54.5 = 136$

$$\Sigma\text{ (negative)} = -54.5 \qquad \frac{(16)(17)}{2} = 136$$

$$T = 54.5$$

$$N = 16$$

$T \leq 29$ is required for rejection at the .05 level. Thus, there is no significant difference in the weight 10 months later.

Interpretation: Put in the most positive language, there is no statistically significant evidence that the participants gained back the weight lost during the workshop.

15.14. The tables show the raw data converted to ranks. The table on the left shows the lowest score ranked 1. If the highest score is ranked 1, the table on the right results. Although the sums at the bottom are not the same, the difference between any two pairs is the same.

Friends	Romans	Countrymen	Friends	Romans	Countrymen
15	9	2	4	10	17
18	13	10	1	6	9
12	4.5	6	7	14.5	13
14	11	1	5	8	18
16	3	8	3	16	11
17	4.5	7	2	14.5	12
$\Sigma = 92$	45	34	$\Sigma = 22$	69	80

	Friends	Romans
Romans	47*	
Countrymen	58**	11

$^*p < .05 \quad {}^{**}p < .01$

For $K = 3$, $N = 6$, the critical values are 43.3 (.05 level) and 53.9 (.01 level).

Interpretation: Romans and countrymen did not differ from each other, but both groups offered significantly less than friends did. So, when you have a plea, ask a friend to lend an ear.

15.15. The table shows the raw scores converted to ranks.

	NO	RE	SE	GE
	6	9	8	15.5
	4	6	18	20
	2	14	17	12.5
	1	6	12.5	19
	3	10	11	15.5
Σ	16	45	66.5	82.5
Σ (reverse ranks)	89	60	38.5	22.5

	NO	RE	SE
RE	29		
SE	50.5*	21.5	
GE	66.5**	37.5	16

$^*p < .05 \quad {}^{**}p < .01$

For $K = 4$, $N = 5$, the critical values are 48.1 (.05 level) and 58.2 (.01 level).
Interpretation: Although group exercise is better than solitary exercise, the difference is not significant. Both of these groups improved significantly when compared to the no-treatment group. The relaxation group was intermediate between no treatment and solitary exercise, but not significantly different from the other three groups. Take-home message: Exercise with others.

15.16.

Authoritarian		Democratic		Laissez-faire	
X	Rank	X	Rank	X	Rank
77	5	90	10.5	50	1
86	9	92	12	62	2
90	10.5	100	16	69	3
97	13	105	21	76	4
100	16	107	22	79	6
102	19	108	23	82	7
120	27.5	110	24	84	8
121	29	118	26	99	14
128	32	125	30	100	16
130	33	131	34	101	18
135	37	132	35.5	103	20
137	38	132	35.5	114	25
141	39.5	146	41	120	27.5
147	42	156	44	126	31
153	43	161	45	141	39.5
Σ	393.5		419.5		222.0
Σ (reverse ranks)	296.5		270.5		468.0

	Authoritarian	Democratic
Democratic	26	
Laissez-faire	171.5*	197.5*

*$p < .05$

For $K = 3$, $N = 15$, a difference in the sum of the ranks of 168.6 is required to reject H_0 at the .05 level.

Interpretation: Both the authoritarian leadership and the democratic leadership resulted in higher personal satisfaction scores than laissez-faire leadership, but the authoritarian and democratic types of leadership did not differ significantly from each other.

15.17. Arranging the data into a summary table gives:

Groups	1	2	3	4	5
2	85				
3	31	54			
4	18	67	13		
5	103	188*	134	121	
6	31	116	62	49	72

*$p < .05$

For $K = 6$, $N = 8$, differences of 159.6 and 188.4 are required to reject the null hypothesis at the .05 and .01 level, respectively. Thus, the only significant difference is between the means of groups 2 and 5 at the .05 level. (With data such as these, the novice investigator may be tempted to report "almost significant at the .01 level." Resist that temptation.)

15.18. The Mann–Whitney U test is used for ranked data from two independent samples with equal or unequal N's. The Wilcoxon matched-pairs signed-ranks T test is used for ranked data from two related samples. The Wilcoxon–Wilcox multiple-comparisons test makes all possible pairwise comparisons among K independent equal-N groups of ranked data.

15.19.
a. Wilcoxon matched-pairs signed-ranks T test; this is a before-and-after study.
b. Wilcoxon matched-pairs signed-ranks T test; pairs are formed by family.
c. Spearman's r_s; the degree of relationship is desired.
d. Wilcoxon–Wilcox multiple-comparisons test; the experiment has three independent groups.
e. Wilcoxon matched-pairs signed-ranks T test; again, this is a before-and-after design.

15.20.

	Symbol of statistic	Appropriate for what design?	Reject H_0 when statistic is (greater, less) than critical value?
Mann–Whitney test	U	Two independent samples	Less
Wilcoxon matched-pairs signed-ranks test	T	Two paired samples	Less
Wilcoxon–Wilcox multiple-comparisons test	None	More than two independent samples, equal N's	Greater

15.21. A low correlation means the philosophers have different sets of criteria of what is important. A high correlation means that their criteria are about the same.

Candidates	Locke	Kant	D	D^2
A	7	8	−1	1
B	10	10	0	0
C	3	5	−2	4
D	9	9	0	0
E	1	1	0	0
F	8	7	1	1
G	5	3	2	4
H	2	4	−2	4
I	6	6	0	0
J	4	2	2	4
				$\Sigma D^2 = 18$

$$r_s = 1 - \frac{6\Sigma D^2}{N(N^2-1)} = 1 - \frac{6(18)}{10(99)} = 1 - .109 = .891$$

Interpretation: The two professors seem to be in pretty close agreement on the selection criteria.

15.22.

Self-esteem scores	Rank	Self-judged attractiveness	Rank	D^2	Other-judged attractiveness	Rank	D^2
40	1	97	3	4	37	6	25
39	2	99	2	0	43	4	4
37	3	100	1	4	29	9	36
36	4	93	6	4	25	12	64
35	5	59	12	49	60	1	16
32	6	88	8	4	54	2	16
28	7	83	9	4	28	10	9
24	8	96	4	16	26	11	9
20	9	90	7	4	40	5	16
16	10	94	5	25	49	3	49
12	11	70	11	0	31	8	9
7	12	78	10	4	34	7	25
			$\Sigma D^2 =$	118			278

a. $r_s = 1 - \frac{6\Sigma D^2}{N(N^2-1)} = 1 - \frac{6(118)}{12(143)}$

$= 1 - \frac{708}{1716} = 1 - .4126 = .587$

In Table L, an r_s of .587 is required for statistical significance at the .05 level for a two-tailed test. Thus, the correlation coefficient of .587 is significantly different from .00 at the .05 level.

b. $r_s = 1 - \frac{6\Sigma D^2}{N(N^2-1)} = 1 - \frac{6(278)}{12(143)}$

$= 1 - \frac{1668}{1716} = 1 - .9720 = .028$

Interpretation: The greater the participant's self-esteem, the greater that person's judgment of his or her own attractiveness ($r_s = .59$). However, when other people judged their attractiveness, the correlation between self-esteem and attractiveness dropped to about zero ($r_s = .03$). Thus, the attractiveness of those with high self-esteem is apparent to themselves but not to others.

15.23.

$$r_s = 1 - \frac{6\Sigma D^2}{N(N^2 - 1)} = 1 - \frac{6(308)}{16(16^2 - 1)} = .55$$

From Table L, a coefficient of .503 is required for significance (two-tailed test). Thus, an r_s = .55 is significantly different from .00.

15.24. With only four pairs of scores there is no possible way to reject the hypothesis that the population correlation is .00 (see Table L). Advise your friend that more data must be obtained before any inference can be made about the population.

What Would You Recommend? Chapters 11–15

a. A Mann–Whitney *U* test should be used to compare systematic desensitization and flooding. The two groups are independent samples—there is no reason to pair the score of a trainee who experienced systematic desensitization with the score of a trainee who experienced flooding. A nonparametric test is called for because the data were skewed.

b. The proper analysis of these data is with a chi square test of independence. The score that a participant received was counted as a frequency in a category (comply or not comply).

c. These data should be analyzed with a factorial ANOVA. The two independent variables are the appearance and intentionality of the defendant. The dependent variable is the likelihood estimate (a quantitative measure). In Berry and McArthur's study (1986) of these variables there was a significant interaction. Baby-faced defendants were less likely to be convicted of an intentional offense than mature-faced defendants, but more likely to be convicted of a negligent offense.

d. A Spearman r_s gives the degree of relationship between two variables. A Spearman r_s is more appropriate than a Pearson *r* because one variable is rank scores.

e. The theory, which predicts the number of participants on subsequent occasions of the exercise, can be evaluated using a chi square goodness-of-fit test. (The expected values are 50, 25, and 12.5.)

f. This experiment has one independent variable that has three levels. Because the variance of the "no cues" condition was much greater than that of the other two conditions, a Wilcoxon–Wilcox multiple-comparisons test is appropriate. See Winter, Uleman, and Cunniff (1985).

g. A Wilcoxon matched-pairs signed-ranks *T* test is appropriate for these skewed weight data. This is a paired-samples design because a before-meal score is paired with an after-meal score. (Both scores were obtained from the same dog.)

h. The problem gives no indication that the data are not normally distributed or have unequal variances. Thus, to compare each pair of recipes, use a Tukey HSD test. If, however, you reasoned that tasty recipes would produce a positively skewed distribution (that is, many waste scores = 0), then the answer is Wilcoxon–Wilcox multiple-comparison tests of the pairs.

i. A one-factor repeated-measures ANOVA will help evaluate the program. Tukey HSD tests can determine the immediate effect of the program (posttest vs. pretest), the long-term effect (follow-up vs. pretest), and the effect of time on retention (follow-up vs. posttest).

CHAPTER 16

Set A

16.1. A weighted mean is needed, and to find it you need to know the number of dollars invested in each division.

16.2. Wilcoxon matched-pairs signed-ranks *T* test. You have evidence that the population of dependent-variable scores (reaction time) is not normally distributed.

16.3. A one-factor repeated-measures ANOVA

16.4. A median is appropriate because the scores are skewed.

16.5. Assume that these measures are normally distributed and use a normal curve to find the proportion (.0228 is the answer).

16.6. Either an independent-measures *t* test or a Mann–Whitney *U* test

16.7. A 95 or 99 percent confidence interval about the mean reading achievement of those 50 sixth-graders

16.8. χ^2 goodness-of-fit test

16.9. Mode

16.10. Neither will do. The relationship described is nonlinear—one that first increases and then decreases. A statistic, eta, is appropriate for curved relationships.

16.11. Paired-samples t test or Wilcoxon matched-pairs signed-ranks T test

16.12. A regression equation will provide the predictions, and a correlation coefficient will indicate how accurate the predictions will be.

16.13. A 2×2 factorial ANOVA

16.14. The student's wondering may be translated into a question of whether a correlation of .20 is statistically significant. Use a t test to find whether $r = .20$ for that class is significantly different from $r = .00$. Or, look in Table A in Appendix C.

16.15. This is a χ^2 problem but it cannot be worked using the techniques in this text because these before-and-after data are correlated, not independent. Intermediate texts describe appropriate techniques for such χ^2 problems.

16.16. A confidence interval about the sample mean. Does the CI include the manufacturer's claim of 1000 hours?

16.17. Pearson product-moment correlation coefficient

16.18. Wilcoxon–Wilcox multiple-comparisons test (if the reaction-time scores are skewed) or one-way analysis of variance

16.19. A χ^2 test of independence may determine whether a decision to report shoplifting is influenced by gender (one χ^2 test) and by the dress of the shoplifter (a second χ^2 test).

16.20. A line graph with serial position on the X axis and number of errors on the Y axis will illustrate this relationship.

16.21. A Mann–Whitney U test is preferred because the dependent variable is skewed. The question is whether there is a relationship between diet and cancer. (Note that if the two groups differ, two interpretations are possible. For example, if the incidence is higher among the red meat cultures, it might be because of the red meat or because of the lack of grains.)

16.22. An r_s will give the degree of relationship for these two ranked variables.

16.23. For each species of fish, a set of z scores may be calculated. The fish with the largest z score should be declared the overall winner.

16.24. Calculating an effect size for the experiments mentioned in the first part of this question will tell you how big a difference psychotherapy makes.

Set B

16.25. $t_{.001}$ (40 df) = 3.55; therefore, reject the null hypothesis and conclude that extended practice *improved* performance. Because this is just the opposite of the theory's prediction, conclude that the theory does not explain the serial position effect.

16.26. An effect size of 0.10 is quite small. Effect sizes of 0.20, 0.50, and 0.80 are classified as small, medium, and large. In the case of a company offering instruction, a large number of clients provides a large N, which can ensure a "statistically significant improvement," even if the effect size is quite small.

16.27. The screening process is clearly worthwhile in making each batch profitable. Of course, the cost of screening has to be taken into account.

16.28. The critical values are 377.6 at the .05 level and 457.6 at the .01 level. Conclude that Herbicide B is significantly better than A, C, or D at the .01 level and that D is better than C at the .05 level.

16.29. Because $\chi^2_{.05}$ (2 df) = 5.99, the null hypothesis for this goodness-of-fit test is retained. The data do fit the theory; the theory is adequate.

16.30. The overall F is significant ($p < .01$), so the differences should not be attributed to chance. The low dose is significantly better than the placebo ($HSD_{.01} = 4.70$) and significantly better than the high dose (which has a smaller mean than the placebo). The placebo and high dose are not significantly different. A low dose of Cymbalta reduced depression.

16.31. The critical value of U for a two-tailed test with $\alpha = .01$ is 70. The difference between the two methods is statistically significant at the .01 level. You cannot, from the information supplied, tell which of the two methods is superior.

16.32. The critical value at the .001 level of an r_s with $N = 18$ is .708 (Table A, $df = 16$). Attitudes of college students and people in business toward the 18 groups are similar. College students and businesspeople have similar attitudes.

16.33. $t_{.05}$ (60 *df*) = 2.00; $F_{.01}$ (30, 40 *df*) = 2.20; the poetry unit appears to have no significant effect on mean attitudes toward poetry. There is a very significant effect on the variability of the attitudes of those who studied poetry. It appears that the poetry unit turned some students on and some students off, thus causing a large variance in attitudes toward poetry.

16.34. $F_{.05}$ (1, 44 *df*) = 4.06; $F_{.01}$ (1, 44 *df*) = 7.24; only the interaction is significant. A graph, as always, helps in interpretation. Whether to present one or both sides to get the most attitude change depends on the level of education of the audience. If the members of a group have less than a high school education, present one side. If they have some college education, present both sides.

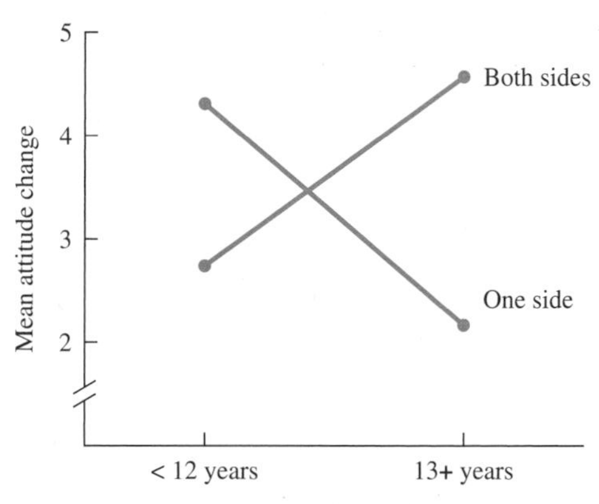

16.35. The tabled value of $F_{.05}$ (2, 60 *df*) = 3.15, so the null hypothesis is retained. There are no significant differences in the verbal ability scores of third-graders who are left- or right-handed or have mixed handedness.

16.36. Because $\chi^2_{.02}$ (2 *df*) = 7.82, the null hypothesis for this test of independence is rejected at the .02 level. Alcoholism and toilet-training are not independent; they are related. Experiments such as this do not allow you to make cause-and-effect statements, however.

APPENDIX A

Part 1 (Pretest)

A.1. About 16	**A.21.** 1.50	**A.41.** 0.1225
A.2. About 15	**A.22.** 3.11	**A.42.** 3.05
A.3. About 15	**A.23.** 6	**A.43.** 0.96
A.4. 6.1	**A.24.** −22	**A.44.** 0.30
A.5. 0.4	**A.25.** −7	**A.45.** 4.20
A.6. 10.3	**A.26.** −6	**A.46.** 4.00
A.7. 11.424	**A.27.** 25	**A.47.** 4.50
A.8. 27.141	**A.28.** −24	**A.48.** 0.45
A.9. 19.85	**A.29.** 3.33	**A.49.** 15.25
A.10. 2.065	**A.30.** −5.25	**A.50.** 8.00
A.11. 49.972	**A.31.** 33%	**A.51.** 16.25
A.12. 0.0567	**A.32.** 22.68	**A.52.** 1.02
A.13. 3.02	**A.33.** 0.19	**A.53.** 12.00
A.14. 17.11	**A.34.** 60 birds	**A.54.** 10, 30
A.15. 1.375	**A.35.** 5	**A.55.** 12, 56
A.16. 0.83	**A.36.** 4	**A.56.** 13.00
A.17. 0.1875	**A.37.** 6, 10	**A.57.** 11.00
A.18. 0.056	**A.38.** 4, 22	**A.58.** 2.50
A.19. 0.60	**A.39.** 16	**A.59.** −2.33
A.20. 0.10	**A.40.** 6.25	

Part 2 (Review)

A.60. **a.** A sum is the answer to an addition problem.
b. A quotient is the answer to a division problem.
c. A product is the answer to a multiplication problem.
d. A difference is the answer to a subtraction problem.

A.61. **a.** About 10 **b.** About 7 **c.** About 100
d. About 0.2 **e.** About 0.16 **f.** About 60
g. Just less than 1 **h.** About 0.1

A.62. **a.** 14 **b.** 126 **c.** 9 **d.** 0 **e.** 128
f. 13 **g.** 12 **h.** 13 **i.** 9

A.63. **a.** 6.33 **b.** 13.00 **c.** 0.05 **d.** 0.97 **e.** 2.61
f. 0.34 **g.** 0.00 **h.** 0.02 **i.** 0.99

A.64.
$$\begin{array}{r} 0.001 \\ 10.000 \\ 3.652 \\ \underline{2.500} \\ 16.153 \end{array}$$

A.65.
$$\begin{array}{r} 14.20 \\ \underline{-7.31} \\ 6.89 \end{array}$$

A.66.
$$\begin{array}{r} 1.26 \\ \times\ 0.04 \\ \hline 0.0504 \end{array}$$

A.67.
$$\begin{array}{r} 143.300 \\ 16.920 \\ 2.307 \\ 8.100 \\ \hline 170.627 \end{array}$$

A.68.
$$\begin{array}{r} 76.5 \\ 0.04\overline{)3.06} \\ \underline{2\,8} \\ 26 \\ \underline{24} \\ 20 \\ \underline{20} \end{array}$$

A.69.
$$\begin{array}{r} 2.04 \\ 11.75\overline{)24.00} \\ \underline{23\,50} \\ 5000 \\ \underline{4700} \\ 300 \end{array}$$

A.70.
$$\begin{array}{r} 152.12 \\ -127.40 \\ \hline 24.72 \end{array}$$

A.71.
$$\begin{array}{r} 0.07 \\ \times\ 0.5 \\ \hline 0.035 \end{array}$$

A.72. $\frac{9}{10} + \frac{1}{2} + \frac{2}{5} = 0.90 + 0.50 + 0.40 = 1.80$

A.73. $\frac{9}{20} \div \frac{19}{20} = 0.45 \div 0.95 = 0.47$

A.74. $\left(\frac{1}{3}\right)\left(\frac{5}{6}\right) = 0.333 \times 0.833 = 0.28$

A.75. $\frac{4}{5} - \frac{1}{6} = 0.80 - 0.167 = 0.63$

A.76. $\frac{1}{3} \div \frac{5}{6} = 0.333 \div 0.833 = 0.40$

A.77. $\frac{3}{4} \times \frac{5}{6} = 0.75 \times 0.833 = 0.62$

A.78. $18 \div \frac{1}{3} = 18 \div 0.333 = 54$

A.79. **a.** $(24) + (-28) = -4$ **b.** -23
c. $(-8) + (11) = 3$ **d.** $(-15) + (8) = -7$

A.80. -40 **A.81.** 24

A.82. 48 **A.83.** -33

A.84. $(-18) - (-9) = -9$

A.85. $14 \div (-6) = -2.33$

A.86. $12 - (-3) = 15$

A.87. $(-6) - (-7) = 1$

A.88. $(-9) \div (-3) = 3$ **A.89.** $(-10) \div 5 = -2$

A.90. $4 \div (-12) = -0.33$ **A.91.** $(-7) - 5 = -12$

A.92. $6 \div 13 = 0.46$

A.93. $.46 \times 100 = 46$ percent

A.94. $18 \div 25 = 0.72$

A.95. $85 \div 115 = 0.74 \times 100 = 74$ percent

A.96. 25.92 **A.97.** 0.31

A.98. 128 **A.99.** 187 coupons

A.100. $|-31| = 31$

A.101. $|21 - 25| = |-4| = 4$

A.102. $12 \pm (2)(5) = 12 \pm 10 = 2, 22$

A.103. $\pm(5)(6) + 10 = \pm 30 + 10 = -20, 40$

A.104. $\pm(2)(2) - 6 = \pm 4 - 6 = -10, -2$

A.105. $(2.5)^2 = 2.5 \times 2.5 = 6.25$

A.106. $9^2 = 9 \times 9 = 81$ **A.107.** $0.3 \times 0.3 = 0.09$

A.108. $(\frac{1}{4})^2 = (0.25)(0.25) = 0.0625$

A.109. **a.** 25.00 **b.** 2.50 **c.** 0.79
d. 0.25 **e.** 4.10 **f.** 0.0548
g. 426.00 **h.** 0.50 **i.** 4.74

A.110.
$$\frac{(4-2)^2 + (0-2)^2}{6} = \frac{2^2 + (-2)^2}{6} = \frac{4+4}{6} = \frac{8}{6} = 1.33$$

A.111.
$$\frac{(12-8)^2 + (8-8)^2 + (5-8)^2 + (7-8)^2}{4-1} = \frac{4^2 + 0^2 + (-3)^2 + (-1)^2}{3} = \frac{16+0+9+1}{3} = \frac{26}{3} = 8.67$$

A.112.
$$\left(\frac{5+6}{3+2-2}\right)\left(\frac{1}{3} + \frac{1}{2}\right) = \left(\frac{11}{3}\right)(0.333 + 0.500) = (3.667)(0.833) = 3.05$$

A.113.
$$\left(\frac{13+18}{6+8-2}\right)\left(\frac{1}{6} + \frac{1}{8}\right) = \left(\frac{31}{12}\right)(0.167 + 0.125) = (2.583)(0.292) = 0.75$$

A.114.
$$\frac{8[(6-2)^2 - 5]}{(3)(2)(4)} = \frac{8[4^2 - 5]}{24} = \frac{8[16-5]}{24} = \frac{8[11]}{24} = \frac{88}{24} = 3.67$$

A.115.
$$\frac{[(8-2)(5-1)]^2}{5(10-7)} = \frac{[(6)(4)]^2}{5(3)} = \frac{[24]^2}{15} = \frac{576}{15} = 38.40$$

A.116.
$$\frac{6}{1/2} + \frac{8}{1/3} = (6 \div 0.5) + (8 \div 0.333) = 12 + 24 = 36$$

A.117.
$$\left(\frac{9}{2/3}\right)^2 + \left(\frac{8}{3/4}\right)^2 = (9 \div 0.667)^2 + (8 \div 0.75)^2 = 13.5^2 + 10.667^2 = 182.25 + 113.785 = 296.03$$

A.118. $\frac{10 - (6^2/9)}{8} = \frac{10 - (36/9)}{8}$

$= \frac{10 - 4}{8} = \frac{6}{8} = 0.75$

A.119. $\frac{104 - (12^2/6)}{5} = \frac{104 - (144/6)}{5}$

$= \frac{104 - 24}{5} = \frac{80}{5} = 16.00$

A.120. $\frac{x - 4}{2} = 2.58, \quad x - 4 = 5.16, \quad x = 9.16$

A.121. $\frac{x - 21}{6.1} = 1.04, \quad x - 21 = 6.344, \quad x = 27.34$

A.122. $x = \frac{14 - 11}{2.5} = \frac{3}{2.5} = 1.20$

A.123. $x = \frac{36 - 41}{8.2} = \frac{-5}{8.2} = -0.61$

APPENDIX B

B.1.

Test scores (class interval)	Tally marks	f
63–67	/	1
58–62	/	1
53–57	///	3
48–52	////	4
43–47	////	4
38–42	~~////~~ //	7
33–37	~~////~~ //	7
28–32	~~////~~ //	7
23–27	~~////~~ //	7
18–22	////	4
13–17	///	3
8–12	/	1
3–7	/	1
		$\Sigma = 50$

B.2.

Weights	Tally marks	f
189–191	/	1
186–188		0
183–185		0
180–182	/	1
177–179		0
174–176	//	2
171–173	/	1
168–170	//	2
165–167	/	1
162–164	/	1
159–161	///	3
156–158	~~////~~	5
153–155	////	4
150–152	~~////~~ /	6
147–149	////	4
		$\Sigma = 31$

The data are positively skewed.

B.3. $\bar{X} = \frac{\Sigma X}{N} = \frac{1720}{50} = 34.40$

Median location $= \frac{N + 1}{2} = \frac{50 + 1}{2} = 25.5.$

There are 23 frequencies in the intervals below 33–37 and 7 scores in the interval. The median is 35, the midpoint of the interval 33–37.

Because four intervals share the largest number of scores, 7, the mode is the midpoint of these four intervals. Mode = 32.5.

B.4. $\bar{X} = \frac{\Sigma X}{N} = \frac{4939}{31} = 159.32$ pounds

Median location $= \frac{N + 1}{2} = \frac{31 + 1}{2} = 16$. There are 14 frequencies below the interval 156–158, which has 5 measurements. The median is 157 pounds.

The mode is 151 pounds because the interval 150–152 has 6 frequencies, more than any other interval.

Note that the mean is larger than the median, confirming your judgment that the distribution is positively skewed.

References

Aiken, L. S., & West, S. G. (2005). Interaction effects. In B. S. Everitt & D. C. Howell (Eds.), *Encyclopedia of statistics in behavioral science* (pp. 929–933). West Sussex, UK: Wiley.

Ambady, N., LaPlante, D., Nguyen, T., Rosenthal, R., Chaumeton, N., & Levinson, W. (2002). Surgeons' tone of voice: A clue to malpractice history. *Surgery, 132,* 5–9. doi: 10.1067/msy.2002.124733

American Psychological Association. (2010). *Publication manual of the American Psychological Association* (6th ed.). Washington, DC: Author.

Aron, A., Aron, E. N., & Coups, E. J. (2009). *Statistics for psychology* (5th ed.). Upper Saddle River, NJ: Pearson Prentice-Hall.

Aronson, E., & Mills, J. (1959). The effect of severity of initiation on liking for a group. *Journal of Abnormal and Social Psychology, 59,* 177–181.

Bahrick, H. P., Bahrick, L. E., Bahrick, A. S., & Bahrick, P. E. (1993). Maintenance of foreign language vocabulary and the spacing effect. *Psychological Science, 4,* 316–321.

Bandura, A., Blanchard, E. B., & Ritter, B. (1969). The relative efficacy of desensitization and modeling approaches for inducing behavioral, affective, and attitudinal changes. *Journal of Personality and Social Psychology, 64,* 173–199.

Barber, T. X. (1976). Suggested ("hypnotic") behavior: The trance paradigm versus an alternative paradigm. In T. X. Barber (Ed.), *Advances in altered states of consciousness & human potentialities* (Vol. 1, pp. 175–259). New York: Psychological Dimensions.

Barker, R. B., Dembo, T., & Lewin, K. (1941). Frustration and regression: An experiment with young children. *University of Iowa Studies in Child Welfare, 18,* No. 1.

Baumeister, R. F., Campbell, J. D., Krueger, J. I., & Vohs, K. D. (2003). Does high self-esteem cause better performance, interpersonal success, happiness, or healthier lifestyles? *Psychological Science in the Public Interest, 4,* 1– 44.

Benjamin, L. T., Cavell, T. A., & Shallenberger, W. R. (1984). Staying with initial answers on objective tests: Is it a myth? *Teaching of Psychology, 11,* 133–141.

Berry, D. S., & McArthur, L. Z. (1986). Perceiving character in faces: The impact of age-related craniofacial changes on social perception. *Psychological Bulletin, 100,* 3–18. doi: 10.1037/0033-2909.100.1.3

Biffen, R. H. (1905). Mendel's laws of inheritance and wheat breeding. *Journal of Agricultural Science, 1,* 4–48.

Birch, H. G., & Rabinowitz, H. S. (1951). The negative effect of previous experience on productive thinking. *Journal of Experimental Psychology, 41,* 121–125.

Blair, R. C., & Higgins, J. J. (1985). Comparison of the power of the paired samples *t* test to that of Wilcoxon's signed-ranks test under various population shapes. *Psychological Bulletin, 97,* 119–128. doi: 10.1037/0033-2909.97.1.119

Blair, R. C., Higgins, J. J., & Smitley, W. D. S. (1980). On the relative power of the *U* and *t* tests. *British Journal of Mathematical and Statistical Psychology, 33,* 114–120.

Boehner, C., & Howe, S. (1996, June). *Statistical graphics in psychological research.* Poster session presented at the annual meeting of the American Psychological Society, San Francisco, CA.

Box, J. F. (1981). Gosset, Fisher, and the *t* distribution. *American Statistician, 35,* 61–66.

Bradley, D. R., Bradley, T. D., McGrath, S. G., & Cutcomb, S. D. (1979). Type I error rate of the chi-square test of independence in R × C tables that have small expected frequencies. *Psychological Bulletin, 86,* 1290–1297.

Bransford, J. D., & Franks, J. J. (1971). The abstraction of linguistic ideas. *Cognitive Psychology, 2,* 331–350.

Brehm, J. W., & Cohen, A. R. (1962). *Explorations in cognitive dissonance.* New York: Wiley.

Bureau of Labor Statistics. (1994, December). Union wages and hours of motor truck drivers and helpers. *Monthly Labor Review,* July 1, 1944.

Byrnes, J. P., Miller, D. C., & Schafer, W. D. (1999). Gender differences in risk taking: A meta-analysis. *Psychological Bulletin, 125,* 367–383.

Camilli, G., & Hopkins, K. D. (1978). Applicability of chi-square to 2 × 2 contingency tables with small expected cell frequencies. *Psychological Bulletin, 85,* 163–167.

Cleveland, W. S. (1994). *The elements of graphing data* (Rev. ed). Murray Hill, NJ: AT&T Bell Laboratories.

Cohen, I. B. (1984, March). Florence Nightingale. *Scientific American, 250,* 128–137.

Cohen, J. (1969). *Statistical power analysis for the behavioral sciences.* New York: Academic.

Cohen, J. (1988). *Statistical power analysis for the behavioral sciences* (2nd ed.). Hillsdale, NJ: Erlbaum.

Cohen, J. (1992). A power primer. *Psychological Bulletin, 112,* 155–159. doi: 10.1037/0033-2909.112.1.155

Cohen, S., & Williamson, G. M. (1991). Stress and infectious disease in humans. *Psychological Bulletin, 109,* 5–24. doi: 10.1037/0033-2909.109.1.5

Coren, S., & Halpern, D. F. (1991). Left-handedness: A marker for decreased survival fitness. *Psychological Bulletin, 109,* 90–106. doi: 10.1037/0033-2909.109.1.90

Cumming, G., & Finch, S. (2005). Inference by eye: Confidence intervals and how to read pictures of data. *American Psychologist, 60,* 170–180. doi: 10.1037/ 0003-066X.60.2.170

David, H. A. (1995). First (?) occurrence of common terms in mathematical statistics. *American Statistician, 49,* 121–133.

Diener, E., Emmons, R. A., & Larsen, R. J. (1985). The Satisfaction with Life Scale. *Journal of Personality Assessment, 49,* 71–75.

Diener, E., & Seligman, M. E. P. (2004). Beyond money: Toward an economy of well-being. *Psychological Science in the Public Interest, 5,* 1–31.

Diener, E., Suh, E. M., Lucas, R. E., & Smith, H. L. (1999). Subjective well-being: Three decades of progress. *Psychological Bulletin, 125,* 276–302. doi: 10.1037/0033-2909.125.2.276

Diener, E., Wolsic, B., & Fujita, F. (1995). Physical attractiveness and subjective well-being. *Journal of Personality and Social Psychology, 68,* 653–663. doi: 10.1037/0022-3514.69.1.120

Dillon, K. M. (1999). I am 95% confident that the world is round: An interview about statistics with Chris Spatz. *Teaching of Psychology, 26,* 232–234.

Doll, R. (1955). Etiology of lung cancer. *Advances in Cancer Research, 3,* 1–50.

Durkheim, E. (1951, reprint). *Suicide.* New York: Free Press. (Original work published 1897)

Edwards, A. W. F. (2001). Fisher, Ronald A. (1890–1962). In N. J. Smelser & P. B. Baltes (Eds.), *International encyclopedia of the social & behavioral sciences* (Vol. 8, pp. 5698–5701). Oxford: Elsevier.

Elgin, J., & Pritchard, M. (2006). Age differences in eating disordered behavior and its correlates. *Psi Chi Journal of Undergraduate Research, 11,* 63–70.

Ellis, D. (1938). *A source book of Gestalt psychology.* London: Routledge & Kegan Paul.

Erceg-Hurn, D. M., & Mirosevich, V. M. (2008). Modern robust statistical methods: An easy way to maximize the accuracy and power of your research. *American Psychologist, 63,* 591–601. doi: 10.1037/0003-066X.63.7.591

Faber, D. (2005). Quetelet, Adolphe. In B. S. Everitt & D. C. Howell (Eds.), *Encyclopedia of statistics in behavioral science* (pp. 1650–1651). West Sussex, UK: Wiley.

Fancher, R. E. (2009). Scientific cousins: The relationship between Charles Darwin and Francis Galton. *American Psychologist, 64,* 84–92. doi: 10.1037/a0013339

Field, A. P. (2005a). Eta and eta squared. In B. S. Everitt & D. C. Howell (Eds.), *Encyclopedia of statistics in behavioral science* (pp. 553–554). West Sussex, UK: Wiley.

Field, A. P. (2005b). Fisher, Sir Ronald Aylmer. In B. S. Everitt & D. C. Howell (Eds.), *Encyclopedia of statistics in behavioral science* (pp. 658–659). West Sussex, UK: Wiley.

Fisher, R. A. (1925). *Statistical methods for research workers.* London: Oliver and Boyd.

Fisher, R. A., & Yates, F. (1963). *Statistical tables for biological, agricultural, and medical research* (6th ed.). Edinburgh: Oliver and Boyd.

Flynn, J. R. (1987). Massive IQ gains in 14 nations: What IQ tests really measure. *Psychological Bulletin, 101,* 171–191. doi: 10.1037/h0090408

Forbs, R., & Meyer, A. B. (1955). *Forestry handbook.* New York: Ronald.

Galton, F. (1869). *Hereditary genius.* London: Macmillan.

Galton, F. (1889). *Natural inheritance.* London: Macmillan.

Galton, F. (1901). Biometry. *Biometrika, 1,* 7–10.

Garcia, S. M., & Tor, A. (2009). The *N*-Effect: More competitors, less competition. *Psychological Science, 20,* 871–877.

Gigerenzer, G., Gaissmaier, W., Kurz-Milcke, E., Schwartz, L. M., & Woloshin, S. (2007). Helping doctors and patients make sense of health statistics. *Psychological Science in the Public Interest, 8,* 53–96.

Guinness World Records 2009. (2008). Enfield, UK: Guinness World Records.

Hacking, I. (1984). Trial by number. *Science, 84,* 5, 69–70.

Hald, A. (1998). *A history of mathematical statistics from 1750 to 1930.* New York: Wiley.

Harter, H. L. (1960). Tables of range and studentized range. *Annals of Mathematical Statistics, 31,* 1122–1147.

Hasler, A. D. (1966). *Underwater guideposts.* Madison: University of Wisconsin Press.

Hasler, A. D., & Scholz, A. T. (1983). *Olfactory imprinting and homing in salmon.* New York: Springer-Verlag.

Hasler, A. D., Scholz, A. T., & Horrall, R. M. (1978). Olfactory imprinting and homing in salmon. *American Scientist, 66,* 347–355.

Hedges, L. V., & Nowell, A. (1995). Sex differences in mental test scores, variability, and numbers of high-scoring individuals. *Science, 269,* 41–45.

Hemphill, J. F. (2003). Interpreting the magnitudes of correlation coefficients. *American Psychologist, 58,* 78–80. doi: 10.1037/0003-066X.58.1.78

Hogan, T. P., & Evalenko, K. (2006). The elusive definition of outliers in introductory statistics textbooks for behavioral sciences. *Teaching of Psychology, 33,* 252–256.

Hollon, S. D., Thrase, M. E., & Markowitz, J. C. (2002). Treatment and prevention of depression. *Psychological Science in the Public Interest, 3,* 39–77.

Howell, D. C. (2008). *Fundamental statistics for the behavioral sciences* (6th ed.). Belmont, CA: Thomson Wadsworth.

Howell, D. C. (2010). *Statistical methods for psychology* (7th ed.). Pacific Grove, CA: Cengage Wadsworth.

Hygge, S., Evans, G. W., & Bullinger, M. (2002). A prospective study of some effects of aircraft noise on cognitive performance in schoolchildren. *Psychological Science, 13,* 469–474.

Ingham, A. G., Levinger, G., Graves, J., & Peckham, V. (1974). The Ringelmann effect: Studies of group size and group performance. *Journal of Experimental Social Psychology, 10,* 371–384.

Irion, A. L. (1976). A survey of the introductory course in psychology. *Teaching of Psychology, 3,* 3–8.

Jacoby, W. (1997). *Statistical graphics for univariate and bivariate data.* Thousand Oaks, CA: Sage.

Jenkins, J. G., & Dallenbach, K. M. (1924). Obliviscence during sleep and waking. *American Journal of Psychology, 35,* 605–612.

Johnson, R. C., McClearn, G. E., Yuen, S., Nagoshi, C. T., Ahern, F. M., & Cole, R. E. (1985). Galton's data a century later. *American Psychologist, 40,* 875–892. doi: 10.1037/ 0003-066X.40.8.875

Killeen, P. R. (2005). An alternative to null-hypothesis significance tests. *Psychological Science, 16,* 345–353.

Kirk, R. E. (1984). *Elementary statistics* (2nd ed.). Pacific Grove, CA: Brooks/Cole.

Kirk, R. E. (1995). *Experimental design: Procedures for the behavioral sciences* (3rd ed.). Pacific Grove, CA: Brooks/Cole.

Kirk, R. E. (2005). Effect size measures. In B. S. Everitt & D. C. Howell (Eds.), *Encyclopedia of statistics in behavioral science* (pp. 532–542). West Sussex, UK: Wiley.

Kirk, R. E. (2008). *Statistics: An introduction* (5th ed.). Belmont, CA: Thompson-Wadsworth.

Kline, R. B. (2004). *Beyond significance testing: Reforming data analysis methods in behavioral research.* Washington, DC: American Psychological Association.

Korbrin, J. L., Patterson, B. F., Shaw, E. J., Mattern, K. D., & Barbuti, S. M. (2008). *Validity of the SAT for predicting first-year college grade point average.* Research Report No. 2008–5. Retrieved from http://professionals.collegeboard.com/profdownload/Validity_of_the_SAT_for_Predicting_First_Year_College_Grade_Point_Average.pdf

Kramer, C. Y. (1956). Extension of multiple range tests to group means with unequal numbers of replications. *Biometrics, 12,* 307–310.

Lewin, K. (1958). Group decision and social change. In E. E. Maccoby, T. M. Newcomb, & E. L. Hartley (Eds.), *Readings in social psychology* (3rd ed.). New York: Holt, Rinehart & Winston.

Lilienfeld, S. O., Wood, J. M., & Garb, H. N. (2000). The scientific status of projective techniques. *Psychological Science in the Public Interest, 1,* 27–66.

Loftus, E. F. (1979). *Eyewitness testimony.* Cambridge, MA: Harvard University Press.

Lovie, S. (2005). Exploratory data analysis. In B. S. Everitt & D. C. Howell (Eds.), *Encyclopedia of statistics in behavioral science* (pp. 586–588). West Sussex, UK: Wiley.

Mackowiak, P. A., Wasserman, S. S., & Levine, M. M. (1992). A critical appraisal of 98.6°F, the upper limit of the normal body temperature, and other legacies of Carl Reinhold August Wunderlich. *Journal of the American Medical Association, 268,* 1578–1580.

Maindonald, J., & Richardson, A. M. (2004). This passionate study: A dialogue with Florence Nightingale. *Journal of Statistics Education, 12.* Retrieved September 27, 2006, from www.amstat.org/publications/jse/v12n1/maindonald.html

Mann, H. B., & Whitney, D. R. (1947). On a test of whether one or two random variables is stochastically larger than the other. *Annals of Mathematical Statistics, 18,* 50–60.

Mayo, E. (1946). *The human problems of an industrial civilization.* Boston: Harvard University Press.

McKeachie, W. J., Pollie, D., & Speisman, J. (1955). Relieving anxiety in classroom examinations. *Journal of Abnormal and Social Psychology, 50,* 93–98.

McKeown, T., & Gibson, J. R. (1951). Observation on all births (23,970) in Birmingham, 1947. IV. "Premature birth." *British Medical Journal, 2,* 513–517.

McMullen, L., & Pearson, E. S. (1939). William Sealy Gosset, 1876–1937. *Biometrika, 30,* 205–253.

Micceri, T. (1989). The unicorn, the normal curve, and other improbable creatures. *Psychological Bulletin, 105,* 156–166. doi: 10.1037/ 0033-2909.105.1.156

Milgram, S. (1969). Note on the drawing power of crowds of different size. *Journal of Personality and Social Psychology, 13,* 79–82.

Milnor, W. R. (1990). *Cardiovascular physiology.* New York: Oxford.

Minium, E. W., & King, B. M. (2002). *Statistical reasoning in psychology and education* (4th ed.). New York: Wiley.

Mischel, H. N. (1974). Sex bias in the evaluation of professional achievements. *Journal of Educational Psychology, 66,* 157–166.

Mosteller, F., & Wallace, D. L. (1989). Deciding authorship. In J. M. Tanur, F. Mosteller, W. H. Kruskal, R. F. Link, R. S. Picters, G. R. Rising, & E. L. Lehmann (Eds.), *Statistics: A guide to the unknown* (3rd ed.). Pacific Grove, CA: Wadsworth & Brooks/Cole.

Mrazek, D. A., Schuman, W. B., & Klinnert, M. (1998). Early asthma onset: Risk of emotional and behavioral difficulties. *Journal of Child Psychology and Psychiatry and Allied Disciplines, 39,* 247–254. doi: 10.1111/1469-7610.00318

Nelson, N., Rosenthal, R., & Rosnow, R. L. (1986). Interpretation of significance levels and effect sizes by psychological researchers. *American Psychologist, 41,* 1299–1301. doi: 10.1037/ 0003-066X.41.11.1299

Nickerson, R. S. (2000). Null hypothesis statistical testing: A review of an old and continuing controversy. *Psychological Methods, 5,* 241–301. doi: 10.1037/1082-989X.5.2.241

Nicol, A., & Pexman, P. (2003). *Displaying your findings: A practical guide for creating figures, posters, and presentations.* Washington, DC: American Psychological Association.

Nijsse, M. (1988). Testing the significance of Kendall's τ and Spearman's r_s. *Psychological Bulletin, 103,* 235–237. doi: 10.1037/0033-2909.103.2.235

Overall, J. E. (1980). Power of chi-square tests for 2 × 2 contingency tables with small expected frequencies. *Psychological Bulletin, 87,* 132–135.

Pagano, R. R. (2007). *Understanding statistics in the behavioral sciences* (8th ed.). Belmont, CA: Wadsworth/Thomson.

Pearson, E. S. (1949). W. S. Gosset. In L. G. Wickham Legg (Ed.), *Dictionary of national biography: 1931–1940.* London: Oxford University Press.

Pearson, K., & Lee, A. (1903). Inheritance of physical characters. *Biometrika, 2,* 357–462.

Peden, B. F. (2001). Correlational analysis and interpretation: Graphs prevent gaffes. *Teaching of Psychology, 28,* 129–131.

Peden, B. F., & Hausmann, S. E. (2000). Data graphs in introductory and upper-level psychology textbooks: A content analysis. *Teaching of Psychology, 27,* 93–97.

Porter, T. M. (1986). *The rise of statistical thinking.* Princeton, NJ: Princeton University Press.

Powers, E., & Witmer, H. (1951). *An experiment in the prevention of delinquency: The Cambridge-Somerville Youth Study.* New York: Columbia University Press.

Pryor, J. H., Hurtado, S., DeAngelo, L., Sharkness, J., Romero, L. C., Korn, W. S., & Tran, S. (2008). *The American freshman: National norms for Fall 2008.* Los Angeles: Higher Education Research Institute.

Rokeach, M., Homant, R., & Penner, L. (1970). A value analysis of the disputed Federalist papers. *Journal of Personality and Social Psychology, 16,* 245–250.

Rosnow, R. L., & Rosenthal, R. (2005). *Beginning behavioral research: A conceptual primer* (5th ed.). Upper Saddle River, NJ: Pearson Prentice Hall.

Runyon, R. P., Coleman, K. A., & Pittenger, D. J. (2000). *Fundamentals of behavioral statistics* (9th ed.). Boston: McGraw-Hill.

Ryan, J. J. (2008). Intelligence. In S. F. Davis & W. Buskist (Eds.), *21st century psychology: A reference handbook* (Vol. 1, pp. 413–421). Thousand Oaks, CA: Sage.

Salsburg, D. (2001). *The lady tasting tea: How statistics revolutionized science in the twentieth century.* New York: Henry Holt.

Schachter, S., & Gross, L. P. (1968). Manipulated time and eating behavior. *Journal of Personality and Social Psychology, 10,* 98–106.

Schumacher, E. F. (1979). *Good work.* New York: Harper & Row.

Shedler, J., & Block, J. (1990). Adolescent drug use and psychological health. *American Psychologist, 45,* 612–630. doi: 10.1037/0003-066X.45.5.612

Sherif, M. (1935). A study of some social factors in perception. *Archives of Psychology.* No. 187.

Sherman, M. (1927). The differentiation of emotional responses in infants: The ability of observers to judge the emotional characteristics of the crying infants, and of the voice of an adult. *Journal of Comparative Psychology, 7,* 335–351.

Smith, L. D., Best, L. A., Stubbs, D. A., Archibald, A. B., & Roberson-Nay, R. (2002). Constructing knowledge: The role of graphs and tables in hard and soft psychology. *American Psychologist, 57,* 749–761. doi: 10.1037/0003-066X.57.10.749

Smith, M. L., & Glass, G. V. (1977). Meta-analysis of psychotherapy outcome studies. *American Psychologist, 32,* 752–760.

Smith, R. A. (1971). The effect of unequal group size on Tukey's HSD procedure. *Psychometrika, 36,* 31–34.

Snedecor, G. W., & Cochran, W. G. (1980). *Statistical methods* (7th ed.). Ames: Iowa State University Press.

Spatz, C. (2000, November–December). Our changing statistical methods: Controversies about the null hypothesis. *Psychology Teacher Network,* 3–4.

Spatz, C. (2008). Statistical techniques and analysis. In S. F. Davis & W. Buskist (Eds.), *21st century psychology: A reference handbook* (Vol. 1, pp. 46–54). Thousand Oaks, CA: Sage.

Spatz, T. S. (1991). Improving breast self-examination training by using the 4MAT instructional model. *Journal of Cancer Education, 6,* 179–183.

Spearman, C. (1930). Autobiography. In C. Murchison (Ed.), *History of psychology in autobiography.* New York: Russell & Russell.

Sprent, P., & Smeeton, N. C. (2007). *Applied nonparametric statistical methods* (4th ed.). Boca Raton, FL: CRC Press.

Sprinthall, R. C. (2007). *Basic statistical analysis* (8th ed.). Boston: Allyn and Bacon.

Sternberg, R. J., & Williams, W. M. (1997). Does the Graduate Record Examination predict meaningful success in the graduate training of psychologists? A case study. *American Psychologist, 52,* 630–641. doi: 10.1037/0003-066X.52.6.630

Stevens, S. S. (1946). On the theory of scales of measurement. *Science, 103,* 677–680.

Strayer, D. L., Drews, F. A., & Johnston, W. A. (2003). Cell phone–induced failures of visual attention during simulated driving. *Journal of Experimental Psychology: Applied, 9,* 23–32. doi: 10.1037/1076-898X.9.1.23

Tanur, J. M., Mosteller, F., Kruskal, W. H., Lehmann, E. L., Link, R. F., Pieters, R. S., & Rising, G. R. (Eds.). (1989). *Statistics: A guide to the unknown* (3rd ed.). Pacific Grove, CA: Wadsworth & Brooks/Cole.

Thomas, R. (2005). Francis Galton. In B. S. Everitt & D. C. Howell (Eds.), *Encyclopedia of statistics in behavioral science* (pp. 687–688). West Sussex, UK: Wiley.

Tufte, E. R. (2001). *The visual display of quantitative information* (2nd ed.). Cheshire, CT: Graphics.

U.S. Census Bureau. (2008). *Statistical abstract of the United States: 2009* (128th ed.). Washington, DC: Author.

von Hippel, P. T. (2005). Mean, median, and skew: Correcting a textbook rule. *Journal of Statistics Education, 13.* Retrieved September 27, 2006, from www.amstat.org/publications/jse/v13n2/vonhippel.html

Wainer, H. (1984). How to display data badly. *American Statistician, 38,* 137–147.

Wainer, H., & Velleman, P. F. (2001). Statistical graphics: Mapping the pathways of science. *Annual Review of Psychology, 52,* 305–335.

Walker, H. M. (1929). *Studies in the history of statistical method.* Baltimore: Williams & Wilkins.

Walker, H. M. (1940). Degrees of freedom. *Journal of Educational Psychology, 31,* 253–269.

Walker, H. M. (1968). Karl Pearson. In D. L. Sills (Ed.), *International encyclopedia of the social sciences.* New York: Macmillan and Free Press.

Wallace, R. K., & Benson, H. (1972). The physiology of meditation. *Scientific American, 226,* 84–90.

Waller, J. C. W. (2001). Sir Francis Galton (1822–1911). In N. J. Smelser & P. B. Baltes (Eds.), *International encyclopedia of the social & behavioral sciences* (Vol. 9, pp. 5860–5863). Oxford: Elsevier.

Watson, J. B. (1924). *Psychology from the standpoint of a behaviorist* (2nd ed.). Philadelphia: Lippincott.

White, E. B. (1970). *The trumpet of the swan.* New York: Harper & Row.

Wight, R. D., & Gable, P. A. (2005). Gauss, Johann Carl Friedrich. In B. S. Everitt & D. C. Howell (Eds.), *Encyclopedia of statistics in behavioral science* (pp. 694–696). West Sussex, UK: Wiley.

Wilcox, R. R. (2005a). Outliers. In B. S. Everitt & D. C. Howell (Eds.), *Encyclopedia of statistics in behavioral science* (pp. 1497–1498). West Sussex, UK: Wiley.

Wilcox, R. R. (2005b). *Introduction to robust estimation and hypothesis testing* (2nd ed.). San Diego: Academic Press.

Wilcoxon, F. (1945). Individual comparisons by ranking methods. *Biometrics, 1,* 80–83.

Wilcoxon, F., & Wilcox, R. A. (1964). *Some rapid approximate statistical procedures* (Rev. ed.). Pearl River, NY: Lederle Laboratories.

Wilkinson, L., and Task Force on Statistical Inference. (1999). Statistical methods in psychology journals: Guidelines and explanations. *American Psychologist, 54,* 594–604. doi: 10.1037/0003-066X.54.8.594

Winer, B. J., Brown, D. R., & Michels, K. M. (1991). *Statistical principles in experimental design* (3rd ed.). New York: McGraw-Hill.

Winter, L., Uleman, J. S., & Cunniff, C. (1985). How automatic are social judgments? *Journal of Personality and Social Psychology, 49,* 904–917. doi: 10.1037/0022-3514.49.4.904

Wittrock, M. C. (1991). Generative teaching of comprehension. *Elementary School Journal, 92,* 169–184.

Wolraich, M. L., Lindgren, S. D., Stumbo, P. J., Stegink, L. D., Applebaum, M. I., & Kiritsy, M. C. (1994). Effects of diets high in sucrose or aspartame on the behavior and cognitive performance of children. *New England Journal of Medicine, 330,* 301–307.

Wood, T. B., & Stratton, F. J. M. (1910). The interpretation of experimental results. *Journal of Agricultural Science, 3,* 417–440.

Woodworth, R. S. (1926). Introduction. In H. E. Garrett (Ed.), *Statistics in psychology and education.* New York: Longmans, Green.

Wuensch, K. L. (2005). Scales of measurement. In B. S. Everitt & D. C. Howell (Eds.), *Encyclopedia of statistics in behavioral science* (pp. 1785–1787). West Sussex, UK: Wiley.

Yates, F. (1981). Sir Ronald Aylmer Fisher. In E. T. Williams & C. S. Nichols (Eds.), *Dictionary of national biography 1961–1970.* Oxford: Oxford University.

Youden, W. J. (1962). *Experimentation and measurement.* Washington, DC: National Science Teachers Association.

Zajonc, R. B. (2001). The family dynamics of intellectual development. *American Psychologist, 56,* 490–496. doi: 10.1037/ 0003-066X.56.6-7.490

Zajonc, R. B., & Bargh, J. (1980). Birth order, family size, and decline of SAT scores. *American Psychologist, 35,* 662–668.

Zwick, R. (1993). Pairwise comparison procedures for one-way analysis of variance designs. In G. Keren & C. Lewis (Eds.), *A handbook for data analysis in the behavioral sciences: Statistical issues* (pp. 43–71). Hillsdale, NJ: Erlbaum.

Index